Theory of Degrees with Applications to Bifurcations and Differential Equations

CANADIAN MATHEMATICAL SOCIETY SERIES OF MONOGRAPHS AND ADVANCED TEXTS

Monographies et Études de la Société Mathématique du Canada

Theory of Degrees with Applications to Bifurcations and Differential Equations

WIESLAW KRAWCEWICZ

University of Alberta

JIANHONG WU

York University

A Wiley-Interscience Publication

JOHN WILEY & SONS, INC.

New York • Chichester • Brisbane • Toronto • Singapore • Weinheim

This text is printed on acid-free paper.

Library of Congress Cataloging in Publication Data:

Krawcewicz, Wiesław.
Theory of degrees, with applications to bifurcations and differential equations / Wiesław Krawcewicz, Jianhong Wu.
p. cm. -- (Canadian Mathematical Society series of monographs and advanced texts)
"A Wiley-Interscience publication."
Includes bibliographical references (p.) and index.
ISBN 0-471-15740-6 (cloth : alk. paper)
1. Topological degree. 2. Bifurcation theory. 3. Differential equations. I Wu, Jianhong. II. Series.
QA612.K73 1996 96-33302
515′.355--dc20 CIP

Printed in the United States of America

10 9 8 7 6 5 4 3 2 1

To Grażyna, Ming,
Karma and Veronica

Contents

Preface

The purpose of this book is to provide an introduction to degree theory and its applications to nonlinear differential equations. Two principles have guided the presentation and organization of this book. First, we intend to address those readers who are applications-oriented and have basic knowledge of functional analysis, general topology, and differential equations. We pay special attention to analytic constructions, normal approximations, computational formulae, and illustrative examples/applications. Second, we intend to provide a unified treatment of the classical Brouwer degree (and its extensions to compact or set-condensing fields in infinite-dimensional spaces), the recently developed S^1-degree [see Dylawerski et al. (1991)], and the Dold–Ulrich degree for equivariant mappings [Ulrich (1980)], as well as their applications to bifurcation problems. This being so, we begin with an introduction to the fundamental concepts and results relating to the transversality theorem, Weierstrass–Sard approximations, differential manifolds and bundles. Of course, when we enter the field of equivariant topology, we cannot avoid the use of basic algebraic and topological terminology and facts regarding compact Lie groups and their representations. This preliminary algebraic and topological material will be provided in the appropriate sections in order to make the book self-contained and suitable for the readers we have in mind.

We now summarize the contents of this book. In Chapter 1, some elementary concepts and results from differential and algebraic topology are introduced. In particular, the Weierstrass–Sard theorem and the transversality theorem are described. Several other notions and related results on manifolds and vector bundles are also presented. They provide the necessary background material for the development and analytic construction of the degree, especially the equivariant degree, in subsequent chapters.

The Brouwer degree in finite-dimensional spaces is the subject of Chapter 2. This degree is developed along axiomatic lines, and an analytic approach is used for its construction. The fundamental axioms are formulated, basic properties are derived, and computation formulae are given.

In Chapter 3, the Brouwer degree is extended to compact perturbations of the identity in infinite-dimensional spaces. The principal technique employed here is the Schauder approximation of a compact map by mappings with finite-dimensional range. The definition of the Leray–Schauder degree for compact perturbations of the identity is introduced and its application to the existence of fixed points is discussed. Also, a computational formula for the Leray–Schauder degree for linear compact fields is obtained. Among several other results, the Schauder fixed-point theorem, Borsuk's theorem, and the Birkhoff–Kellog theorem are established in this chapter.

The Leray–Schauder degree is extended to condensing fields in Chapter 4. The concepts of measure of noncompactness and of condensing fields play a vital role in the development of this subject. The bijection theorem that connects condensing fields to compact mappings is established. This theorem is used to formulate a constructive definition of a degree for condensing fields. Furthermore, several useful fixed-point theorems and nonlinear alternatives are presented. In addition, some topological and analytical results, such as the invariance of domain theorem and the Fredholm alternative theorem, are also established. The homotopic properties of the class of linear condensing fields together with Sard–Smale theorem enable us to develop regular approximations and the regular value formula for C^1-condensing fields. Finally, a composite coincidence degree for certain coincidence problems is introduced and applied to deal with boundary value problems for neutral functional differential equations.

Chapter 5 is concerned with the application of degree theory to the investigation of the structure of the solution set for a large class of eigenvalue problems in a Banach space. A local bifurcation theorem due to Krasnosel'skii, which relates an eigenvalue of odd algebraic multiplicity to the occurrence of a bifurcation point, is presented. Further, a global bifurcation theorem of Rabinowitz, which shows that bifurcation resulting from eigenvalues of odd multiplicity is a global phenomenon, is discussed. The general theory is illustrated by an application to the Sturm–Liouville problem for second-order ordinary differential equations.

The central object of Chapter 6 is the development of the S^1-equivariant degree theory for a class of nonlinear mappings preserving the symmetry given by an action of the circle group S^1. In order to make the presentation self-contained, a brief introduction to representation theory is provided. The equivariant generic approximation theorem established in this chapter plays a crucial role in the construction of the S^1-degree for S^1-equivariant mappings [cf. Dylawerski, et al. (1991)]. It is interesting to note that this degree is a sequence of integers, each of which "measures" the number of orbits of zeros of a certain type. It is shown that such a degree has all the standard properties of a topological degree discussed in earlier chapters. Finally, an equivariant bijection theorem is presented that enables one to extend the S^1-degree to the class of condensing fields and to treat a class of equivariant composite coincidence problems.

Chapter 7 provides some applications of the S^1-equivariant degree to the global Hopf bifurcation problem for neutral functional differential equations. First, an abstract equivariant global bifurcation theorem is established for an S^1-equivariant composite coincidence problem involving two parameters. Next, this theorem is employed to obtain a global Hopf bifurcation theorem for neutral equations. This theorem provides detailed information on the maximal continuation of periodic solutions. An example from the lossless transmission lines problem is presented. This application demonstrates the power of the global bifurcation theorem for establishing the existence of large amplitude periodic solutions.

The final Chapter 8 of this book deals with an equivariant degree called G-degree. It is inspired by the equivariant fixed-point index introduced by Dold (1983) and Ulrich (1980). The G-degree is defined through the use of regular normal approximations of equivariant mappings. The main goal of Chapter 8 is to present a variety of examples to illustrate the computation of G-degree for a number of simple classical compact Lie groups G. Applications to time-reversible systems involving functional differential equations of mixed type and to symmetry breaking steady-state bifurcations are included in this chapter.

We would like to pay special homage to our teacher and friend Professor Kazimierz Gęba who gave us the inspiration and strength to undertake the task of writing this book. Words cannot express our gratitude for his kindness and our admiration for his insight, erudition, and patience. We would like to express our deepest thanks to our colleagues Zalman Balanov and Yuli Billig for their help in the preparation of this book, in particular for their expertise and commentaries in the preparation of the last part of the book dealing with the equivariant degree. Our thanks also go to Professors H. Steinlein and E. N. Dancer for their suggestions and comments. We thank our colleagues Lynn Erbe, I. Kleiner, A. Shenitzer, and Joseph So for their unfailing support and inspiration. We owe a special debt to V. Sree Hari Rao, who encouraged us to undertake this work and critiqued and reviewed some of the manuscript. We also thank Huaxing Xia and Assia Barabanova for their critical remarks and suggestions. We are grateful to our students at the University of Alberta and York University, Zaoyang Guo, Paola Vivi, and Xingfu Zou, who struggled with evolving versions of the manuscript and helped us with their questions and insight. The first author renders his heartfelt thanks to Professor B. G. Sidharth and the Birla Science Center in Hyderabad for the support and hospitality during his visits in 1992 and 1993, when part of the manuscript was written. Special thanks go to Peter Habala whose technical knowledge and good humor helped us in our struggle with computers, to Elizabeth Leonard for her excellent typing, and to Christian Labadie for his help in the preparation of PostScript figures for this book. Finally, we would like to thank H. W. Knobloch, V. Dragalin, J. de Tibeiro, M. Pohl, T. Spanily, and W. Marzantowicz for their encouragement, advice, and scientific exchanges.

It is our pleasure to acknowledge the financial support of the Natural Sciences and Engineering Research Council of Canada, the Alexander von Humboldt Foundation (W. Krawcewicz), the Faculty of Arts Fellowship at York University (J. Wu), and the Central Research Fund at the University of Alberta (W. Krawcewicz). We are obliged to Professor J. Borwein and the staff of John Wiley and Sons, in particular, to S. Quigley, E. Singletary, P. Brecht and A. Volan for their interest and cooperation in transforming the manuscript into a book.

Alberta, Canada WIESLAW KRAWCEWICZ
Ontario, Canada JIANHONG WU

Introduction

Many problems in nonlinear differential equations lead to a fixed-point equation

$$(*) \qquad x = f(x)$$

for a mapping $f: E \to E$ satisfying a certain compactness or condensing property with respect to some measure of noncompactness on an abstract space E. If the system under consideration possesses a symmetry, or when special solutions such as periodic solutions or waves are sought, then the associated mapping f will have the additional property of being equivariant with respect to a compact Lie group acting on E.

A natural question related to the fixed-point equation is the existence, the uniqueness or multiplicity, and the distribution of fixed points. Of equal importance is the problem of the change of the set of fixed points when a certain parameter is varied, such as the occurrence and global continuation of fixed points bifurcating from known ones. In the case when f is equivariant with respect to a compact Lie group, it is also important to describe the symmetries of its fixed points, the interaction of the symmetries of multiple fixed points as well as the change of symmetries along a specific branch of fixed points.

Degree theory has been introduced to provide a topological approach to these problems. Roughly speaking, the degree of a mapping f, denoted by $\deg(\mathrm{Id}-f, \Omega)$, for f belonging to a certain class of mappings defined on an open, bounded subset Ω of a Banach space such that $x \neq f(x)$ for $x \in \partial\Omega$, is an integer satisfying, among others, the following properties: (i) if $\deg(\mathrm{Id} - f, \Omega) \neq 0$, then $x = f(x)$ has a solution in Ω; (ii) $\deg(\mathrm{Id}-f, \Omega) = \deg(\mathrm{Id} - f, \Omega_1) + \deg(\mathrm{Id}-f, \Omega_2)$, provided Ω_1 and Ω_2 are disjoint open subsets of Ω such that $x \neq f(x)$ for $x \in \overline{\Omega \backslash (\Omega_1 \cup \Omega_2}$; and (iii) if f can be "continuously deformed" into a mapping g such that no fixed points "escape" from $\partial\Omega$ during the deformation, then $\deg(\mathrm{Id}-f, \Omega) = \deg(\mathrm{Id} - g, \Omega)$. The first two properties indicate that $\deg(\mathrm{Id}-f, \Omega)$ provides an algebraic count of the number of fixed points of f in Ω. The third property, referred to as homotopy invariance, enables us to compute the degree of f by deforming f to a possibly simpler map g. When f is equivariant with respect to a compact

Lie group G, the equivariant degree $G\text{-Deg}(\mathrm{Id} - f, \Omega)$ is a family of integers, to distinguish fixed points of different symmetries. Degree theory for the fixed-point equation (*) when f has no symmetry has been well developed and widely applied to various problems of nonlinear differential equations. The monographs by Cronin (1964), Deimling (1984), Lloyd (1978), Rothe (1986), and Zeidler (1986), to name few, provide excellent expositions of this subject. More recently, equivariant degrees have been introduced to treat (*) when f is equivariant with respect to a compact Lie group. Though still at an early stage of development, equivariant degree theory has found interesting applications to the study of symmetry breaking, Hopf bifurcations, and various wave solutions of nonlinear differential equations.

Theory of Degrees with Applications to Bifurcations and Differential Equations

Chapter One

Elements of Differential Topology

In this chapter we introduce some elementary concepts and results from differential and algebraic topology. Three important theorems—Sard's theorem, Weierstrass' theorem and the transversality theorem—will be presented together with some necessary notions such as manifolds, vector bundles, etc. These will provide essential technical tools for a systematic development and analytic construction of the degree theory.

1.1 DIFFERENTIAL MANIFOLDS AND SUBMANIFOLDS

Throughout this section we assume that all topological spaces are *Hausdorff*, *paracompact*, and have *countable bases*, unless otherwise stated.

Definition 1.1.1 A topological space M is called an *n-dimensional manifold of class* C^r, $r = 1, 2, \ldots$ or $r = \infty$, or simply *C^r-manifold of dimension* $\dim M = n$, if there is an open cover $\mathcal{U} = \{U_i\}_{i \in \Lambda}$ of M such that:

- **(i)** For each $i \in \Lambda$, there is a map $\varphi_i : U_i \to \mathbb{R}^n$ that is a homeomorphism from U_i onto an open subset of $\mathbb{R}^n$.
- **(ii)** For every pair of such homeomorphisms $\varphi_i : U_i \to \mathbb{R}^n$ and $\varphi_j : U_j \to \mathbb{R}^n$, the *coordinate change* $\varphi_j \circ \varphi_i^{-1} : \varphi_i(U_i \cap U_j) \to \varphi_j(U_i \cap U_j)$ is a differentiable map of class C^r.

In the above definition, the pair (φ_i, U_i) is called a *chart* or a *coordinate system* with domain U_i, and the set $\Phi = \{(\varphi_i, U_i)\}_{i \in \Lambda}$ is called an *atlas* on M. Further, in case (ii) we say that (φ_i, U_i) and (φ_j, U_j) have *C^r-overlap*. In this case, the atlas Φ on M is called a *C^r-atlas*. For every C^r-atlas Φ on M, there exists a unique maximal C^r-atlas α in M that contains Φ, with respect to the inclusion relation. A *C^r-differential structure* on M is a maximal C^r-atlas α

on M. Consequently, a manifold M of class C^r is a pair (M, Φ), where Φ is a C^r-atlas on M. We will also use (M, α) to denote a C^r-manifold, where α denotes a C^r-differential structure on M. A C^∞-manifold will be called a *smooth manifold*.

To determine a C^r-differential structure, it suffices to give a single C^r-atlas. Moreover, every C^s-manifold may be regarded as a C^r-manifold for $1 \le r < s$ in the following sense: If α is a C^s-differential structure on M and $1 \le r < s$, then α is also a C^r-atlas that is contained in a unique C^r-differential structure on M obtained by adding to α all charts having C^r-overlap with every chart from α.

We say that two ordered bases $(b_1, \ldots, b_n)$ and $(b'_1, \ldots, b'_n)$ of a finite-dimensional vector space V determine the *same orientation* of V if $b'_i = \sum_{j=1}^{n} a_{ij} b_j$, $j = 1, \ldots, n$, with $\det[a_{ij}] > 0$. Otherwise, we say that they determine *opposite orientations* of V. The vector space $\mathbb{R}^n$ has *standard* orientation corresponding to the basis $e_1 = (1, 0, \ldots, 0)$, $e_2 = (0, 1, \ldots, 0)$, $\ldots$, $e_n = (0, \ldots, 0, 1)$.

Definition 1.1.2 Let (M, α) be a C^r-manifold. We say that M is *orientable* if there exists a C^r-atlas $\Phi = \{(\varphi_i, U_i)\}$ such that $\Phi \subset \alpha$ and for every two overlapping charts (φ_i, U_i) and (φ_j, U_j) of Φ, the coordinate change map $\varphi_j \circ \varphi_i^{-1} : \varphi_i(U_i \cap U_j) \to \varphi_j(U_i \cap U_j)$ satisfies $\det D(\varphi_j \circ \varphi_i^{-1})(x) > 0$ for all $x \in \varphi_i(U_i \cap U_j)$. Such an atlas Φ will be called *oriented*. We will also say that two oriented atlases $\Phi = \{(\varphi_i, U_i)\}$ and $\Psi = \{(\psi_j, V_j)\}$ are *equivalent* or *determine the same orientation* on M if for every two overlapping charts (φ_i, U_i) of Φ and (ψ_j, V_j) of Ψ, we have $\det D(\psi_j \circ \varphi_i^{-1})(x) > 0$ for all $x \in \varphi_i(U_i \cap V_j)$. The above relation is an equivalence relation and its equivalence class will be called an *orientation* on M. The manifold M together with a fixed orientation will be called an *oriented manifold*.

It can be shown that if (M, α) is a connected orientable manifold, then there exist exactly two orientations on M.

Example 1.1.3

(i) The space $\mathbb{R}^n$ has a C^r-differential structure, for $r = 1, 2, \ldots$ or $r = \infty$, containing the atlas $\Phi = \{(\mathrm{Id}, \mathbb{R}^n)\}$.

(ii) Every open set $U \subseteq \mathbb{R}^n$ has a C^r-differential structure, $r = 1, 2, \ldots$ or $r = \infty$, containing the atlas $\Phi = \{(i, U)\}$, where $i : U \to \mathbb{R}^n$ is the inclusion map.

(iii) *The unit n-sphere* $S^n := \{x \in \mathbb{R}^{n+1}; |x| = 1\}$, where $|x| = \sqrt{\sum_{i=1}^{n+1} x_i^2}$, $x = (x_1, \ldots, x_{n+1}) \in \mathbb{R}^{n+1}$, is a C^∞-manifold. In fact, the atlas $\Phi := \{(\varphi_+, U_+), (\varphi_-, U_-)\}$ gives a C^∞-differential structure on S^n, where $U_+ = \{x = (x_1, \ldots, x_{n+1}) \in S^n; \quad x_{n+1} < 1\}$, $U_- = \{x =$

$(x_1, \ldots, x_{n+1}) \in S^n; x_{n+1} > -1\}$ and $\varphi_+ : U_+ \to \mathbb{R}^n$ and $\varphi_- : U_- \to \mathbb{R}^n$ are the stereographic projections from the North Pole $p_+ = (0, \ldots, 0, 1)$ and the South Pole $p_- = (0, \ldots, 0, -1)$, respectively, that is

$$\varphi_{\pm}(x_1, \ldots, x_{n+1}) = \left(\frac{x_1}{1 \mp x_{n+1}}, \ldots, \frac{x_n}{1 \mp x_{n+1}} \right)$$

$$(x_1, \ldots, x_{n+1}) \in U_{\pm}$$

Since $\varphi_- \circ \varphi_+^{-1}(y) = \dfrac{y}{|y|^2}$ and $\varphi_+ \circ \varphi_-^{-1}(y) = \dfrac{y}{|y|^2}$ for $y \in \mathbb{R}^n \backslash \{0\}$, Φ gives a C^∞-differential structure on S^n. The atlas Φ is oriented and thus S^n is orientable.

Definition 1.1.4 Let (M, Φ) be a C^∞-manifold. The manifold (M, Φ) is called *analytic* if for every two charts (φ_i, U_i) and (φ_j, U_j), the coordinate change $\varphi_j \circ \varphi_i^{-1} : \varphi_i(U_i \cap U_j) \to \varphi_j(U_i \cap U_j)$ is an analytic map.

Remark 1.1.5 There are several ways of constructing manifolds from the given ones, among which we mention:

(i) *Product Manifold*: Let (M, Φ) and (N, Ψ) be two manifolds, with dimensions m and n, and of classes C^r and C^s, respectively. Their *Cartesian product* is the manifold $(M \times N, \Theta)$ of class C^ℓ, $\ell = \min(r, s)$, where Θ is a C^ℓ-differential structure containing all charts of the form $(\varphi \times \psi, U \times V)$, $(\varphi, U) \in \Phi$, $(\psi, V) \in \Psi$, and the map $\varphi \times \psi$ is simply the product of maps φ and ψ from $U \times V$ into $\mathbb{R}^{m+n} = \mathbb{R}^m \times \mathbb{R}^n$.

(ii) *Restricted Manifold*: A manifold (M, α), where α is a maximal atlas on M, induces a differential structure $\alpha|_W$ on every open subset $W \subseteq M$. The maximal atlas $\alpha|_W$ is defined by

$$\alpha|_W := \{(\varphi, U) \in \alpha; U \subseteq W\}$$

(iii) *Induced Manifold*: Let M be a topological space, (N, Ψ) a C^r-manifold and $h : M \to N$ a homeomorphism of M onto N. The *induced differential structure* on M is the maximal C^r-atlas on M containing the *induced* C^r-atlas $h^*\Psi$ on M defined by

$$h^*\Psi := \{(\psi \circ h, h^{-1}(U)); (\psi, U) \in \Psi\}$$

Definition 1.1.6 A subset A of an n-dimensional C^r-manifold (M, α) is called a C^r-*submanifold* of (M, α) if for some $k \in \mathbb{N}$, $0 \leq k \leq n$, each point of A belongs to the domain of a chart (φ, U) from the maximal atlas α such that $U \cap A = \varphi^{-1}(\mathbb{R}^k \cap \varphi(U))$, where $\mathbb{R}^k \subseteq \mathbb{R}^n$ is the set of vectors whose last $n-k$ coordinates are zero. We call such a chart (φ, U) a *submanifold* chart for $A \subseteq M$.

Clearly, if A is a submanifold of M, then the maps $\varphi_{|U\cap A} : U \cap A \to \mathbb{R}^k$, where (φ, U) is a submanifold chart, form a C^r-atlas for A. Thus, A is a C^r-manifold of dimension k. The *codimension* of A in M is $n-k$.

Remark 1.1.7 It can be verified that the property of being a C^r-submanifold is *local in character*, in the sense that a subset A of M is a submanifold of M if and only if A_i is a submanifold of M_i for each i, where $\{A_i\}$ is an open cover of A and each M_i is an open subset of M containing A_i.

Let $W \subseteq \mathbb{R}^n$ be an open set and $f : W \to \mathbb{R}^q$ a C^r-map, $1 \le r \le \infty$. We say $z \in \mathbb{R}^q$ is a *regular value* of f if either $f^{-1}(z) = \varnothing$ or for every $p \in f^{-1}(z)$, the derivative $Df(p)$ of f at p has rank q, that is, $Df(p)$ is surjective.

Theorem 1.1.8

Let $W \subseteq \mathbb{R}^n$ be an open set and $f : W \to \mathbb{R}^q$ a C^r-map, $1 \le r \le \infty$. If z is a regular value of f, then $f^{-1}(z) = \varnothing$ or $f^{-1}(z)$ is an orientable C^r-submanifold of $\mathbb{R}^n$ of codimension q.

Proof. We represent the space $\mathbb{R}^n$ as the product $\mathbb{R}^{n-q} \times \mathbb{R}^q$ and denote by $j_1 : \mathbb{R}^{n-q} \to \mathbb{R}^{n-q} \times \mathbb{R}^q$ and $j_2 : \mathbb{R}^q \to \mathbb{R}^{n-q} \times \mathbb{R}^q$ the inclusions of $\mathbb{R}^{n-q}$ and $\mathbb{R}^q$ onto the first and the second component, respectively. Let $p \in f^{-1}(z)$ and $A : \mathbb{R}^n \to \mathbb{R}^{n-q}$ be a linear map such that $\det(j_1A + j_2Df(p)) > 0$. Since $j_1A + j_2Df(p) : \mathbb{R}^n \to \mathbb{R}^n$ is an isomorphism, by the implicit function theorem, there exists a neighborhood U of p such that the map $\psi : U \to \psi(U) \subset \mathbb{R}^n$ given by $\psi(x) := j_1Ax + j_2(f(x) - z)$ is a C^r-diffeomorphism. Since $j_1A + j_2Df(x) : \mathbb{R}^n \to \mathbb{R}^n$ is an isomorphism, we may assume, without loss of generality, that $\det(j_1A + j_2Df(x)) > 0$ for all $x \in U$. It is clear that $\psi(x) \in \mathbb{R}^{n-q}$ if and only if $x \in f^{-1}(z) \cap U$. Therefore, the pair $(\psi, U \times V)$ is a submanifold chart of class C^r for $f^{-1}(z)$. In order to show that $f^{-1}(z)$ is orientable, suppose that (ψ', U') is another chart such that $\psi'(x) = j_1A'x + j_2(f(x) - z)$, $x \in U'$, and $\det(j_1A' + j_2Df(x)) > 0$ for all $x \in U'$, which overlaps with (ψ, U). Let $x \in U \cap U'$, we denote by $V := \operatorname{Ker} Df(x)$, $A_1 = A_{|V}$, $A_2 := A_{|V^\perp}$, $A_1' = A'_{|V}$, $A_2' := A'_{|V_\perp}$, where $V^\perp$ is the orthogonal complement of V in $\mathbb{R}^n$. Then we have the following matrix representation of $D(\psi' \circ \psi^{-1})(\psi(x)) = D\psi'(x) \circ [D\psi(x)]^{-1}$:

$$\begin{bmatrix} A_1' & A_2' \\ 0 & Df(x)_{|V^\perp} \end{bmatrix} \circ \begin{bmatrix} A_1^{-1} & -A_1^{-1}A_2(Df(x)_{|V^\perp})^{-1} \\ 0 & (Df(x)_{|V^\perp})^{-1} \end{bmatrix}$$

$$= \begin{bmatrix} A_1'A_1^{-1} & -A_1'A_1^{-1}A_2(Df(x)_{|V^\perp})^{-1} + A_2'(Df(x)_{|V^\perp})^{-1} \\ 0 & \mathrm{Id} \end{bmatrix}$$

Since both determinants $\det D\psi(x)$ and $\det D\psi'(x)$ are positive, we have that $\det A_1' A_1^{-1} > 0$, and consequently the atlas $\{(\varphi, \mathcal{U})\}$ on $f^{-1}(z)$, where $\mathcal{U} := U \cap f^{-1}(z)$ and $\varphi := \psi|_{\mathcal{U}}$, is oriented. $\square$

Example 1.1.9 As an application of Theorem 1.1.8, we consider the space $M_n(\mathbb{R})$ of all $n \times n$ matrices that can be identified with $\mathbb{R}^{n^2}$ and let $\det : M_n(\mathbb{R}) \to \mathbb{R}$ denote the determinant function. We define $SL(n; \mathbb{R}) := \{A \in M_n(\mathbb{R}); \det(A) = 1\}$, that is, $SL(n; \mathbb{R}) = \det^{-1}(1)$. It is clear that det is C^∞-differentiable, since it is a polynomial. We will show that 1 is a regular value of det. Indeed, it suffices to show that if $\det(A) = 1$, then $D(\det)(A)$: $M_n(\mathbb{R}) \to \mathbb{R}$ is surjective, that is, the gradient of det at A, denoted by $\nabla(\det)(A)$, is different from 0. However, by a direct computation $\dfrac{\partial \det}{\partial x_{ij}}(A) = M_{ij}$, where M_{ij} denotes the *minor* of the (i, j)th element of the matrix A, that is, $M_{ij} = (-1)^{i+j} \det(A_{ij})$, where A_{ij} denotes the matrix obtained from A by deleting the row and column containing the (i, j)th element of the matrix A. Consequently, $\nabla(\det)(A) \neq 0$, since $\det(A) = 1$, $\det(A_{ij}) \neq 0$ for some $i, j \in \{1, \ldots, n\}$. Therefore, the set $SL(n, \mathbb{R})$ is a C^∞-submanifold of $M_n(\mathbb{R}) = \mathbb{R}^{n^2}$.

EXERCISES

1.1.1 Show that the hyperboloid in $\mathbb{R}^3$, defined by $x^2 + y^2 - z^2 = 4$, is manifold. Is the set $\{(x, y, z); x^2 + y^2 - z^2 = 0\}$ a manifold?

1.1.2 Let $f : \mathbb{R}^n \to \mathbb{R}^m$ be a map of class C^r. Show that the *graph* of f defined by $\text{graph}(f) = \{(x, f(x)); x \in \mathbb{R}^n\}$ is a C^r-manifold.

1.1.3 Show that the set $A = \{(x_1, x_2, x_3, x_4) \in \mathbb{R}^4; x_1^2 + x_2^2 = 1, x_3^2 + x_4^2 = 1\}$ is a submanifold of $\mathbb{R}^4$.

1.1.4 Show that the set $O(n)$ of all $n \times n$ orthogonal matrices is a C^∞-submanifold of $M_n(\mathbb{R})$. What is the dimension of $O(n)$? *Hint:* Denote by $S(n)$ the vector space of all real $n \times n$ symmetric matrices and consider the map $\varphi : M_n(\mathbb{R}) \to S(n)$ given by $\varphi(A) = A^t A$, where $A \in M_n(\mathbb{R})$. Show that $\text{Id} \in S(n)$ is a regular value of φ and $\varphi^{-1}(\text{Id}) = O(n)$.

1.1.5 Show that the set of all complex unitary matrices $U(n)$ is a manifold. *Hint:* Follow similar steps as in Exercise 1.1.4.

1.1.6 Let M be C^r-manifold. Show that the *diagonal* $\Delta(M) := \{(x, x) \in M \times M; x \in M\}$ is a submanifold of $M \times M$.

1.1.7 On $\mathbb{R}^{n+1} \backslash \{0\}$ we define $x \sim y$ if there is a nonzero real constant λ such that $x = \lambda y$. The relation $\sim$ is an equivalence relation. The *real projective space* $\mathbb{RP}^n$ is defined as the quotient space of $(\mathbb{R}^{n+1} \backslash \{0\} / \sim$.

Show that $\mathbb{RP}^n$ is a compact smooth manifold. *Hint:* Use the fact that $\mathbb{RP}^n = S^n/\sim$.

1.1.8 Show that neither S^n nor $\mathbb{RP}^n$ can be covered by a single chart. *Hint:* Both S^n and $\mathbb{RP}^n$ are compact.

1.1.9 Show that $\mathbb{RP}^1 \simeq S^1$.

1.1.10 Show that $\mathbb{RP}^2$ is not orientable.

1.2 DIFFERENTIABLE MAPS, LIE GROUPS, AND TANGENT BUNDLES

From now on, when discussing a manifold (M, α), we will frequently suppress the notation α of a differentiable structure on the manifold M.

Let $f: M \to N$ be a map between two C^r-manifolds. A pair of charts (φ, U) for M and (ψ, V) for N is said to be *adjusted* to f if $f(U) \subseteq V$. In this case, the map $\psi f \varphi^{-1}: \varphi(U) \to \psi(V)$ is defined and is called the *local representation* of f in the given charts. We will also say that $\psi f \varphi^{-1}: \varphi(U) \to \psi(V)$ is a *local representation* of f at every point $x \in U$. It is clear that if f is continuous at a point $x \in M$, then f admits a local representation at the point x.

Definition 1.2.1 A map $f: M \to N$ between two C^r-manifolds is called *differentiable* at a point $x \in M$ if f has a local representation $\psi f \varphi^{-1}: \varphi(U) \to \psi(V)$, which is differentiable at $\varphi(x)$. f is called a C^k-map or a (*differentiable*) *map of class* C^k, $1 \leq k \leq r$, if it has local representations at every point of M that have continuous kth derivatives. If f is of class C^∞, then it is called *smooth*.

One can show that if f is of class C^k, $1 \leq k \leq r$, then every local representation of f is of class C^k. Moreover, if $f: M \to N$ and $g: N \to P$ are C^k-maps between C^r-manifolds, then the composition $g \circ f: M \to P$ is also a C^k-map.

Definition 1.2.2 A *C^k-diffeomorphism* $f: M \to N$ is a C^k-map between C^r-manifolds M and N, $1 \leq k \leq r$, which is a homeomorphism and whose inverse $f^{-1}: N \to M$ is also class C^k. If such a diffeomorphism exists, we call M and N *C^k-diffeomorphic manifolds* (or M and N are *equivalent*) and write $M \approx N$.

Remark 1.2.3 There is no essential difference between C^r and C^s-manifolds for $1 \leq r < s < \infty$ or even $s = \infty$. In fact, Whitney (1944) showed that every C^r-manifold is C^r-diffeomorphic to a C^∞-manifold.

Definition 1.2.4 A *Lie group* is a smooth manifold G that is also a group such that the group multiplication $\cdot : G \times G \to G$ and the inverse map $\nu : G \to G$, $\nu(g) = g^{-1}$, are smooth. A *homomorphism* of Lie groups is a smooth group homomorphism between Lie groups.

The following theorem is a fundamental result in the theory of Lie groups [see, e.g., Adams (1969), Bröcker and tom Dieck (1985), Kawakubo (1991), Zhlelobenko and Shtern (1983)].

Theorem 1.2.5

A closed subgroup H of a Lie group G is also a Lie group.

Example 1.2.6

(i) The unit circle $S^1 = \{z \in \mathbb{C};\ |z| = 1\}$, viewed as a multiplicative subgroup of $\mathbb{C}$, is a compact Lie group.

(ii) We denote by $GL(n, \mathbb{R})$ the set of all invertible $n \times n$ real matrices. In the literature, $GL(n, \mathbb{R})$ is called the *general linear group* of $\mathbb{R}^n$. $GL(n, \mathbb{R})$ is an open subset in the linear space $M_n(\mathbb{R})$ of all $n \times n$ matrices. The multiplication of matrices is evidently a smooth map from $GL(n, \mathbb{R}) \times GL(n, \mathbb{R})$ to $GL(n, \mathbb{R})$. Moreover, the map $\nu : GL(n, \mathbb{R}) \to GL(n, \mathbb{R})$ given by $\nu(A) = A^{-1} = (\det A)^{-1}[M_{ij}]^t$, where $M_{ij} = (-1)^{i+j} \det(A_{ij})$ denotes the minor of the (i, j)th element of the matrix A, is smooth. Consequently, the group $GL(n, \mathbb{R})$ is a Lie group.

(iii) Since the subgroup $SL(n; \mathbb{R}) = \{A \in GL(n, \mathbb{R});\ \det(A) = 1\}$ of $GL(n, \mathbb{R})$ is closed, it is a Lie group.

(iv) The subgroup $O(n)$ of $GL(n, \mathbb{R})$ consisting of all orthogonal matrices is closed in $GL(n, \mathbb{R})$; thus, it is a compact Lie group. Similarly, the group $SO(n) := O(n) \cap SL(n; \mathbb{R})$ is also a compact Lie group.

(v) The group $U(n)$ of all complex unitary $n \times n$ matrices, which is a closed subgroup of $GL(2n, \mathbb{R})$, is a compact Lie group.

(vi) The torus $T^n := \underbrace{S^1 \times \cdots \times S^1}_{n \text{ times}}, n = 2, 3, \ldots$, is a compact Lie group.

(vii) Let $\mathbb{H}$ denote the *quaternion algebra*, which may be described as the $\mathbb{R}$-algebra of 2×2 complex matrices of the form $\begin{bmatrix} a & b \\ -\bar{b} & \bar{a} \end{bmatrix}$ with matrix addition and multiplication. As $\mathbb{H}$ has a standard (complex) *basis* comprised of two elements $1 = \begin{bmatrix} 1 & 0 \\ 0 & 1 \end{bmatrix}$ and $j = \begin{bmatrix} 0 & 1 \\ -1 & 0 \end{bmatrix}$ with the rules for multiplication $zj = j\bar{z}$ for $z \in \mathbb{C}$ and $j^2 = -1$, an element $h \in \mathbb{H}$ can be represented as $h = a + jb$, where $a, b \in \mathbb{C}$. We define $N(a + jb) = |a|^2 + |b|^2$ and the *quaternion group* $Sp(1) = \{a + jb \in$

$\mathbb{H}$; $N(a+jb)=1\}$. Since $\mathbb{H}$ is $\mathbb{R}$-isomorphic to $\mathbb{R}^4$, the group $Sp(1)$ can be identified with S^3.

The dominating concept of the local theory of differentiable manifolds is that of the tangent space at a point x of a manifold M. If the manifold M is embedded in $\mathbb{R}^n$, then it is quite obvious that to every point $x \in M$ there is assigned a certain linear subspace of $\mathbb{R}^n$, the space of tangent vectors to M at x. Since, in general, such an embedding is not canonically given, we must describe the tangent space by intrinsic properties of the manifold.

Let (M, α) be an n-dimensional C^{r+1}-manifold, $1 \leq r \leq \infty$, with $\alpha = \{(\varphi_i, U_i)\}_{i\in\Lambda}$. Define a relation in $\{(x, i, v) \in M \times \Lambda \times \mathbb{R}^n;\ x \in U_i\}$ by $(x, i, v) \sim (y, j, w)$ if and only if $x = y$, and $w = D(\varphi_j\varphi_i^{-1})(\varphi_i(x))v$. We can verify that this relation is an equivalence relation. An equivalence class $[x, i, v]$ of triples (x, i, v) under this equivalence relation $\sim$ is called a *tangent vector* to M at the point x, and the set of all tangent vectors, denoted by TM, is called the *tangent bundle* of M.

The map $p_M : TM \to M$, given by $p_M([x, i, v]) = x$, is well defined. For any subset $A \subseteq M$ we put $p_M^{-1}(A) := T_AM$. If $U \subseteq M$ is open, then $(U, \alpha|_U)$ is also a C^{r+1}-manifold, and we make the harmless identification $T_UM = TU$.

The topological and differential structure on TM is defined as follows. For every chart $(\varphi_i, U_i) \in \alpha$, there is a well-defined bijective map $T\varphi_i : TU_i \to \varphi_i(U_i) \times \mathbb{R}^n \subseteq \mathbb{R}^n \times \mathbb{R}^n$ given by $T\varphi_i([x, i, v]) = (\varphi_i(x), v)$. As the map $(T\varphi_j)(T\varphi_i)^{-1} : \varphi_i(U_i \cap U_j) \times \mathbb{R}^n \to \varphi_j(U_i \cap U_j) \times \mathbb{R}^n$ is clearly the homeomorphism

$$(y, v) \to (\varphi_j\varphi_i^{-1}(y), D(\varphi_j\varphi_i^{-1})(y)v)$$

we can endow TM with the unique topology that makes each $T\varphi_i$ a homeomorphism. Moreover, since $(T\varphi_j)(T\varphi_i)^{-1}$ is a C^r-diffeomorphism, the set of charts $\{(T\varphi_i, TU_i)\}_{i\in\Lambda}$ is a C^r-atlas on TM. Consequently, TM is a C^r-manifold and the projection map $p_M : TM \to M$ is a C^r-map. The charts $(T\varphi_i, TU_i)$ are called *natural charts* on TM.

For every $x \in M$, let $T_xM := p_M^{-1}(\{x\})$. For a given chart $(\varphi_i, U_i) \in \alpha$, if $x \in U_i$, then the map $T_x\varphi_i : T_xM \to \mathbb{R}^n$ defined as the composition

$$T_xM \xrightarrow{\subseteq} TU_i \xrightarrow{T\varphi_i} \varphi_i(U_i) \times \mathbb{R}^n \xrightarrow{\pi_2} \mathbb{R}^n$$

where π_2 denotes the projection onto $\mathbb{R}^n$, is a bijection. Hence, $T_x\varphi_i$ induces on T_xM a structure of an n-dimensional vector space. This structure is independent of i. Indeed, for $x \in U_j$,

$$(T_x\varphi_j)(T_x\varphi_i)^{-1} = D(\varphi_j\varphi_i^{-1})(\varphi_i(x))$$

is a linear isomorphism of $\mathbb{R}^n$. T_xM equipped with the above vector space structure is called the *tangent space* to M at x. It is clear that $TM = \bigcup_{x\in M} T_xM$ and $T_xM \cap T_yM = \varnothing$ for $x \neq y$.

Let $f: M \to N$ be a C^{r+1}-map, $1 \le r \le \infty$. Assume $\varphi_i : U_i \to \mathbb{R}^m$, $\psi_j : V_j \to \mathbb{R}^n$ are charts for M and N, respectively, with $f(U_i) \subseteq V_j$. Define the C^r-map $(Tf)_{ij} : TU_i \to TV_j$ by

$$(Tf)_{ij}([x, i, v]) = [f(x), j, D(\psi_j f \varphi_i^{-1})(\varphi_i(x))v]$$

By the chain rule this map is independent of the choice of charts for M and N. So, there is a well-defined map $Tf : TM \to TN$ that coincides with $(Tf)_{ij}$ on TU_i. That is, a local representation of Tf in natural charts of TM and TN is the derivative of the corresponding local representation of f.

If $f(x) = y$, then Tf maps T_xM into T_yN, and the restriction of Tf to the tangent space T_xM is a linear map denoted by $T_xf : T_xM \to T_yN$. In the natural chart this is just the derivative of x of the corresponding local representation of f. Thus, T_xf may be thought of as the derivative of f at x, and Tf is called the *tangent* of f.

Using natural charts, we can easily show the diagram

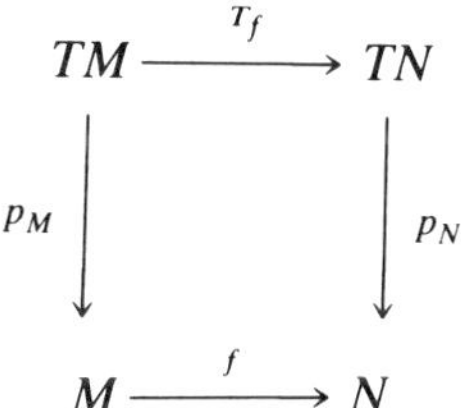

commutes, that is, $f \circ p_M = p_N \circ Tf$. Moreover, if $f : M \to N$ and $g : N \to P$ are C^{r+1}-maps, then $T(g \circ f) = (Tg) \circ (Tf)$.

The tangent vector bundle TM can be described in another, but equivalent way. We say that a C^{r+1}-map $\sigma : J \to M$ from an open interval J into M is a *curve of class* C^{r+1} in M. Two curves $\sigma_1 : J_1 \to M$ and $\sigma_2 : J_2 \to M$ are called *related* at x if:

(i) $J_1 \cap J_2 \ni 0$.
(ii) $\sigma_1(0) = \sigma_2(0) = x$.

For two related at x curves σ_1 and σ_2, we write $\sigma_1 \sim \sigma_2$ if $\frac{d}{dt}(\varphi \circ \sigma_1)(0) = \frac{d}{dt}(\varphi \circ \sigma_2)(0)$ for every chart (φ, U) such that $x \in U$. The relation $\sim$ is an equivalence relation and the set of equivalence classes of this relation can be shown to be in one-to-one correspondence with T_xM defined above.

Let M be an oriented C^r-manifold. Then the orientation of M determines an orientation of the tangent space T_xM for every $x \in M$. Indeed, suppose that the orientation of M is given by an oriented atlas $\Phi = \{(\varphi_i, U_i)\}$

and $x \in U_i$. Then the map $T_x\varphi_i : T_xM \to \mathbb{R}^n$ defines an oriented basis $(v_1, \ldots, v_n)$ in T_xM such that $T_x\varphi_i(v_j) = e_j$ for $j = 1, \ldots, n$.

Example 1.2.7 Let $f : \mathbb{R}^n \to \mathbb{R}^q$ be a smooth map such that zero is a regular value of f. Assume that $M = f^{-1}(0) \neq \emptyset$. Then M is a submanifold of $\mathbb{R}^n$. The tangent bundle of M can be identified with the set $\{(x, v) \in M \times \mathbb{R}^n;\ v \in \operatorname{Ker} Df(x)\}$. We can define the set $N(M) := \{(x, v) \in M \times \mathbb{R}^n;\ v \perp T_xM\}$, which is called the *normal bundle* of M in $\mathbb{R}^n$. It can be verified (see Exercise 1.2.2) that $N(M)$ is a smooth manifold and the map $\pi : N(M) \to M$ defined by $\pi(x, v) = x$ for $(x, v) \in N(M)$ is smooth.

Definition 1.2.8 Let M be an n-dimensional C^r-manifold, $r \geq 2$. A *Riemannian metric* on M is a function that associates to every point $x \in M$ a positive definite and symmetric bilinear form $\langle \cdot, \cdot \rangle_x : T_xM \times T_xM \to \mathbb{R}$ such that for every chart (φ, U) on M, the map $g : U \to M_n(\mathbb{R})$ defined by $g(x)(v, w) = \langle (T_x\varphi)^{-1}v, (T_x\varphi)^{-1}w \rangle_x$, $x \in U$, $v, w \in \mathbb{R}^n$, is of class C^{r-1}. We will denote $|v|_x = \sqrt{\langle v, v \rangle_x}$ and we call a manifold endowed with a Riemannian metric a *Riemannian manifold*.

Let us notice that if M is a submanifold of $\mathbb{R}^n$, then $TM \subset T\mathbb{R}^n = \mathbb{R}^n \times \mathbb{R}^n$, and since the space $\mathbb{R}^n$ is equipped with the standard inner product $\langle \cdot, \cdot \rangle$, its restriction to the tangent space T_xM defines a positive and symmetric bilinear form on T_xM. That implies M is a Riemannian manifold. In what follows, we will show that every C^r-manifold can be equipped with a Riemannian metric. To prove this fact, we need the following technical tool.

Definition 1.2.9 Let X be a topological space. A *partition of unity* for X is a family of continuous real-valued functions $\{\lambda_\alpha\}_{\alpha \in \Lambda}$ defined in X such that:

(i) $\lambda_\alpha(x) \geq 0$ for all $x \in X$, $\alpha \in \Lambda$

(ii) $\{\operatorname{supp}(\lambda_\alpha)\}_{\alpha \in \Lambda}$ is a locally finite covering of X

(iii) $\sum_{\alpha \in \Lambda} \lambda_\alpha(x) = 1$ for all $x \in X$

where $\operatorname{supp}(\lambda_\alpha) = \overline{\{x \in X;\ \lambda_\alpha(x) \neq 0\}}$ is the *support* of λ_α. We say that a partition of unity $\{\lambda_\alpha\}_{\alpha \in \Lambda}$ on X is *subordinate* to a cover $\{U_i\}_{i \in \Gamma}$ of X, if for every $\alpha \in \Lambda$, there is $i \in \Gamma$ such that $\operatorname{supp}(\lambda_\alpha) \subset U_i$.

Let M be a C^r-manifold, $r \geq 1$. A *partition of unity* $\{\lambda_\alpha\}_{\alpha \in \Lambda}$ on M is said to be of class C^r if each of the functions λ_α is of class C^r. The following result says that for every cover $\{U_i\}_{i \in \Gamma}$ of a C^r-manifold M, there always exists a subordinate to $\{U_i\}_{i \in \Gamma}$ partition of unity of class C^r.

Lemma 1.2.10

Let M be a C^r-manifold and $\{U_\beta\}_{\beta\in B}$ an open covering of M. Then there is a C^r-partition of unity subordinate to $\{U_\beta\}_{\beta\in B}$.

Proof. The manifold M is paracompact with countable basis. Thus, we may assume, without loss of generality, that $\{U_\beta\}_{\beta\in B}$ is a locally finite countable covering of M. Let $\{V_\alpha\}_{\alpha\in\Lambda}$ be a refinement of $\{U_\beta\}_{\beta\in B}$ such that $\overline{V_\alpha}$ is compact, and $\{\overline{V_\alpha}\}_{\alpha\in\Lambda}$ is also a locally finite countable refinement of $\{U_\beta\}_{\beta\subset D}$.

Let $0<r_1<r_2<r$. We define

$$\theta(t)=\begin{cases}\exp\left[\dfrac{1}{(t+r_1)(t+r_2)}\right] & \text{if } -r_2<t<-r_1\\ 0 & \text{otherwise}\end{cases}$$

See Figure 1.2.1.

$$\mu(t)=\frac{\displaystyle\int_{-r_2}^{t}\theta(s)\,ds}{\displaystyle\int_{-r_2}^{-r_1}\theta(s)\,ds},\qquad t\in\mathbb{R}$$

and

$$\gamma(t)=\begin{cases}\mu(t) & \text{if } t<0\\ \mu(-t) & \text{if } t\geq 0\end{cases}$$

See Figures 1.2.2 and 1.2.3.

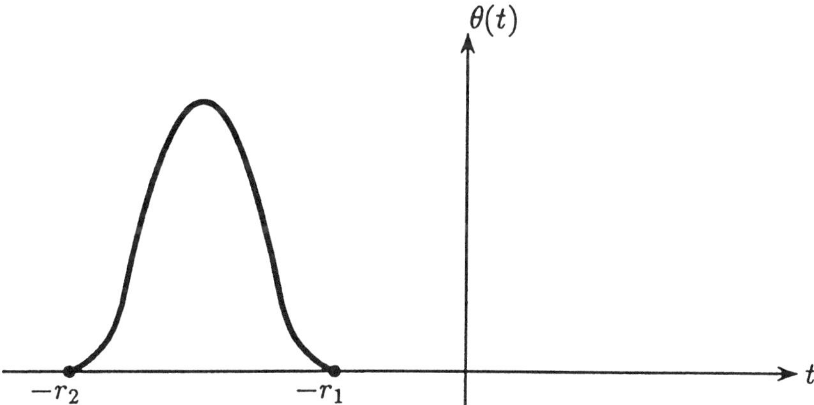

FIGURE 1.2.1. The graph of θ function.

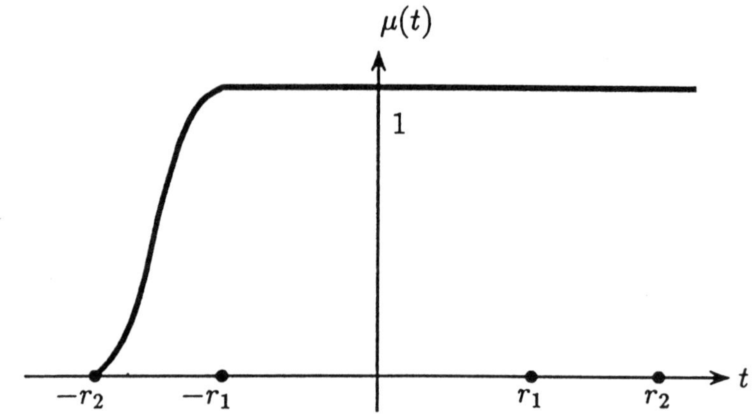

FIGURE 1.2.2. The graph of μ function.

We can identify, using the local charts, a subset U_β with an open subset of $\mathbb{R}^n$, $n = \dim M$. For every $\alpha \in \Lambda$, we can find $\beta \in B$ such that $\overline{V_\alpha} \subseteq U_\beta$ and $\overline{V_\alpha}$ is a compact subset. By the compactness of $\overline{V_\alpha}$, there exists an $\varepsilon > 0$ and a finite set $\{x_1, \ldots, x_N\} \subset \overline{V_\alpha}$ such that $\overline{V_\alpha} \subseteq \bigcup_{i=1}^{N} B_\varepsilon(x_i)$ and $\bigcup_{i=1}^{N} \overline{B_{2\varepsilon}(x_i)} \subset U_\beta$. We put $r_1 = \varepsilon^2$, $r_2 = (2\varepsilon)^2$, and define $\tilde{\lambda}_{i,\alpha}(x) = \gamma(|x - x_i|^2)$ for $i = 1, 2, \ldots, N$. Finally, let

$$\tilde{\lambda}_\alpha(x) = \sum_{i=1}^{N} \tilde{\lambda}_{i,\alpha}(x)$$

It can be verified that $\tilde{\lambda}_\alpha$ is a smooth function on M, $\{\operatorname{supp}(\tilde{\lambda}_\alpha)\}$ is a locally finite compact covering of M such that $V_\alpha \subseteq \operatorname{supp}(\tilde{\lambda}_\alpha) \subseteq U_\beta$. Define

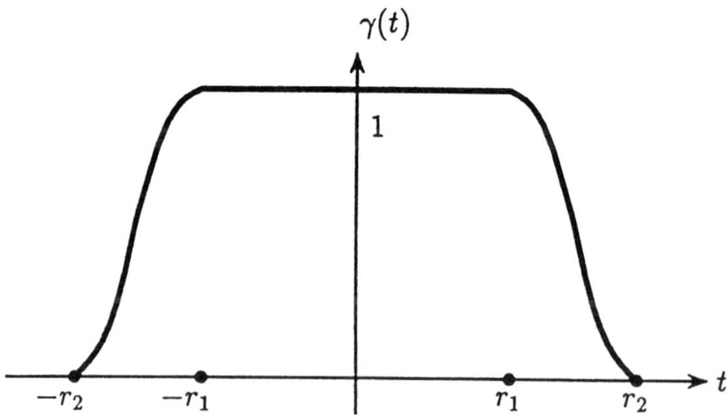

FIGURE 1.2.3. The graph of γ function.

$$\lambda_\alpha(x) = \frac{\tilde{\lambda}_\alpha(x)}{\sum_{\alpha'} \tilde{\lambda}_{\alpha'}(x)}$$

It is clear that $\{\lambda_\alpha\}$ is a C^r-partition of unity subordinate to $\{U_\beta\}_{\beta \in B}$. □

Corollary 1.2.11

Let M be a C^r-manifold and $A, B \subseteq M$ be two closed subsets such that $A \cap B = \varnothing$. Then there exists a C^r-function $\alpha: M \to \mathbb{R}$ separating A and B, that is, $\alpha(x) \in [0, 1]$ for $x \in M$, $A \subseteq \alpha^{-1}(\{0\})$, and $B \subseteq \alpha^{-1}(\{1\})$.

Proof. We consider the covering of M composed of two subsets $\{(M \setminus A), (M \setminus B)\}$. Let $\{\lambda_\alpha\}_{\alpha \in \Lambda}$ be a C^r-partition of unity subordinate to this covering. Let $\Lambda' = \{\alpha \in \Lambda;\ \text{supp}(\lambda_\alpha) \subseteq M \setminus A\}$ and define

$$\alpha(x) = \sum_{\alpha \in \Lambda'} \lambda_\alpha(x)$$

The function α is a well-defined C^r-function on M. Clearly, if $x \in A$, then $\alpha(x) = 0$, and if $x \in B$, then $\alpha(x) = 1$. □

Theorem 1.2.12

Every C^r-manifold M has a Riemannian metric.

Proof. Let $\{(\varphi_i, U_i)\}$ be a C^r-atlas on M and let $\{\lambda_\alpha\}$ be a subordinate to $\{U_i\}$ partition of unity of class C^r. In particular, the set $\{(\psi_\alpha, V_\alpha)\}$, where $V_\alpha = \{x \in M;\ \lambda_\alpha(x) > 0\} \subseteq U_i$, and $\psi_\alpha := \varphi_i|_{V_\alpha}$ for some i, is a C^r-atlas on M. Denote by $\langle \cdot, \cdot \rangle$ the standard inner product on $\mathbb{R}^n$, $n = \dim M$. For $x \in M$ and $v, w \in T_x M$, $\langle v, w \rangle_x := \sum_\alpha \lambda_\alpha(x) \langle (T_x \varphi_\alpha) v, (T_x \varphi_\alpha) w \rangle$. A direct verification shows that $\langle \cdot, \cdot \rangle_x$ is a well-defined Riemannian metric on M. □

Definition 1.2.13 Let M be a Riemannian manifold and let $\sigma: [a, b] \to M$ be a curve of class C^1. The *length* of σ is defined by

$$L(\sigma) = \int_a^b |\dot{\sigma}(t)|_{\sigma(t)}\, dt = \int_a^b \sqrt{\langle \dot{\sigma}(t), \dot{\sigma}(t) \rangle_{\sigma(t)}}\, dt$$

where $\dot{\sigma}(t) = T_t \sigma(1)$, $t \in [a, b]$.

Let $x, y \in M$; we denote by $\mathscr{C}_x^y$ the set of all C^1-curves from x to y, that is, $\sigma : [a, b] \to M$ belongs to $\mathscr{C}_x^y$ if σ is of class C^1, $\sigma(a) = x$, and $\sigma(b) = y$. If M is a connected Riemannian manifold, then the set $\mathscr{C}_x^y$ is nonempty for all

$x, y \in M$. We define the *geodesic distance* $d(x, y)$ from x to y by

$$d(x, y) := \inf\{L(\sigma); \sigma \in \mathscr{C}_x^y\}$$

The geodesic distance is a metric on M and this metric topology coincides with the topology on M. Consequently, every C^r-manifold is metrizable.

EXERCISES

1.2.1 Consider the C^∞-differential structures on $\mathbb{R}$ given by (i) the atlas $\{(\text{Id}, \mathbb{R})\}$, and (ii) the atlas $\{(\varphi, \mathbb{R})\}$, where $\varphi(t) = t^3$. Show that these manifolds are diffeomorphic, whereas the identity map is not a diffeomorphism.

1.2.2 Show that the normal bundle $N(M)$ defined in Example 1.2.7 is a smooth manifold.

1.2.3 Let M be a smooth manifold and $\pi : S^n \to \mathbb{RP}^n$ denote the natural projection. Show that $f : \mathbb{RP}^n \to M$ is smooth if and only if the map $f \circ \pi : S^n \to M$ is smooth.

1.2.4 Let $f : \mathbb{R} \to \mathbb{R}$ be a local diffeomorphism. Prove that the image of f is an open interval and that f maps $\mathbb{R}$ diffeomorphically onto this interval.

1.2.5 Construct $f : \mathbb{R}^2 \to \mathbb{R}^2$ that is a local diffeomorphism but not one-to-one.

1.2.6 Let M be a C^r-manifold and A a nonempty compact submanifold of M. Show that there exists a C^r-function $a : M \to \mathbb{R}$ such that $\alpha^{-1}(0) = A$.

1.2.7 Prove that C^r-differential structures (M, α) and (M, β) on the space M are equivalent if and only if the identity map Id: $(M, \alpha) \to (M, \beta)$ is a diffeomorphism.

1.2.8 Show that the bump function $\gamma : \mathbb{R} \to \mathbb{R}$ used in the proof of Lemma 1.2.10 can also be defined by

$$\gamma(t) = \frac{\varphi(r_2^2 - t^2)}{\varphi(r_2^2 - t^2) + \varphi(t^2 - r_1^2)}$$

where

$$\varphi(t) = \begin{cases} e^{-1/t}, & \text{for } t > 0 \\ 0, & \text{for } t \le 0 \end{cases}$$

1.3 EMBEDDINGS, IMMERSIONS, AND TRANSVERSALITY

In this section we explain the notions of embedding and immersion. Our main goal is to present the well-known theorems due to Sard and Weierstrass. In addition, we discuss a very useful and important result known as the transversality theorem.

Definition 1.3.1 Let $f : M \to N$ be a C^1-map between C^r-manifolds M and N, $r \geq 1$. We say that:

- **(i)** f is *immersive* at $x \in M$ if the linear map $T_x f : T_x M \to T_{f(x)} N$ is injective.
- **(ii)** f is *submersive* at $x \in M$ if $T_x f$ is surjective.
- **(iii)** f is an *immersion* if f is immersive at every point of M.
- **(iv)** f is a *submersion* if f is submersive at every point of M.
- **(v)** f is an *embedding* if f is an immersion and maps M homeomorphically onto its image. We will write $f : M \hookrightarrow N$ to denote that f is an embedding.

In what follows, we provide several useful methods to construct submanifolds. First, we point out that the property of being a C^r-submanifold is *preserved under C^r-diffeomorphisms*, that is, $A \subseteq M$ is a C^r-submanifold of M if and only if $g(A) \subseteq N'$ is a C^r-submanifold of N', where $g : N \to N'$ is a C^r-diffeomorphism.

Theorem 1.3.2

Let N be a C^r-manifold, $r \geq 1$. A subset $A \subseteq N$ is a C^r-submanifold of N if and only if A is the image of a C^r-embedding.

Proof. Suppose that A is a C^r-submanifold. Then A has a natural C^r-differential structure derived from a covering by submanifold charts. With respect to this differential structure, the inclusion of A into N is a C^r-embedding.

Conversely, suppose $f : M \hookrightarrow N$ is a C^r-embedding and $A = f(M)$. Let $\Psi = \{(\psi_i, V_i)\}_{i \in \Lambda}$ be a family of charts $\psi_i : V_i \to \mathbb{R}^n$ in N that covers A. Then we can find an atlas $\Phi = \{(\varphi_i, W_i)\}_{i \in \Lambda}$ for M, $\varphi_i : W_i \to \mathbb{R}^m$, such that $f(W_i) \subseteq V_i$ (reindexing Ψ if necessary). Since f is an embedding, Φ and Ψ can be chosen so that $f(W_i) = A \cap V_i$. By the local character (Remark 1.1.7) and the preservation under diffeomorphisms of submanifolds, it suffices to show that $\psi_i(f(W_i)) \subseteq \mathbb{R}^n$ is a C^r-submanifold. Set $U_i := \varphi_i(W_i) \subseteq \mathbb{R}^m$ and $f_i := \psi_i \circ f \circ \varphi_i^{-1} : U_i \to \mathbb{R}^n$. Then f_i is a C^r-embedding and $f_i(U_i) = \psi_i(f(W_i))$. Thus, we have reduced the verification to the special case where $N = \mathbb{R}^n$, M is an open set $U \subseteq \mathbb{R}^n$, and $f : U \hookrightarrow \mathbb{R}^n$ is a C^r-embedding. In this case, the implicit function theorem implies that there is a C^r-submanifold chart for $(\mathbb{R}^n, f(U))$ at each point of $f(U)$. □

Definition 1.3.3 Let $f: M \to N$ be a C^1-map. We say that $x \in M$ is a *regular point* of f if f is submersive at x. A point x is called a *critical point* of f if it is not a regular point of f. In this case, we call $f(x)$ a *critical value* of f. If $y \in N$ is not a critical value of f, it is called a *regular value*. If $y \in f(M)$ is a regular value, $f^{-1}(y)$ is called a *regular level surface*.

Employing arguments similar to that used in Theorem 1.1.8 and 1.3.2, we can prove the following *regular value theorem*.

Theorem 1.3.4

Let $f: M \to N$ be a C^r-map, $r \geq 1$. If $y \in f(M)$ is a regular value, then the regular level surface $f^{-1}(y)$ is a C^r-submanifold of M. Moreover, $\dim f^{-1}(y) = \dim M - \dim N$.

The above result can be extended to a more general situation. Before stating this extension, we introduce the following important definition.

Definition 1.3.5 A C^1-map $f: M \to N$ is *transverse* to a submanifold $A \subseteq N$ on $B \subseteq M$ if $T_{f(x)}A + T_x f(T_x M) = T_{f(x)}N$ for every $x \in f^{-1}(A) \cap B$. That is, the tangent space to N at $f(x)$ is spanned by the tangent space to A at $f(x)$ and the image of the tangent space to M at x. If f is transverse to A on B, we will write $f \pitchfork_B A$. If $f \pitchfork_M A$, then we will call f *transverse to* A and we will simply write $f \pitchfork A$.

If $f: \mathbb{R}^n \to \mathbb{R}^q$ is a C^1-map and $z \in \mathbb{R}^q$, then $f \pitchfork \{z\}$ simply means that z is a regular value of f.

Theorem 1.3.6

Let $f: M \to N$ be a C^r-map, $r \geq 1$, and $A \subseteq N$ be a C^r-submanifold. If f is transverse to A, then $f^{-1}(A)$ is a C^r-submanifold of M. Moreover, the codimension of $f^{-1}(A)$ in A is the same as the codimension of A in N, that is, $\dim f^{-1}(A) = \dim M - \dim N + \dim A$.

Proof. Due to the local character and the preservation under diffeomorphisms of submanifolds, it suffices to prove the theorem locally. Without loss of generality, we can replace the pair (N, A) by $(U \times V, U \times \{0\})$, where $U \times V \subseteq \mathbb{R}^p \times \mathbb{R}^q$ is an open neighborhood of $(0, 0)$. It is easy to see that the map $f: M \to N$ is transverse to $U \times \{0\}$ if and only if the composition map $g: M \to V$ defined by $g = \pi \circ f$, where $\pi: U \times V \to V$ is the projection onto V, has 0 as a regular value. As $f^{-1}(U \times \{0\}) = g^{-1}(0)$, the theorem follows from Theorem 1.1.8. □

Lemma 1.3.7

Let $U \subset \mathbb{R}^n$ be an open subset and $f : U \to \mathbb{R}^n$ a mapping of class C^1. If $X \subset U$ has a Lebesgue measure zero, then $f(X)$ also has Lebesgue measure zero.

Proof. Let $I^n \subset U$ be a cube of length $a > 0$. By the assumption that f is a C^1-mapping, the norms $\|Df(x)\|$ are uniformly bounded for $x \in I^n$. Therefore, by Taylor's formula, there is a positive constant β such that

$$|f(x) - f(y)| \leq \beta |x - y| \qquad \text{for all } x, y \in I^n$$

Consequently, $f(I^n)$ is contained in a cube of length $\sqrt{n}\beta a =: Ca$. It follows that the Lebesgue measure $\lambda(f(I^n))$ of the image $f(I^n)$ satisfies

$$\lambda(f(I^n)) \leq C^n \lambda(I^n)$$

Let $\varepsilon > 0$ be arbitrarily small. Then the set X can be represented as $X = \bigcup_{i=1}^{\infty} X_i$, where every subset X_i is contained in a cube I_i^n such that $\sum_{i=1}^{\infty} \lambda(I_i^n) < \varepsilon$. Since a countable union of sets of measure zero is a set of measure zero, we can assume, without loss of generality, that all the cubes I_i^n, $i = 1, 2, \ldots$ are contained in I^n. Consequently, $f(X) \subset \bigcup_{i=1}^{\infty} f(I_i^n)$ and

$$\lambda(f(X)) \leq \sum_{i=1}^{\infty} \lambda(f(I_i^n)) \leq C^n \sum_{i=1}^{\infty} \lambda(I_i^n) < C^n \varepsilon$$

Since ε was chosen arbitrarily, the Lebesgue measure of $f(X)$ is zero. □

Theorem 1.3.8

Let M and N be two C^r-manifolds, $r \geq 1$, such that $\dim M < \dim N$. *If $f : M \to N$ is a mapping of class C^1, then the set $f(M)$ is nowhere dense (i.e., its closure does not contain any open set) and therefore $N \backslash f(M)$ is a dense open subset of N.*

Proof. It is sufficient to show that for a representation $f : \overline{U} \to \mathbb{R}^n$ of f in local coordinates, where $\overline{U} \subset \mathbb{R}^m$, $m = \dim M$ and $n = \dim N$, the set $f(\overline{U})$ has Lebesgue measure zero. Indeed, the manifold M can be represented as a union $M = \bigcup_{i=1}^{\infty} \overline{U_i}$, where $\overline{U_i}$ is compact and is contained in some local coordinate set W_i such that $f(W_i)$ is also contained in a local coordinate set V_i of N. Since $f(\overline{U_i})$ is closed and nowhere dense in N, by Baire's category theorem $f(M) = \bigcup_{i=1}^{\infty} f(\overline{U_i})$ is a nowhere dense set of N. Consequently, we can assume that $f : \overline{U} \to \mathbb{R}^n$, where $U \subset \mathbb{R}^m$ is an open set, is a C^1-mapping such that $m < n$. We consider the following composition:

$$\overline{U} \times \{0\} \subset \overline{U} \times \mathbb{R}^{n-m} \xrightarrow{\pi} \overline{U} \xrightarrow{f} \mathbb{R}^n$$

where $\pi : \overline{U} \times \mathbb{R}^{n-m} \to \overline{U}$ is the canonical projection on $\overline{U}$. Since $f \circ \pi(U \times \{0\}) = f(U)$ and $\overline{U} \times \{0\}$, as a subset of $\mathbb{R}^n$, has Lebesgue measure zero, the conclusion follows from Lemma 1.3.7. □

Theorem 1.3.9 (Morse–Sard Theorem)

Let M and N be two C^r-manifolds of dimensions m and n, respectively, and let $f : M \to N$ be a mapping of class C^r, where $r > \max\{0, m - n\}$. We denote by Σ_f the set of all critical points of f. Then the set $f(\Sigma_f)$ is nowhere dense. In particular, the set of all regular values of f is a Baire's second category.

This is an immediate consequence of the following lemma.

Lemma 1.3.10 (Sard's Theorem)

Let $U \subset \mathbb{R}^m$ be an open set and $f : U \to \mathbb{R}^n$ be a mapping of C^r such that $r > \max\{0, m - n\}$. Denote by Σ_f the set of all critical points of f. Then the set $f(\Sigma_f)$ of all critical values of f has Lebesgue measure zero.

Proof. We will present the proof of Sard's theorem (for the sake of simplicity) only for the mapping f of class C^∞. For $m = 0$ and/or $n = 0$, the theorem is clearly true. We may assume that $m, n \geq 1$. The following proof is based on the application of the principle of mathematical induction with respect to m.

Let $\alpha = (\alpha_1, \ldots, \alpha_m)$ denote a multiindex with $\alpha_i \in \mathbb{N} \cup \{0\}$. Define $|\alpha| = \alpha_1 + \cdots + \alpha_n$ and $D^\alpha = \dfrac{\partial^{|\alpha|}}{\partial x_1^{\alpha_1} \cdots \partial x_m^{\alpha_m}}$. Let $f = (f_1, f_2, \ldots, f_n)$. We introduce the following sequence of sets $\Sigma_f \supset \Sigma^1 \supset \Sigma^2 \supset \cdots$ by

$$\Sigma^i := \{x \in U;\, D^\alpha f_j(x) = 0 \text{ for } 1 \leq |\alpha| \leq i \text{ and } j = 1, \ldots, n\}, \quad i = 1, 2, \ldots$$

In the proof, we will employ the following form of Fubini's theorem: *If $A \subseteq \mathbb{R}^n = \mathbb{R}^1 \times \mathbb{R}^{n-1}$ is a subset such that the intersection of the set A with every hyperplane $\{t\} \times \mathbb{R}^{n-1}$, $t \in \mathbb{R}^1$, has $(n-1)$-dimensional Lebesgue measure zero, then A has n-dimensional Lebesgue measure zero.* For details, see Abraham et al. (1988).

Step 1. *The set $f(\Sigma_f \backslash \Sigma^1)$ has measure zero.* Since for $n = 1$ we have $\Sigma_f = \Sigma^1$, we may assume that $n \geq 2$. Suppose that $x' \in \Sigma_f \backslash \Sigma^1$. To complete the proof, it suffices to show that there exists an open neighborhood $V \subseteq \mathbb{R}^m$ of x' such that $f(V \cap \Sigma_f)$ has Lebesgue measure zero, as the set $\Sigma_f \backslash \Sigma^1$ may be covered by a countable collection of such sets V.

Since $x' \not\in \Sigma^1$, at least one of the partial derivatives, say, $\dfrac{\partial f_1}{\partial x_1}$, is not equal to zero. Define a mapping $h : U \to \mathbb{R}^m$ by $h(x) = (f_1(x), x_2, \ldots, x_m)$. Since the derivative $Dh(x')$ is an isomorphism, the implicit function theorem implies that a neighborhood $V \subset \mathbb{R}^m$ of x' may be chosen such that h maps V diffeomorphically onto a neighborhood V' of $h(x')$. Consider the composition $g := f \circ h^{-1} : V' \to \mathbb{R}^n$. The set $\tilde{\Sigma}$ of critical points of g coincides with $h(V \cap \Sigma_f)$, that is, $g(\tilde{\Sigma}) = f(V \cap \Sigma_f)$ is the set of critical values for g. Figure 1.3.1 illustrates the various mappings involved.

It is easy to see that for each $(t, x_2, \ldots, x_n) \in V'$, $g(t, x_2, \ldots, x_n)$ lies on the hyperplane $\{t\} \times \mathbb{R}^{n-1}$; accordingly, g maps the hyperplane $\{t\} \times \mathbb{R}^{m-1}$ into the hyperplane $\{t\} \times \mathbb{R}^{n-1}$. We consider the family of smooth maps $g^t : (\{t\} \times \mathbb{R}^{m-1}) \cap V' \to \{t\} \times \mathbb{R}^{n-1}$. Observe that a point $p \in \{t\} \times \mathbb{R}^{m-1}$ is a critical point of g^t if and only if p is a critical point of g. Indeed, we have the following relation of Jacobian matrices:

$$\left[\frac{\partial g_i}{\partial x_j}\right] = \begin{bmatrix} 1 & 0 \\ * & \left[\dfrac{\partial g_i^t}{\partial x_j}\right] \end{bmatrix}$$

Notice that for $m = 0$, the statement is obvious. We suppose that the statement holds for $m - 1$. Now, since the Lebesgue measure of $g(\tilde{\Sigma} \cap (\{t\} \times \mathbb{R}^{m-1})) = g^t(\tilde{\Sigma} \cap (\{t\} \times \mathbb{R}^{m-1}))$ is zero for all t, we see that the conclusion for m follows from Fubini's theorem.

STEP 2. *The set $f(\Sigma^i \backslash \Sigma^{i+1})$ has Lebesgue measure zero for $i \geq 1$.* We shall sketch the proof only, as it is analogous to that of Step 1. For every point

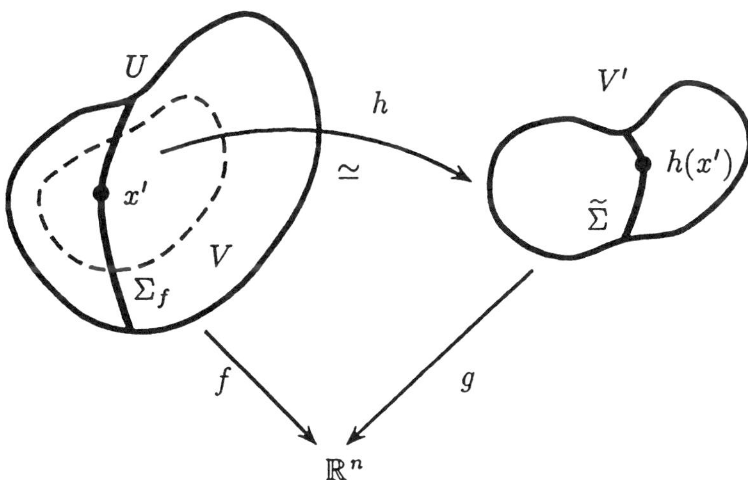

FIGURE 1.3.1. Diagram of mappings involved in the proof of Sard's Theorem.

$x' \in \Sigma^i \backslash \Sigma^{i+1}$, all the partial derivatives of f of order $\leq i$ are zero, but there is a multi-index α such that $|\alpha| = i$ and $\frac{\partial}{\partial x_j} D^\alpha f_k(x') \neq 0$. We denote by w the function $w(x) = D^\alpha f_k(x)$. Then $w(x') = 0$ and $\frac{\partial w}{\partial x_j} \neq 0$. We may assume for simplicity that $j = 1$. Define $h : U \to \mathbb{R}^m$ by $h(x) = (w(x), x_2, \ldots, x_m)$. Then h is a diffeomorphism from a neighborhood V of x' onto a neighborhood V' of $h(x')$ in $\mathbb{R}^m$. Consider the set $h(\Sigma^i \cap V)$. Since $w(x) = D^\alpha f_k(x)$ is one of the coordinate functions of h that is zero on $\Sigma^i \cap V$, the set $h(\Sigma^i \cap V)$ is contained in a hyperplane $\{0\} \times \mathbb{R}^{m-1}$. Consequently, h maps $\Sigma^i \cap V$ into $\{0\} \times \mathbb{R}^{m-1}$. Similarly as in Step 1, we consider the composition $g := f \circ h^{-1} : V' \to \mathbb{R}^n$ and its restriction $g' : (\{0\} \times \mathbb{R}^{m-1}) \cap V' \to \mathbb{R}^n$. Let $\tilde{\Sigma}$ denote the set of critical points of g'. By the induction assumption on m, the Lebesgue measure of $g'(\tilde{\Sigma})$ is zero. Since every point x of the set $h(\Sigma^i \cap V)$ is critical for g' and all the partial derivatives of f at x of order $\leq i$ are zero (in particular, the dimension of the range of $Df(x)$ is smaller than n), we have $g' \circ h(\Sigma^i \cap V) = f(\Sigma^i \cap V)$. Consequently, the Lebesgue measure of $f(\Sigma^i \cap V)$ in $\mathbb{R}^n$ is equal to zero.

STEP 3. *The set $f(\Sigma^r)$ has measure zero for sufficiently large r.* We assume that $r > m/n$ and we set $\nu := r + 1$. Let $I^m \subseteq U$ be a cube of length $a > 0$. It follows from the assumption that f is of class C^{r+1} and from Taylor's formula that

$$|f(x) - f(y)| \leq \beta |x - y|^\nu \qquad \text{for } x \in I^m \cap \Sigma^r, \quad y \in I^m$$

where β is a constant independent of x and y. Let $k \geq 1$ be an integer. Subdivide the cube I^m onto k^m cubes $J_1, \ldots, J_{k^m}$, each of which has length a/k. By the above estimate, for each $i = 1, \ldots, k^m$

$$|f(x) - f(y)| \leq \beta (a\sqrt{m})^\nu \frac{1}{k^\nu}, \qquad x \in J_i \cap \Sigma^r, \quad y \in J_i$$

This says that $f(J_i \cap \Sigma^r)$ lies in a cube of length $2\beta(a\sqrt{m}/k)^\nu$ and, thus, $f(\Sigma^r \cap I^m)$ lies in the union of k^m cubes whose sum of volumes is

$$k^m \left[2\beta \left(\frac{a\sqrt{m}}{k} \right)^\nu \right]^n = k^{m-n\nu} [2\beta a^\nu m^{n\nu/2}]^n = k^{m-n\nu} C$$

Since $\nu > m/n$, $m - n\nu < 0$ and thus $Ck^{m-n\nu} \to 0$ as $k \to \infty$. This shows that the set $f(\Sigma^r \cap I^m)$ has the Lebesgue measure zero. Since $f(\Sigma^r)$ is a countable union of sets $f(\Sigma^r \cap I_j^m)$, where I_j^m are cubes in U, the conclusion that $f(\Sigma^r)$ has measure zero follows. This completes the proof. □

Let M and N be two C^r-manifolds. We denote by $C^r(M, N)$, $0 \leq r \leq \infty$, the set of all C^r-mappings $f : M \to N$. We define the *weak topology* in $C^r(M, N)$, $r < \infty$, as follows:

Let $f \in C^r(M, N)$, (φ, U), (ψ, V) be two charts on M and N, respectively, $K \subseteq U$ a compact subset such that $f(K) \subset V$. Also let $\varepsilon > 0$ be a given real number. We define a basic *weak neighborhood* $\mathcal{N}^r(U, V, K, \varepsilon)$ of f in $C^r(M, N)$ as the set of all C^r-mappings $g : M \to N$ such that $g(K) \subset V$ and for all $x \in \varphi(K)$, $k = 0, \ldots, r$

$$\|D^\alpha(\psi f \varphi^{-1})(x) - D^\alpha(\psi g \varphi^{-1})(x)\| < \varepsilon \,, \qquad |\alpha| \leq k$$

The weak topology on $C^r(M, N)$ is generated by the basis $\{\mathcal{N}^r(U, V, K, \varepsilon)\}$ of weak neighborhoods. The space $C^r(M, N)$ endowed with the weak topology will be denoted by $C^r_w(M, N)$.

Theorem 1.3.11 (WEIERSTRASS THEOREM)

Let M and N be two smooth manifolds. Then the set $C^\infty(M, N)$ of all smooth mappings from M to N is dense in $C^r_w(M, N)$ for $r = 0, 1, \ldots$.

Proof. Let $\varepsilon > 0$, $f \in C^r(M, N)$, and let (φ, U), (ψ, V) be two charts on M and N, $K \subseteq U$ a compact subset such that $f(K) \subset V$. We need to show that there exists $g \in \mathcal{N}^r(U, V, K, \varepsilon) \cap C^\infty(M, N)$. Assume for simplicity that $f(U) \subseteq V$. We consider

$$\psi f \varphi^{-1} : U' \to V' \,, \qquad U' = \varphi(U) \,, \qquad V' = \psi(V)$$

Put $K' = \varphi(K)$. By Corollary 1.2.11, there is a C^∞-function $\alpha : U' \to \mathbb{R}$ with compact support such that $K' \subset \alpha^{-1}(\{1\})$. Then by the classical Weierstrass theorem, there is $\hat{g} \in \mathcal{N}^r(U', V', K', \varepsilon) \cap C^\infty(U', V')$. We put

$$\tilde{g}(v) = \psi f \varphi^{-1}(v) + \alpha(v)[\hat{g}(v) - \psi f \varphi^{-1}(v)] \,, \qquad v \in U'$$

and we define $g' : M \to N$ by

$$g'(x) = \begin{cases} \psi^{-1} \circ \tilde{g} \circ \varphi(x) & \text{if } x \in U \\ f(x) & \text{if } x \notin U \end{cases}$$

It is clear that $g' \in \mathcal{N}^r(U, V, V, K, \varepsilon)$. Let $L = \operatorname{supp}(\alpha)$ be the support of α and put $\Omega = \varphi^{-1}(\operatorname{Int} L)$. Then $g'|_\Omega$ is a C^∞-mapping. In order to "correct" g' to a C^∞-mapping on M, we use countable atlases (φ_i, U_i), (ψ_i, V_i) on M and N such that $g'(U_i) \subset V_i$, and we apply again, locally, the classical Weierstrass theorem, to approximate g' by C^∞-mappings and successively modify it, in the same way as it was done above with f on U, to a C^∞-mapping on other elements of the covering $\{U_i\}$. This construction leads to a C^∞-mapping $g \in \mathcal{N}^r(U, V, K, \varepsilon)$. $\square$

Let M and N be C^r-manifolds. For a submanifold $A \subseteq N$ and a subset $L \subseteq M$, we define

$$\pitchfork^r_L (M,N;A)=\{f\in C^r(M,N);\ f\pitchfork_L A\}$$

We are now in a position to state the transversality theorem.

Theorem 1.3.12 (TRANSVERSALITY THEOREM)

Let M and N be two smooth manifolds, A a compact submanifold of N, L a compact subset of M, and $1\le r<\infty$ a given integer. Then the set $\pitchfork^r_L(M,N;A)$ is open and dense in $C^r_w(M,N)$.

To simplify the proof, we introduce the following notation for a fixed $1\le r<\infty$ and a fixed compact submanifold A of N. For open sets $U\subseteq M$ and $V\subseteq N$ and a compact subset $L\subseteq U$, we let

$$\mathfrak{T}_L(U,V):=\pitchfork^r_L (U,V;V\cap A)\subseteq C^r(U,V)$$

Before we proceed with the proof of the transversality theorem, we establish the following lemma.

Lemma 1.3.13

There exist open coverings $\{U_\alpha\}$ and $\{V_\beta\}$ of the manifolds M and N, respectively, such that if $L\subseteq U_\alpha$ is a compact subset, then $\mathfrak{T}_L(U_\alpha,V_\beta)$ is dense and open in $C^r_w(U_\alpha,V_\beta)$.

Proof. Let $\{(\psi_\beta,V_\beta)\}$ be an atlas on N such that $\psi_\beta(A\cap V_\beta)=\mathbb{R}^k\cap V'_\beta$, where $\dim A=k$, $V'_\beta\subseteq\mathbb{R}^n$ is an open subset. Therefore, it is enough to show that $\mathfrak{T}_L(U_\alpha,V'_\beta)$ is dense and open in $C^r_w(U_\alpha,V'_\beta)$. Since $C^r_w(U_\alpha,V'_\beta)$ is open in $C^r_w(U_\alpha,\mathbb{R}^n)$, we can assume, without loss of generality, that $V'_\beta=\mathbb{R}^n$. Let $\pi:\mathbb{R}^n\to\mathbb{R}^n/\mathbb{R}^k$ be the natrual projection. We remark that if $f\in C^r(U_\alpha,\mathbb{R}^n)$ and $f\pitchfork_x\mathbb{R}^k$ for some $x\in U_\alpha$, then either $f(x)\notin\mathbb{R}^k$ or $f(x)\in\mathbb{R}^k$ and x is a regular point of $\pi\circ f:U_\alpha\to\mathbb{R}^n/\mathbb{R}^k$.

We now show the openness of $\mathfrak{T}_L(U_\alpha,\mathbb{R}^n)$. Let $f\in\mathfrak{T}_L(U_\alpha,\mathbb{R}^n)$. Then every point $y\in L$ has a neighborhood $L_y\subseteq U_\alpha$ such that for all $x\in L_y$, either $f(x)\notin\mathbb{R}^k$ or $f(x)\in\mathbb{R}^k$ and x is a regular point of $\pi\circ f$. Then it is easy to obtain a weak neighborhood of f that is contained in $\mathfrak{T}_L(U_\alpha,\mathbb{R}^n)$. Therefore, the set $\pitchfork^r_L(U_\alpha,\mathbb{R}^n;\mathbb{R}^k)$ is open in $C^r_w(U_\alpha,\mathbb{R}^n)$ and the conclusion follows.

In order to show that the set $\mathfrak{T}_L(U_\alpha,\mathbb{R}^n)$ is dense in $C^r_w(U_\alpha,\mathbb{R}^n)$, we fix $f\in C^r_w(U_\alpha,\mathbb{R}^n)$. By the Weierstrass theorem, C^∞-mappings are dense in $C^r_w(U_\alpha,\mathbb{R}^n)$. Therefore, we can assume, without loss of generality, that f is a C^∞-mapping. By Sard's theorem the set of regular values of $\pi\circ f|_L$ is dense in $\mathbb{R}^n$. Let $\{y_j\}_{j=1}^\infty$ be a sequence of points in $\mathbb{R}^n$ such that $\lim_{j\to\infty}y_j=0$ and $\pi(y_j)$ is regular value of $\pi\circ f:U_\alpha\to\mathbb{R}^n/\mathbb{R}^k$ for each j. We consider the mappings $f_j:U_\alpha\to\mathbb{R}^n$ given by $f_j(x)=f(x)-y_j$, $j=1,2,\ldots$. Then $f_j\to f$ in

$C^r_w(U_\alpha, \mathbb{R}^n)$. Since $f_j \pitchfork \mathbb{R}^k$, it follows that $\pitchfork^r_L(U_\alpha, \mathbb{R}^n; \mathbb{R}^k)$ is dense in $C^r_w(U_\alpha, \mathbb{R}^n)$. This completes the proof. □

We now prove the transversality theorem.

Proof. We fix a mapping $f \in C^r(M, N)$. Consider an atlas $\mathfrak{U} = \{U_i, \varphi_i\}$ on M such that $\{U_i\}$ is a countable locally finite covering of M. Let $\{L_{i_1}, \ldots, L_{i_k}\}$ be compact sets such that $L_{i_j} \subset U_{i_j}$, $L = \bigcup_{j=1}^{k} L_{i_j}$. We put $L_i = \varnothing$ if $i \neq i_j$ for some $j = 1, \ldots, k$. Let $\mathcal{B} = \{\psi_i, V_i\}$ be an atlas on N such that $f(L_i) \subset V_i$. By Lemma 1.3.13, the sets $\mathfrak{T}_{L_i}(U, V_i)$, where $L_i \subseteq U \subseteq U_i$ and U is an open subset of U_i, are open and dense in $C^r_w(U, V_i)$.

We define $\mathcal{M} \subset C^r(M, N)$ to be the set of all $g \in C^r(M, N)$ such that

$$g|_{U_i} \in \mathfrak{T}_{L_i}(U_i, V_i) \qquad \text{for all } i$$

It is clear that $\mathcal{M} \subset \pitchfork^r_L(M, N; A)$. Conversely, if $f \in \pitchfork^r_L(M, N; A)$, then by assumption, $f|_{U_i} \in \pitchfork^r_{L_i}(U_i, V_i; A \cap V_i)$, and therefore $f \in \mathcal{M}$. Since for all i $\mathfrak{T}_{L_i}(U_i, V_i)$ is open in $C^r_w(U_i, V_i)$, it follows from the compactness of L that $\mathcal{M}$ is open in $C^r_w(M, N)$ and, therefore, the openness of $\pitchfork^r_L(M, N; A)$ follows.

In order to prove that $\pitchfork^r_L(M, N; A)$ is dense, we arbitrarily fix a mapping $f \in C^r(M, N)$. Let $\varepsilon_j > 0$, $j = 1, \ldots, k$. We define the following basic neighborhood $\mathcal{N}$ of f:

$$\mathcal{N} := \bigcap_{j=1}^{k} \mathcal{N}^r(U_{i_j}, V_{i_j}, L_{i_j}, \varepsilon_j)$$

Choose $i = i_j$ for $j \in \{1, \ldots, k\}$ and put $E = U_i \cap f^{-1}(V_i)$. Then $L_i \subseteq E$. It follows from Lemma 1.3.13 that $\mathfrak{T}_{L_i}(E, V_i)$ is dense in $C^r_w(E, V_i)$. In what follows, we identify V_i, via the chart ψ_i, with an open subset V'_i of $\mathbb{R}^n$ so that we can use the vector operations in V_i. Let $\lambda : E \to [0, 1]$ be a C^r-mapping with compact support so that $\lambda \equiv 1$ on L_i. Let $g \in C^r_w(E, V_i)$ be sufficiently close to $f|_E$ such that the following mapping $\Gamma(g) = h \in C^r(M, N)$ is well defined

$$h(x) = \begin{cases} f(x) + \lambda(x)[g(x) - f(x)] & \text{for } x \in E \\ f(x) & \text{for } x \in M \backslash E \end{cases}$$

Let us notice that if g converges to $f|_E$ in weak topology, then $\Gamma(g)$ converges to f in weak topology. Since $\mathfrak{T}_{L_i}(E, V_i)$ is dense in $C^r_w(E, V_i)$, we can choose $g \in \mathfrak{T}_{L_i}(E, V_i)$ sufficiently close to $f|_E$ so that $h = \Gamma(g) \in \mathcal{N}$. Since $h \equiv g$ in a neighborhood of L_i, it follows that $h \in \mathfrak{T}_{L_i}(M, N)$. Finally, as $\mathfrak{T}_L(M, N) = \bigcap_{j=1}^{k} \mathfrak{T}_{L_{i_j}}(M, N)$, the density of $\pitchfork_L(M, N; A) = \mathfrak{T}_L(M, N)$ follows from Baire's theorem. □

Proposition 1.3.14

Let M be a smooth manifold, A a compact submanifold of $\mathbb{R}^n$, and L a compact subset of M. Let $f \in C^r_w(M; \mathbb{R}^n)$ be such that $f \in \pitchfork^r_{L_0}(M, \mathbb{R}^n; A)$ for some compact $L_0 \subseteq L$. Then for every neighborhood $\mathcal{N}$ of f in $C^r_w(M; \mathbb{R}^n)$, there exists $h \in \mathcal{N} \cap \pitchfork^r_L(M, \mathbb{R}^n; A)$ such that $h|_{L_0} = f|_{L_0}$.

Proof. It is sufficient to prove the statement for $\mathcal{N} = \mathcal{N}^r(U, V; K, \varepsilon)$, where $L \subseteq K$. Since $f \in \pitchfork^r_{L_0}(M, \mathbb{R}^n; A)$, there is a compact neighborhood B of L_0 such that $f \in \pitchfork^r_B(M, \mathbb{R}^n; A)$. Let $\alpha : M \to [0, 1]$ be a C^∞-function separating K and $M \backslash B$, that is, $\alpha \equiv 1$ on L_0, $\alpha \equiv 0$ on $M \backslash B$. For $g \in \pitchfork^r_L(M, \mathbb{R}^n; A)$, we define

$$h(x) := g(x) + \alpha(x)[f(x) - g(x)] , \qquad x \in M$$

Since the multiplication by $\alpha(x)$ is a continuous operation on $C^r_w(M, \mathbb{R}^n)$, it follows that if g is sufficiently close to f, then $h \in \pitchfork^r_B(M, \mathbb{R}^n; A)$. However, by the definition, $h \in \pitchfork^r_{L \backslash B}(M, \mathbb{R}^n; A)$ and $h|_{L_0} = f|_{L_0}$. Thus, $h \in \pitchfork^r_L(M, \mathbb{R}^n; A)$ and $h \in \mathcal{N}$. □

Corollary 1.3.15

Let M and N be two smooth manifolds, A a compact submanifold of N, and $L_0 \subseteq L$ a pair of compact sets of M. Let $f \in C^r(M, N) \cap \pitchfork^r_{L_0}(M, N; A)$; then for every neighborhood $\mathcal{N}$ of f in $C^r(M, N)$, there exists $g \in \mathcal{N} \cap \pitchfork^r_L(M, N; A)$ such that $g|_{L_0} = f|_{L_0}$.

EXERCISES

1.3.1 Give an example of an immersion that is one-to-one but is not an embedding.

1.3.2 Define explicitly, using elementary functions, an embedding of the torus $S^1 \times S^1$ into $\mathbb{R}^3$.

1.3.3 Let M be a smooth manifold and $\Phi = \{(\varphi_i, U_i)\}$ an atlas on M. For every $x \in U_i$, we define $z(x) := [(x, i, 0)] \in T_xM$. Show that $z : M \to TM$ is a well-defined embedding of M such that $p_M \circ z = \mathrm{Id}_M$ (the map z is called the *zero section* of TM).

1.3.4 Give an example of a continuous map $\sigma : \mathbb{R} \to \mathbb{R}^2$ such that the Lebesgue measure of $\sigma(\mathbb{R})$ is positive.

1.3.5 Show that every C^{r+1}-map $f : S^n \to S^{n+r}$ can be extended to a C^{r+1}-map $\bar{f} : \bar{B}^{n+1} \to S^{n+r}$, where $\bar{B}^{n+1} = \{x \in \mathbb{R}^{n+1}; \|x\| \leq 1\}$.

1.3.6 Let M be a smooth manifold of dimension n and $f : M \to \mathbb{R}^m$ a one-to-one smooth immersion. For a nonzero vector $v \in \mathbb{R}^m$, we denote by V_v the subspace orthogonal to $span(v)$, that is, $V_v := \{x \in \mathbb{R}^m;\ \langle x, v \rangle = 0\}$, and by $P_v : \mathbb{R}^m \to V_v$ the orthogonal projection onto V_v. Show that if $m > 2n + 1$, then there exists a nonzero vector $v \in \mathbb{R}^m$ such that $P_v \circ f : M \to V_v \simeq \mathbb{R}^{m-1}$ is again a one-to-one immersion. *Hint:* Consider the composition

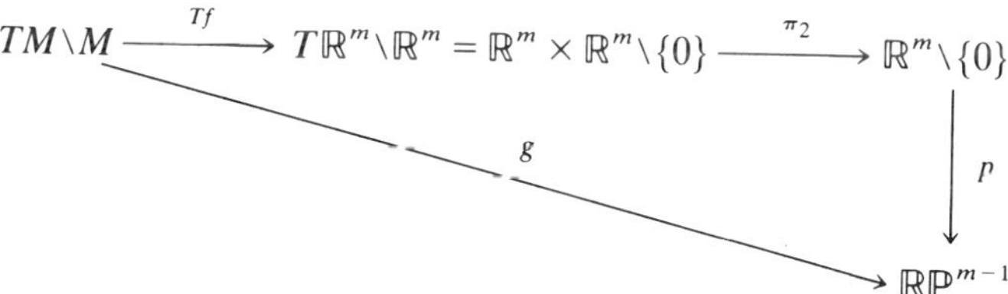

where we identify M (respectively, $\mathbb{R}^m$) with the zero section $z(M)$ of TM [respectively, $z(\mathbb{R}^m)$ of $T\mathbb{R}^m$], $\pi_2 : \mathbb{R}^m \times \mathbb{R}^m \backslash \{0\} \to \mathbb{R}^m \backslash \{0\}$ is a projection on the second component, and $p : \mathbb{R}^m \backslash \{0\} \to \mathbb{RP}^{m-1}$ is the natural projection. Show that the map g is well defined, and that if $p(v) \not\in \mathrm{Im}(g)$, then $P_v \circ f$ is an immersion. Next, consider the map h defined by the following diagram:

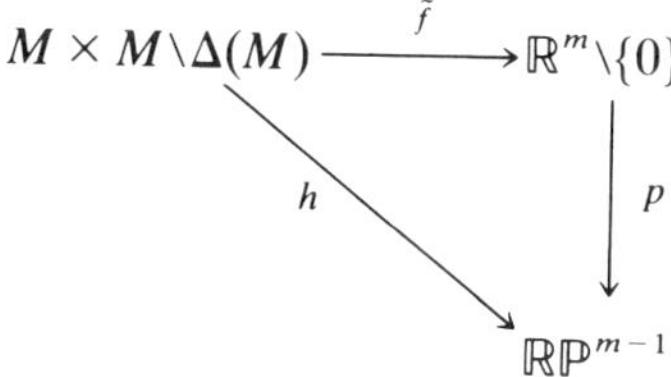

where $\Delta(M) = \{(x, x) \in M \times M;\ \ x \in M\}$ and $\tilde{f}(x, y) = f(x) - f(y)$. Show that if $p(v) \not\in \mathrm{Im}(h)$, then $P_v \circ f$ is one-to-one.

1.3.7 Show that a compact manifold M of dimension n may be embedded into $\mathbb{R}^{2n+1}$ and there exists an immersion from M into $\mathbb{R}^{2n}$. *Hint:* Use a finite atlas on M to construct an embedding $f : M \to \mathbb{R}^N$ and next apply the results from Exercise 1.3.6.

1.3.8 Show that if $f : \mathbb{R}^n \to \mathbb{R}^n$ is a smooth map, then there exists $y \in \mathbb{R}^n$ such that the set $f^{-1}(y)$ is empty or countable.

1.3.9 Construct an embedding of $S^2 \times S^2$ into $\mathbb{R}^5$. *Hint:* Find an open subset in $\mathbb{R}^5$ homeomorphic to $S^2 \times \mathbb{R}^3$.

1.4 VECTOR BUNDLES

The central object in this section is a topological-algebraic mixture called a vector bundle. Roughly speaking, a vector bundle is a topological space with a vector space attached to each point. The most commonly encountered vector bundle is the tangent bundle TM.

Definition 1.4.1 Let E and B be two topological spaces and let $p : E \to B$ be a continuous map. An n-dimensional *vector bundle* is a triple $\xi = (p, E, B)$, where $p : E \to B$ is a surjective map, and for every $x \in B$, the space $E_x := p^{-1}(x)$ has a structure of an n-dimensional real vector space such that the following property is satisfied:

Local Triviality: For every point x in B, there are a neighborhood U and a homeomorphism $\varphi : p^{-1}(U) \to U \times \mathbb{R}^n$ such that the diagram

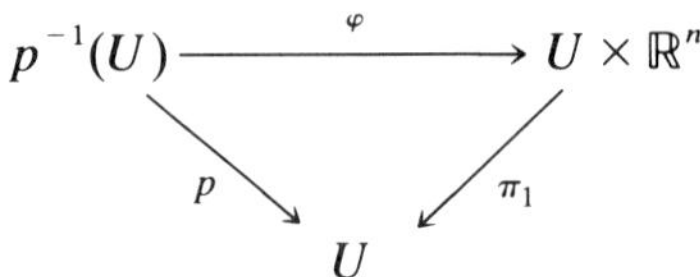

commutes, where $\pi_1(x, y) = x$ for $(x, y) \in U \times \mathbb{R}^n$, and the map $\varphi_x := f_{|E_x} : E_x \to \mathbb{R}^n$ defined as the composition

$$p^{-1}(x) \xrightarrow{\varphi} \{x\} \times \mathbb{R}^n \longrightarrow \mathbb{R}^n$$

is a vector space isomorphism.

Then the space E is called the *total space*, B the *base*, p the *projection*, and E_x the *fiber* of the vector bundle ξ. The pair (φ, U) is called a *vector bundle chart*.

Definition 1.4.2 Let (p, E, B) be a vector bundle of dimension n. A *vector bundle atlas* Φ on (p, E, B) is a family of vector bundle charts $\{(\varphi_i, U_i)\}$ on (p, E, B), where $\{U_i\}$ is a covering of B. A maximal vector bundle atlas α is called a *vector bundle structure* on (p, E, B).

Let Φ be a vector bundle atlas. Whenever (φ, U) and (ψ, V) are in Φ and $x \in U \cap V$, $\psi_x \varphi_x^{-1} : \mathbb{R}^n \to \mathbb{R}^n$ is an isomorphism and the map $x \in U \cap V \mapsto \psi_x \varphi_x^{-1} \in GL(n, \mathbb{R})$ is continuous. In what follows, we will frequently abuse notation and use ξ and E to denote the vector bundle or its total space.

Definition 1.4.3 Let $\xi = (p, E, B)$ be an n-dimensional vector bundle and $E' \subset E$ a subset such that for every point $x \in B$, there exists a bundle chart (φ, U) such that

$$\varphi(p^{-1}(U) \cap E') = U \times \mathbb{R}^k \subset U \times \mathbb{R}^n$$

Then $(p_{|E'}, E', B)$ is a vector bundle, called a k-dimensional *subbundle* of ξ.

If $\Phi = \{(\varphi_i, U_i)\}_{i\in\Lambda}$ is an atlas of a vector bundle ξ, then we obtain a family of functions, called *transition functions*, $g_{ij} : U_i \cap U_j \to GL(n, \mathbb{R})$ defined by $g_{ij}(x) = (\varphi_j)_x(\varphi_i)_x^{-1}$. These functions satisfy the following identities:

(i) $g_{ij}(x)g_{jk}(x) = g_{ik}(x)$ for $x \in U_i \cap U_j \cap U_k$.
(ii) $g_{ii}(x) = \mathrm{Id} \in GL(n, \mathbb{R})$ for $x \in U_i$.
(iii) $g_{ij}(x) = g_{ji}^{-1}(x)$ for $x \in U_i \cap U_j$.

Theorem 1.4.4

Let B be a topological space and $\{U_i\}_{i\in\Lambda}$ an open cover of B. Suppose that there is a system of continuous functions $\{g_{ij}\}$, $i, j \in \Lambda$, $g_{ij} : U_i \cap U_j \to GL(n, \mathbb{R})$ satisfying the conditions (i), (ii) and (iii) above. Then there exists a vector bundle $\xi = (p, E, B)$ having transition functions $\{g_{ij}\}$.

Proof. We construct the vector bundle $\xi = (p, E, B)$ as follows: Consider the space $\mathscr{E} := \{(x, i, v) \in B \times \Lambda \times \mathbb{R}^n;\ x \in U_i\}$ with the topology induced from $B \times \Lambda \times \mathbb{R}^n$, where Λ is endowed with the discrete topology. We introduce on $\mathscr{E}$ the following equivalence relation:

$$(x, i, v) \sim (y, j, w) \qquad \text{if and only if } x = y,\ g_{ij}(x)v = w$$

Let $E = \mathscr{E}/\sim$ be the set of all equivalence classes of $\sim$ equipped with the quotient topology. We notice that there exists a unique continuous map $p : E \to B$ such that the following diagram:

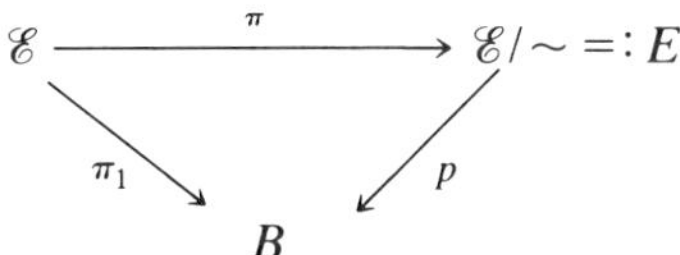

commutes, where $\pi : \mathscr{E} \to \mathscr{E}/\sim$ is the natural projection and $\pi_1 : \mathscr{E} \to B$ denotes the projection onto B, that is, $\pi_1(x, i, v) = x$. Moreover, for every $i \in \Lambda$, the inclusion $U_i \times \mathbb{R}^n \simeq U_i \times \{i\} \times \mathbb{R}^n \hookrightarrow \mathscr{E}$ composed with the projection $\pi : \mathscr{E} \to \mathscr{E}/\sim = E$ is a homeomorphism, denoted by φ_i^{-1}, from $U_i \times \mathbb{R}^n$ onto an open subset of E. It is easy to check that $\{(\varphi_i, U_i)\}_{i\in\Lambda}$ is a vector bundle atlas on $\xi = (p, E, B)$ and the transition functions of ξ are exactly g_{ij}, $i, j \in \Lambda$. □

Example 1.4.5

(i) $\varepsilon_B^n = (p, B \times \mathbb{R}^n, B)$ is a vector bundle, where $p : B \times \mathbb{R}^n \to B$ is the natural projection. This is called a *product* vector bundle.

(ii) Let $\xi = (p, E, B)$ be a vector bundle and $B_o \subset B$. Then the triple $(p_{|p^{-1}(B_o)}, p^{-1}(B_o), B_o)$ is a vector bundle, which is usually denoted by $\xi_{|B_o}$ and called the *restriction* of ξ to B_o.

(iii) Let $\xi = (p, E, B)$ be a vector bundle and let $f: X \to B$ be a continuous map. We define the space $E_X := \{(e, x) \in E \times X;\ p(e) = f(x)\}$ and the projection $\pi: E_X \to X$ by $\pi(e, x) = x$. Then it can be verified that the triple (π, E_X, X) is a vector bundle of the same dimension as ξ, which is usually written as $f^*(\xi)$ and is called the *induced bundle* of ξ by f.

(iv) Let $\xi = (p, E, B)$ and $\xi' = (p', E', B)$ be two vector bundles over the same total space B. Then there exist vector bundle atlases Φ and Φ', respectively, on ξ and ξ' that are both associated with the same covering $\{U_i\}_{i \in \Lambda}$ of B. Let $\{g_{ij}\}$ and $\{g'_{ij}\}$, respectively, denote the transition functions of ξ and ξ'. By using the transition functions $g_{ij}: U_i \cap U_j \to GL(n, \mathbb{R})$ and $g'_{ij}: U_i \cap U_j \to GL(m, \mathbb{R})$, we can construct a new vector bundle, called the *Whitney sum* $\xi \oplus \xi'$ of ξ and ξ', which is the vector bundle with transition functions $\widetilde{g_{ij}}(x) := g_{ij}(x) \oplus g'_{ij}(x) \in GL(n + m, \mathbb{R})$.

(v) If $\xi' = (p_{|E'}, E', B)$ is a subbundle of ξ of dimension k, then by taking the transition functions $\hat{g}_{ij}(x) := [g_{ij}(x)]: \mathbb{R}^{n-k} \to \mathbb{R}^{n-k}$, where $\mathbb{R}^{n-k} := \mathbb{R}^n/\mathbb{R}^k$ and $[g_{ij}(x)]$ is the isomorphism of the quotient space $\mathbb{R}^n/\mathbb{R}^k$ induced by $g_{ij}(x)$, we can construct a new vector bundle, denoted by ξ/ξ', which is called the *quotient bundle*.

Definition 1.4.6 Let $\xi_i = (p_i, E_i, B_i)$ be a vector bundle, $i = 0, 1$. A *morphism* of vector bundles, $F: \xi_0 \to \xi_1$ is a continuous map $F: E_0 \to E_1$ which covers some continuous map $f: B_0 \to B_1$, that is, the following diagram:

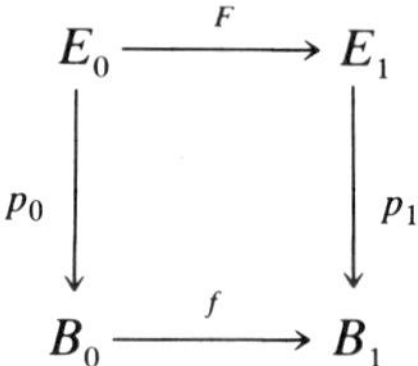

commutes and the restriction $F_x: F_{|E_{0x}}: E_{0x} \to E_{1f(x)}$ is a linear map. A morphism F of vector bundles is called:

(i) *Monomorphism*, if each F_x is injective.

(ii) *Epimorphism*, if each F_x is surjective.

(iii) *Bimorphism*, if each F_x is bijective.

If F is a bimorphism covering a homeomorphism $f: B_0 \to B_1$, then F is called an *equivalence*. If $B = B_0 = B_1$ and $f = \mathrm{Id}_B$, then a morphism F is called a bundle *homomorphism*. If, in addition, the bundle homomorphism F is a bimorphism, then we call it *isomorphism* and we write $\xi_0 \approx \xi_1$. A vector bundle ξ over B is called *trivial* if it is isomorphic to ε_B^n for some n. Such an isomorphism is called a *trivialization* of ξ.

Example 1.4.7 Let $F: E \to E$ be a bundle homomorphism from $\xi = (p, E, B)$ into itself. If for every $x \in B$, the image $F_x(E_x)$ has a constant dimension, then we can define $\mathrm{Ker}(F) := \{v \in E;\, v \in \mathrm{Ker}\, F_{p(v)}\} = \bigcup_{x \in B} \mathrm{Ker}\, F_x$ and $\mathrm{Im}(F) := \{v \in E;\; v \in \mathrm{Im}\, F_{p(v)}\} = \bigcup_{x \in B} \mathrm{Im}\, F_x$. Then it can be verified [Husemoller (1966) or Bröcker and Jänich (1982)] that both subsets $\mathrm{Ker}(F)$ and $\mathrm{Im}(F)$ are subbundles of ξ.

Definition 1.4.8 By a *section* of a vector bundle $\xi = (p, E, B)$ we mean a continuous map $\sigma: B \to E$ such that $p \circ \sigma = \mathrm{Id}_{|B}$. The section, which to every $x \in B$ assigns the zero vector in E_x, is called the *zero section* of ξ and is denoted by $z: B \to E$.

Definition 1.4.9. Two mappings $f, g: X \to Y$ are called *homotopic*, if there exists a continuous mapping $h: [0, 1] \times X \to Y$ such that $h(0, x) = f(x)$ and $h(1, x) = g(x)$ for all $x \in X$.

We state the following results without proofs. Interested readers may get more details from Husemoller (1966).

Proposition 1.4.10

Let $\xi = (p, E, B)$ be a vector bundle:

(i) *If $f: Y \to X$ and $g: X \to B$ are two continuous mappings, then the vector bundle $f^*(g^*(\xi))$ is isomorphic to the vector bundle $(g \circ f)^*(\xi)$.*

(ii) *If $f, g: X \to B$ are two homotopic mappings, then the induced vector bundles $f^*(\xi)$ and $g^*(\xi)$ are isomorphic.*

(iii) *If $f: X \to B$ is a mapping homotopic to a constant mapping, then the induced vector bundle $f^*(\xi)$ is isomorphic to trivial vector bundle. Consequently, every vector bundle over a contractible topological space is isomorphic to a trivial vector bundle.*

For a differentiability class C^r, $1 \le r \le \infty$, the above definitions make sense if all spaces involved are required to be C^r-manifolds, and maps are required to be C^r-maps. In this case, we talk about *C^r-vector bundles*, *C^r-morphisms*, *C^r-bimorphisms*, or *C^r-isomorphisms*. For $r = \infty$, we will simply say *smooth vector bundles*, *smooth morphisms*, *smooth bimorphisms*, or *smooth isomorphisms*. We also interpret a C^0-vector bundle as a vector bundle originally defined, likewise C^0-morphisms, etc.

Example 1.4.11

(a) The tangent bundle $p : TM \to M$ of a C^{r+1}-manifold M is a C^r-vector bundle. For each chart $\varphi : U \to \mathbb{R}^n$ of the manifold M, we define a vector bundle chart $p^{-1}(U) \to U \times \mathbb{R}^n$ by sending a tangent vector $v \in T_xM$ to $(x, D\varphi(x)v)$. If $f : M \to N$ is a C^{r+1}-map between two C^{r+1}-manifolds, then $Tf : TM \to TN$ is a C^r-vector bundle morphism. We can easily show that Tf is a monomorphism, epimorphism, or equivalence if f is an immersion, submersion, or diffeomorphism, respectively.

(b) If $A \subseteq M$ is a C^{r+1}-submanifold, then TA is a C^r-subbundle of $TM|_A$.

(c) Let $A \subset \mathbb{R}^n$ be a k-dimensional submanifold. Then TA is a k-dimensional vector subbundle of $T\mathbb{R}^n_{|A} = A \times \mathbb{R}^n$. We can define the *normal bundle* of A in $\mathbb{R}^n$ as $\nu(A) := (\pi, N(A), A)$, where $N(A) = \{(x, v) \in A \times \mathbb{R}^n;\ v \perp T_xA\}$ and $\pi(x, v) = x$ for $(x, v) \in N(A)$. It can be verified that $\nu(A)$ is a vector bundle. It is also possible to check that the bundle $\nu(A)$ is isomorphic to the quotient bundle $T\mathbb{R}^n_{|A}/TA$. In what follows, we will denote by $N_x(A)$ the fiber of the normal bundle over the point $x \in A$.

In what follows, we fix a differentiability class C^r, $0 < r \leq \infty$, and write "bundle" for "C^r-vector bundle." We will usually not mention the atlas of a vector bundle explicitly.

In the remaining part of this section, we will prove the existence of tubular neighborhoods of a submanifold in $\mathbb{R}^n$. In the study of a submanifold M in $\mathbb{R}^n$, we must often handle problems that are not local, and thus we can not use local coordinates in $\mathbb{R}^n$. It is very useful to know that there are neighborhoods of M in $\mathbb{R}^n$ in which the "position" of a point is determined by its position in the subspace orthogonal to the submanifold M, that is, in a fiber of the normal bundle. More precisely, let $\nu(M) = (\pi, N(M), M)$ denote the normal bundle of the compact submanifold M in $\mathbb{R}^n$, where $N(M) = \{(x, v) \in M \times \mathbb{R}^n;\ v \perp T_xM\}$. Let us fix $\varepsilon > 0$. We define the set $\mathcal{N}(M, \varepsilon) := \{(x, v) \in N(M);\ |v| < \varepsilon\}$. This set is called the *ε-disk bundle of* $\nu(M)$. Since $\mathcal{N}(M, \varepsilon)$ is an open subset of the manifold $N(M)$, the map $e : \mathcal{N}(M, \varepsilon) \to \mathbb{R}^n$ defined by $e(x, v) = x + v$ is of class C^r. Let $U := e(\mathcal{N}(M, \varepsilon))$. We say that the set U is a *tubular neighborhood* of M in $\mathbb{R}^n$ if the map $e : \mathcal{N}(M, \varepsilon) \to U$ is a diffeomorphism. If U is a tubular neighborhood of a compact submanifold M in $\mathbb{R}^n$, then every point $z \in U$ can be uniquely represented as $z = x + v$, where $x \in M$ and $v \perp T_xM$, via the diffeomorphism.

Theorem 1.4.12

For every compact submanifold M in $\mathbb{R}^n$, there exists a tubular neighborhood.

Proof. Let $x \in M$ be an arbitrary point. The tangent map $T_{(x,0)}e : T_{(x,0)}N(M) = T_xM \times N_x(M) \to T_x\mathbb{R}^n = T_xM \times N_x$ is the identity operator. Consequently, by the implicit function theorem, there exists a neighborhood U_x of $(x, 0)$ in $N(M)$ such that the restricted map $e_{|U_x} : U_x \to e(U_x)$ is a diffeomorphism. By the compactness of M, we may choose a finite covering $\{U_i\}_{i=1}^N$ of the zero section M in $N(M)$ such that for each $i = 1, \ldots, N$, the map $e_{|U_i} : U_i \to e(U_i)$ is a diffeomorphism. Put $U := \bigcup_{i=1}^N U_i$. The restricted map $e_{|U} : U \to e(U)$ is a local homeomorphism. We will show that there exists a neighborhood $V := \mathcal{N}(M, \varepsilon) \subset U$ of the submanifold M in $N(M)$ such that $e_{|V} : V \to e(V)$ is a homeomorphism. For this purpose, it is sufficient to find an open neighborhood V of M such that $e_{|V}$ is injective. Suppose that for any neighborhood V of M in $N(M)$ the restriction $e_{|V}$ is not injective. Then there exist sequences $x_k, y_k \in M$ and $v_k \in N_{x_k}M$ and $w_k \in N_{y_k}M$ such that $x_k \neq y_k$, $|v_k| \to 0$, $|w_k| \to 0$ as $k \to \infty$ and $x_k + v_k = y_k + w_k$. Since the manifold M is compact, we may assume, without loss of generality, that $x_k \to x$ and $y_k \to y$ as $k \to \infty$. By the continuity of e, we see that $x = y$. But then e is not a local homeomorphism at x, which is a contradiction. Consequently, there exists an $\varepsilon > 0$ such that for $V = \mathcal{N}(M, \varepsilon)$, the map $e_{|V} : V \to e(V)$ is a homeomorphism. Therefore, it is a diffeomorphism and the conclusion follows. □

EXERCISES

1.4.1 Show that an n dimensional vector bundle $\xi = (p, E, B)$ is trivial if and only if there exist n sections $s_i : B \to E$, $i = 1, \ldots, n$, of the vector bundle ξ such that for every $x \in B$, the vectors $s_1(x), \ldots, s_n(x)$ form a basis in E_x.

1.4.2 Show that if $\xi = (p, E, B)$ is a trivial bundle, then every induced bundle $f^*(\xi)$, where $f : X \to B$ is a continuous map, is also trivial.

1.4.3 We define the canonical vector bundle γ_n over the projective space $\mathbb{RP}^n$ in the following way: The total space is defined by $E_n := \{(L, v) \in \mathbb{RP}^n \times \mathbb{R}^{n+1};\ v \in L\}$, and the projection $p : E_n \to \mathbb{RP}^n$ is given by $p(L, v) = L$. The space E_1 is called the (unbounded) *Möbius band*. Recall that $\mathbb{RP}^n = (\mathbb{R}^{n+1} \backslash \{0\})/\sim$, where $x \sim y$ if and only if $x = \lambda y$ for a nonzero real number λ. Consequently, the equivalence class L of a point $x \in \mathbb{R}^{n+1} \setminus \{0\}$ can be identified with *span*(x).

(**i**) Show that the vector bundle γ_1 is not isomorphic to a trivial vector bundle.

(**ii**) Show that the natural embedding $i : \mathbb{RP}^1 \hookrightarrow \mathbb{RP}^2$ induces a vector bundle $i^*(\gamma_2)$ that is isomorphic to γ_1.

(**iii**) Show that the vector bundle γ_n is not isomorphic to a trivial bundle for $n \geq 1$.

1.4.4 Show that every one-dimensional vector bundle over $S^1 \simeq \mathbb{RP}^1$ is either trivial or isomorphic to the bundle γ_1.

1.4.5 Show that TS^1 is isomorphic to $S^1 \times \mathbb{R}$.

1.4.6 Let ξ, ζ be two vector bundles over the same base space B. Show that the Whitney sum $\xi \oplus \zeta$ of ξ and ζ is equivalent to the induced vector bundle $\Delta^*(\xi \times \zeta)$, where $\xi \times \zeta$ denotes the product of the vector bundles ξ and ζ over $B \times B$, and $\Delta : B \to B \times B$ is the diagonal map $\Delta(x) = (x, x)$.

1.4.7 Let $\xi = (p, E, B)$ be a vector bundle and $F : \xi \to \xi$ be a vector bundle homomorphism such that $F \circ F = F$. Show that $\mathrm{Ker}(F)$ and $\mathrm{Im}(F)$ are vector subbundles of ξ.

1.4.8 Let $\xi = (p, E, B)$ be a vector bundle and $F : \xi \to \xi$ be a vector bundle homomorphism such that $F \circ F = \mathrm{Id}_E$. Show that $Fix(F) := \{v \in E;\ F(v) = v\}$ is a vector subbundle of ξ.

1.5 ELEMENTS OF HOMOTOPY THEORY

In this section we recall some elementary notions and basic results of the homotopy theory. These will be occasionally used in later chapters, in particular, in those chapters dealing with the equivariant degree theory. For all details and proofs, we refer the reader to Spanier (1966), Switzer (1975), or Greenberg and Harper (1981).

A *topological pair* (X, A) consists of topological space X and a subspace $A \subseteq X$. The pair $(X, \emptyset)$ is identified with the space X. A *subpair* $(X_0, A_0) \subset (X, A)$ is a topological pair with $X_0 \subset X$ and $A_0 \subset A$. By a *pointed space* (X, x_0) we mean a nonempty topological space X with a *base point* $x_0 \in X$. Such a pointed space (X, x_0) may be considered as a topological pair $(X, \{x_0\})$.

A map $f : (X, A) \to (Y, B)$ between two pairs is a continuous map f from X to Y such that $f(A) \subseteq B$. In what follows, we will always assume that all the involved maps are continuous. Given a pair (X, A), we write $[0, 1] \times (X, A)$ to denote the pair $([0, 1] \times X, [0, 1] \times A)$. Let $X_0 \subset X$ and suppose that f_0, $f_1 : (X, A) \to (Y, B)$ are such that $f_{0|X_0} = f_{1|X_0}$, in which case we will say that f_0 and f_1 *agree* on X_0. Then f_0 is *homotopic* to f_1 *relative* to X_0, denoted by $f_0 \sim f_1 \text{ rel } X_0$, if there exists a map $h : [0, 1] \times (X, A) \to (Y, B)$ such that $h(0, x) = f_0(x)$, $h(1, x) = f_1(x)$ for all $x \in X$, and $h(t, x) = f_0(x)$ for $x \in X_0$, $t \in [0, 1]$. Such a map h is called a *homotopy relative* to X_0 from f_0 to f_1 and is denoted by $h : f_0 \sim f_1 \text{ rel } X_0$. If $X_0 = \emptyset$, then we will simply say that h is a *homotopy* from f_0 to f_1. The homotopy relation relative to X_0 is an equivalence relation in the set of maps from (X, A) to (Y, B), which gives a partition of the set into disjoint equivalence classes. These equivalence classes are called homotopy classes relative to X_0. We will use $[X, A; Y, B]_{X_0}$ to denote this set of homotopy classes and we will use $[f]_{X_0}$ (or simply $[f]$) to denote the element of $[X, A; Y, B]_{X_0}$ determined by $f : (X, A) \to (Y, B)$. In the case where $X_0 = \emptyset$, we will denote the set of homotopy classes by $[X, A; Y, B]$.

A map $f : (X, A) \to (Y, B)$ is called a *homotopy equivalence* if there exists a map $g : (Y, B) \to (X, A)$, called a *homotopy inverse* of f, such that $[g \circ f] = [\text{Id}_{(X,A)}]$ and $[f \circ g] = [\text{Id}_{(Y,B)}]$, where $\text{Id}_{(X,A)} : (X, A) \to (X, A)$ and $\text{Id}_{(Y,B)} : (Y, B) \to (Y, B)$ are the identity maps. The pairs (X, A) and (Y, B) are said to be *homotopy equivalent* or to have the same *homotopy type* if there exists a homotopy equivalence $f : (X, A) \to (Y, B)$. In the case where $A = \emptyset$ and $B = \emptyset$, we will say that the spaces X and Y are *homotopy equivalent* or have the same *homotopy type*. A topological space X is said to be *contractible* if the identity map Id_X of X is homotopic to some constant map of X to itself. A homotopy from Id_X to the constant map of X to $x_0 \in X$ is called a *contraction* of X to x_0.

A subspace A of X is called a *retract* of X if the inclusion map $i : A \hookrightarrow X$ has a left inverse, that is, there exists a map $r : X \to A$ such that $r \circ i = \text{Id}_A$. Such a map r is called a *retraction* of X to A.

Let Y be a topological space and (X, A) a topological pair. A map $f : A \to Y$ is said to be *extendable* (respectively, *neighborhood extendable*) over X if there exists a map $\tilde{f} : X \to Y$ (respectively, $\tilde{f} : U \to Y$, where U is a neighborhood of A in X) such that $\tilde{f}_{|A} = f$.

Definition 1.5.1 A metrizable space Y is called an *absolute retract* (respectively, *absolute neighborhood retract*), or simply *AR-space* (respectively, *ANR-space*), if for any metric space X and a closed subset $A \subseteq X$, every map $f : A \to Y$ is extendable (respectively, neighborhood extendable) over X.

Let X and Y be two topological spaces and $r : X \to Y$ and $s : Y \to X$ two continuous maps such that $rs = \text{Id}_Y$, where Id_Y denotes the identity map on Y. Then Y is said to be *r-dominated* by X.

As every ANR-space can be imbedded isometrically as a closed subset of a normed linear space [see, e.g., Dugundji (1966)], we have the following result.

Theorem 1.5.2

A metric space Y is an ANR-space if and only if it is r-dominated by an open subset of a normed linear space.

Definition 1.5.3 A metric space Y is called a *Euclidean neighborhood retract* (or simply *ENR-space*) if it is r-dominated by an open subset in a Euclidean space $\mathbb{R}^n$.

Clearly, every ENR-space is an ANR-space. The condition on Y to be an ANR is not as restrictive as it seems to be at a first glance. For example, the following result implies that every convex subset C of a normed linear space is an AR-space.

Proposition 1.5.4 (DUGUNDJI'S EXTENSION THEOREM)

Let X and Y be normed linear spaces, $A \subseteq X$ closed, and $F : A \to Y$ continuous. Then F has a continuous extension $\tilde{F} : X \to Y$ such that $\tilde{F}(X) \subseteq$ $\operatorname{conv}(F(A))$.

Proof. We will use the following notations: For two subsets A_0 and B_0 of X and $x \in X$, we will denote $d(A_0, B_0) := \inf\{\|x - y\|;\ x \in A_0, y \in B_0\}$, $d(x, A_0) := d(\{x\}, A_0)$, and $\operatorname{diam}(A_0) = \sup\{\|x - y\|; x, y \in A_0\}$. For every $x \in X \setminus A$, let $B_{r_x}(x)$ be the ball with its center at x and the radius $r_x = d(x, A)/6$. Then evidently, $2r_x = \operatorname{diam} B_{r_x}(x) \leq d(B_{r_x}(x), A)$. Since $\{B_{r_x}(x)\}_{x \in X \setminus A}$ is a covering of $X \setminus A$ and $X \setminus A$ is paracompact, $\{B_{r_x}(x)\}_{x \in X \setminus A}$ admits a locally finite refinement $\{U_\lambda\}_{\lambda \in \Lambda}$. Since $U_\lambda \subseteq B_{r_x}(x)$ for some $x \in X \setminus A$, $d(U_\lambda, A) \geq d(B_{r_x}, A) > 0$. Therefore, we can choose $a_\lambda \in A$ such that $d(a_\lambda, U_\lambda) < 2d(U_\lambda, A)$ for every $\lambda \in \Lambda$. We define, for $x \in X \setminus A$,

$$\varphi_\lambda(x) = \begin{cases} 0 & \text{if } x \notin U_\lambda \\ d(x, \partial U_\lambda) & \text{if } x \in U_\lambda \end{cases}$$

and $\psi_\lambda(x) = \dfrac{\varphi_\lambda(x)}{\sum\limits_{\mu \in \Lambda} \varphi_\mu(x)}$. Notice that $\sum\limits_{\mu} \varphi_\mu(x) > 0$ in $X \setminus A$ and ψ_λ is continuous. Moreover, $0 \leq \psi_\lambda(x) \leq 1$ and $\sum\limits_{\lambda} \psi_\lambda(x) = 1$ for $x \in X \setminus A$.

We let

$$\tilde{F}(x) = \begin{cases} F(x) & \text{for } x \in A \\ \sum\limits_{\lambda} \psi_\lambda(x) F(a_\lambda) & \text{for } x \notin A \end{cases}$$

Obviously, $\tilde{F}$ is an extension of F with $\tilde{F}(x) \in \operatorname{conv} F(A)$, $\tilde{F}$ is continuous in $X \setminus A$ and in the interior of A. Moreover,

$$\|\tilde{F}(x) - F(x_0)\| \le \sum_{\lambda} \psi_\lambda(x)\|F(a_\lambda) - F(x_0)\| \qquad \text{for } x \notin A, \quad x_0 \in A$$

Assume $x_0 \in \partial A$. Given $\varepsilon > 0$, we then find $\delta > 0$ such that $\|F(z) - F(x_0)\| < \varepsilon$ in $A \cap B_\delta(x_0)$. Hence, to prove the continuity of $\tilde{F}$ at x_0, it suffices to show that if $\psi_\lambda(x) \neq 0$ (i.e., $x \in U_\lambda$) and $\|x - x_0\|$ is sufficiently small, then the corresponding a_λ must be in $B_\delta(x_0)$, since then

$$\|\tilde{F}(x) - F(x_0)\| \le \varepsilon \sum_{\lambda} \psi_\lambda(x) = \varepsilon$$

Suppose $\|x - x_0\| < \delta/4$ and $\psi_\lambda(x) \neq 0$, that is, $x \in U_\lambda \subseteq B_{r_z}(z)$ for some $z \in X \setminus A$. Then

$$\begin{aligned} \|x - a_\lambda\| &\le d(a_\lambda, U_\lambda) + \operatorname{diam} U_\lambda \\ &\le 2d(U_\lambda, A) + \operatorname{diam} B_{r_z}(z) \\ &\le 3d(U_\lambda, A) \le 3\|x - x_0\| \le 4\|x - x_0\| < \delta \end{aligned}$$

This completes the proof. □

Corollary 1.5.5
Every closed convex set C in a normed linear space X is a retract of X.

Every open subset of an ANR-space is also ANR-space. It follows that every open subset of C is an ANR-space. Moreover, every space r-dominated by an ANR-space is also ANR-space, and every topological product or finite collection of ANR-spaces is an ANR-space. Furthermore, if Y_1 and Y_2 are subspaces of Y such that Y_1, Y_2, and $Y_1 \cap Y_2$ are ANR-spaces, then the union $Y_1 \cup Y_2$ is also an ANR-space.

Definition 1.5.6 Let (X, x_0) be a pointed space. We denote by $\pi_1(X, x_0)$ the set of all homotopy classes relatively $\{0, 1\}$ of the continuous mappings $\gamma : ([0,1], \{0,1\}) \to (X, \{x_0\})$. We denote by $[\gamma]$ the homotopy class of the map γ. The set $\pi(X, x_0)$ is called the *fundamental group* of X (based at x_0).

The name fundamental group is justified as follows: Let $\alpha, \beta : ([0,1], \{0,1\}) \to (X, \{x_0\})$ be two continuous maps. We define a new map $\gamma : ([0,1], \{0,1\}) \to (X, \{x_0\})$, called the *composition* of α and β, by

$$\gamma(t) = \begin{cases} \alpha(2t) & \text{if } 0 \le t \le \frac{1}{2} \\ \beta(2t-1) & \text{if } \frac{1}{2} \le t \le 1 \end{cases}$$

and we put $[\alpha][\beta] := [\gamma]$. This formula defines an operation of multiplication in $\pi_1(X, x_0)$, which does not depend on the choice of representatives α and β. The set $\pi_1(X, x_0)$ together with this operation of multiplication is a group. It is clear that the above definition depends on the choice of the base point x_0. However, in the case of an arc-connected space X, all the fundamental groups $\pi_1(X, x)$, where $x \in X$, are isomorphic. If $f : X \to Y$ is a continuous map, then we can define the map $f_* : \pi_1(X, x_0) \to \pi_1(Y, f(x_0))$ by $f_*([\alpha]) := [f \circ \alpha]$ for $\alpha : ([0, 1], \{0, 1\}) \to (X, \{x_0\})$, which is a group homomorphism.

The fundamental group is also called the *first homotopy group*. It is clear that the group $\pi_1(X, x_0)$ can be described as the set of all homotopy classes of continuous maps $\alpha : (S^1, \{s_0\}) \to (X, \{x_0\})$ (where the homotopy relation is taken relative to the set $\{s_0\}$) for a fixed $s_0 \in S^1$.

It is also possible to define *higher homotopy groups* $\pi_n(X, x_0)$ as the sets $[S^n, \{s_0\}; X, \{x_0\}]_{\{s_0\}}$. The set $\pi_n(X, x_0)$ has also a natural structure of a group, which is called the nth *homotopy group* (based at x_0) of the space X. However, in order to describe this group structure, we will need some additional notations.

Let (X, x_0) and (Y, y_0) be pointed spaces. We denote by $(X \vee Y, \{(x_0, y_0)\})$ the *wedge product* of (X, x_0) and (Y, y_0), which is a pointed space with $X \vee Y := \{(x, y) \in X \times Y;\ x = x_0 \text{ or } y = y_0\}$. The *smash product* $(X \wedge Y, *)$ of (X, x_0) and (Y, y_0) is a pointed space $X \wedge Y := X \times Y / X \vee Y$ with the base point $* = p(X \vee Y)$, where $p : X \times Y \to X \times Y / X \vee Y$ is the natural projection. We denote by $S^1 = \{z \in \mathbb{C};\ |z| = 1\}$ the unit circle in $\mathbb{C}$ and let $s_0 = 1$. Sometimes, it is convenient to identify the circle S^1 with the quotient space $[0, 1]/\{0, 1\}$. The smash product $(S^1 \wedge X, *)$ of (S^1, s_0) and (X, x_0) is called the *suspension* of (X, x_0) and is denoted by $(SX, *)$. An important property of the operation of suspension is that the suspension of a sphere is again a sphere. More precisely, we have that SS^n is naturally homeomorphic to S^{n+1} [see Switzer (1975) and Spanier (1966)]. If we use the identification $S^1 = [0, 1]/\{0, 1\}$, then the suspension SX is the quotient space

$$p : [0, 1] \times X \to SX = ([0, 1] \times X)/(\{0\} \times X \cup [0, 1] \times \{x_0\} \cup \{1\} \times X)$$

where p is the natural projection. For a point $(t, x) \in [0, 1] \times X$, we denote by $[t, x]$ the point $p(t, x)$ of SX. The suspension SX is illustrated in Figure 1.5.1.

We can define a map $\mu : SX \to SX \vee SX$ by

$$\mu([t, x]) = \begin{cases} ([2t, x], x_0)\,, & \text{if } 0 \le t \le \frac{1}{2} \\ (x_0, [2t - 1, x])\,, & \text{if } \frac{1}{2} \le t \le 1 \end{cases}$$

(see Figure 1.5.2). Since

$$\begin{aligned} \pi_n(X, x_0) &= [S^n, \{s_0\}; X, \{x_0\}]_{\{s_0\}} \\ &= [SS^{n-1}, \{*\}; X, \{x_0\}]_{\{*\}} \end{aligned}$$

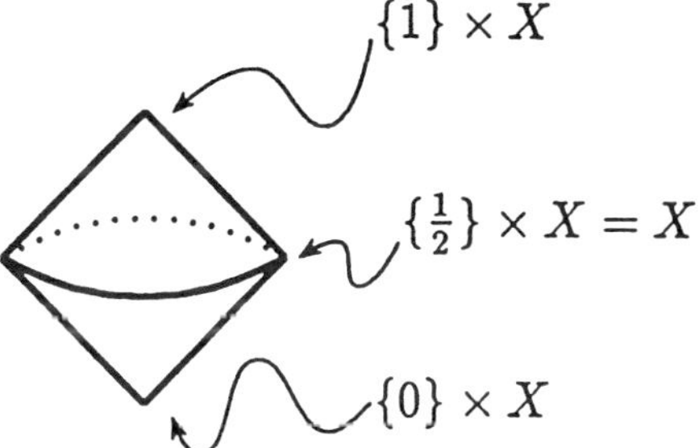

FIGURE 1.5.1. The suspension SX of a space X.

we may define the group operation in $\pi_n(X, x_0)$ as follows: For two maps $\alpha, \beta : (S^n, \{s_0\}) \to (X, \{x_0\})$, we define the product $[\alpha][\beta] := [(\alpha \vee \beta) \circ \mu]$, where

$$(\alpha \vee \beta)(x, y) = \begin{cases} \beta(y) & \text{if } x = s_0 \\ \alpha(x), & \text{if } y = s_0' \end{cases}$$

If $n > 1$, then the group $\pi_n(X, x_0)$ is abelian. As in the case of the fundamental group, any map $f : X \to Y$ induces group homomorphisms $f^* : \pi_n(X, x_0) \to \pi_n(Y, f(x_0))$ for $n = 2, 3, \ldots$, which are called the *induced homomorphisms*.

Let us consider, as an example, the homotopy groups of sphere $\pi_n(S^k, s_0)$. Since both S^n and S^k are smooth manifolds, it is clear (by the Weierstrass theorem) that the homotopy group $\pi_n(S^k, s_0)$ is generated by smooth mappings $\alpha : (S^n, \{s_0\}) \to (S^k, \{s_0'\})$. If $k > n$, then by the transversality theorem α cannot be surjective, and therefore there is a point $x \in S^k$ that

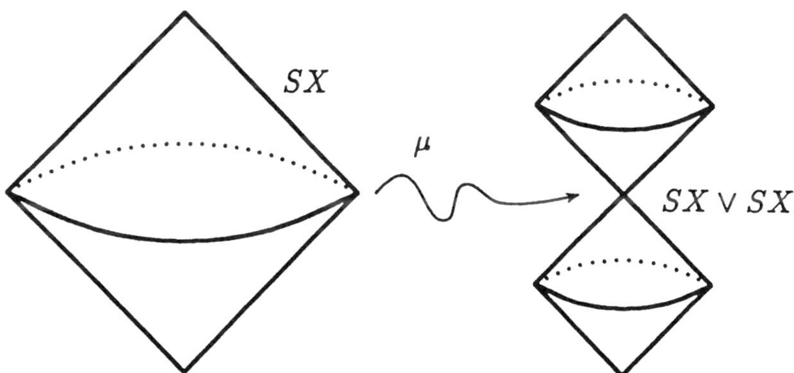

FIGURE 1.5.2. A map from SX into $SX \vee SX$.

does not belong to $\text{Im}(\alpha)$. Since $S^k \setminus \{x\}$ is contractible, the map α is homotopic to a constant map. Therefore, $\pi_n(S^k, s_0) = \{0\}$. If $n = k$, then it can be shown that $\pi_n(S^n, s_0) = \mathbb{Z}$ [see Greenberg and Harper (1981)]. Computations of other homotopy groups of spheres are difficult and involve complex methods of algebraic topology.

Let X and Y be two topological spaces. A map $f : X \to Y$ is called a *weak homotopy equivalence* if f induces a one-to-one correspondence between the path components of X and Y, and if for every $x \in X$, the induced homomorphisms $f_* : \pi_n(X, x) \to \pi_n(Y, f(x))$ are isomorphisms for all $n = 1, 2, 3, \ldots$. If a map $f : X \to Y$ is a weak homotopy equivalence, then there exists a map $g : Y \to X$, which is called a *weak homotopy inverse* of f, such that $[g \circ f]_* = \text{Id} : \pi_n(X, x) \to \pi_n(X, x)$ and $[f \circ g]_* = \text{Id} : \pi_n(Y, f(x)) \to \pi_n(Y, f(x))$ for $n = 1, 2, 3, \ldots$. If there exists a weak homotopy equivalence $f : X \to Y$, then we say that the spaces X and Y are *weakly homotopy equivalent* or X and Y have the same *weak homotopy type*.

One can easily verify that if two spaces X and Y are homotopy equivalent, then they are also weak homotopy equivalent. It is not true, in general, that weak homotopy equivalence implies homotopy equivalence. However, there is a large class of topological spaces for which this implication is true.

Theorem 1.5.7

If $f : X \to Y$ is a weak homotopy equivalence between two ANRs X and Y, then f is a homotopy equivalence.

We refer to Hu (1965), Milnor (1959), Spanier (1966), or Murayama (1983) for some detailed discussions and proofs.

An important problem in the homotopy theory, called the lifting problem, leads to the concept of fibration. Let $p : E \to B$ and $f : X \to B$ be maps. The *lifting problem* for f is to determine whether there is a continuous map $\tilde{f} : X \to E$ such that $f = p \circ \tilde{f}$, that is, the following diagram commutes:

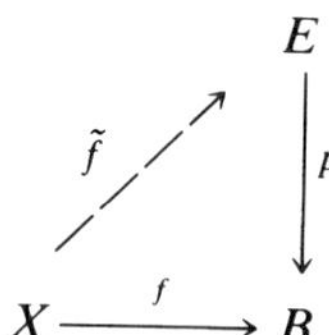

Definition 1.5.8 Let E and B be two topological spaces and $p : E \to B$ a continuous map. The triple (p, E, B) is called *fibration* (or *Hurewicz fiber space*) if for every space X and for every commuting rectangle of continuous maps

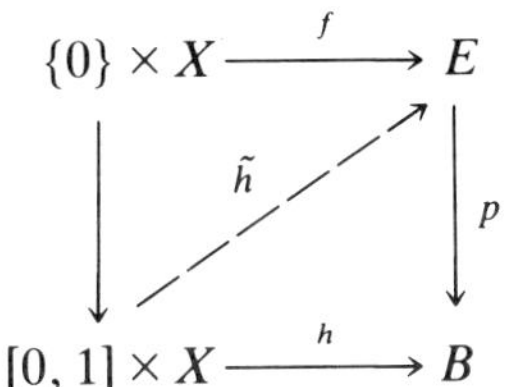

there exists a lift $\tilde{h}$ of h making the diagram commute. We say that the map $p: E \to B$ has the *homotopy lifting property* (*HLP*) with respect to all homotopies $h: [0, 1] \times X \to B$ with an initial lift $f: \{0\} \times X \to E$. For a point $b_0 \in B$, the set $F := p^{-1}(b_0)$ is called the *fiber* over b_0.

Example 1.5.9 A *covering space* $p: \hat{X} \to X$ is a space such that for every $x \in X$, there is an open neighborhood U of x such that $p^{-1}(U)$ is the disjoint union of open subsets of $\hat{X}$, each of which is mapped homeomorphically onto U by p. A covering space is a fibration. See Spanier (1966) for details.

Let G be a compact Lie group and M be a smooth manifold. We say that the group G acts on M if there is a continuous map $\varphi: G \times M \to M$ that satisfies the following properties:

(i) $\varphi(g, \varphi(h, x)) = \varphi(gh, x)$ for all $x \in M$ and $g, h \in G$.
(ii) $\varphi(1, x) = x$ for $x \in M$, where 1 denotes the neutral element of G.

The map φ is called an *action* of G on M. The action φ is called *free* if $\varphi(g, x) \neq x$ for all $g \neq 1$, $g \in G$ and $x \in M$. We should point out that if $\varphi(g, \cdot)$ is a diffeomorphism for all $g \in G$, the continuity of the action φ implies that φ is, in fact, a C^∞-map, that is, the action of G is smooth [Montgomery and Zippin (1955) or Bredon (1972)]. In order to simplify the notations, we will write gx instead of $\varphi(g, x)$ for $g \in G$ and $x \in M$.

Assume that a compact Lie group G acts on a smooth manifold M; then we can define the equivalence relation $x \sim y$ if and only if there is a $g \in G$ such that $gx = y$, where $x, y \in M$. The quotient space $M/\sim$ is denoted by M/G and is called the *orbit space* of M.

We refer the reader to Bredon (1972) for the proof of the following result.

Theorem 1.5.10

Assume that a compact Lie group G acts freely on a smooth manifold M. Then the triple $(p, M, M/G)$, where $p: M \to M/G$ is the quotient map, is a fibration with the fiber homeomorphic to G. In particular, if H is a closed subgroup of G, then the triple $(\pi, G, G/H)$, where $\pi: G \to G/H$ is the quotient map, is a fibration with the fiber homeomorphic to H.

We conclude with the following result, whose proof can be found in Spanier (1966).

Theorem 1.5.11

Suppose that (p, E, B) *is a fibration,* $e_0 \in E$ *and* $b_0 \in B$ *are such that* $p(e_0) = b_0$. *Then we have the following long exact sequence of homotopy groups*:

$$\cdots \longrightarrow \pi_{n+1}(E, e_0) \xrightarrow{p_*} \pi_{n+1}(B, b_0) \xrightarrow{\partial} \pi_n(F, e_0) \xrightarrow{i_*} \pi_n(E, e_0) \xrightarrow{p_*} \cdots$$

where $i : F \rightarrow B$ *denotes the inclusion, and the boundary homomorphism* $\partial \colon \pi_{n+1}(B, b_0) \rightarrow \pi_n(F, e_0)$ *is defined by* $\partial[f] = [f|_{\{0\} \times S \simeq S^n}] \in \pi_n(F, e_0)$ *for every* $[f] \in \pi_{n+1}(B, b_0)$.

1.6 BIBLIOGRAPHICAL NOTES

There are many excellent books on differential manifolds, submanifolds, differentiable maps, tangent bundles, and vector fields. We cite Abraham et al. (1988), Arnol'd (1978), Bröcker and Jänich (1982), Dubrovin et al. (1986), Duda (1986), Guillemin and Pollack (1974), Hirsch (1976), Lang (1985), Matsushima (1966), Milnor (1959), Mischenko and Fomenko (1980), Narasimhan (1968), Pontriagin (1976), to name a few.

The proof of Lemma 1.3.10 (Sard's theorem) is standard and can be found in Milnor (1959), Golubitsky and Guillemin (1973), and Dubrovin et al. (1986); the proof of Theorem 1.3.12 (the transversality theorem) is taken from Hirsch (1976). For more details on transversality, see also Abraham et al. (1988), Guillemin and Pollack (1974), Abraham and Robbin (1967), and Mischenko and Fomenko (1980). For more information on Lie groups and smooth actions, we refer the reader to Adams (1969), Bredon (1972), Bröcker and tom Dieck (1985), Kawakubo (1991), Montgomery and Zippin (1955), and Zhelobenko (1970).

Our presentstion of the results in Section 1.4 regarding vector bundles is along the lines of Husemoller (1966) and Bröcker and Jänich (1982). More materials on vector bundles and related notions may be found in Abraham et al. (1988), Hirsch (1976), and Husemoller (1966).

For more information on the basic concepts of the homotopy theory, we refer the reader to Dold (1972), Fucks et al. (1969), Greenberg and Harper (1981), Palais (1966), Spanier (1966), and Switzer (1975).

Chapter Two

Degree in Finite-Dimensional Spaces

In this chapter we introduce a degree, called the Brouwer degree, in finite-dimensional spaces. We formulate its fundamental axioms, derive basic properties and computational formulae, and give its analytic construction. Some elementary examples will be discussed to demonstrate the applications of this degree. The degree is developed along the axiomatic lines, and an analytic approach is followed for construction.

2.1 AXIOMS AND BASIC PROPERTIES

The main purpose of this section is to formulate fundamental axioms of a degree and to derive some useful properties.

For a subset $D \subseteq \mathbb{R}^n$, we denote by ∂D the boundary and by $\overline{D}$ the closure of D. Let $\Omega \subset \mathbb{R}^n$ be an open, nonempty, bounded set and let $f : \mathbb{R}^n \to \mathbb{R}^n$ be a continuous mapping. We say that f is Ω-*admissible* if $f(x) \neq 0$ for all $x \in \partial\Omega$, and we will call the pair (f, Ω) *admissible*. We denote by $\mathcal{M}$ the set of all admissible pairs (f, Ω).

Our goal is to define an integer-valued function $\deg : \mathcal{M} \to \mathbb{Z}$, called the *Brouwer degree*, satisfying the following properties:

(P1) *Normalization*: Let $\Omega \subseteq \mathbb{R}^n$ be an open, nonempty, bounded set, $x_0 \in \mathbb{R}^n$ such that $x_0 \notin \partial\Omega$. Then

$$\deg(\mathrm{Id} - x_0, \Omega) = \begin{cases} 1 & \text{if } x_0 \in \Omega \\ 0 & \text{if } x_0 \notin \Omega \end{cases}$$

where Id: $\mathbb{R}^n \to \mathbb{R}^n$ is the identity map and x_0 denotes the constant mapping with the value x_0.

(P2) *Additivity*: If $(f, \Omega) \in \mathcal{M}$, then $\deg(f, \Omega) = \deg(f, \Omega_1) + \deg(f, \Omega_2)$

whenever Ω_1 and Ω_2 are two disjoint open nonempty subsets of Ω such that $f^{-1}(0) \cap \bar{\Omega} \subset \Omega_1 \cup \Omega_2$.

(P3) *Homotopy Invariance*: Let $\Omega \subset \mathbb{R}^n$ be an open, nonempty, bounded set and $h : [0, 1] \times \mathbb{R}^n \to \mathbb{R}^n$ be a continuous mapping such that $h(t, x) \neq 0$ for $x \in \partial\Omega$ and $t \in [0, 1]$. Then $\deg(h(t, \cdot), \Omega)$ is independent of $t \in [0, 1]$.

Note that property (P1) is simply a normalization, while (P2) is an abstract formulation of the desire that $\deg(f,\Omega)$ should yield information on the location of zeros of f in Ω in the sense that if Ω_1 and Ω_2 are disjoint open subsets of Ω and f has finitely many zeros in $\Omega_1 \cup \Omega_2$ but no zero in $\bar{\Omega}\backslash(\Omega_1 \cup \Omega_2)$, then the number of zeros of f in Ω is, in some sense, a sum of the numbers of zeros of f in Ω_1 and Ω_2.

Property (P3) reflects the desire that for a complicated f, the integer $\deg(f, \Omega)$ can be calculated by $\deg(g, \Omega)$ with simpler g, at least if f can be continuously deformed into g so that at no stage of the deformation we get zeros on the boundary of Ω. Let, $f, g : \mathbb{R}^n \to \mathbb{R}^n$ be two Ω-admissible mappings. If there is a continuous mapping $h : [0, 1] \times \mathbb{R}^n \to \mathbb{R}^n$ such that $h(t, x) \neq 0$ for $(t, x) \in [0, 1] \times \partial\Omega$, $h(0, x) = f(x)$, and $h(1, x) = g(x)$ for $x \in \mathbb{R}^n$, then we say that f and g are *Ω-homotopic*, h is an *Ω-admissible homotopy* (between f and g), and we write $f \overset{\Omega}{\sim} g$. It is easy to check that $\overset{\Omega}{\sim}$ is an equivalence relation. Property (P3) says that deg is constant on every equivalence class of the relation $\overset{\Omega}{\sim}$.

For a continuous mapping $f : \bar{\Omega} \to \mathbb{R}^n$ defined only on $\bar{\Omega}$, we can apply the Tietze extension theorem* to get a continuous mapping $\hat{f} : \mathbb{R}^n \to \mathbb{R}^n$ so that $\hat{f}|_{\bar{\Omega}} = f$. If we further assume that $f(x) \neq 0$ for $x \in \partial\Omega$, then we can define $\deg(f, \Omega) = \deg(\hat{f}, \Omega)$. It is an easy exercise to apply the homotopy invariance to show that the above definition $\deg(f, \Omega)$ is independent of the choice of the extension $\hat{f}$ of f. So, in what follows, we will always assume the considered mapping is defined in the whole space.

As (P1–P3) involve only topological concepts such as open sets, continuous mappings and boundaries, and the group $\mathbb{Z}$ of integers, it is not surprising that the above degree can be constructed by using various techniques in algebraic topology. We refer interested readers to Bott and Tu (1982), Dold (1983), Dugundji and Granas (1982), Greenberg and Harper (1981), Postnikov (1984), Spanier (1966), and Switzer (1975). Our approach in constructing the degree is analytic in order to better suit the inclinations of analysts and applied mathematicians. The treatment is elementary in the sense that only some basic analytic tools such as the approximation theorem of Weierstrass and Sard's lemma will be employed. Before constructing such

* Tietze Extension Theorem: *Let X be a normal Hausdorff space and $A \subset X$ a closed subset. Then every continuous function $f : A \to \mathbb{R}$ has a continuous extension $\tilde{f} : X \to \mathbb{R}$.* See Dugundi (1951).

a degree, we derive a few useful properties of the degree from the fundamental properties (P1–P3).

Proposition 2.1.1

Assume that $\deg : \mathcal{M} \to \mathbb{Z}$ *is a function satisfying the properties (P1–P3). Then it also satisfies:*

- **(P4)** *Existence: For every* $(f, \Omega) \in \mathcal{M}$, *if* $\deg(f, \Omega) \neq 0$, *then* $f^{-1}(0) \cap \Omega \neq \emptyset$, *that is, there exits a solution* $x \in \Omega$ *of the equation* $f(x) = 0$.
- **(P5)** *Excision: For every* $(f, \Omega) \in \mathcal{M}$, $\deg(f, \Omega) = \deg(f, \Omega_1)$ *whenever* Ω_1 *is an open subset of* Ω *such that* $f^{-1}(0) \cap \Omega \subset \Omega_1$.

Proof. We first prove (P4). Let Ω_1, Ω_2, Ω_3, Ω_4 be disjoint nonempty open subsets of Ω, and assume that $f : \mathbb{R}^n \to \mathbb{R}^n$ is such that $f^{-1}(0) \cap \overline{\Omega} = \emptyset$. Then by the additivity property, we have

$$\deg(f, \Omega) = \deg(f, \Omega_1) + \deg(f, \Omega_2)$$

and

$$\deg(f, \Omega) = \deg(f, \Omega_3) + \deg(f, \Omega_4)$$

On the other hand, application of the additivity property also leads to

$$\begin{aligned}\deg(f, \Omega) &= \deg(f, \Omega_1 \cup \Omega_2) + \deg(f, \Omega_3 \cup \Omega_4)\\ &= \deg(f, \Omega_1) + \deg(f, \Omega_2) + \deg(f, \Omega_3) + \deg(f, \Omega_4)\end{aligned}$$

Putting the above four equalities together, we get $\deg(f, \Omega) = 2\deg(f, \Omega)$, from which it follows that $\deg(f, \Omega) = 0$.

To prove the excision property, we assume that $\Omega_1 \subset \Omega$ and $f^{-1}(0) \cap \Omega \subset \Omega_1$. If $\Omega_1 \neq \Omega$, then we take as Ω_2 the interior of $\Omega \backslash \Omega_1$, which is evidently nonempty. It follows from the additivity property and (P4) that

$$\deg(f, \Omega) = \deg(f, \Omega_1) + \deg(f, \Omega_2) = \deg(f, \Omega_1) \quad \square$$

Let Ω be an open, bounded, nonempty subset of $\mathbb{R}^n$. We denote by $C(\overline{\Omega}; \mathbb{R}^n)$ the space of all continuous mappings $f : \overline{\Omega} \to \mathbb{R}^n$. $C(\overline{\Omega}; \mathbb{R}^n)$, equipped with the supremum norm $\|f\|_\infty := \sup\{|f(x)|;\ x \in \overline{\Omega}\}$ is a Banach space. Let $C(\overline{\Omega}; \partial\Omega)$ be the subset of $C(\overline{\Omega}; \mathbb{R}^n)$ of all mappings $f : \overline{\Omega} \to \mathbb{R}^n$ satisfying $f(x) \neq 0$ for $x \in \partial\Omega$. Then we get a well-defined function $\deg : C(\overline{\Omega}; \partial\Omega) \to \mathbb{Z}$.

Proposition 2.1.2

Assume that $\deg : \mathcal{M} \to \mathbb{Z}$ *satisfies the properties (P1–P3). Then we have:*

(P6) *Continuity: For every nonempty, bounded, open set* $\Omega \subseteq \mathbb{R}^n$, *the associated degree map* $\deg: C(\overline{\Omega}; \partial\Omega) \to \mathbb{Z}$ *is continuous. More precisely, if* $\|f - g\|_\infty < \min\{|f(x)|;\ x \in \partial\Omega\}$, *then* $\deg(f, \Omega) = \deg(g, \Omega)$.

(P7) *Boundary Values Dependence: Let* Ω *be a nonempty, bounded, open set in* $\mathbb{R}^n$. *Then for every* $f, g \in C(\overline{\Omega}; \partial\Omega)$ *satisfying* $f(x) = g(x)$ *for* $x \in \partial\Omega$, *we have* $\deg(f, \Omega) = \deg(g, \Omega)$.

Proof. We first prove (P6). Let $f \in C(\overline{\Omega}; \partial\Omega)$. Since $\partial\Omega$ is compact, $\min\{|f(x)|;\ x \in \partial\Omega\} = \varepsilon > 0$. Assume that $\|f - g\|_\infty < \varepsilon$. For every $x \in \partial\Omega$, we have

$$\begin{aligned} |g(x)| &\geq |f(x)| - |f(x) - g(x)| \geq |f(x)| - \|f - g\|_\infty \\ &\geq \min\{|f(y)|;\ y \in \partial\Omega\} - \|f - g\|_\infty = \varepsilon - \|f - g\|_\infty > 0 \end{aligned}$$

Consequently, $g \in C(\overline{\Omega}; \partial\Omega)$ and $\deg(g, \Omega)$ is well defined. We define the following homotopy h: $[0, 1] \times \overline{\Omega} \to \mathbb{R}^n$, which we will call *linear homotopy*, by

$$h(t, x) = tg(x) + (1 - t)f(x)\,, \qquad t \in [0, 1), \quad x \in \overline{\Omega}$$

Since for $x \in \partial\Omega$ and $t \in [0, 1]$,

$$|h(t, x)| \geq |f(x)| - t|f(x) - g(x)| \geq \varepsilon - t\|f - g\|_\infty > 0$$

the homotopy h is Ω-admissible. Hence, (P3) implies $\deg(f, \Omega) = \deg(g, \Omega)$.

To prove (P7), we note that $f|_{\partial\Omega} = g|_{\partial\Omega}$ guarantees that the linear homotopy $h : [0, 1] \times \overline{\Omega} \to \mathbb{R}^n$ defined above is Ω-admissible and, hence, $\deg(f, \Omega) = \deg(g, \Omega)$. □

We now present the following Brouwer's fixed-point theorem, assuming the Brouwer degree exists.

Theorem 2.1.3

Let $\Omega \subseteq \mathbb{R}^n$ *be a nonempty, open, bounded, and convex set. Then every continuous mapping* $F : \overline{\Omega} \to \overline{\Omega}$ *has a fixed point. That is, there exists* $x \in \overline{\Omega}$ *such that* $F(x) = x$.

Proof. We may assume that $F(x) \neq x$ for all $x \in \partial\Omega$, for otherwise the statement is trivial. We can also assume, without loss of generality, that $0 \in \Omega$. Define a homotopy $h : [0, 1] \times \overline{\Omega} \to \mathbb{R}^n$ as follows:

$$h(t, x) = x - tF(x)\,, \qquad x \in \overline{\Omega}\,, \quad t \in [0, 1]$$

Evidently, h is continuous and for $x \in \partial\Omega$ the point $tF(x)$ belongs to the segment joining $F(x)$ to 0. If $t=1$, then by the assumption $h(1,x)=x-F(x)\neq 0$. If $t\in[0,1)$, then by the assumption that Ω is convex and $0\in\Omega$, $tF(x)\in\Omega$ and consequently for all $x\in\partial\Omega$, we have $tF(x)\neq x$, that is, $h(t,x)\neq 0$. This shows h is an Ω-admissible homotopy. Therefore, the homotopy invariance implies that $\deg(h(t,\cdot),\Omega)$ does not depend on $t\in[0,1]$. In particular,

$$\deg(\mathrm{Id}-F,\Omega)=\deg(\mathrm{Id},\Omega)=1$$

Now, we apply the existence property to conclude that there is $x\in\Omega$ such that $x-F(x)=0$, that is, x is a fixed point of F. □

EXERCISES

2.1.1 Use Brouwer's fixed-point theorem to show that every $n\times n$ matrix with positive entries has a positive eigenvalue.

2.1.2 Let $\Omega\subset\mathbb{R}^n$ be the open unit ball and $f, g:\mathbb{R}^n\to\mathbb{R}^n$ be two Ω-admissible maps satisfying $\lambda f(x)\neq -g(x)$ for all $x\in\partial\Omega=S^{n-1}$ and $\lambda>0$. Show that $\deg(f,\Omega)=\deg(g,\Omega)$.

2.1.3 Let Ω be an open, bounded neighborhood of zero in $\mathbb{R}^n$, and $f=\mathrm{Id}-F:\mathbb{R}^n\to\mathbb{R}^n$ be a continuous map. Show that either F has a fixed point $x\in\overline{\Omega}$, or there is $\lambda>1$ and $y\in\partial\Omega$ such that $\lambda y=F(y)$.

2.1.4 Let $\Omega\subset\mathbb{R}^n$ be the open unit ball and $f:\mathbb{R}^n\to\mathbb{R}^n$ be a continuous map such that $f(x)=x$ for all $x\in S^n$. Show that $\Omega\subset f(\Omega)$.

2.2 COMPUTATION OF THE DEGREE

In this section we assume that a degree $\deg:\mathcal{M}\to\mathbb{Z}$ exists and satisfies fundamental axioms (P1–P3). We want to derive a few computational formulae for the degree when f is a linear or a smooth mapping satisfying a certain regularity condition. The computation of the degree for general mappings will be developed in the next section. We will also give a degree-theoretical proof of the fundamental theorem of algebra to demonstrate how the degree theory is applied.

We will use the following notation:

$$GL^+(n,\mathbb{R})=\{A\in GL(n,\mathbb{R});\ \det A>0\}$$

$$GL^-(n,\mathbb{R})=\{A\in GL(n,\mathbb{R});\ \det A<0\}$$

Note that $GL(n,\mathbb{R})$ can be regarded as a subset of $\mathbb{R}^{n\times n}$, and hence $GL(n,\mathbb{R})$ is a topological space with the topology induced from $\mathbb{R}^{n\times n}$.

Lemma 2.2.1

The sets $GL^+(n, \mathbb{R})$ and $GL^-(n, \mathbb{R})$ are two open, connected components of $GL(n, \mathbb{R})$.

Proof. As $\det : GL(n, \mathbb{R}) \to \mathbb{R}$ is a continuous function, the openness of $GL^{\pm}(n, \mathbb{R})$ is obvious. To prove the connectness of $GL^{\pm}(n, \mathbb{R})$, we will show that every matrix $T \in GL(n, \mathbb{R})$ can be connected by a path in $GL(n, \mathbb{R})$ either to the identity matrix Id or to the matrix J, where

$$J = \begin{bmatrix} -1 & 0 & \cdots & 0 \\ 0 & 1 & \cdots & 0 \\ \vdots & \vdots & \ddots & \vdots \\ 0 & 0 & \cdots & 1 \end{bmatrix}$$

It follows from the spectral theorem [see, e.g., Dunford and Schwartz (1958), and Schwartz (1965)] for linear operators in $\mathbb{R}^n$ that for each fixed $T \in GL(n, \mathbb{R})$, there exist two invariant generalized eigenspaces X^- and X^0 of $\mathbb{R}^n$ such that $\mathbb{R}^n = X^- \oplus X^0$, $\sigma(T^-) = \{\lambda \in \sigma(T);\ \lambda < 0\}$, and $\sigma(T^0) = \sigma(T) \backslash \sigma(T^-)$, where $\sigma(\cdot)$ denotes the set of all (complex) eigenvalues of a linear operator defined in a finite-dimensional vector space, $T^- = T|_{X_-}$ and $T^0 = T|_{X^0}$. Let us write $T : \mathbb{R}^n \to \mathbb{R}^n$ in the form of a block diagonal matrix associated with the decomposition $\mathbb{R}^n = X^- \oplus X^0$:

$$T = \begin{bmatrix} T^- & 0 \\ 0 & T^0 \end{bmatrix}$$

Define the following deformation:

$$T_t = \begin{bmatrix} (1-t)T^- - t\,\mathrm{Id} & 0 \\ 0 & (1-t)T^0 + t\,\mathrm{Id} \end{bmatrix}$$

It is clear that $T_t \in GL(n, \mathbb{R})$ for all $t \in [0, 1]$. Evidently, T_t defines a path in $GL(n, \mathbb{R})$ between the matrix T and the matrix

$$T_1 = \begin{bmatrix} -\mathrm{Id} & 0 \\ 0 & \mathrm{Id} \end{bmatrix}$$

We now fix a basis in X^-. If $\dim X^-$ is even (say, $2k$), we can deform $T_1|_{X^-} = -\mathrm{Id}$ to the identity operator Id by the deformation

$$\begin{bmatrix} A(\theta) & & & \mathbf{0} \\ & A(\theta) & & \\ & & \ddots & \\ \mathbf{0} & & & A(\theta) \end{bmatrix} : X^- \to X^-$$

where

$$A(\theta) = \begin{bmatrix} -\cos\theta & -\sin\theta \\ \sin\theta & -\cos\theta \end{bmatrix}, \qquad \theta \in [0, \pi]$$

If X^- has an odd dimension (say, $2k+1$), the above deformation can be applied to the $2k \times 2k$ block of the matrix $-\mathrm{Id}$ corresponding to the last $2k$ vectors of the basis of X^-. This defines a path in $GL(n, \mathbb{R})$ between T_1 and the matrix J. □

Lemma 2.2.2

Let Ω be an open, bounded neighborhood of zero in $\mathbb{R}^n$ and $T \in GL(n, \mathbb{R})$. Then $\deg(T, \Omega) = \text{sign}\det T$. *Here and in what follows, we will use* sign det T *to denote a real number with absolute value* 1 *and* sign *being determined by the* sign *of* det T.

Proof. By the excision property we can assume, without loss of generality, that Ω is the ball $B_2(0)$ of radius 2 centered at zero. It follows from Lemma 2.2.1 that T can be connected by a path T_t in $GL(n, \mathbb{R})$ either to $T_1 = \mathrm{Id}$ (if $\text{sign}\det T = 1$) or to the matrix $T_1 = J$ (if $\text{sign}\det T = -1$). Evidently, such a path T_t is an Ω-admissible homotopy. Therefore, by the homotopy invariance, $\deg(T, \Omega) = \deg(T_1, \Omega)$. If $\text{sign}\det T > 0$, then $T_1 = \mathrm{Id}$ and the normalization property implies that $\deg(T, \Omega) = \deg(\mathrm{Id}, \Omega) = 1$. If $\det T < 0$, then $\deg(T, \Omega) = \deg(J, \Omega)$. We now define a mapping $f : \mathbb{R}^n \to \mathbb{R}^n$ by

$$f(x_1, x_2, \ldots, x_n) = (1, x_2, \ldots, x_n), \qquad x = (x_1, \ldots, x_n) \in \mathbb{R}^n$$

Since $f(x) \neq 0$ for all $x \in \mathbb{R}^n$, it follows from the existence property that $\deg(f, \Omega) = 0$. Define the following homotopy:

$$h(t, x) = ((1-t) + t\varphi(x_1), x_2, \ldots, x_n), \qquad x = (x_1, \ldots, x_n) \in \mathbb{R}^n$$

where

$$\varphi(t) = \begin{cases} 1 & \text{if } 1 < t \text{ or } t < -2 \\ -t-1 & \text{if } -2 \leq t < -\frac{1}{2} \\ t & \text{if } -\frac{1}{2} \leq t \leq 1 \end{cases}$$

See Figure 2.2.1. Clearly, h is an Ω-admissible homotopy and $h_1(\cdot) := h(1, \cdot)$ satisfies

$$h_1^{-1}(0) = \{(-1, 0, \ldots, 0), (0, 0, \ldots, 0)\}$$

Let Ω_1 and Ω_2 be two disjoint open balls centered at $x_0 = (-1, 0, \ldots, 0)$

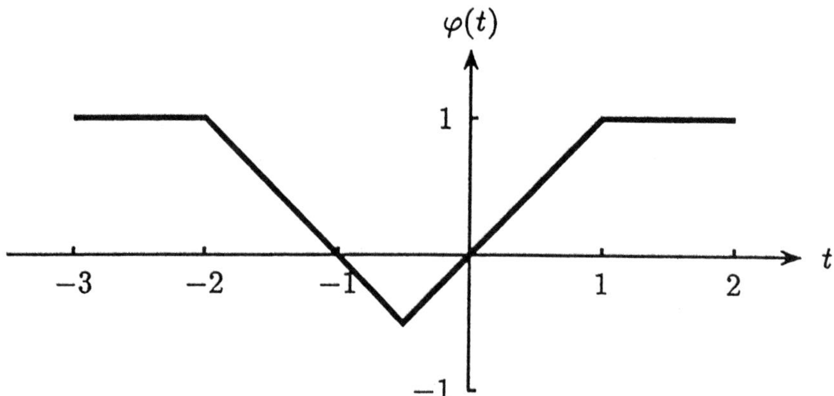

FIGURE 2.2.1. The piecewise linear function φ.

and 0, respectively. Then, since $h_1|_{\Omega_2} = \mathrm{Id}$, we have

$$\begin{aligned}0 &= \deg(f, \Omega) = \deg(h_1, \Omega)\\ &= \deg(h_1, \Omega_1) + \deg(h_1, \Omega_2) = \deg(h_1, \Omega_1) + 1\\ &= \deg(J + x_0, \Omega) + 1\end{aligned}$$

It is trivial to see that $J + tx_0$ is an Ω-admissible homotopy between $J + x_0$ and J. So $\deg(J + x_0, \Omega) = \deg(J, \Omega)$. Therefore, $\deg(T, \Omega) = \deg(J, \Omega) = -1$. □

To state the next result, we recall that zero is a regular value of a C^1-map $g : \Omega \to \mathbb{R}^n$, where Ω is an open, nonempty, bounded subset of $\mathbb{R}^n$, if either $g^{-1}(0) = \varnothing$, or for each $x \in g^{-1}(0)$ we have $\det Dg(x) \neq 0$.

Theorem 2.2.3

Let Ω be a nonempty, open, bounded subset of $\mathbb{R}^n$ and f an Ω-admissible C^1-mapping such that zero is a regular value of $f|_\Omega$. Then

$$\deg(f, \Omega) = \sum_{x \in f^{-1}(0) \cap \Omega} \operatorname{sign} \det Df(x)$$

where $\sum$ over an empty set is defined as zero.

Proof. Let $B_\delta(x_0) = \{x \in \mathbb{R}^n; |x - x_0| < \delta\}$ denote the ball of radius $\delta > 0$ centered at $x_0 \in \mathbb{R}^n$. It is easy to show that the set $f^{-1}(0) \cap \Omega$ is finite. Therefore, we can find $\delta > 0$ such that for every $x \in f^{-1}(0) \cap \Omega$, $B_{2\delta}(x) \cap$

$f^{-1}(0) = \{x\}$ and $B_{2\delta}(x) \subseteq \Omega$. Consequently, by the additivity property, we have

$$\deg(f, \Omega) = \sum_{x \in f^{-1}(0) \cap \Omega} \deg(f, B_\delta(x))$$

For every $x \in f^{-1}(0) \cap \Omega$, we define $g_x(v) = Df(x)(v - x)$. Let $h_x(t, v) = tf(v) + (1-t)Df(x)(v - x)$ denote the linear homotopy. Note that

$$\begin{aligned} h_x(t, v) &= t[f(v) - Df(x)(v - x)] + Df(x)(v - x) \\ &= t \cdot o(|v - x|) + Df(x)(v - x) \end{aligned}$$

It follows that if $\delta > 0$ is sufficiently small, then $|h_x(t, v)| > 0$ for all $t \in [0, 1]$ and for all v such that $|v - x| = \delta$. Consequently, h_x is a $B_\delta(x)$-admissible homotopy between f and g_x. By the homotopy property, we get

$$\deg(f, \Omega) = \sum_{x \in f^{-1}(0) \cap \Omega} \deg(g, B_\delta(x))$$

Let $R > 0$ be such that $\Omega \subseteq B_R(0)$. Then the excision property and the homotopy property imply that

$$\deg(g_x, B_\delta(x)) = \deg(g_x, B_R(0)) = \deg(Df(x), B_R(0)) = \text{sign det } Df(x)$$

from which the conclusion of the theorem follows. □

Example 2.2.4 We can now give a degree-theoretical proof of the fundamental theorem of algebra: *Every polynomial*

$$p(z) = z^n + a_{n-1}z^{n-1} + \cdots + a_1 z + a_0$$

has at least one complex zero if $n \geq 1$, where a_i, $0 \leq i \leq n-1$, are complex numbers. To prove this, we first identify the complex plane with $\mathbb{R}^2$. Then p defines a mapping from $\mathbb{R}^2$ into $\mathbb{R}^2$. Let $\varepsilon > 0$ be given and set

$$H(t, z) = z^n - \varepsilon + t[a_{n-1}z^{n-1} + \cdots + a_1 z + a_0 + \varepsilon]$$

for $t \in [0, 1]$. It is easy to show that if ε is sufficiently small and $R > 0$ is sufficiently large, then H is Ω-admissible with $\Omega = B_R(0)$. Consequently,

$$\deg(p, \Omega) = \deg(H(1, \cdot), \Omega) = \deg(H(0, \cdot), \Omega)$$

Clearly, $H(0, \cdot)\colon \mathbb{R}^2 \to \mathbb{R}^2$ is a C^1-map

$$H(0, \cdot)^{-1}(0) = \{\varepsilon^{1/n} e^{i(2\pi/n)j};\ 0 \leq j \leq n-1\}$$

and $\det DH(0,z)|_{z=\varepsilon^{1/n}e^{i(2\pi/n)j}}>0$. Therefore, zero is a regular value of $H(0,\cdot)|_\Omega$ and

$$\deg(H(0,\cdot),\Omega)=\sum_{j=0}^{n-1}\operatorname{sign}\det DH(0,z)|_{z=\varepsilon^{1/n}e^{i(2\pi/n)j}}=n$$

Consequently, $\deg(p,\Omega)=n>0$ and the conclusion follows from the existence property of the degree.

The above example demonstrates how a degree is used in analysis. First, an existence problem is associated with the zero of a mapping f. Second, the existence property of the degree implies that it is sufficient to show the nontriviality of the degree of the mapping f relative to an open set where the expected zero is located. Third, we construct an admissible homotopy between f and a new mapping g such that the degree of g can be easily computed (or at least estimated).

However, the existence of zeros of f in Ω does not necessarily imply the nontriviality of $\deg(f,\Omega)$. For example, consider the mapping $f:\mathbb{R}\to\mathbb{R}$ defined by $f(x)=x^2-1$. Let $\Omega=(-2,2)$. Clearly, f has two zeros ± 1 in Ω and zero is a regular value of $f|_\Omega$. But, since $\operatorname{sign} f'(1)>0$ and $\operatorname{sign} f'(-1)<0$, we have $\deg(f,\Omega)=\operatorname{sign} f'(1)+\operatorname{sign} f'(-1)=0$.

EXERCISES

2.2.1 Suppose that $f:S^n\to S^n$ is homotopic to a constant map. Show that there must exist $x,y\in S^n$ such that $x=f(x)$ and $y=-f(y)$.

2.2.2 Let Ω be the unit ball in $\mathbb{C}$, and $f:\mathbb{C}\to\mathbb{C}$ be given by $f(z)=\bar{z}^5$. Compute $\deg(f,\Omega)$.

2.2.3 Show that for odd n the antipodal mapping $-\mathrm{Id}:S^n-S^n$ is homotopic to Id.

2.2.4 Let Ω be an open subset of $\mathbb{R}^n$ such that $0\in\Omega$ and $F:\mathbb{R}^n\to\mathbb{R}^n$ a differentiable *contraction* (i.e., there exists $k\in[0,1)$ such that $\|F(x)\|\le k\|x\|$ for all $x\in\mathbb{R}^n$) such that $|\deg(\mathrm{Id}-F,\Omega)|>1$. Show that zero is a critical value of $\mathrm{Id}-F$.

2.2.5 Let $F:\mathbb{R}^n\to\mathbb{R}^n$ be a C^1-mapping such that there is a nonsingular $n\times n$ matrix A such that $\lim_{|x|\to\infty}\dfrac{|F(x)-Ax|}{|x|}=0$, and let $x_0\in\mathbb{R}^n$ be a regular zero of F, that is, $F(x_0)=0$ and $\det DF(x_0)\neq 0$. Show that if $\operatorname{sign}\det DF(x_0)\neq\operatorname{sign}\det A$, then F has at least two zeros in $\mathbb{R}^n$.

2.2.6 Let X,Y be topological spaces. A continuous mapping $f:X\to Y$ is called *universal* if for any continuous mapping $g:X\to Y$, there exists $x\in X$ such that $f(x)=g(x)$. Show that if a continuous mapping

$f : \overline{B} \to \overline{B}$, where B denotes the open unit ball in $\mathbb{R}^{n+1}$, is such that $\deg(f, B) \neq 0$, then f is universal. *Hint:* Consider the homotopy $h(t, x) = f(x) - tg(x)$.

2.2.7 Let B denote the open unit ball in $\mathbb{R}^{n+1}$ and $f, g : \overline{B} \to \mathbb{R}^{n+1}$ be two continuous mappings such that $\lambda f(x) \neq -g(x)$ [respectively, $\lambda f(x) \neq g(x)$] for all $x \in S^n$ and $\lambda > 0$. Show that $\deg(f, B) = \deg(g, B)$ [respectively, $\deg(f, B) = (-1)^{n+1} \deg(g, B)$].

2.2.8 Let B denote the open unit ball in $\mathbb{R}^{n+1}$. Let $f : \mathbb{R}^{n+1} \to \mathbb{R}^{n+1}$ be B-admissible. We say that f is *B-essential* if for any given $g : \mathbb{R}^{n+1} \to \mathbb{R}^{n+1}$ satisfying $g|_{S^n} = f|_{S^n}$, there is a point $x_0 \in B$ such that $g(x_0) = 0$. Show that f is B-essential if and only if $\deg(f, B) \neq 0$.

2.3. CONSTRUCTION OF THE DEGREE

In this section we present an analytic construction of the degree.

Let $\Omega \subseteq \mathbb{R}^n$ be a nonempty, bounded, open set and $f : \mathbb{R}^n \to \mathbb{R}^n$ an Ω-admissible map. We define $\deg(f, \Omega)$ as follows:

Let $\varepsilon = \min\{|f(x)|;\ x \in \partial\Omega\}$. By the Weierstrass theorem (Theorem 1.3.11), there exists $\tilde{g} \in C^\infty(\mathbb{R}^n; \mathbb{R}^n)$ such that $\max\{|f(x) - \tilde{g}(x)|;\ x \in \overline{\Omega}\} < \varepsilon/2$. By Sard's theorem (Lemma 1.3.10), there exists a regular value $y_0 \in \mathbb{R}^n$ of the mapping $\tilde{g}_{|\Omega}$ such that $|y_0| < \varepsilon/3$. We define $g : \mathbb{R}^n \to \mathbb{R}^n$ by $g(x) = \tilde{g}(x) - y_0$. It is clear that $g \in C^\infty(\mathbb{R}^n; \mathbb{R}^n)$, $\max\{|\tilde{g}(x) - g(x)|;\ x \in \overline{\Omega}\} < \varepsilon/2$, and 0 is a regular value of $g_{|\Omega}$. We then define

$$\deg(f, \Omega) = \sum_{x \in g^{-1}(0) \cap \Omega} \operatorname{sign} \det Dg(x) \tag{2.3.1}$$

Note that the right-hand side of (2.3.1) is well defined. In fact, $\max\{|f(x) - g(x)|;\ x \in \overline{\Omega}\} < \varepsilon$. Therefore, $g : \mathbb{R}^n \to \mathbb{R}^n$ is Ω-admissible. In what follows, we will call g a *regular approximation* of f.

In order to prove that formula (2.3.1) does not depend on the regular approximation of g, we remark that if g' is another regular approximation of f, then the homotopy

$$h(t, x) = tg(x) + (1 - t)g'(x), \qquad t \in [0, 1], \quad x \in \mathbb{R}^n$$

is a well-defined Ω-admissible homotopy between g and g'. Indeed, since

$$\max\{|f(x) - g(x)|; x \in \overline{\Omega}\} < \varepsilon \quad \text{and} \quad \max\{|f(x) - g'(x)|; x \in \overline{\Omega}\} < \varepsilon$$

we have, for every $t \in [0, 1]$,

$$\begin{aligned}
&\min\{|h(t, x)|; x \in \partial\Omega\} \\
&\quad \geq \min\{|f(x)|; x \in \partial\Omega\} - \max\{|f(x) - (tg(x) + (1-t)g'(x))|; x \in \partial\Omega\} \\
&\quad \geq \varepsilon - t\max\{|f(x) - g(x)|; x \in \partial\Omega\} - (1-t)\max\{|f(x) - g'(x)|; x \in \partial\Omega\} \\
&\quad > \varepsilon - t\varepsilon - (1-t)\varepsilon = 0
\end{aligned}$$

In order to show that formula (2.3.1) does not depend on the regular approximation, we only need the following.

Lemma 2.3.1

Let $h : [0, 1] \times \mathbb{R}^n \to \mathbb{R}^n$ be a Ω-admissible homopoty of class C^1 between $g_1 : \mathbb{R}^n \to \mathbb{R}^n$ and $g_2 : \mathbb{R}^n \to \mathbb{R}^n$. We assume that both mappings $g_1|_\Omega$ and $g_2|_\Omega$ have zero as a regular value. Then

$$\sum_{x \in g_1^{-1}(0) \cap \Omega} \operatorname{sign} \det Dg_1(x) = \sum_{x \in g_2^{-1}(0) \cap \Omega} \operatorname{sign} \det Dg_2(x)$$

Proof. We may assume, without loss of generality, that $h : \mathbb{R} \times \mathbb{R}^n \to \mathbb{R}^n$. Let $\alpha : \mathbb{R} \to [0, 1]$ be a function of class C^∞ such that

$$\alpha(t) = \begin{cases} 0 & \text{if } t \leq \frac{1}{3} \\ 1 & \text{if } t \geq \frac{2}{3} \end{cases}$$

We define a new homotopy h^* between g and g' by

$$h^*(t, x) = h(\alpha(t), x)\,; \qquad t \in \mathbb{R}, \quad x \in \mathbb{R}^n$$

We denote $L = [0, 1] \times \overline{\Omega}$ and $L_0 = ([0, \frac{1}{3}] \cup [\frac{2}{3}, 1]) \times \overline{\Omega}$. From the definition of h^* and the assumption that $g_1|_{\overline{\Omega}}$ and $g_2|_{\overline{\Omega}}$ are transverse to $\{0\}$, it follows that $h^*|_{L_0}$ is transverse to $\{0\}$. Let $\delta = \min\{|h^*(t, x)|;\ (t, x) \in [0, 1] \times \partial\Omega\}$. From Corollary 1.3.15, there exists $\tilde{h} \in ⋔(\mathbb{R}^{n+1}, \mathbb{R}^n; \{0\})$ such that $\tilde{h}|_{L_0} \equiv h^*|_{L_0}$ and

$$\max\{|\tilde{h}(t, x) - h^*(t, x)|;\ (t, x) \in [0, 1] \times \overline{\Omega}\} < \delta$$

It is clear that $\tilde{h}$ is an Ω-admissible homotopy between g_1 and g_2, and zero is a regular value of $\tilde{h}|_L$.

In view of Theorem 1.3.6, there exists a neighborhood U of L such that $M := \tilde{h}^{-1}(0) \cap U$ is a one-dimensional submanifold of U. We remark that M is naturally oriented by the mapping $\tilde{h}$ (see Theorem 1.1.8). Indeed, the tangent space $T_{(t,x)}M$ is exactly the kernel $\operatorname{Ker} D\tilde{h}(t, x)$. Let $P_{(t,x)} : \mathbb{R} \times \mathbb{R}^n \to T_{(t,x)}M$ be the orthogonal projection onto $T_{(t,x)}M$. It is clear that P depends continuously on (t, x). We choose a linear operator

$A_{(t,x)} : T_{(t,x)}M \to \mathbb{R}$ such that the linear mapping $B_{(t,x)} : \mathbb{R} \times \mathbb{R}^n \to \mathbb{R} \times \mathbb{R}^n$, defined by $B_{(t,x)}(v) = (A_{(t,x)}P_{(t,x)}(v),\ D\tilde{h}(t,x)v) \in \mathbb{R} \times \mathbb{R}^n$ for $v \in \mathbb{R} \times \mathbb{R}^n$, preserves the orientation of $\mathbb{R} \times \mathbb{R}^n$, that is, $\det B_{(t,x)} > 0$. The mapping $A^{-1}_{(t,x)}$ determines the orientation of $T_{(t,x)}M$.

Denote $\tilde{h}_t : \mathbb{R}^n \to \mathbb{R}^n$ by $\tilde{h}_t(x) = \tilde{h}(t,x)$. Let $x \in \tilde{h}_t^{-1}(0) \cap \Omega$ be a regular point of $\tilde{h}_t$. Then $D\tilde{h}_t(x) : \mathbb{R}^n \to \mathbb{R}^n$ is an isomorphism. If we further assume $t \in [0, \frac{1}{3}] \cup [\frac{2}{3}, 1]$, then the definition of the function α implies that $T_{(t,x)}M = \mathbb{R} \times \{0\} = \mathbb{R}$. Therefore, we have the following block matrix decomposition of $D\tilde{h}(t,x) : \mathbb{R} \times \mathbb{R}^n \to \mathbb{R}^n$:

$$D\tilde{h}(t,x) = [0,\quad D\tilde{h}_t(x)]$$

Let $A : T_{(t,x)}M \to \mathbb{R}$ be any isomorphism; then the linear operator $B(v) = (AP_{(t,x)}(v), D\tilde{h}(t,x)(v)) : \mathbb{R} \times \mathbb{R}^n \to \mathbb{R} \times \mathbb{R}^n$ has the block matrix decomposition

$$B = \begin{bmatrix} A & 0 \\ 0 & D\tilde{h}_t(x) \end{bmatrix}$$

and it is clear that B preserves the orientation of $\mathbb{R} \times \mathbb{R}^n$ if and only if $\det D\tilde{h}_t(x) \det A > 0$. Consequently, if $\det D\tilde{h}_t(x) > 0$, then the orientation of M at (t,x) is the direction of increasing t, and if $\det D\tilde{h}_t(x) < 0$, then the orientation of M at (t,x) is the direction of decreasing t. See Figure 2.3.1.

Since $\tilde{h}$ is an Ω-admissible homotopy, the only points belonging to $M \cap \partial L$ are the points $(0, x_1)$, $x_1 \in g_1^{-1}(0) \cap \Omega$, or $(1, x_2)$, $x_2 \in g_2^{-1}(0) \cap \Omega$. Every oriented curve of solutions to $\tilde{h}(t,x) = 0$ that starts at $(0, x_1)$ or $(1, x_2)$ is either entering the region L or exiting L. The portion of that curve inside L has to be connected to another solution of $\tilde{h}(t,x) = 0$ of type $(0, x_1')$ or $(1, x_2')$. See Figure 2.3.2.

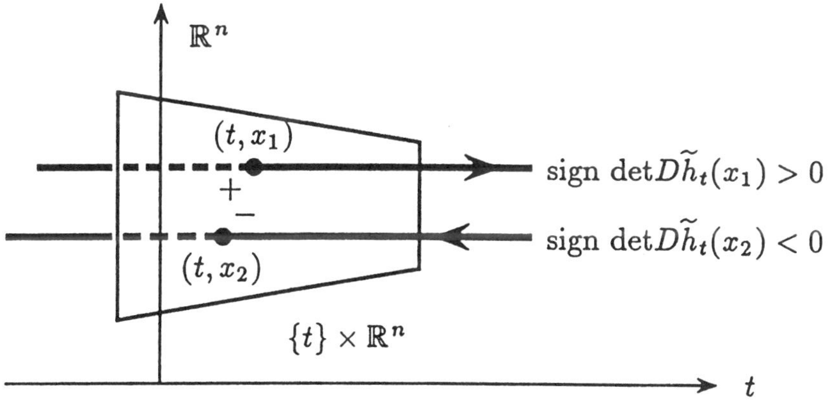

FIGURE 2.3.1. The orientation of M.

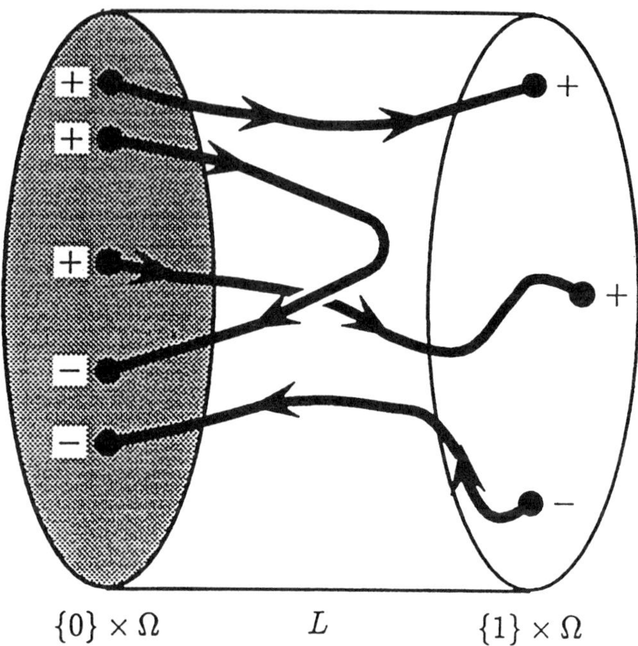

FIGURE 2.3.2. Prototypes of oriented curves of zeros of a regular homotopy.

It is now evident that any oriented curve of $M \cap L$ must be of one of the following types:

A: It contains points $(0, x_0)$ and $(1, x_1)$ with $x_0 \in g_1^{-1}(0) \cap \Omega$, $x_1 \in g_2^{-1}(0) \cap \Omega$, and sign det $Dg_1(x_0) =$ sign det $Dg_2(x_1)$.

B: It contains points $(0, x_0)$ and $(0, x_0^*)$ with $x_0, x_0^* \in g_1^{-1}(0) \cap \Omega$, and sign det $Dg_1(x_0) = -$sign det $Dg_1(x_0^*)$.

C: It contains points $(1, x_1)$ and $(1, x_1^*)$ with $x_1, x_1^* \in g_2^{-1}(0) \cap \Omega$ and sign det $Dg_2(x_1) = -$sign det $Dg_2(x_1^*)$.

D: It is contained in $(0, 1) \times \Omega$ and, hence, it does not contain any point $(0, x_0)$ or $(1, x_1)$.

Accordingly, we obtain the following equality:

$$\sum_{x \in g_1^{-1}(0) \cap \Omega} \text{sign det } Dg_1(x) = \sum_{x \in g_2^{-1}(0) \cap \Omega} \text{sign det } Dg_2(x) \quad \square$$

Theorem 2.3.2

The degree defined by formula (2.3.1) satisfies properties (P1–P3).

Proof. Property (P1) is a trivial consequence of the definition.

In order to prove the additivity, we consider an Ω-admissible mapping $f:\mathbb{R}^n \to \mathbb{R}^n$ such that there are two nonempty, disjoint, open subsets Ω_1 and Ω_2 of Ω satisfying $f^{-1}(0) \cap \overline{\Omega} \subset \Omega_1 \cup \Omega_2$.

Let $\varepsilon = \min\{|f(x)|;\ x \in \overline{\Omega} \setminus (\Omega_1 \cup \Omega_2)\} > 0$ and let $g:\mathbb{R}^n \to \mathbb{R}^n$ be a regular approximation of f such that $\max\{|g(x) - f(x)|;\ x \in \overline{\Omega}\} < \varepsilon$. Then it can be easily verified that $g^{-1}(0) \cap \overline{\Omega} \subseteq \Omega_1 \cup \Omega_2$ and, therefore,

$$\begin{aligned}\deg(f,\Omega)) &= \sum_{x \subset g^{-1}(0)\cap\Omega} \text{sign}\det Dg(x)\\ &= \sum_{x \in g^{-1}(0)\cap\Omega_1} \text{sign}\det Dg(x) + \sum_{x \in g^{-1}(0)\cap\Omega_2} \text{sign}\det Dg(x)\\ &= \deg(f,\Omega_1) + \deg(f,\Omega_2)\end{aligned}$$

Finally, we prove the homotopy property (P3). It is easy to prove that if f_1 and f_2 are Ω-homotopic and if g_1 and g_2 are two regular approximations of f_1 and f_2, respectively, then g_1 and g_2 are also Ω-homotopic. Applying Weierstrass theorem, we may assume that the Ω-admissible homotopy between g_1 and g_2 is of class C^1. Then, by Lemma 2.3.1, we get $\deg(g_1,\Omega) = \deg(g_2,\Omega)$, from which it follows that $\deg(f_1,\Omega) = \deg(f_2,\Omega)$. □

EXERCISES

2.3.1 Let $\xi = (p, E, B)$ denote a vector bundle over an arc-connected base space B and let $U \subset E$ be an open set such that $U \cap E_x$ is bounded for all $x \in B$. Let $f: E \to E$ be a continuous mapping such that $p(e) = f(p(e))$ for all $e \in E$, and $f(e)$ does not belong to the zero section of ξ for all $e \in \partial U$. Then, for every $x \in B$, we can define $f_x: E_x \to E_x$ by $f_x := f_{|E_x}$. Show that $\deg(f_x, U_x) = \text{constant}$ for all $x \in B$, where $U_x = U \cap E_x$.

2.3.2 Let M be a compact manifold and N a connected manifold, both of dimension n. For every C^1-mapping $f: M \to N$, we define the *mod* (2) *degree* of f with values in $\mathbb{Z}_2$ by $\deg_2(f, M) = \#f^{-1}(y) \pmod 2$, where y is a regular value of f and $\#f^{-1}(y)$ denotes the number of elements in the set $f^{-1}(y)$. Show that

- **(i)** The above definition does not depend on the choice of the regular value of f.
- **(ii)** One may define the (mod 2) degree of a continuous mapping $g: M \to N$ by using differentiable approximations of g.
- **(iii)** The mod (2) degree satisfies the existence and homotopy properties of a degree.

2.4 FURTHER PROPERTIES OF THE DEGREE

This section deals with properties of the Brouwer degree that are consequences of the analytic definition. We begin this section with the following.

Theorem 2.4.1 (BORSUK'S THEOREM)

Let $\Omega \subseteq \mathbb{R}^n$ be an open, bounded, symmetric set (i.e., $x \in \Omega$ if and only if $-x \in \Omega$) such that $0 \in \Omega$. Assume that $f : \mathbb{R}^n \to \mathbb{R}^n$ is an Ω-admissible mapping such that $f|_{\partial\Omega}$ is odd, that is, $f(-x) = -f(x)$ for all $x \in \partial\Omega$. Then $\deg(f, \Omega)$ *is an odd integer number. In particular,* $\deg(f, \Omega) \neq 0$.

Proof. The main technical difficulty in the following proof is to construct a regular odd mapping g (i.e., $g|_\Omega \pitchfork \{0\}$) such that f and g are Ω-homotopic.

STEP 1. *Averaging of f.* We put

$$f_1(x) = \tfrac{1}{2}[f(x) - f(-x)] , \qquad x \in \mathbb{R}^n$$

Since f is odd on $\partial\Omega$, we have

$$f_1(x) = \tfrac{1}{2}[f(x) - f(-x)] = f(x) , \qquad x \in \partial\Omega$$

It follows from the boundary values' dependence that

$$\deg(f, \Omega) = \deg(f_1, \Omega)$$

STEP 2. *Normal approximation of f.* Let $\delta > 0$ be a number such that $\overline{B_{2\delta}(0)} \subset \Omega$. Let $\alpha : [0, 2\delta] \to [0, 1]$ be a continuous function such that

$$\alpha(t) = \begin{cases} 1 & \text{if } 0 \leq t \leq \delta \\ 0 & \text{if } \frac{3}{2}\delta \leq t \leq 2\delta \end{cases}$$

Define a mapping $f_2 : \mathbb{R}^n \to \mathbb{R}^n$ by

$$f_2(x) = \begin{cases} f_1(x) + \alpha(|x|)(x - f_1(x)) & \text{if } x \in B_{2\delta}(0) \\ f_1(x) & \text{if } x \not\in B_{2\delta}(0) \end{cases}$$

Since $f_2|_{\partial\Omega} = f_1|_{\partial\Omega}$, f_2 is an Ω-admissible mapping and

$$\deg(f_2, \Omega) = \deg(f_1, \Omega) = \deg(f, \Omega)$$

We will call f_2 a *zero-normal approximation* of f_1.

STEP 3. *Localization of the degree.* Since $f_1|_{B_\delta(0)} = \mathrm{Id}|_{B_\delta(0)}$, applying the normalization and additivity to $\Omega_1 = B_\delta(0)$ and $\Omega^* = \Omega\backslash\overline{B_\delta(0)}$, we get

$$\begin{aligned}\deg(f, \Omega) = \deg(f_2, \Omega) &= \deg(f_2, \Omega_1) + \deg(f_2, \Omega^*)\\ &= 1 + \deg(f_2, \Omega^*)\end{aligned}$$

STEP 4. *Smooth odd approximation of* f_2. Let $\varepsilon = \min\{|f_2(x)|; x \in \partial\Omega^*\}$ and let g^* be a smooth mapping such that $g^*|_{\Omega^*} \pitchfork \{0\}$ and

$$\max\{|g^*(x) - f_2(x)|; x \subset \overline{\Omega}^*\} < \varepsilon$$

Then g^* is Ω^*-admissible. Define $f_3 : \mathbb{R}^n \to \mathbb{R}^n$ by $f_3(x) = \frac{1}{2}[g^*(x) - g^*(-x)]$, $x \in \mathbb{R}^n$. For all $x \in \overline{\Omega}^*$, we have

$$\begin{aligned}|f_3(x) - f_2(x)| &= |\tfrac{1}{2}[g^*(x) - g^*(-x)] - \tfrac{1}{2}[f_2(x) - f_2(-x)]|\\ &\le \tfrac{1}{2}|g^*(x) - f_2(x)| + \tfrac{1}{2}|g^*(-x) - f_2(-x)|\\ &< \tfrac{1}{2}\varepsilon + \tfrac{1}{2}\varepsilon = \varepsilon\end{aligned}$$

Therefore, f_3 is Ω^*-admissible and $\deg(f_2, \Omega^*) = \deg(f_3, \Omega^*)$.

STEP 5. *Regular odd approximation of* f_3. We define the sets

$$\Omega_k^* := \{x \in \Omega^*; x_i \neq 0 \text{ for some } i \le k\}\ ,\ \ k = 1, 2, \ldots, n$$

It is clear that $\Omega_1^* \subseteq \Omega_2^* \subseteq \cdots \subseteq \Omega_n^* = \Omega^*$. Let $\varphi : \mathbb{R} \to \mathbb{R}$ denote the function $\varphi(t) = t^3$. We define $\tilde{g}_1(x) = f_3(x)/\varphi(x_1)$, $x \in \Omega_1^*$. Let y^1 be a regular value of $\tilde{g}_1$ with $|y^1|$ as small as necessary. Then, zero is a regular value for $g_1(x) = f_3(x) - \varphi(x_1)y^1$ on Ω_1^*. Indeed, if $g_1(x) = f_3(x) - \varphi(x_1)y^1 = 0$, then

$$\begin{aligned}Dg_1(x) &= D(\varphi(x_1)\tilde{g}_1(x)) - D(\varphi(x_1)y^1)\\ &= \varphi(x_1)D\tilde{g}_1(x) + \tilde{g}_1(x)\varphi'(x_1)P_1 - y^1\varphi'(x_1)P_1\\ &= \varphi(x_1)D\tilde{g}_1(x) + y^1\varphi'(x_1)P_1 - y^1\varphi'(x_1)P_1\\ &= \varphi(x_1)D\tilde{g}_1(x)\end{aligned}$$

where $P_1 : \mathbb{R}^n \to \mathbb{R}^n$ is the projection onto the first coordinate, that is, $P_1(x) = x_1$. Consequently, x is a regular point of g_1 and g_1 is odd on Ω_1^*. We can extend g_1 to the set $\overline{\Omega}^*$ by putting $g_1(x) = f_3(x)$ if $x \in \overline{\Omega}^*\backslash\Omega_1^*$. Since $\varphi'(0) = 0$, the mapping g_1 is also differentiable at points $x \in \overline{\Omega}^*$ such that $P_1(x) = x_1 = 0$. Moreover, the extended mapping g_1 is of class C^1 on $\overline{\Omega}^*$.

We now assume, for the purpose of induction, that for $k < n$ we have already constructed a C^1-mapping $g_k : \overline{\Omega}^* \to \mathbb{R}^n$ which is odd and $g_k|_{\Omega_k^*} \pitchfork \{0\}$. We consider $\tilde{g}_k(x) = g_k(x)/\varphi(x_{k+1})$ for $x_{k+1} \neq 0$, and let y^{k+1} be a regular value of $\tilde{g}_k$ on the set $\{x \in \Omega^*;\ x_{k+1} \neq 0\}$ with $|y^{k+1}|$ as small as necessary. Then we define $g_{k+1}(x) = g_k(x) - \varphi(x_{k+1})y^{k+1}$, $x \in \{x \in \Omega^*; x_{k+1} \neq 0\}$. An argument similar to that for g_1 shows that zero is a regular value of g_{k+1} on $\{x \in \Omega^*, x_{k+1} \neq 0\}$. We can also extend g_{k+1} to the set $\overline{\Omega}^*$ by simply putting $g_{k+1}(x) = g_k(x)$ if $x \in \overline{\Omega}^*$ is not in the set $\{x \in \Omega^*; x_{k+1} \neq 0\}$. Again, since $\varphi'(0) = 0$, this extension is of class C^1 on the set $\overline{\Omega}^*$. It is also clear that g_{k+1} is odd.

Note that if $x \in \Omega^*_{k+1}$ and $x_{k+1} = 0$, then $x \in \Omega^*_k$ and $g_{k+1}(x) = g_k(x)$ as well as $Dg_{k+1}(x) = Dg_k(x)$. Therefore, by the induction assumption, if $g_{k+1}(x) = 0$, then x is a regular point of g_{k+1}. This implies that zero is a regular value of $g_{k+1}|_{\Omega^*_{k+1}}$. Since $\Omega^*_n = \Omega^*$, the construction leads when $n = k$ to an odd regular approximation of f on $\overline{\Omega}^*$. We denote this approximation by g.

Evidently, as g can be made sufficiently close to f, we have $\deg(f_3, \Omega^*) = \deg(g, \Omega^*)$.

STEP 6. *Computation of $\deg(g, \Omega^*)$.* The mapping g is of class C^1, zero is a regular value of g, and g is odd, that is, $g(-x) = -g(x)$. Therefore, if $x \in g^{-1}(0) \cap \Omega^*$, then $-x \in g^{-1}(0) \cap \Omega^*$. On the other hand, the identity $-g(-x) = g(x)$ for all $x \in \Omega^*$ implies $Dg(-x) = Dg(x)$. Consequently, all solutions in Ω^* of the equation $g(x) = 0$ occur in pairs $\{x, -x\}$ and $\operatorname{sign}\det Dg(x) = \operatorname{sign}\det Dg(-x)$. Therefore, the degree $\deg(g, \Omega^*)$ is an even integer number, say, $2N$. By using the result in Step 3, we get

$$\deg(f, \Omega) = 1 + \deg(g, \Omega^*) = 1 + 2N$$

from which the conclusion follows. □

Note that in the above argument, we established the following result that will be used in later chapters.

Corollary 2.4.2

Let $\Omega \subseteq \mathbb{R}^n$ be an open, bounded, symmetric set and let $f : \mathbb{R}^n \to \mathbb{R}^n$ be an Ω-admissible odd mapping. Then for every $\varepsilon > 0$, there exists an odd mapping $g : \mathbb{R}^n \to \mathbb{R}^n$ of class C^1 such that zero is a regular value of $g|_{\overline{\Omega}}$ and

$$\max\{|f(x) - g(x)|; x \in \Omega\} < \varepsilon$$

Theorem 2.4.3 (PRODUCT FORMULA)

Suppose that $\Omega \subseteq \mathbb{R}^n \times \mathbb{R}^m$ is a nonempty, bounded, open set and let $g : \mathbb{R}^n \times \mathbb{R}^m \to \mathbb{R}^n \times \mathbb{R}^m$ be a mapping defined by

$$g(x, y) = (f(x, y), y)\,, \qquad x \in \mathbb{R}^n, \quad y \in \mathbb{R}^m$$

where $f : \mathbb{R}^n \times \mathbb{R}^m \to \mathbb{R}^n$ is a continuous mapping. Assume that g is Ω-

admissible; then $f|_{\mathbb{R}^n}$ *is* $(\Omega \cap \mathbb{R}^n)$*-admissible and*

$$\deg(g, \Omega) = \deg(f|_{\mathbb{R}^n}, \Omega \cap \mathbb{R}^n)$$

Proof. We can assume, without loss of generality, that f is of class C^1, zero is a regular value of $f|_{\Omega}$, and zero is a regular value of $f_0 = f|_{\mathbb{R}^n}$ on $\Omega_0 = \Omega \cap \mathbb{R}^n$. Evidently, $g^{-1}(0) \cap \Omega = f_0^{-1}(0) \cap \Omega_0$. Let $x \in \Omega_0$ be such that $f_0(x) = 0$. Then $Dg(x, 0) : \mathbb{R}^n \times \mathbb{R}^m \to \mathbb{R}^n \times \mathbb{R}^m$ has the following block matrix representation:

$$Dg(x, 0) = \begin{bmatrix} Df_0(x) & * \\ 0 & \mathrm{Id} \end{bmatrix}$$

It is clear that sign det $Dg(x, 0) =$ sign det $Df_0(x)$. Therefore,

$$\deg(g, \Omega) = \deg(f|_{\mathbb{R}^n}, \Omega \cap \mathbb{R}^n) \quad \square$$

Proposition 2.4.4

Let $\Omega \subseteq \mathbb{C}$ *be a nonempty, bounded, open set and let* $f : \overline{\Omega} \to \mathbb{C}$ *be a* Ω*-admissible mapping such that* $f|_{\Omega}$ *is analytic. Then* $\deg(f, \Omega)$ *is equal to the number of all zeros of* f *in* Ω *counted with multiplicity.*

Proof. In what follows, we identify $\mathbb{C}$ with $\mathbb{R}^2$. If $z_0 \in \Omega$ is a zero of f of multiplicity k, then there is a neighborhood $\mathcal{V}$ of z_0 and an analytic function $g(z)$ such that $g(z) \neq 0$ and $f(z) = (z - z_0)^k g(z)$ for $z \in \bar{\mathcal{V}}$. By applying the additivity property of degree, it is sufficient to show that $\deg(f, \mathcal{V}) = k$. Let $\varepsilon > 0$ be a sufficiently small number. Choose distinct numbers $a_1, \ldots, a_k \in \mathbb{C}$ such that $|a_i| < \varepsilon$, $i = 1, \ldots, k$. Define

$$f_\varepsilon(z) = (z - z_1) \cdots (z - z_k) g(z)$$

$$z_i = z_0 + a_i\,, \qquad i = 1, \ldots, k$$

It is clear that for sufficiently small ε, f_ε is a regular approximation of f on $\mathcal{V}$. Therefore, $\deg(f, \mathcal{V}) = \deg(f_\varepsilon, \mathcal{V})$. Since the function f_ε has only simple zeros in $\mathcal{V}$, it is sufficient, by the additivity property, to show that if $k = 1$, then $\deg(f, \mathcal{V}) = 1$. This is obvious, since $f(z) = (z - z_0) g(z)$, $f'(z_0) = g(z_0) \neq 0$, and $\det Df(z_0) = |f'(z_0)|^2 > 0$. Therefore, $\deg(f, \mathcal{V}) = 1$. This completes the proof. $\square$

We now discuss the degree of composition functions.

Theorem 2.4.5 (Multiplication Formula)

Let $\Omega \subseteq \mathbb{R}^n$ *be a nonempty, bounded, open set. Assume that* $f : \overline{\Omega} \to \mathbb{R}^n$ *and*

$g : \mathbb{R}^n \to \mathbb{R}^n$ are continuous mappings such that $0 \notin (g \circ f)(\partial\Omega)$. Then

$$\deg(g \circ f, \Omega) = \sum_i \deg(f - p_i, \Omega) \deg(g, U_i) \tag{2.4.1}$$

where $\{U_i\}$ denotes the bounded, connected components of $\mathbb{R}^n \setminus f(\partial\Omega)$ and p_i is an arbitrary element $p_i \in U_i$.

Proof. It can be shown by using the standard properties of the degree and regular approximations of f and g that we may assume, without loss of generality, that both maps f and g are of class C^1 and zero is a regular value of $g \circ f_{|\Omega}$. Denote by $\{x_1, \ldots, x_m\}$ the set $(g \circ f)^{-1}(0) \cap \Omega$ and let $\{y_1, \ldots, y_r\} = f(\{x_1, \ldots, x_m\})$. Then, it follows from the definition of the degree that

$$\begin{aligned}\deg(g \circ f, \Omega) &= \sum_{s=1}^{m} \operatorname{sign} \det Df(x_s) \cdot \operatorname{sign} \det Dg(f(x_s)) \\ &= \sum_{j=1}^{r} \deg(f - y_j, \Omega) \cdot \operatorname{sign} \det Dg(y_j)\end{aligned}$$

For every bounded connected component U_i, $i = 1, \ldots, \ell$, of $\mathbb{R}^n \setminus f(\partial\Omega)$, we choose a point $p_i \in U_i$, and for the unbounded component U_0 of $\mathbb{R}^n \setminus f(\partial\Omega)$, we choose a point $p_0 \in \mathbb{R}^n$ such that $p_0 \notin f(\overline{\Omega})$. Suppose that y_j belongs to U_i, where i is one of the numbers $0, 1, \ldots, \ell$. Then there exists a continuous curve $\sigma : [0, 1] \to U_i$ such that $\sigma(0) = y_j$ and $\sigma(1) = p_i$. Define the homotopy $h : [0, 1] \times \mathbb{R}^n \to \mathbb{R}^n$ by $h(t, x) = f(x) - \sigma(t)$. It is clear that h is Ω-admissible and thus $\deg(f - y_j, \Omega) = \deg(f - p_i, \Omega)$. If the point y_j belongs to the unbounded connected component U_0, that is, $p_i = p_0$ and $f^{-1}(p_0) \cap \Omega = \emptyset$, then $\deg(f - p_0, \Omega) = 0$. Consequently, we obtain for every bounded connected component U_i,

$$\begin{aligned}\sum_{y_j \in U_i} \deg(f - y_j, \Omega) \cdot \operatorname{sign} \det Dg(y_j) &= \deg(f - p_i, \Omega) \sum_{y_i \in U_i} \operatorname{sign} \det Dg(y_i) \\ &= \deg(f - p_i, \Omega) \cdot \deg(g, U_i)\end{aligned}$$

from which formula (2.4.1) follows. □

Remark 2.4.6 If $\Omega \subseteq \mathbb{R}^n$ is an open and bounded subset of $\mathbb{R}^n$, $f : \overline{\Omega} \to \mathbb{R}^n$ is continuous, and $p \in \mathbb{R}^n \setminus f(\partial\Omega)$, then the proof of Theorem 2.4.5 shows that $\deg(f - p, \Omega)$ is a constant as a function of p belonging to a connected component of $\mathbb{R}^n \setminus f(\partial\Omega)$.

Remark 2.4.7 Due to the compactness, $g^{-1}(0) \cap f(\overline{\Omega})$ is covered by finitely many sets U_i. Consequently, the existence property of the degree implies that $\deg(f-p_i,\Omega)\deg(g,U_i)=0$ except for a finite number of i. This shows that the summation in Theorem 2.4.5 is finite.

As a simple application of the above result, we will show that there is no free continuous action of a nontrivial group on an even-dimensional sphere except for the group $\mathbb{Z}_2:=\{-1,1\}\subseteq\mathbb{R}$. Let us recall that a group G *acts* on a set X if there is a map $\varphi: G\times X\to X$ satisfying the properties:

(i) $\varphi(g,\varphi(h,x))=\varphi(gh,x)$ for all $g, h\in G$, and $x\in X$.
(ii) $\varphi(e,x)=x$ for all $x\in X$, where $e\in G$ denotes the neutral element of G.

It is clear from properties (i) and (ii) that for every $g\in G$, the map $\varphi_g: X\to X$, given by $\varphi_g(x)=gx$, is a bijection of X such that $(\varphi_g)^{-1}=\varphi_{g^{-1}}$. Assume, in addition, that X is a topological space. We say that an action $\varphi: G\times X\to X$ is *continuous* if all the maps $\varphi_g: X\to X$, $g\in G$, are continuous and thus homeomorphisms. An action φ of G on X is said to be *free* if for all $x\in X$ and $g\neq e$, we have $\varphi(g,x)\neq x$.

We have the following.

Proposition 2.4.8

Let S^n be an even-dimensional sphere. Assume that there is a free continuous action φ of a group G on S^n. Then G is either the trivial group $\{e\}$ or is isomorphic to $\mathbb{Z}_2=\{-1,1\}$.

Proof. For every $g\in G$, we define the map $\tilde{\varphi}_g:\mathbb{R}^{n+1}\to\mathbb{R}^{n+1}$by

$$\tilde{\varphi}_g(x):=\begin{cases}|x|\varphi\left(g,\dfrac{x}{|x|}\right), & \text{for } x\neq 0\\ 0, & \text{if } x=0\end{cases}$$

By the continuity of the action φ, the maps $\tilde{\varphi}_g$, $g\in G$, are well-defined continuous maps such that $\tilde{\varphi}_g\circ\tilde{\varphi}_h=\tilde{\varphi}_{gh}$. Denote by Ω the unit ball in $\mathbb{R}^{n+1}$. Then all the maps $\tilde{\varphi}_g$, $g\in G$, are Ω-admissible, and by the assumption that φ is a free action, we have for every $g\neq e$, $\tilde{\varphi}_g(x)\neq\lambda x$ for all $x\in\partial\Omega=S^n$ and for all $\lambda>0$. Consequently, we obtain (see Exercise 2.2.7)

$$\deg(\tilde{\varphi}_g,\Omega)=(-1)^{n+1}\deg(\mathrm{Id},\Omega)=-1 \tag{2.4.2}$$

On the other hand, it follows from Theorem 2.4.5 that if $g,h\in G$ are two elements of G such that $g\neq e$ and $h\neq e$, then

$$\deg(\tilde{\varphi}_{gh},\Omega)=\deg(\tilde{\varphi}_g,\Omega)\cdot\deg(\tilde{\varphi}_h,\Omega)=1$$

and, thus, by (2.4.2) we obtain $gh = e$. In particular, by taking $g = h$, we see that for every $g \in G$, we have $g^2 = e$. Suppose now that there exist two different elements g and h in $G \setminus \{e\}$. Since $g^2 = e$ and $h^2 = e$, $gh \neq e$, a contradiction. Consequently, the group G can contain at most two elements. □

EXERCISES

2.4.1 Let $\Omega \subset \mathbb{R}^n$ be a nonempty, bounded, open set and let $f : \mathbb{R}^n \to \mathbb{R}^n$ be a Ω-admissible mapping. Show that if $\deg(f, \Omega) \neq 0$, then $0 \in \text{Int } f(\Omega)$.

2.4.2 Let $U \subset \mathbb{R}^n$ be a nonempty, open set such that $0 \in U$ and let $f : U \to \mathbb{R}^n$ be a continuous, one-to-one mapping such that $f(0) = 0$. Show that there is a small ball Ω centered at zero such that

(i) f is Ω-admissible;

(ii) there is an Ω-admissible homotopy between f and an odd mapping on Ω. *Hint:* Use the homotopy $h(t, x) = f\left(\frac{x}{1+t}\right) - f\left(\frac{-tx}{1+t}\right)$

2.4.3 Let $U \subset \mathbb{R}^n$ be an open, nonempty set and $f : U \to \mathbb{R}^n$ a continuous, one-to-one mapping. Show that $f(U)$ is open and $f : U \to f(U)$ is a homeomorphism.

2.4.4 Prove the following Borsuk–Ulam theorem: Every continuous mapping $f : S^n \to \mathbb{R}^n$ sends at least one pair of antipodal points to the same point, that is, there exists $p \in S^n$ such that $f(p) = f(-p)$.

2.4.5 Suppose that $A_1, A_2, \ldots, A_n$ are bounded measurable subsets of $\mathbb{R}^n$. Show that there exists a hyperplane L in $\mathbb{R}^n$ and $y \in \mathbb{R}^n$ such that the $(n-1)$-dimensional flat $L + y$ simultaneously bisects each one of the sets $A_1, A_2, \ldots, A_n$. (This result has a nice practical interpretation, namely, it says that a sandwich composed of bread, cheese, and ham can always be cut with a single blow of a knife into two pieces, each of them containing exactly the same amount of bread, cheese, and ham.) *Hint:* Let L_x denote the n-dimensional subspace of $\mathbb{R}^{n+1}$ perpendicular to $x \in S^n$. Construct a mapping $f : S^n \to \mathbb{R}^n$, where $f_i(x)$ denotes the measure of the part of $A_i \subset \mathbb{R}^n \subset \mathbb{R}^{n+1}$ that lies on the same side of $L_x + (0, \ldots, 0, 1)$ as does the point $x + (0, \ldots, 0, 1)$.

2.4.6 Prove that $\mathbb{R}^m$ is not homeomorphic to $\mathbb{R}^n$ whenever $n \neq m$.

2.4.7 Assume that $f : S^n \to S^n$ is a continuous mapping homotopic to a constant mapping. Show that (i) there is a fixed point of f, that is, there is $x_0 \in S^N$ such that $f(x_0) = x_0$; (ii) there is a point $x_1 \in S^n$ such that $f(x_1) = -x_1$.

2.5 APPLICATION: EXISTENCE OF EQUILIBRIA IN DIFFERENTIAL EQUATIONS

In this section we describe a result due to Hofbauer and Sigmund (1988) that applies the Brouwer degree to establish the existence of equilibria in an ecological system.

Consider the system

$$\dot{x}_i(t) = x_i(t)f_i(x(t)), \qquad 1 \le i \le n \tag{2.5.1}$$

where $f_i : \mathbb{R}^n \to \mathbb{R}$ is continuously differentiable, $1 \le i \le n$. This system includes many important models in the theoretical ecology. Clearly, if $x \in \mathbb{R}^n_+ := \{x \in \mathbb{R}^n;\ x = (x_1, \ldots, x_n),\ x_i \ge 0 \text{ for } i = 1, \ldots, n\}$, then Eq. (2.5.1) has a unique solution, denoted by $\varphi(t, x)$, satisfying $\varphi(0, x) = x$. This solution satisfies $\varphi(t, x) \in \mathbb{R}^n_+$ for all $t \ge 0$, provided that $\varphi(t, x)$ exists. Moreover, if $x_i > 0$, then $\varphi_i(t, x) > 0$ for $t \ge 0$.

We say system (2.5.1) is *permanent* if:

- **(H1)** For any $x \in \mathbb{R}^n_+$, $\varphi(t, x)$ is defined for all $t \ge 0$.
- **(H2)** There exists $\delta > 0$ such that if $x_i > 0$ for some $1 \le i \le n$, then $\liminf_{t \to \infty} \varphi_i(t, x) > \delta$.
- **(H3)** There exists $D > 0$ such that if $x \in \operatorname{Int} \mathbb{R}^n_+$, then $\limsup_{t \to \infty} \varphi_i(t, x) \le D$ for all $1 \le i \le n$.

(H1) ensures that $\varphi : [0, \infty) \times \mathbb{R}^n_+ \to \mathbb{R}^n_+$ is a continuous mapping. (H2) is related to the important question of extinction in theoretical ecology and (H3) is the usual dissipativeness assumption. Under these assumptions, for each $x \in \operatorname{Int} \mathbb{R}^n_+$, the so-called *$\omega$-limit set* of x defined by $\omega(x) := \{y;$ there exists $t_n \to \infty$ such that $\varphi(t_n, x) \to y$ as $n \to \infty\}$ is nonempty, compact, connected, and invariant in the sense that if $y \in \omega(x)$, then $\varphi(t, y)$ exists and $\varphi(t, y) \in \omega(x)$ for all $t \in \mathbb{R}$. Moreover, $\omega(x) \subseteq \operatorname{Int} \mathbb{R}^n_+$.

Note that if x is an equilibrium, then $\omega(x) = \{x\}$, and if x is a periodic point, that is, $\varphi(p, x) = x$ for some $p > 0$, then $\omega(x) = \bigcup_{0 \le t \le p} \{\varphi(t, x)\}$.

Theorem 2.5.1

If (2.5.1) is permanent, then for any open, bounded set U such that

$$\bigcup_{x \in \operatorname{Int} \mathbb{R}^n_+} \omega(x) \subseteq U \subseteq \bar{U} \subseteq \operatorname{Int} \mathbb{R}^n_+$$

$\deg(h, U) = (-1)^n$, *where $h : \mathbb{R}^n \to \mathbb{R}^n$ is defined by $h_i(x) = x_i f_i(x)$, $1 \le i \le n$, $x \in \mathbb{R}^n$. In particular, there exists an equilibrium of (2.5.1) in* $\operatorname{Int} \mathbb{R}^n_+$ *(i.e., a zero of h).*

Proof. Let $K \subseteq \mathrm{Int}\,\mathbb{R}^n_+$ be a compact set containing $\bigcup_{x \in \mathrm{Int}\,\mathbb{R}^n_+} \omega(x)$. Define $\tau(x) = \inf\{t \geq 0;\ \varphi(s, x) \in \mathrm{Int}\,K \text{ for all } s \geq t\}$. It can be easily shown that τ is well defined, locally bounded, and the maximum value of $\tau(x)$ over $x \in K$ can be attained. Let $T = \max_{x \in K} \tau(x)$, $K^+ = \{\varphi(t, x);\ x \in K,\ 0 \leq t \leq T\}$. Then K^+ is compact and $\varphi(t, x) \in K^+$ for all $t \geq 0$ and $y \in K^+$.

Now, assume that U is a convex open bounded set such that $\overline{U} \subseteq \mathrm{Int}\,\mathbb{R}^n_+$ and $K^+ \subseteq U$. Define $s = \max_{x \in \overline{U}} \tau(x)$. Again, such a maximum value can be attained. Consider the following homotopy $H : [0, s] \times \bar{U} \to \mathbb{R}^n$ defined by

$$H(t, x) = \begin{cases} h(x) & \text{if } t = 0 \\ \dfrac{\varphi(t, x) - x}{t} & \text{if } t \in (0, s] \end{cases}$$

Clearly, $H(t, x) \neq 0$ for $t \in [0, s]$ and $x \in \partial U$. Since ∂U contains no equilibria or periodic points, $\deg(h, U) = \deg(H(s, \cdot\,), U)$. On the other hand, if $x \in \partial U$, then the definition of s implies that $\varphi(s, x) \in K^+ \subseteq U$. This implies that there exists no $\theta \in [0, 1]$ so that $\frac{\theta x_0 + (1-\theta)\varphi(s, x) - x}{s} = 0$, where $x_0 \in U$. Consequently, $H(s, \cdot)$ is U-homotopic to $-\mathrm{Id} + x_0$. This ensures $\deg(H(x, \cdot), U) = \deg(-\mathrm{Id} + x_0, U) = (-1)^n$. □

EXERCISES

2.5.1 An equilibrium of (2.5.1) is said to be *saturated* if $f_i(p) \leq 0$ whenever $p_i = 0$. Clearly, every equilibrium in $\mathrm{Int}\,\mathbb{R}^n_+$ is saturated. Prove that if (H1) and (H2) in Theorem 2.5.2 are satisfied, then Eq. (2.5.1) has a saturated equilibrium.

2.5.2 Consider the following system of ordinary differential equations $\dot{u}(t) = f(t, u(t))$, where $f : \mathbb{R} \times \mathbb{R}^n \to \mathbb{R}^n$ is continuously differentiable and there exists $\omega > 0$ such that $f(t + \omega, u) = f(t, u)$ for $(t, u) \in \mathbb{R} \times \mathbb{R}^n$. Assume that for any $x \in \mathbb{R}^n$, there exists a unique solution $u(t, x)$ to the initial value problem $\dot{u}(t) = f(t, u(t))$, $u(0) = x$, on $[0, \infty)$. Define $P_\omega : \mathbb{R}^n \to \mathbb{R}^n$ by $P_\omega(x) = u(\omega, x)$, the so-called *Poincaré operator*. We say a point $x \in \mathbb{R}^n$ is ω-irreversible if $u(t, x) \neq x$ for $t \in (0, \omega]$. Suppose that $\Omega \subseteq \mathbb{R}^n$ is nonempty, open, and bounded, $0 \notin f(0, \partial\Omega)$ and every $x \in \partial\Omega$ is ω-irreversible. Then $\deg(\mathrm{Id} - P_\omega, \Omega) = \deg(-f(0, \cdot\,), \Omega)$. *Hint:* Consider the homotopy

$$h(t, x) = \begin{cases} (x - u(\omega t, x))\left(\dfrac{1-t}{t\omega} + t\right), & t \neq 0 \\ -f(0, x)\,, & t = 0 \end{cases}$$

2.6 BIBLIOGRAPHICAL NOTES

The presentation of the Brouwer degree and its properties in this chapter is standard and follows the usual analytical construction. In particular, we take the approach based on the application of the transversality theory. This can be compared to Deimling (1984), Hirsch (1976), or Milnor (1978). We refer the reader to the original papers of Brouwer (1912a,b) and the monographs of Cronin (1964), Lloyd (1978), Krasnosiel'skii (1965), Rothe (1986), and Zeidler (1986) for similar or different treatments. The application of the Brouwer degree to the existence of equilibria in differential euations (Section 2.5) is based on the work of Hofbauer and Sigmund (1988).

Chapter Three

Leray–Schauder Degree for Compact Fields

This chapter extends the Brouwer degree to compact perturbations of the identity defined in infinite-dimensional spaces, using the technique of Schauder approximation of a compact map by mappings with finite-dimensional ranges.

The definition of Leray–Schauder degree for compact perturbations of the identity will be introduced and its applications to the existence of fixed points will be presented. We will also develop a computational formula for Leray–Schauder degree of linear compact fields. Some applications to boundary value problems and the extension of the degree to a fixed-point index for ANRs will also be discussed.

Note that in finite-dimensional spaces, Brouwer degree was defined for the class of continuous mappings. For infinite-dimensional spaces, however, some restriction has to be placed on the mappings, as may be seen from the following.

Example Let $X := C([0, 1])$ be the Banach space of continuous functions $x : [0, 1] \to \mathbb{R}$ with the norm $\|x\| = \sup_{0 \le s \le 1} |x(s)|$. Denote by x_0 the constant function $x_0(s) = \frac{1}{2}$, $0 \le s \le 1$, and let $\Omega = \{x; x \in X, \|x - x_0\| < \frac{1}{2}\}$. Clearly, Ω is open and bounded. Fix $\varphi \in X$ such that $\varphi(0) = 0$, $\varphi(1) = 1$, and $0 \le \varphi(s) \le 1$ for $0 \le s \le 1$. Define $\Phi : \overline{\Omega} \to X$ by $\Phi(x) = \varphi \circ x$, $x \in \overline{\Omega}$. Then $\Phi(\overline{\Omega}) \subseteq \overline{\Omega}$. Consider the mapping $H : [0, 1] \times \overline{\Omega} \to X$ defined by

$$H(t, x) = t\Phi(x) + (1 - t)x$$

We claim that $H(t, \partial\Omega) \subseteq \partial\Omega$ for $t \in [0, 1]$. In fact, if $y \in \partial\Omega$, then $\|x_0 - y\| = \frac{1}{2}$, so that $0 \le y(s) \le 1$ for $s \in [0, 1]$, and for some $s_0 \in [0, 1]$, either $y(s_0) = 0$ or $y(s_0) = 1$. In the case where $y(s_0) = 0$, we have $H(t, y)(s_0) = 0$, and in the case where $y(s_0) = 1$, we have $H(t, y)(s_0) = 1$. Moreover, as $0 \le \varphi(s) \le 1$ for $0 \le s \le 1$, we have $0 \le H(t, y)(s) \le 1$ for $0 \le t \le 1$ and $0 \le s \le 1$. Hence,

$H(t, y) \in \partial\Omega$ for $0 \le t \le 1$. Therefore, if $p \in \Omega$, then for every $t \in [0, 1]$ and $y \in \partial\Omega$, we have $H(t, y) - p \neq 0$, that is, $H - p$ is an *Ω-admissible homotopy* from $\Phi - p$ to $\mathrm{Id} - p$.

Suppose now that there is a concept of degree deg for the class $C(\overline{\Omega}; X)$ of continuous functions from $\overline{\Omega}$ into X that satisfies at least the homotopy invariance, existence, and normalization properties. Then for any $p \in \Omega$, we have

$$\deg(\Phi - p, \Omega) = \deg(\mathrm{Id} - p, \Omega) = 1$$

and consequently, there exists $x \in \Omega$ so that $\Phi(x) = p$. However, this conclusion is erroneous. To see this, define functions p and φ as follows:

$$p(s) = \tfrac{1}{4} + \tfrac{1}{2}s; \qquad \varphi(s) = \begin{cases} s, & 0 \le s \le \tfrac{1}{2} \\ 1 - s, & \tfrac{1}{2} < s \le \tfrac{5}{8} \\ \tfrac{5}{3}(s - 1) + 1, & \tfrac{5}{8} < s \le 1 \end{cases}$$

If x is a solution of $\Phi(x) = p$, we have $\varphi(x(0)) = \tfrac{1}{4}$, hence $x(0) = \tfrac{1}{4}$. Thus, $\varphi(x(s))$ can increase from $\tfrac{1}{4}$ by at most $\tfrac{1}{4}$ before it starts to decrease. But $p(s)$ increases from $\tfrac{1}{4}$ to $\tfrac{3}{4}$. Therefore, no suitable function x can be found in Ω.

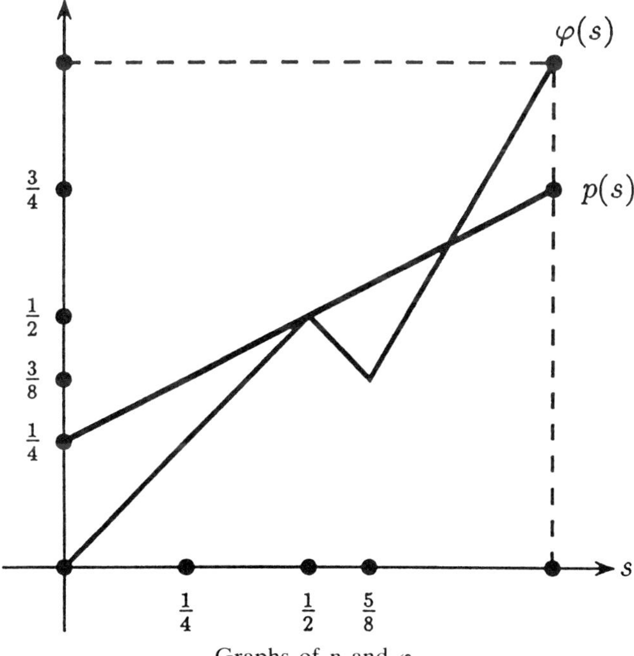

Graphs of p and φ.

Thus, it is impossible to satisfactorily define a concept of degree for the class of all continuous functions from $\overline{\Omega}$ into X.

3.1 SCHAUDER'S APPROXIMATIONS AND LERAY–SCHAUDER DEGREE

In this section we introduce the notion of Schauder's approximation and define Leray–Schauder's degree for compact fields. We begin with the following.

Definition 3.1.1 Let X, Y be topological spaces and $F : X \to Y$ be a continuous mapping:

(i) F is called *compact* if $F(X)$ is contained in a compact subset of Y.

(ii) F is called *completely continuous* if X is a metric space and the image of each bounded set in X is contained in a compact subset of Y.

(iii) A compact mapping $F : X \to Y$ is called *finite-dimensional* if Y is a linear topological space and $F(X)$ is contained in a finite-dimensional linear subspace of Y.

(iv) F is called *bounded* if Y is a metric space and $F(X)$ is a bounded subset of Y.

We denote by $\mathcal{K}(X, Y)$ the set of all compact mappings from X to Y. If (Y, d) is a metric space, $\mathcal{K}(X, Y)$ is a subspace of all bounded mappings $\mathcal{B}(X, Y)$ from X to Y, equipped with the metric

$$\rho(F, G) := \sup_{x \in X} d(F(x), G(x))$$

Proposition 3.1.2

Let X be a topological space and Y be a complete metric space. Then the space $\mathcal{K}(X, Y)$ equipped with the metric ρ, is also complete.

Proof. Since $\mathcal{B}(X, Y)$ is complete, it is sufficient to show that $\mathcal{K}(X, Y)$ is closed in $\mathcal{B}(X, Y)$. Let $\{F_n\}$ be a sequence in $\mathcal{K}(X, Y)$ and $\rho(F_n, F) \to 0$ for some $F \in \mathcal{B}(X, Y)$. We need to show $F \in \mathcal{K}(X, Y)$. In particular, we will show that F is continuous and $F(X)$ is totally bounded, that is, for all $\varepsilon > 0$, there is a finite ε-net for $F(X)$.

Let $\varepsilon > 0$ be a given number. By assumption, there is an integer k such that $\rho(F, F_k) = \sup_{x \in X} d(F(x), F_k(x)) \le \varepsilon/2$. Let $N = \{y_1, \ldots, y_n\}$ be an $\varepsilon/2$ net for the totally bounded set $F_k(X)$. Since for every $x \in X$, there is $y_i \in N$ such that $d(F_k(x), y_i) < \varepsilon/2$, we have

$$d(F(x), y_i) = d(F(x), F_k(x)) + d(F_k(x), y_i) < \varepsilon$$

which shows that N is an ε-net for the set $F(X)$.

In order to establish the continuity of F, assume again that $\varepsilon > 0$ is a given number and $x \in X$ a fixed point. By assumption, there is an integer m such that $\rho(F, F_m) \leq \varepsilon/3$, and there is a neighborhood U of x such that for all $x' \in U$, $d(F_m(x), F_m(x')) < \varepsilon/3$. Then, for all $x' \in U$,

$$\begin{aligned} d(F(x), F(x')) &\leq d(F(x), F_m(x)) + d(F_m(x), F_m(x')) + d(F_m(x'), F(x')) \\ &< \frac{\varepsilon}{3} + \frac{\varepsilon}{3} + \frac{\varepsilon}{3} = \varepsilon \end{aligned}$$

and the statement follows. □

Proposition 3.1.3

Let D be a bounded, closed subset of a normed linear space X and $F : D \to X$ be a compact map. Then, we have the following:

(i) *$(\mathrm{Id} - F)(A)$ is closed for every closed subset $A \subseteq D$.*
(ii) *$(\mathrm{Id} - F)^{-1}(K)$ is compact for every compact subset K of X.*

Proof. (i) Let $\{x_k\} \subset A$ be a sequence such that $y_k = x_k - F(x_k)$ converges to $y_0 \in X$. We need to show that $y_0 \in (\mathrm{Id} - F)(A)$. By the compactness of F, the sequence $\{F(x_k)\}$ is relatively compact. Therefore, we may assume, without loss of generality, that $F(x_k) \to y_1$ as $k \to \infty$ and thus $x_k = y_k + F(x_k) \to x_0 := y_0 + y_1$ as $k \to \infty$. Since A is closed, $x_0 \in A$, and by the continuity of F we obtain $x_0 - F(x_0) = y_0$; hence, $y_0 \in (\mathrm{Id} - F)(A)$.

(ii) Assume that $K \subset X$ is a compact subset and let $\{x_k\} \subset (\mathrm{Id} - F)^{-1}(K)$. We need to show that the sequence $\{x_k\}$ contains a convergent subsequence. We put $y_k := x_k - F(x_k)$, $k = 1, 2, 3, \ldots$. Since $\{y_k\} \subseteq K$, we may assume, without loss of generality, that $y_k \to y_0$ as $k \to \infty$. By the compactness of F, the sequence $\{F(x_k)\}$ contains a convergent subsequence $F(x_{k_l}) \to y_1$ as $l \to \infty$. Therefore, $x_{k_l} = y_{k_l} + F(x_{k_l}) \to y_0 + y_1$ as $l \to \infty$ and, thus, $\{x_{k_l}\}$ is a convergent subsequence of $\{x_k\}$. □

Let $N = \{y_1, \ldots, y_n\}$ be a finite set in a normed linear space Y, and let $\varepsilon > 0$ be a fixed number. We set $B(\varepsilon, N) := \bigcup_{i=1}^{n} B_\varepsilon(y_i)$. For each $i = 1, 2, \ldots, n$, let $\lambda_i : B(\varepsilon, N) \to \mathbb{R}$ be the function $\lambda_i(x) = \max\{0, \varepsilon - \|x - y_i\|\}$. The *Schauder projection* $p_\varepsilon : B(\varepsilon, N) \to \mathrm{conv}(N)$ is given by $p_\varepsilon(x) =$

$\dfrac{1}{\sum_{i=1}^{n} \lambda_i(x)} \sum_{i=1}^{n} \lambda_i(x) y_i$, where conv$(N)$ denotes the convex hull of the set N.

Proposition 3.1.4

Let C be a convex set of the normed space Y, $N = \{y_1, \ldots, y_n\}$ a finite subset of C, and $\varepsilon > 0$ a given number. Let $p_\varepsilon : B(\varepsilon, N) \to \mathrm{conv}(N) \subseteq C$ be the Schauder projection. Then we have the following:

(i) *p_ε is a finite-dimensional mapping.*
(ii) *For all $x \in B(\varepsilon, N)$, $\|x - p_\varepsilon(x)\| < \varepsilon$.*
(iii) *If N is symmetric with respect to zero, that is, $N = \{y_1, \ldots, y_k, -y_1, \ldots, -y_k\}$, then $B(\varepsilon, N)$ is a symmetric set and $p_\varepsilon(-x) = -p_\varepsilon(x)$ for all $x \in B(\varepsilon, N)$, that is, p_ε is odd.*

Proof. The proof of statement (i) is obvious and (ii) follows from the following inequality:

$$\begin{aligned}\|x - p_\varepsilon(x)\| &= \frac{1}{\sum_{i=1}^{n} \lambda_i(x)} \left\| \sum_{i=1}^{n} \lambda_i(x)[x - y_i] \right\| \\ &\leq \frac{1}{\sum_{i=1}^{n} \lambda_i(x)} \sum_{i=1}^{n} \lambda_i(x) \|x - y_i\| < \varepsilon\end{aligned}$$

To prove the statement (iii), we denote $-y_i$ by y_{-i}. Then (iii) follows from the definition of λ_i and the fact $\lambda_i(x) = \lambda_{-i}(-x)$, $x \in B(\varepsilon, N)$. □

Theorem 3.1.5 (SCHAUDER'S APPROXIMATION THEOREM)

Let X be a topological space, Y a normed linear space, $C \subseteq Y$ a closed convex subset, and $F : X \to C$ a compact map from X into C. Then for each $\varepsilon > 0$, there is a finite set $N = \{y_1, \ldots, y_n\} \subseteq F(x) \subseteq C$ and a finite-dimensional mapping $F_\varepsilon : X \to C$ satisfying:

(i) *$\|F_\varepsilon(x) - F(x)\| < \varepsilon$ for all $x \in X$.*
(ii) *$F_\varepsilon(X) \subseteq \mathrm{conv}(N) \subseteq C$.*

Proof. Since $F : X \to C$ is a compact map, there is a finite set $N = \{y_1, \ldots, y_n\} \subseteq F(X)$ with $F(X) \subseteq B(\varepsilon, N)$. Define $F_\varepsilon : X \to C$ by

$$F_\varepsilon(x) = p_\varepsilon(F(x)), \qquad x \in X$$

where $p_\varepsilon : B(\varepsilon, N) \to \text{conv}(N)$ is the Schauder's projection. It is clear that F_ε has the required properties. □

In what follows, when we say a sct A in a linear topological space is symmetric, we mean the set A is symmetric with respect to 0, that is, $x \in A$ if and only if $-x \in A$.

Corollary 3.1.6

Let $X_0 \subseteq X$ be symmetric subsets of a linear topological space, Y a normed linear space, $C \subseteq Y$ a closed convex symmetric subset, and let $F : X \to C$ be a compact map from X into C. Assume that $F(-x) = -F(x)$ for $x \in X_0 \subseteq X$. Then for each $\varepsilon > 0$, there is a finite set $N = \{y_1, \dots, y_n\} \subset F(X) \subseteq C$ and a finite-dimensional mapping $F_\varepsilon : X \to C$ satisfying:

(i) $\|F_\varepsilon(x) - F(x)\| < \varepsilon$ *for all* $x \in X$.
(ii) $F_\varepsilon(X) \subseteq \text{conv}(N) \subseteq C$.
(iii) $F_\varepsilon(-x) = -F_\varepsilon(x)$ *for all* $x \in X_0$.

Proof. As $\overline{F(X)} \cup \overline{(-F(X))}$ is a compact symmetric subset of C, we can find a symmetric ε-net N for $\overline{F(X)} \cup \overline{(-F(X))}$. Define $F_\varepsilon(x) = p_\varepsilon(F(x))$, where $p_\varepsilon : B(\varepsilon, N) \to \text{conv}(N)$ is the Schauder's projection. It is easy to verify that F_ε satisfies all required properties. □

Definition 3.1.7 Let X be a normed linear space:

(i) For a subset $A \subseteq X$, a mapping $f : A \to X$ is called a *compact field* on A if $f(x) = x - F(x)$, $x \in A$, where $F : A \to X$ is a compact mapping. We say that $h : [0, 1] \times A \to X$ is a *homotopy* of *compact fields* if $h(t, x) = x - H(t, x)$, $x \in A$, $t \in [0, 1]$, where $H : [0, 1] \times A \to X$ is a compact mapping.
(ii) Let $\Omega \subseteq X$ be an open, bounded, nonempty subset of X. A mapping $f : \overline{\Omega} \to X$ is called an Ω-*admissible compact field* if f is a compact field on $\overline{\Omega}$ and $f(x) \neq 0$ for all $x \in \partial\Omega$. A homotopy $h : [0, 1] \times \overline{\Omega} \to X$ of compact fields is called Ω-*homotopy* (or is said to be an Ω-*admissible homotopy*) if $h(t, x) \neq 0$ for all $(t, x) \in [0, 1] \times \partial\Omega$.

Proposition 3.1.8

Let $\Omega \subseteq X$ be an open, bounded, nonempty subset of X, and let $h : [0, 1] \times \overline{\Omega} \to X$ be an Ω-admissible homotopy of compact fields. Then

$$\inf\{\|h(t, x)\|; x \in \partial\Omega, t \in [0, 1]\} > 0$$

Proof. Assume that $h(t, x) = x - H(t, x)$ with $H : [0, 1] \times \overline{\Omega} \to X$ being compact. Suppose that $\{(t_n, x_n)\} \subseteq [0, 1] \times \partial\Omega$ is a sequence such that $h(t_n, x_n) = x_n - H(t_n, x_n) \to 0$ as $n \to \infty$. Since H is a compact mapping, we can assume, without loss of generality, that $H(t_n, x_n) \to y_0$ as $n \to \infty$. Since $\{t_n\} \subseteq [0, 1]$, we can also assume, without loss of generality, that $t_n \to t_0$ as $n \to \infty$. Then it follows that $(t_n, x_n) \to (t_0, y_0) \in [0, 1] \times \partial\Omega$ as $n \to \infty$ and $h(t_0, y_0) = \lim_{n\to\infty} h(t_n, x_n) = 0$. This contradicts the assumption that h is an Ω-admissible homotopy. □

We are now in a position to define the Leray–Schauder degree for compact fields.

Assume that $\Omega \subset X$ is an open, bounded, nonempty subset. Let $f : \overline{\Omega} \to X$ be an Ω-admissible compact field. Put $f(x) = x - F(x)$, $x \in \overline{\Omega}$. Let $\varepsilon = \inf\{\|f(x)\|;\ x \in \partial\Omega\}$. From Proposition 3.1.8 it follows that $\varepsilon > 0$. By Schauder's approximation theorem, there exists a finite-dimensional mapping $F_\varepsilon : \overline{\Omega} \to X$, called an *$\varepsilon/3$-approximation of F*, such that

$$\|F_\varepsilon(x) - F(x)\| < \frac{\varepsilon}{3} \qquad \text{for } x \in \overline{\Omega}$$

Then $f_\varepsilon : \overline{\Omega} \to X$ defined by $f_\varepsilon(x) = x - F_\varepsilon(x)$ for $x \in \overline{\Omega}$ is also an Ω-admissible compact field. Indeed, it follows from the definition that

$$\|f_\varepsilon(x) - f(x)\| = \|F_\varepsilon(x) - F(x)\| < \varepsilon$$

for $x \in \overline{\Omega}$. Therefore, $\|f_\varepsilon(x)\| > 0$ for all $x \in \partial\Omega$.

Let X_0 be a finite-dimensional subspace of X such that (i) $\Omega \cap X_0 \neq \varnothing$ and (ii) $F_\varepsilon(\overline{\Omega}) \subseteq X_0$. We define the Leray–Schauder degree $\deg(f, \Omega)$ of f on Ω as

$$\deg(f, \Omega) \stackrel{\text{def}}{=} \deg(f_0, \Omega_0) \tag{3.1.1}$$

where

$$f_0 := f_\varepsilon|_{\overline{\Omega \cap X_0}} : \overline{\Omega \cap X_0} \to X_0\,, \qquad \Omega_0 = \Omega \cap X_0$$

Theorem 3.1.9

The integer $\deg(f, \Omega)$, *defined by formula (3.1.1), does not depend on the choice of the finite-dimensional approximation F_ε and the choice of X_0. Moreover, this integer satisfies the following properties:*

(P1) *Normalization: Let $\Omega \subseteq X$ be a nonempty, open, bounded subset, $x_0 \in X$ such that $x_0 \notin \partial\Omega$. Then*

$$\deg(\mathrm{Id} - x_0, \Omega) = \begin{cases} 1 & \text{if } x_0 \in \Omega \\ 0 & \text{if } x_0 \notin \Omega \end{cases}$$

(P2) *Additivity: Let $\Omega \subseteq X$ be an open, bounded, nonempty subset and let $\Omega_1, \Omega_2 \subseteq \Omega$ be two disjoint, nonempty, open subsets. Let $f : \overline{\Omega} \to X$ be an Ω-admissible compact field such that $f^{-1}(0) \subseteq \Omega_1 \cup \Omega_2$. Then*

$$\deg(f, \Omega) = \deg(f, \Omega_1) + \deg(f, \Omega_2)$$

(P3) *Homotopy: Let $\Omega \subseteq X$ be a nonempty, open, bounded subset and let $h : [0, 1] \times \overline{\Omega} \to X$ be an Ω-admissible homotopy of compact fields. Then $\deg(h(t, \cdot), \Omega)$ is independent of $t \in [0, 1]$.*

Proof. In order to show that the definition of $\deg(f, \Omega)$ does not depend on the finite-dimensional $\varepsilon/3$ approximation F_ε of $F = \mathrm{Id} - f$ and the choice of X_0, we consider another finite-dimensional $\varepsilon/3$ approximation $F'_\varepsilon : \overline{\Omega} \to X$ of F such that $F'_\varepsilon(\overline{\Omega}) \subseteq X'_0$ with $\dim X'_0 < \infty$. Put $f'_\varepsilon(x) = x - F'_\varepsilon(x)$, $x \in \overline{\Omega}$. We denote by X_1 the subspace $X'_0 + X_0$ and define $f_0 = f_\varepsilon|_{\overline{X_0 \cap \Omega}}$, $f'_0 := f'_\varepsilon|_{\overline{X'_0 \cap \Omega}}$, $f_1 := f_\varepsilon|_{\overline{X_1 \cap \Omega}}$, $f'_1 := f'_\varepsilon|_{\overline{X_1 \cap \Omega}}$, and $\Omega_1 = \Omega \cap X_1$, $\Omega'_0 = \Omega \cap X'_0$, $\Omega_0 = \Omega \cap X_0$. It can be easily shown that $h_1(t, x) := tf_1(x) + (1 - t)f'_1(x)$ is an Ω_1-admissible homotopy. Therefore,

$$\deg(f_1, \Omega_1) = \deg(f'_1, \Omega_1)$$

On the other hand, it follows from the product formula of the Brouwer degree that

$$\deg(f_1, \Omega_1) = \deg(f_0, \Omega_0) \qquad \text{and} \qquad \deg(f'_1, \Omega_1) = \deg(f'_0, \Omega'_0)$$

Consequently, $\deg(f_0, \Omega_0) = \deg(f'_0, \Omega'_0)$. That is, the definition of $\deg(f, \Omega)$ does not depend on the approximation F_ε and the choice of the finite-dimensional subspace X_0.

In order to prove the normalization property, let $x_0 \notin \partial\Omega$. We consider a finite-dimensional subspace X_0 containing x_0 and such that $\Omega_0 = \Omega \cap X_0 \neq \varnothing$. Let $f(x) = x - x_0$, $x \in \overline{\Omega}$. Then $f_0 = f|_{\overline{\Omega_0}}$ maps $\overline{\Omega_0}$ into X_0, $f_0 = \mathrm{Id} - x_0$. Therefore, from the normalization property of the Brouwer degree, it follows that

$$\deg(f, \Omega) = \deg(f_0, \Omega_0) = \begin{cases} 1 & \text{if } x_0 \in \Omega_0 \\ 0 & \text{if } x_0 \notin \Omega_0 \end{cases}$$

In order to prove the additivity property, we consider a compact field $f : \overline{\Omega} \to X$, $f(x) - F(x)$, $x \in \overline{\Omega}$, such that $f^{-1}(0) \subseteq \Omega_1 \cup \Omega_2 \subseteq \Omega$, where Ω_1 and Ω_2 are two disjoint open subsets. Let

$$\varepsilon = \inf\{\|f(x)\|;\ x \in \overline{\Omega} \setminus (\Omega_1 \cup \Omega_2)\} > 0$$

and let F_ε be a finite-dimensional $\varepsilon/3$-approximation of F. Then $f_\varepsilon(x) := x - F_\varepsilon(x)$, $\overline{x} \in \overline{\Omega}$, satisfies the property $f_\varepsilon^{-1}(0) \subseteq \Omega_1 \cup \Omega_2$. Let X_0 be a finite-dimensional subspace of X such that $F_\varepsilon(\overline{\Omega}) \subseteq X_0$, $\Omega_0 = \Omega \cap X_0 \neq \emptyset$, $\Omega_{01} = \Omega_1 \cap X_0 \neq \emptyset$, and $\Omega_{02} = \Omega_2 \cap X_0 \neq \emptyset$. Then by the additivity property of the Brouwer degree, we get

$$\begin{aligned}\deg(f, \Omega) &= \deg(f_0, \Omega_0)\\ &= \deg(f_0, \Omega_{01}) + \deg(f_0, \Omega_{02})\\ &= \deg(f, \Omega_1) + \deg(f, \Omega_2)\end{aligned}$$

Assume now that $h : [0, 1] \times \overline{\Omega} \to X$ is an Ω-admissible homotopy of compact fields. Let $h(t, x) = x - H(t, x)$, where $H : [0, 1] \times \overline{\Omega} \to X$ is a compact mapping. Put

$$\varepsilon = \inf\{\|h(t, x)\|; (t, x) \in [0, 1] \times \partial\Omega\} > 0$$

We consider a finite-dimensional ε-approximation H_ε of H. Put $h_\varepsilon(t, x) = x - H_\varepsilon(t, x)$ and let $X_0 \subseteq X$ be a finite-dimensional subspace such that $h_\varepsilon([0, 1] \times \overline{\Omega}) \subseteq X_0$ and $\Omega_0 = \Omega \cap X_0 \neq \emptyset$. It is clear that $h_0 := h_\varepsilon|_{[0,1]\times\overline{\Omega}_0} : [0, 1] \times \overline{\Omega}_0 \to X_0$ is an Ω-homotopy. Thus, by the homotopy property of the Brouwer degree, we see that $\deg(h(t, \cdot), \Omega) = \deg(h_0(t, \cdot), \Omega_0)$ does not depend on $t \in [0, 1]$. □

Remark 3.1.10 Assume that $\Omega \subseteq X$ is an open but not necessarily bounded subset of X. Let $f : \Omega \to X$ be a continuous map such that (i) $\mathrm{Id} - f$ is completely continuous; (ii) the set $f^{-1}(0)$ is a compact subset of Ω. We will simply say that $f : \Omega \to X$ is Ω-admissible in a *larger sense*. The Leray–Schauder degree $\deg(f, \Omega)$ for an Ω-admissible map f in a larger sense can be defined by the formula

$$\deg(f, \Omega) := \deg(f, \Omega_0)$$

where Ω_0 is a bounded, open subset of Ω such that $f^{-1}(0) \subset \Omega_0 \subset \overline{\Omega}_0 \subset \Omega$. It is clear from the excision property below that the above definition does not depend on the choice of the subset Ω_0.

We shall now list other properties of the Leray–Schauder degree.

Theorem 3.1.11

Let $\Omega \subset X$ be an open, bounded, and nonempty subset, and $f = \mathrm{Id} - F : \overline{\Omega} \to X$ be an Ω-admissible compact field. The Leray–Schauder degree $\deg(f, \Omega)$ defined by formula (3.1.1) satisfies the following properties:

- **(P4)** *Existence: If $\deg(f, \Omega) \neq 0$, then there is $x \in \Omega$ such that $f(x) = 0$.*
- **(P5)** *Excision: If Ω_1 is an open subset of Ω such that $f^{-1}(0) \subseteq \Omega_1$, then $\deg(f, \Omega) = \deg(f, \Omega_1)$.*

(P6) *Continuity: Let $g\colon \overline{\Omega} \to X$ be a compact field such that*

$$\|f-g\|_\infty := \sup\{\|f(x)-g(x)\|;\quad x \in \overline{\Omega}\} < \inf\{\|f(x)\|; x \in \partial\Omega\}$$

then $\deg(f,\Omega) = \deg(g,\Omega)$.

(P7) *Boundary Values Dependence: If $g: \overline{\Omega} \to X$ is a compact field such that $f(x) = g(x)$ for $x \in \partial\Omega$, then $\deg(f,\Omega) = \deg(g,\Omega)$.*

(P8) *Multiplication Formula: Let $U = X \setminus f(\partial\Omega)$ and $\{U_\lambda\}_\lambda$ be the bounded components of the set U. If $G: X \to X$ is compact, $g = \mathrm{Id} - G$ and $0 \notin g \circ f(\partial\Omega)$, then $\deg(g \circ f, \Omega) = \sum_\lambda \deg(f - y_\lambda, \Omega)\deg(g, U_\lambda)$, where $y_\lambda \in U_\lambda$ is any given element, and only finitely many terms in the sum are nonzero.*

(P9) *Reduction Formula: If X_0 is a closed subspace of X and $F(\overline{\Omega}) \subseteq X_0$, then $\deg(f,\Omega) = \deg(f|_{\overline{X_0\cap\Omega}}, \Omega \cap X_0)$.*

(P10) *Commutativity:* Let Ω' *be an open and nonempty subset of a Banach space X' and assume that $F: \Omega \to X'$ and $G: \Omega' \to X$ are continuous maps such that one of them is compact. If one of the mappings $\mathrm{Id} - G \circ F : U = F^{-1}(\Omega') \to X$ and $\mathrm{Id} - F \circ G : U' = G^{-1}(\Omega) \to X'$ is an admissible compact field, then so is the other and*

$$\deg(\mathrm{Id} - F \circ G, U') = \deg(\mathrm{Id} - G \circ F, U)^*$$

Proof. The proofs of the properties (P4–P7) are similar to those in the case of the Brouwer degree, and the reduction formula (P9) is an immediate consequence of the definition of the Leray–Schauder degree. So we will only give the proofs of the multiplication formula (P8) and the commutativity property (P10).

We begin with the proof of the multiplication formula (P8). By Proposition 3.1.3, the set $f(\partial\Omega)$ is closed and bounded. Therefore, there is one unbounded connected component of the set $U = X \setminus f(\partial\Omega)$. We denote by $\{U_\lambda\}_\lambda$ the family of all bounded connected components of U and let $U_0 = \bigcup_\lambda U_\lambda$. Since G is compact, by Proposition 3.1.3, the set $K := U_0 \cap (\mathrm{Id} - G)^{-1}(0)$ is compact. The family $\{U_\lambda\}$ is an open covering of K; thus, it contains a finite subcovering $\{U_{\lambda_1}, \ldots, U_{\lambda_n}\}$ of K. We put $K_i = K \cap U_{\lambda_i}$ for $i = 1, \ldots, n$. For every $i = 1, \ldots, n$, there exists (see Exercise 3.1.1) a closed arc-connected subset $A_i \subset U_{\lambda_i}$ such that $K_i \subset A_i$ and $\inf\{\|x-y\|;\ x \in \partial U_{\lambda_i},\ y \in A_i\} =: \delta_i > 0$. Let $\varepsilon := \frac{1}{2}\min\{\delta_1, \ldots, \delta_n\}$ and F_ε be a finite-dimensional ε-approximation of F. Then it is clear that $A_i \subset X \setminus (\mathrm{Id} - F_\varepsilon)(\partial\Omega)$ for all $i = 1, \ldots, n$, and thus each of the sets K_i is again contained in a single connected component $W_{j(i)}$ of the set $X \setminus (\mathrm{Id} - F_\varepsilon)(\partial\Omega)$. We may assume, without loss of generality, that $W_{j(i)}$ is bounded. We construct a

* Note that U and U' are not necessarily bounded and the degrees are in a larger sense.

finite-dimensional δ-approximation G_δ of G, where $\delta = \frac{1}{3}\inf\{\|x - G(x)\|;\ x \in \partial W_{j(i)}$ for all $j(i)\}$. Next, we consider the finite-dimensional subspace $X_0 := \text{span}\{F_\varepsilon(\overline{\Omega}), G_\delta(X)\}$ and we obtain

$$\begin{aligned}\deg(g \circ f, \Omega) &= \deg(g_\delta \circ f_\varepsilon|_{X_0}, \Omega \cap X_0)\\ &= \sum_i \deg(f_\varepsilon|_{X_0} - y_i, \Omega \cap X_0)\deg(g_\delta|_{X_0}, U_{\lambda_i} \cap X_0)\\ &= \sum_i \deg(f_\varepsilon - y_i, \Omega)\deg(g_\delta, U_{\lambda_i})\\ &= \sum_i \deg(f - y_i, \Omega)\deg(g, U_{\lambda_i})\end{aligned}$$

where $y_i \in K_i$ is a fixed point for each connected component U_{λ_i}, $f_\varepsilon := \text{Id} - F_\varepsilon$ and $g_\delta := \text{Id} - G_\delta$. This completes the proof of the multiplication property.

Now we prove the commutativity property (P10). For a given mapping $\Phi : V \to Y$, $V \subset Y$, we denote by $\text{Fix}(\Phi)$ the set of fixed points of Φ in V, that is, $\text{Fix}(\Phi) := \{x \in V;\ x = \Phi(x)\}$. We note that the maps $F : \text{Fix}(G \circ F) \to \text{Fix}(F \circ G)$ and $G : \text{Fix}(F \circ G) \to \text{Fix}(G \circ F)$ are inverse to each other and, hence, the fixed point sets $\text{Fix}(G \circ F)$ and $\text{Fix}(F \circ G)$ are homeomorphic. Thus if one of them is compact, then so is the other. The proof of the commutativity property will be done in two steps.

Step 1. We first assume that both mappings F and G are compact. We define $\Phi : U \times U' \to X \times X'$ by $\Phi(x, y) = (G(y), F(x))$, $(x, y) \in U \times U'$. We also define the homotopies

$$\begin{aligned}H_t, H'_t &: U \times U' \to X \times X'\\ \Psi_t &: U \times X' \to X \times X'\\ \Psi'_t &: X \times U' \to X \times X'\end{aligned}$$

by the following formulae:

$$\begin{aligned}H_t(x, y) &= (tG \circ F(x) + (1 - t)G(y), F(x))\\ H'_t(x, y) &= (G(y), tF \circ G(y) + (1 - t)F(x))\\ \Psi_t(x, y) &= (G \circ F(x), (1 - t)F(x))\\ \Psi'_t(x, y) &= ((1 - t)G(y), F \circ G(y))\end{aligned}$$

We have $\Phi = H_0 = H'_0$. Since F and G are compact, the mapping Φ and all the above homotopies are compact. On the other hand, the fixed-point sets $\text{Fix}(\Phi)$ and $\text{Fix}(G \circ F)$ are homeomorphic under the mappings $x \mapsto (x, F(x))$, and $(x, y) \mapsto x$; thus, $\text{Id} - \Phi$ is $U \times U'$-admissible. Moreover, by direct computation $\mathrm{Fix}(\Phi) = \mathrm{Fix}(H_t) = \mathrm{Fix}(H'_t)$. Therefore, the homotopies H_t and H'_t are also admissible. Similarly, the homotopies Ψ_t and Ψ'_t are admissible. Consequently, by the homotopy property, we obtain

$$\deg(\mathrm{Id} - \Phi, U \times U') = \deg(\mathrm{Id} - H_1, U \times U') = \deg(\mathrm{Id} - H'_1, U \times U')$$

On the other hand, since $H_1 = \Psi_0|_{U \times U'}$ and $H'_1 = \Psi'_0|_{U \times U'}$, we get by excision and homotopy,

$$\deg(\mathrm{Id} - \Phi, U \times U') = \deg(\mathrm{Id} - \Psi_1, U \times X') = \deg(\mathrm{Id} - \Psi'_1, U \times X')$$

As both mappings $\Psi_1 = (G \circ F) \times 0$ and $\Psi'_1 = 0 \times (F \circ G)$ are product maps, we obtain from the reduction formula $\deg(\mathrm{Id} - G \circ F, U) = \deg(\mathrm{Id} - F \circ G, U')$.

STEP 2. We assume now that $F : \Omega \to X'$ is compact and $G : \Omega' \to X$ is continuous. By the excision property, we may assume, without loss of generality, that both F and G are defined, in fact, on $\overline{\Omega}$ and $\overline{\Omega'}$, respectively. We assume that $G \circ F$ is admissible and, thus, $F \circ G$ is also admissible. We can find a bounded $\Omega'_0 \subseteq \Omega'$ such that $\mathrm{Fix}(F \circ G) \subseteq \Omega'_0 \subseteq \overline{\Omega'_0} \subset \Omega'$. Put $\Omega_0 := F^{-1}(\Omega'_0)$. Clearly, $\mathrm{Fix}(G \circ F) \subset \Omega'_0$. By the assumption that F is compact, both $G \circ F : F^{-1}(\Omega'_0) \to X$ and $F \circ G : G^{-1}(\Omega_0) \to X'$ are compact. By the excision property, it is sufficient to show that $\deg(\mathrm{Id} - G \circ F,\ F^{-1}(\Omega'_0)) = \deg(\mathrm{Id} - F \circ G, G^{-1}(\Omega_0))$. Let us put

$$\eta_1 = \inf\{\|x - G \circ F(x)\|; x \in \partial F^{-1}(\Omega'_0)\}$$

$$\eta_2 = \inf\{\|x - F \circ G(x)\|; x \in \partial G^{-1}(\Omega_0)\}$$

$$\eta = \min\{\eta_1, \eta_2\}$$

The number η is positive. Let K be a compact set containing $\overline{F(\Omega)} \subset X'$. The continuity of G implies that for each $y \in K \cap \Omega'_0$, there is $\delta_y > 0$ such that (i) the open ball $B_{\delta_y}(y)$ of radius δ_y and centered at y is contained in Ω'; (ii) if $y_1, y_2 \in B_{\delta_y}(y)$, then $\|G(y_1) - G(y_2)\| < \eta$. From the compactness of $K \cap \overline{\Omega}'$, it follows that we can select a finite number of balls $B_{\delta_{y_1}}(y_1), \ldots, B_{\delta_{y_k}}(y_k)$ to cover $K \cap \overline{\Omega}'$. We put $\delta = \min\{\delta_{y_1}, \ldots, \delta_{y_k}\}$, $B = \bigcup_{i=1}^{k} B_{\delta_{y_i}}(y_i)$, and $\varepsilon = \min\{\delta/2, \eta\}$. It is clear that if $y \in K \cap \overline{\Omega}'$ and $\|y' - y\| < \varepsilon$, then $\|G(y') - G(y)\| < \eta$, and $ty + (1-t)y' \in B$ for all $t \in [0, 1]$.

Now let $F_\varepsilon : \Omega_0 \to X'$ be a finite-dimensional ε-approximation of $F : \Omega_0 \to X'$ and $H_t : \Omega_0 \to X'$ be given by $H_t(x) = tF(x) + (1-t)F_\varepsilon(x)$. Clearly, H_t is a compact homotopy joining F and F_ε. Since $\overline{F^{-1}(\Omega'_0)} \subset F^{-1}(\overline{\Omega'_0})$, the values of H_t are in a compact subset of $B \subset \Omega'$. We may consider on $\overline{F^{-1}(\Omega'_0)}$ the composition $G \circ H_t$. It follows from the choice of the number η that this is a homotopy joining $G \circ F$ with its η-approximation $G \circ F_\varepsilon$.

Moreover, $G \circ H_t$ has no fixed point on $\partial F^{-1}(\Omega_0')$. Thus, $\mathrm{Id} - G \circ H_t : F^{-1}(\Omega_0') \to X$ is an admissible homotopy joining $\mathrm{Id} - G \circ F_\varepsilon$ and $\mathrm{Id} - G \circ F$ on $F^{-1}(\Omega_0')$.

Consequently, by the homotopy property, we have

$$\deg(\mathrm{Id} - G \circ F, F^{-1}(\Omega_0')) = \deg(\mathrm{Id} - G \circ F_\varepsilon, F^{-1}(\Omega_0')) \tag{3.1.2}$$

Next, since F_ε has its range in a finite-dimensional subspace X_0', that is, $F_\varepsilon(\Omega) \subset X_0'$, we consider the following restrictions: $\widetilde{F_\varepsilon} : F^{-1}(\Omega_0') \to X_0' \cap \Omega_0'$ and $\tilde{G} : X_0' \cap \Omega_0' \to X$ of F_ε and G, respectively. On $F^{-1}(\Omega_0')$, we have $G \circ F_\varepsilon = \tilde{G} \circ \tilde{F}_\varepsilon$. Therefore,

$$\deg(\mathrm{Id} - G \circ F_\varepsilon, F^{-1}(\Omega_0')) = \deg(\mathrm{Id} - \tilde{G} \circ \tilde{F}_\varepsilon, F^{-1}(\Omega_0')) \tag{3.1.3}$$

Further, since Ω_0' is bounded and $\tilde{G}$ is the restriction of $G_{|X_0' \cap \overline{\Omega_0'}}$ with $X_0' \cap \overline{\Omega_0'}$ compact, we conclude that $\tilde{G}$ is compact. Thus, both $\tilde{F}_\varepsilon$ and $\tilde{G}$ are compact and we obtain from step 1

$$\deg(\mathrm{Id} - \tilde{G} \circ \tilde{F}_\varepsilon, F^{-1}(\Omega_0')) = \deg(\mathrm{Id} - \tilde{F}_\varepsilon \circ \tilde{G}, \tilde{G}^{-1}(\Omega_0)) \tag{3.1.4}$$

On the other hand, for the composition $H_t \circ G$ on $\overline{G^{-1}(\Omega_0)} \subset G^{-1}(\overline{\Omega_0})$, $H_t \circ G$ is a compact homotopy joining $F \circ G$ with $F_\varepsilon \circ G$ that has no fixed point on $\partial G^{-1}(\Omega_0)$. In other words, $H_t \circ G : G^{-1}(\Omega_0) \to X$ is an admissible compact homotopy joining $F_\varepsilon \circ G$ with $F \circ G$ on $G^{-1}(\Omega_0)$. Consequently, by the homotopy property, we have

$$\deg(\mathrm{Id} - F \circ G, G^{-1}(\Omega_0)) = \deg(\mathrm{Id} - F_\varepsilon \circ G, G^{-1}(\Omega_0)) \tag{3.1.5}$$

Since the values of $F_\varepsilon \circ G$ are in $X_0' \cap \Omega_0'$, by the definition of the Leray–Schauder degree, we get

$$\begin{aligned} \deg(\mathrm{Id} - F_\varepsilon \circ G, G^{-1}(\Omega_0)) &= \deg(\mathrm{Id} - F_\varepsilon \circ G, G^{-1}(\Omega_0) \cap \Omega_0' \cap X_0') \\ &= \deg(\mathrm{Id} - \tilde{F}_\varepsilon \circ \tilde{G}, \tilde{G}^{-1}(\Omega_0)) \end{aligned} \tag{3.1.6}$$

By comparing formulas (3.1.5), (3.1.3), and (3.1.2), we obtain the conclusion. □

EXERCISES

3.1.1 Let U be an open, connected, bounded set in a normed linear space X and $K \subset U$ a compact subset. Show that there exists an arc-connected, closed subset $A \subset U$ such that $K \subset A$ and $\inf\{\|x - y\|;\ x \in \partial U,\ y \in A\} > 0$.

3.1.2 Let A be a subset of a metric space (X, d) and $F : A \to X$ be a compact map. A point $a \in A$ is said to be an *ε-fixed point* of F for $\varepsilon > 0$ if $d(a, F(a)) < \varepsilon$. Show that F has a fixed point if and only if it has an ε-fixed point for each $\varepsilon > 0$.

3.1.3 Following similar steps as in the finite-dimensional case (Chapter 2), complete the proof of Theorem 3.1.11.

3.1.4 Let U be an open, not necessarily bounded, subset of a Banach space X and let $F : U \to X$ be a completely continuous map such that $\text{Fix}(F) := \{x; F(x) = x\}$ is a compact subset of U. Show that $\inf\{\|x - F(x)\|; x \in \partial U\} > 0$.

3.1.5 Show by verifying all the necessary details that the Leray–Schauder degree can be extended to the class of admissible maps in a larger sense, that is, maps $\text{Id} - F : U \to X$, where U is an open (not necessarily bounded) subset of X and $F : U \to X$ is a completely continuous map such that $\text{Fix}(F)$ is a compact subset of U.

3.1.6 Let $N = \{y_1, \ldots, y_n\}$ be a finite set in a locally convex linear space Y and let μ_U be a seminorm corresponding to a convex symmetric neighborhood U of the origin. Let $B(U, N) := \bigcup_{i=1}^{n} (\{y_i\} + U)$, and $\lambda_{i,U} : B(U, N) \to \mathbb{R}$ be defined by $\lambda_{i,U}(x) = \max\{1 - \mu_U(x - y_i), 0\}$. Then one can define the *Schauder projection* $p_U : B(U, N) \to \text{conv}(N)$ by

$$p_U(x) = \frac{1}{\sum_{i=1}^{n} \lambda_{i,U}(x)} \sum_{i=1}^{n} \lambda_{i,U}(x) y_i$$

Show:

(i) p_U is a finite-dimensional mapping.

(ii) For all $x \in B(U, N)$, $x - p_U(x) \in U$.

(iii) If N is symmetric with respect to zero, then $-B(U, N) = B(U, N)$ and $p_U(-x) = -p_U(x)$ for all $x \in B(U, N)$, that is, p_U is odd.

3.1.7 Let X be a topological space, Y a locally convex linear space, $C \subseteq Y$ a convex subset, and $F : X \to C$ a compact map from X into C. Show that for every convex symmetric neighborhood U of the origin in Y, there are a finite set $N = \{y_1, \ldots, y_n\} \subseteq F(X) \subseteq C$ and a finite-dimensional mapping $F_U : X \to C$ satisfying:

(i) $F_U(x) - F(x) \in U$ for all $x \in U$.

(ii) $F_U(x) \subseteq \text{conv}(N) \subseteq C$.

(iii) If, in addition, X is a linear topological space and $F(x)$ is an odd mapping, then $F_U(x)$ is also an odd mapping.

3.1.8 Let $\Omega \subseteq X$ be an open, bounded subset in a locally convex linear space X and $F : \overline{\Omega} \to X$ be an Ω-admissible compact mapping.

(i) Show that there exists a convex symmetric neighborhood U of the origin such that $x - F(x) \not\in U$ for all $x \in \partial\Omega$.

(ii) Using the idea of $\frac{1}{3}U$-*approximations* and an analogue of formula (3.1.1), extend the definition of the degree to the class of compact fields in a locally convex linear space.

(iii) Prove that the degree for compact fields in locally convex linear space satisfies all the properties (P1–P9).

3.2 FIXED-POINT THEOREMS

In this section we develop several fixed-point theorems as applications of the Leray–Schauder degree.

Proposition 3.2.1

Let Ω be a nonempty, convex, open, bounded set of a normed space X and let $F : \overline{\Omega} \to \overline{\Omega}$ be a compact map such that $F(x) \neq x$ for $x \in \partial\Omega$. Then $f = \mathrm{Id} - F$ is an Ω-admissible compact field and $\deg(f, \Omega) = 1$. *In particular, F has a fixed point in Ω.*

Proof. Let $y_0 \in \Omega$ be a fixed element. Define $H : [0, 1] \times \overline{\Omega} \to \overline{\Omega}$ by

$$H(t, x) = ty_0 + (1 - t)F(x) , \qquad t \in [0, 1] , \quad x \in \overline{\Omega}$$

Clearly, H is a compact homotopy and $H(t, x) \neq x$ for $x \in \partial\Omega$ and $t \in [0, 1]$. Therefore, the homotopy invariance and normalization property imply that

$$\deg(f, \Omega) = \deg(\mathrm{Id} - H(t, \cdot\,), \Omega) = \deg(\mathrm{Id} - \{y_0\}, \Omega) = 1$$

The existence of a fixed point of F is an immediate consequence of the existence property of the degree. □

We have the following extension of Brouwer's fixed-point theorem to compact maps.

Theorem 3.2.2 (Schauder's Fixed-Point Theorem)

Let C be a nonempty, closed, convex subset of a normed space X. Then each compact map $F : C \to C$ has a fixed point.

Proof. By Dugundji's extension theorem (Proposition 1.5.4), there exists a continuous extension $\tilde{F} : X \to X$ of F such that $\tilde{F}(X) \subseteq \mathrm{conv}(F(C))$. Since C is a closed convex subset of X, $\tilde{F}(X) \subseteq \overline{\mathrm{conv}(F(C)} \subseteq C$, and thus

$\tilde{F}: X \to X$ is a compact extension of F such that $\overline{\tilde{F}(X)} \subset C$. Let $\Omega \subset X$ be an open ball centered at x_0 for some $x_0 \in C$ such that $\overline{\tilde{F}(X)} \subset \Omega$. Define $h(t, x) = x - (1-t)\tilde{F}(x) - tx_0$ for $t \in [0, 1]$ and $x \in X$. If $x - (1-t)\tilde{F}(x) - tx_0 = 0$ for some $x \subset X$, then $x \in \operatorname{conv}(x_0, \tilde{F}(X)) \subset \Omega \cap C$. Thus, $h: [0, 1] \times X \to X$ is an Ω-admissible homotopy between $\mathrm{Id} - \tilde{F}$ and $\mathrm{Id} - x_0$. Consequently, $\deg(\mathrm{Id} - \tilde{F}, \Omega) = \deg(\mathrm{Id} - x_0, \Omega) = 1$. Therefore, from the existence property it follows that there is $x \in \Omega$ such that $\tilde{F}(x) = x$. On the other hand, since $\tilde{F}(x) \in C$ and $\tilde{F}$ is an extension of $F: C \to C$, $F(x) = \tilde{F}(x)$ and the conclusion follows. □

Theorem 3.2.3 (Borsuk's Theorem)

Let Ω be a bounded, convex, symmetric, open neighborhood of the origin in a normed linear space X. Let $f: \overline{\Omega} \to X$ be a compact Ω-admissible field such that f is odd on $\partial\Omega$. Then $\deg(f, \Omega)$ is an odd integer. In particular, $\deg(f, \Omega) \neq 0$ and therefore there is $x \in \Omega$ such that $f(x) = 0$.

Proof. Let $g(x) = \frac{1}{2}[f(x) - f(-x)]$, $x \in \overline{\Omega}$. Since g is a compact field such that $g|_{\partial\Omega} = f|_{\partial\Omega}$, $\deg(f, \Omega) = \deg(g, \Omega)$. Moreover, g is odd. By Corollary 3.1.5, there is a finite-dimensional mapping $F_\varepsilon: \overline{\Omega} \to X$ such that:

(i) $\|F_\varepsilon(x) - G(x)\| < \varepsilon$ for all $x \in \overline{\Omega}$
(ii) $F_\varepsilon(-x) = -F_\varepsilon(x)$ for $x \in \overline{\Omega}$

where $G(x) = x - g(x)$, $x \in \overline{\Omega}$. Let X_0 be a finite-dimensional subspace of X such that $X_0 \cap \Omega \neq \varnothing$ and $F_\varepsilon(\overline{\Omega}) \subset X_0$. Put $g_\varepsilon(x) = x - F_\varepsilon(x)$, $x \in \overline{\Omega}$. Clearly, $g_\varepsilon(\overline{\Omega} \cap X_0) \subseteq X_0$. Then by the definition of the Leray–Schauder degree, we have

$$\deg(g, \Omega) = \deg(g_\varepsilon|_{X_0 \cap \overline{\Omega}}, X_0 \cap \overline{\Omega})$$

Since $g_\varepsilon|_{X_0 \cap \overline{\Omega}}$ satisfies the hypotheses of Theorem 2.4.1, $\deg(g_\varepsilon|_{X_0 \cap \overline{\Omega}}, X_0 \cap \overline{\Omega})$ is an odd integer and the conclusion follows. □

Theorem 3.2.4 (Birkhoff–Kellogg Theorem)

Let Ω be a bounded, open neighborhood of zero in an infinite-dimensional normed space X and let $F: \overline{\Omega} \to X$ be a compact map such that $\inf_{x \in \partial\Omega} \|F(x)\| > 0$. Then F has an invariant direction on $\partial\Omega$, that is, there is an $x \in \partial\Omega$ and $\mu > 0$ such that $x = \mu F(x)$.

Proof. Assume that $x \neq \mu F(x)$ for all $x \in \partial\Omega$ and $\mu > 0$. Let $\alpha = \inf_{x \in \partial\Omega} \|F(x)\|$. Since $\overline{F(\overline{\Omega})} = K$ is compact, it cannot cover the noncompact ball $B_{\alpha/2}(0)$, so there is some $y_0 \in B_{\alpha/2}(0)\backslash K$. Define a homeomorphism $h:(X, B_{\alpha/2}(0)) \to (X, B_{\alpha/2}(0))$ that is the identity on $X\backslash B_{\alpha/2}(0)$ and maps y_0 to zero. This homeomorphism can be constructed as follows: First express each $z \in B_{\alpha/2}(0)\backslash\{y_0\}$ uniquely as $z=(1-t)y_0+t\hat{z}$, where $\hat{z} \in \partial B_{\alpha/2}(0)$, $0<t<1$, and then define

$$h(z)=\begin{cases} 0 & \text{if } z=y_0 \\ t\hat{z} & \text{if } z\in B_{\alpha/2}(0)\backslash\{y_0\} \\ z & \text{if } z \in X\backslash B_{\alpha/2}(0) \end{cases}$$

We put $G=h\circ F$. Then $G:\overline{\Omega}\to X$ is a compact map such that $G|_{\partial\Omega}=F|_{\partial\Omega}$. By assumption, $f(x):=x-F(x)\neq 0$ for $x\in\partial\Omega$. Therefore, $\deg(f,\Omega)=\deg(g,\Omega)$, where $g(x)=x-G(x)$ for $x\in\overline{\Omega}$.

Now choose an integer N so large that $\overline{\Omega}$ is contained in the interior of $B_N(0)$ and consider the mapping $G_1(x)=N\dfrac{G(x)}{\|G(x)\|}$, $x\in\overline{\Omega}$. It is clear that G_1 is compact. Moreover, $G_1(\overline{\Omega})\cap\overline{\Omega}=\varnothing$. Therefore, for $f_1(x):=x-G_1(x)$, $x\in\overline{\Omega}$, we have $\deg(f_1,\Omega)=0$. Moreover, it follows from the assumption that the mapping

$$h(t,x)=x-tN\frac{G(x)}{\|G(x)\|}, \qquad t\in[0,1], \quad x\in\overline{\Omega}$$

is an Ω-admissible homotopy of compact fields such that $h_0=\mathrm{Id}|_{\overline{\Omega}}$. Therefore, by the normalization property and the homotopy property, we get

$$1=\deg(h_0,\Omega)=\deg(h_1,\Omega)=\deg(f_1,\Omega)=0$$

a contradiction. Consequently, there exist $x\in\partial\Omega$ and $\mu>0$ such that $x=\mu F(x)$. □

EXERCISES

3.2.1 Give an example to show that Birkhoff–Kellogg's theorem is not true in a finite-dimensional space.

3.2.2 Let $X=\mathbb{R}^n_+ :=\{(x_1,\ldots,x_n)\in\mathbb{R}^n;\ x_i\geq 0,\ i=1,\ldots,n\}$ and let $F:X\to X$ be a continuous mapping such that $\|F(x)\|<\|x\|$ for all $x\in X$ with $\|x\|=1$, and $\|F(x)\|>\|x\|$ for all $x\in X$ with $\|x\|=2$. Show that there exists an $x\in X$ such that $1<\|x\|<2$ and $F(x)=x$.

3.2.3 Let $X=\mathbb{R}^n_+$ and let $F:X\to X$ be a continuous mapping such that $\|F(x)\|>\|x\|$ for all $x\in X$ with $\|x\|=1$, and $\|F(x)\|<\|x\|$ for all $x\in X$ with $\|x\|=2$. Show that there exists an $x\in X$ such that $1<\|x\|<2$ and $F(x)=x$.

3.2.4 Let Ω be an open, bounded subset in a locally convex linear space X and $F : \overline{\Omega} \to X$ be a compact map. Show that if for every convex symmetric neighborhood U of the origin, there exists a point $x_U \in \overline{\Omega}$ such that $F(x_U) - x_U \in U$, then F has a fixed point in Ω.

3.2.5 Let C be a closed convex subset of a locally convex linear space X. Show that every compact map $F : C \to C$ has a fixed point. *Hint:* Approximate the map F by finite, dimensional maps and use the result from Exercise 3.2.4.

3.2.6 Let U be a convex, symmetric neighborhood of the origin in a locally convex linear space X and let $F : \bar{U} \to X$ be a compact map such that $F(-x) = -F(x)$ for all $x \in \partial U$. Show that $\deg(\mathrm{Id} - F, U)$ is an odd number and, consequently, F has a fixed point in U.

3.3 COMPACT LINEAR OPERATORS

In this section we establish a relation between the degree $\deg(\mathrm{Id} - T, \Omega)$ and the multiplicities of eigenvalues of T, where X is a real Banach space, Ω is an open and bounded subset of X containing 0, and T is a compact linear operator.

For simplicity, we will denote by $\mathcal{K}(X)$ the set of all compact linear operators from X into itself.

Theorem 3.3.1

Let X be a real Banach space, $T_0 \in \mathcal{K}(X)$, and $L = \mathrm{Id} - T_0$. Then we have the following:

(i) *If L and M are one-to-one, where $M = \mathrm{Id} - S_0$ and $S_0 \in \mathcal{K}(X)$, then $\deg(LM, \Omega) = \deg(L, \Omega) \cdot \deg(M, \Omega)$ for every open and bounded $\Omega \subseteq X$ with $0 \in \Omega$.*

(ii) *If L is one-to-one and X is the topological direct sum of subspaces $X_1, \ldots, X_m$ such that $L(X_i) \subseteq X_i$, then*

$$\deg(L, \Omega) = \prod_{i=1}^{m} \deg(L|_{X_i}, \Omega \cap X_i)$$

Proof. The statement (i) follows immediately from the multiplication formula. To prove statement (ii), it suffices to consider the case $m = 2$. Let $P_i : X \to X_i$, $i = 1, 2$, be chosen projections and consider $M_1 = \mathrm{Id} - T_0P_1$, $M_2 = \mathrm{Id} - T_0P_2$. Then $L = M_1M_2$, M_1 and M_2 are one-to-one, $T_0P_i \in \mathcal{K}(X_i)$, $i = 1, 2$.

Let $\Omega = B_1(0)$. Then by (i), we have $\deg(L, \Omega) = \prod_{i=1}^{2} \deg(M_i, \Omega)$. From the product formula, we have

$$\deg(L, \Omega) = \prod_{i=1}^{2} \deg(M_i|_{\Omega \cap X_i}, \Omega \cap X_i) = \prod_{i=1}^{2} \deg(L|_{\Omega \cap X_i}, \Omega \cap X_i)$$

This completes the proof. □

We now recall the following spectral theorem for compact linear operators in Banach spaces. For a proof, we refer to Rudin (1976).

Theorem 3.3.2

Let X be a Banach space over $K = \mathbb{R}$ or $\mathbb{C}$, $T_0 \in \mathscr{K}(X)$, and $L_\lambda = T_0 - \lambda \,\mathrm{Id}$ for $\lambda \in K$. Denote by Λ the set of all eigenvalues of T_0. Then, we have the following:

(i) $\Lambda \subseteq \{\mu \in K;\ |\mu| \le \|T_0\|\}$, *$\Lambda$ is at most countable, and only $\mu = 0$ may be a cluster point of Λ.*

(ii) *L_λ is a continuous automorphism of X for every $\lambda \notin \Lambda \cup \{0\}$.*

(iii) *For every $\lambda \in \Lambda \backslash \{0\}$, there exists the least natural number $k = k(\lambda)$ such that $\mathrm{Ker}(L_\lambda^n) = \mathrm{Ker}(L_\lambda^k)$ for $n \ge k$, and*
   - (a) $X = L_\lambda^k(X) \oplus \mathrm{Ker}(L_\lambda^k)$, $\dim \mathrm{Ker}(L_\lambda^k) < \infty$, *and $L_\lambda^k(X)$ is closed.*
   - (b) *$L_\lambda^k(X)$ and $\mathrm{Ker}(L_\lambda^k)$ are invariant under T_0 and $L_\lambda|_{L_\lambda^k(X)}$ is a homeomorphism onto $L_\lambda^k(X)$.*
   - (c) $\mathrm{Ker}(L_\mu^{k(\mu)}) \subseteq L_\lambda^{k(\lambda)}(X)$ *whenever $\lambda, \mu \in \Lambda \backslash \{0\}$, and $\lambda \ne \mu$.*

In what follows, the closure Λ is called the spectrum of T_0 and is denoted by $\sigma(T_0)$. Also for $\lambda \in \Lambda \backslash \{0\}$, $\dim \mathrm{Ker}(L_\lambda^{k(\lambda)})$ is called the *algebraic multiplicity* of λ, and $\mathrm{Ker}(L_\lambda^{k(\lambda)})$ and $L_\lambda^{k(\lambda)}(X)$ are denoted by $N(\lambda)$ and $R(\lambda)$, respectively.

Theorem 3.3.3

Let X be a real Banach space and $T \in \mathscr{K}(X)$. Assume that λ is a nonzero real number such that λ^{-1} is not an eigenvalue of T. Then for any open and bounded subset $\Omega \subseteq X$ with $0 \in \Omega$, we have

$$\deg(\mathrm{Id} - \lambda T, \Omega) = (-1)^{m(\lambda)}$$

where $m(\lambda)$ is the sum of the algebraic multiplicities of the real eigenvalues μ satisfying $\mu\lambda > 1$ and $m(\lambda) = 0$ if T has no such eigenvalues μ.

Proof. First, we note that

$$\mathrm{Id} - \lambda T = -\lambda(T - \lambda^{-1}\,\mathrm{Id})$$

is a homeomorphism onto X. Hence, we can, without loss of generality, assume that $\Omega = B_1(0)$.

By (i) of Theorem 3.3.2, there are at most finitely many $\mu \in \Lambda$ such that $\mu\lambda > 1$, that is, sign μ = sign λ and $|\mu| > |\lambda|^{-1}$, say, $\mu_1, \ldots, \mu_p$. Let $V = \bigoplus_{i=1}^{p} N(\mu_i)$ and $W = \bigcap_{i=1}^{p} R(\mu_i)$. Then we claim that $X = V \oplus W$. In fact, if $x \in V \cap W$, then $x = \sum_{j=1}^{p} x_j$ with $x_j \in N(\mu_j)$. By (iii, c) of Theorem 3.3.2, $\sum_{j=2}^{p} x_j \in R(\mu_1)$ and, hence, $x_1 = x - \sum_{j=2}^{p} x_j \in R(\mu_1)$. But $N(\mu_1) \cap R(\mu_1) = \{0\}$. So, $x_1 = 0$. Similarly, we can obtain $x_2 = \cdots = x_p = 0$. This shows that $\{0\} = V \cap W$. Moreover, for any $x \in X$, there exists $x_j \in N(\mu_j)$ and $y_j \in R(\mu_j)$ so that $x - x_j + y_j$. So, by (iii, c) of Theorem 3.3.2 again, for any $k = 1, \ldots, n$, we have $x - \sum_{j=1}^{p} x_j = x - x_k - \sum_{j \neq k} x_j \in R(\mu_k)$. Therefore, $x - \sum_{j=1}^{p} x_j \in W$ and, hence, $X = V \oplus W$.

By Theorem 3.3.1, we have

$$\deg(\mathrm{Id} - \lambda T, \Omega) = \deg(\mathrm{Id} - \lambda T|_V, \Omega \cap \mathrm{V}) \deg(\mathrm{Id} - \lambda T|_W, \Omega \cap W)$$

Note that $T|_W$ has no eigenvalues with $\mu\lambda > 1$. So $0 \notin (\mathrm{Id} - t\lambda T|_W)(\partial\Omega)$ for all $t \in [0, 1]$. Consequently, $\deg(\mathrm{Id} - \lambda T|_W, \Omega \cap W) = \deg(\mathrm{Id}|_W, \Omega \cap W) = 1$. By Theorem 3.3.1 again, we get

$$\deg(\mathrm{Id} - \lambda T|_V, \Omega \cap V) = \prod_{i=1}^{p} \deg(\mathrm{Id} - \lambda T|_{N(\mu_i)}, \Omega \cap N(\mu_i))$$

As μ_j is the only eigenvalue of $T|_{N(\mu_j)}$, $0 \notin ((2t-1)\,\mathrm{Id} - t\lambda T|_{N(\mu_j)})(\partial\Omega)$ for $t \in [0, 1]$. Therefore,

$$\deg(\mathrm{Id} - \lambda T|_{N(\mu_j)}, \Omega \cap N(\mu_j)) = \deg(-\mathrm{Id}|_{N(\mu_j)}, \Omega \cap N(\mu_j)) = (-1)^{\dim N(\mu_j)}$$

Consequently,

$$\deg(\mathrm{Id} - \lambda T, \Omega) = \prod_{j=1}^{p} (-1)^{\dim N(\mu_j)} = (-1)^{m(\lambda)} \quad \square$$

Example 3.3.4 Consider the boundary value problem

$$\begin{cases} x'' + \mu x = 0 & \text{in } J = [0, 1] \\ x(0) = x(1) = 0 \end{cases} \tag{3.3.1}$$

Let X denote the Banach space of all continuous functions from $[0, 1]$ into $\mathbb{R}$, equipped with the supremum norm. It is easy to verify that (3.3.1) is equivalent to the problem of finding $x \in X$ such that

$$x(t) - \mu \int_0^1 k(t,s)x(s)\,ds = 0 \qquad \text{for } t \in J \tag{3.3.2}$$

where

$$k(t,s) = \begin{cases} s(1-t), & 0 \le s \le t \le 1 \\ t(1-s), & 0 \le t \le s \le 1 \end{cases}$$

is the Green's function of (3.3.1). Let $T : X \to X$ be defined by

$$Tx(t) = \int_0^1 k(t,s)x(s)\,ds\,, \qquad t \in J$$

By the well-known Ascoli–Arzela theorem [see, e.g., Coddington and Levinson (1955)], T is compact. To compute $\deg(\mathrm{Id} - \lambda T, B_1(0))$, we first investigate the distribution of eigenvalues of T. For $\mu \le 0$, the general solution of $x'' + \mu x = 0$ is

$$x(t) = \begin{cases} ce^{\sqrt{-\mu}t} + de^{-\sqrt{-\mu}t}\,, & \text{if } \mu < 0 \\ c + dt\,, & \text{if } \mu = 0 \end{cases}$$

Therefore, c and d must be zero for x to be a solution of (3.3.1). Consequently, T has no negative eigenvalue. For $\mu > 0$, the general solution of $x'' + \mu x = 0$ is

$$x(t) = c \sin(\sqrt{\mu}t) + d \cos(\sqrt{\mu}t)$$

Therefore, for x to be a nonzero solution of (3.3.1), $\mu = n^2\pi^2$ for some positive integer n. Therefore, the eigenvalues of T are given by $\lambda_n = (n^2\pi^2)^{-1}$, $n = 1, 2, \ldots$, and $\mathrm{Ker}(T - \lambda_n\,\mathrm{Id})$ is the one-dimensional subspace of X spanned by $x_n(t) = \sin(n\pi t)$.

We now claim that the algebraic multiplicity of λ_n is 1. In fact, if $(T - \lambda_n\,\mathrm{Id})^2 x = 0$, then $y := (T - \lambda_n\,\mathrm{Id})x \in \mathrm{Ker}(T - \lambda_n\,\mathrm{Id})$ and, hence, $(T - \lambda_n\,\mathrm{Id})x = \alpha \sin(n\pi t)$. Therefore, $x'' + \lambda_n^{-1}x = -\alpha\lambda_n^{-1}\sin(n\pi t)$, from which it follows that

$$-\frac{\alpha}{2}\lambda_n^{-1} = -\alpha\lambda_n^{-1}\int_0^1 \sin^2(n\pi t)\,dt = \int_0^1 (x'' + \lambda_n^{-1}x)\sin(n\pi t)\,dt$$

But, by integration by parts, we have

$$\int_0^1 (x'' + \lambda_n^{-1} x) \sin(n\pi t)\, dt = 0$$

Therefore, $\alpha = 0$ and, hence, $x \in \text{Ker}(T - \lambda_n \text{Id})$. This implies that $\text{Ker}(T - \lambda_n \text{Id})^2 = \text{Ker}(T - \lambda_n \text{Id})$. That is, the algebraic multiplicity of λ_n is 1.

We can now apply Theorem 3.3.3 to get

$$\deg(\text{Id} - \lambda T, B_1(0)) = \begin{cases} 1 & \text{if } \lambda \in (-\infty, \pi^2) \\ (-1)^n & \text{if } \lambda \in (n^2\pi^2, (n+1)^2\pi^2) \end{cases}$$

EXERCISES

3.3.1 Let X be a real Banach space, $\Omega \subset X$ a bounded open subset, and $f: X \to X$ an Ω-admissible compact field. Let $A: X \to X$ be an isomorphism such that $A = \text{Id} - T_0$ for some $T_0 \in \mathcal{K}(X)$. Show that $\deg(f, \Omega) = (-1)^{m_-(A)} \deg(Af, \Omega)$, where $m_-(A)$ denotes the sum of the algebraic multiplicities of all negative eigenvalues of A.

3.3.2 Consider the boundary value problem

$$\begin{cases} x'' + \mu x = 0, & \text{in } J = [0, 1] \\ x(0) = x(1),\ x'(0) = x'(1) \end{cases}$$

Following the idea of Example 3.3.4, compute the degree of the compact field associated with this problem.

3.3.3 Let H be a Hilbert space and $T_0: H \to H$ a compact self-adjoint operator. Consider a polynomial $P(t)$ such that $1 \notin \sigma(P(T_0))$. Compute $\deg(\text{Id} - P(T_0), B_1(0))$.

3.4 APPLICATION: BERNSTEIN'S THEOREM FOR BVPs

Let $C^k([0, 1]; \mathbb{R})$, $k \geq 0$, be the Banach space of functions defined on $[0, 1]$ with a continuous kth-derivative, equipped with the norm

$$\|u\|_k := \sum_{j=0}^{k} \sup_{t \in [0,1]} |u^{(j)}(t)|$$

for $u \in C^k([0, 1]; \mathbb{R})$.

We consider the following two-point boundary value problem:

$$\begin{cases} y'' = f(t, y, y') \\ y(0) = 0 = y(1) \end{cases} \tag{3.4.1}$$

where $f : [0, 1] \times \mathbb{R} \times \mathbb{R} \to \mathbb{R}$ is a continuous function satisfying the following conditions:

(A1) *There is a constant M_0 such that $yf(t, y, 0) > 0$ for $|y| \geq M_0$.*

(A2) *There are constants A, $B > 0$ such that for all $(t, y, p) \in [0, 1] \times [-M_0, M_0] \times \mathbb{R}$, we have $|f(t, y, p)| \leq Ap^2 + B$.*

Theorem 3.4.1

Assume that $f : [0, 1] \times \mathbb{R} \times \mathbb{R} \to \mathbb{R}$ is a continuous function satisfying conditions (A1) and (A2). Then there exists a solution $y \in C^2([0, 1]; \mathbb{R})$ of the problem (3.4.1).

Proof. Suppose $\lambda \in [0, 1]$ and consider the following family of boundary value problems:

$$\begin{cases} y'' = \lambda f(t, y, y'), \qquad \lambda \in [0, 1] \\ y(0) = 0 = y(1) \end{cases} \tag{3.4.2$_\lambda$}$$

We put $C_0^2([0, 1]; \mathbb{R}) := \{u \in C^2([0, 1]; \mathbb{R});\ u(0) = u(1) = 0\}$ and define $L : C_0^2([0, 1]; \mathbb{R}) \to C([0, 1]; \mathbb{R})$, $F : C^1([0, 1]; \mathbb{R}) \to C([0, 1]; \mathbb{R})$ by

$$Lu = u'', \qquad u \in C_0^2([0, 1]; \mathbb{R})$$

$$F(u)(t) = f(t, u(t), u'(t)), \qquad u \in C^1([0, 1]; \mathbb{R})$$

Then L is invertible and F is continuous. Let $j : C_0^2([0, 1]; \mathbb{R}) \to C^1([0, 1]; \mathbb{R})$ be the natural inclusion. By the Arzela–Ascoli theorem, j is a compact operator. Clearly, (3.4.2$_\lambda$) can be formulated as the following nonlinear equation:

$$Ly = \lambda F(j(y)), \qquad y \in C_0^2([0, 1]; \mathbb{R}), \quad \lambda \in [0, 1] \tag{3.4.3$_\lambda$}$$

which is equivalent to the following family of fixed-point problems:

$$y = \lambda L^{-1} F(j(y)), \qquad y \in C_0^2([0, 1]; \mathbb{R}), \quad \lambda \in [0, 1] \tag{3.4.4$_\lambda$}$$

It is clear that $\mathcal{F} : C_0^2([0, 1]; \mathbb{R}) \to C_0^2([0, 1]; \mathbb{R})$ defined by $\mathcal{F}(y) = L^{-1}F(j(y))$ is a completely continuous mapping. We consider the set $\Sigma(\mathcal{F}) := \{y \in C_0^2([0, 1]; \mathbb{R});$ there exists $\lambda \in (0, 1)$ such that $y = \lambda\mathcal{F}(y)\}$.

We claim that $\Sigma(\mathscr{F})$ is bounded. Suppose that $y \in C_0^2([0,1];\mathbb{R})$ is a solution of $(3.4.3_\lambda)$ with $\lambda \in (0,1)$. Since $y(0)=y(1)=0$, the function $r(t)=\frac{1}{2}y^2(t)$ must attain its maximum at $t_0 \in (0,1)$. We may assume, without loss of generality, that $r(t_0)=\frac{1}{2}y^2(t_0)>0$. Thus, $y'(t_0)=0$ and $r''(t_0)=y(t_0)\cdot y''(t_0)\le 0$. Since y is a solution to $(3.4.3_\lambda)$, we have

$$0 \ge r''(t_0) = y(t_0)\cdot y''(t_0) - \lambda y(t_0)\cdot f(t_0, y(t_0), 0)$$

By assumption (A1), it follows that $|y(t_0)| \le M_0$. Therefore, $\|y\|_0 \le M_0$.

Now, we establish an a priori bound on y'. Since $y(0)=y(1)=0$, there exists $t_1 \in (0,1)$ such that $y'(t_1)=0$. Let $[a,b]\subset[0,1]$ be an interval such that $y'(t)\ne 0$ for $t\in(a,b)$ and $y'(a)=0$ and/or $y'(b)=0$. For example, suppose $y'(a)=0$ and $y'(t)>0$ for $t\in(a,b)$. Then by (A2)

$$\frac{d}{dt}[\ln(Ay'^2+B)] = \frac{2Ay'y''}{Ay'^2+B} \le 2A\lambda y'\frac{Ay'^2+B}{Ay'^2+B} = 2A\lambda y' \tag{3.4.5}$$

on $[a,b]$. Integrating (3.4.5) from a to t, we get $\ln\left(\frac{Ay'^2+B}{B}\right) \le 4A\lambda M_0 \le 4AM_0$ and, hence, $|y'(t)| \le \{B/A(e^{4AM_0}-1)\}^{1/2} =: M_1$ on $[a,b]$. Since the constant M_1 does not depend on the choice of the interval $[a,b]$, $|y'(t)|\le M_1$ for all $t\in[0,1]$. Finally, we put $M_2=\max\{|f(t,y,p)|;\, t\in[0,1], |y|\le M_0, |p|\le M_1\}$. Then $|y''(t)|\le M_2$ for all $t\in[0,1]$ and, therefore, $\|y\|_2 \le M := M_0+M_1+M_2$. So $\Sigma(\mathscr{F})$ is bounded.

Let $R>0$ be given so that $\Sigma(\mathscr{F})\subseteq B_R(0) := \{u\in C_0^2([0,1];\mathbb{R}); \|u\|_2<R\}$. If $\mathscr{F}$ has no fixed point in $\partial B_R(0)$, then $H:[0,1]\times C_0^2([0,1],\mathbb{R})\to C_0^2([0,1];\mathbb{R})$ given by $H(\lambda,u)=u-\lambda L^{-1}F(j(u))$ is a $B_R(0)$-admissible homotopy. This implies that $\deg(\mathrm{Id}-\mathscr{F}, B_R(0)) = \deg(\mathrm{Id}, B_R(0)) = 1$. Therefore, $\mathscr{F}$ has a fixed point in $B_R(0)$. This completes the proof. □

The following result shows that for the boundary value problem (3.4.1) to have a solution, one may replace (A2) by the following Bernstein–Nagumo growth condition:

(A3) *There is a continuous function $\psi:[0,\infty)\to(0,\infty)$ such that $\int_0^\infty \frac{t\,dt}{\psi(t)} = \infty$ and $|f(t,y,p)|\le\psi(|p|)$ for $(t,y,p)\in[0,1]\times[-M_0,M_0]\times\mathbb{R}$, where the constant M_0 is given in (A1).*

Theorem 3.4.2

Assume that $f:[0,1]\times\mathbb{R}\times\mathbb{R}\to\mathbb{R}$ is a continuous function satisfying conditions (A1) and (A3). Then there exists a solution $y\in C^2([0,1];\mathbb{R})$ to the problem (3.4.1).

Proof. Since the proof is essentially the same as for Theorem 3.4.1, we will use the same notations. We need only to obtain a priori bounds for $\sup_{t\in[0,1]} |y'(t)|$ for any solution y of $(3.4.3_\lambda)$ with $\lambda \in (0, 1)$.

Let $t_0 \in [0, 1]$ be such that $|y'(t_0)|$ attains its maximum. We may assume, without loss of generality, that $y'(t_0) > 0$. By (A3), $y''(t) \le \psi(|y'(t)|)$ for $t \in [0, 1]$. Let $[a, b] \subset [0, 1]$ be an interval, $t_0 \in (a, b)$, $y'(t) > 0$ for $t \in (a, b)$, $y'(a) = 0$. Then by (A3), we have

$$\frac{y''(t)y'(t)}{\psi(y'(t))} \le y'(t)\,, \qquad t \in [a, b] \tag{3.4.6}$$

Integrating (3.4.6) from a to t_0, we get

$$\int_a^{t_0} \frac{y'(t)y''(t)}{\psi(y'(t))}\,dt \le \int_a^{t_0} y'(t)\,dt = y(t_0) - y(a) \le 2M_0$$

On the other hand, by changing the variable $x = y'(t)$, we obtain

$$\int_0^{y'(t_0)} \frac{x\,dx}{\psi(x)} = \int_a^{t_0} \frac{y'(t)y''(t)}{\psi(y'(t))}\,dt \le 2M_0$$

Since $\int_0^\infty \frac{x\,dx}{\psi(x)} = \infty$, there exists a constant $M_1 = M_1(M_0) \le 0$ such that $y'(t_0) \le M_1$. Thus, $|y'(t)| \le M_1$ for $t \in [0, 1]$. The remaining part of the proof is the same as in the proof of Theorem 3.4.1 and, hence, is omitted. □

EXERCISES

3.4.1 Show that the boundary value problem $y'' = 2\tan^{-1}(y'^2 + y^2 + t)$, $y(0) = 0 = y(1)$, has a solution.

3.4.2 Show that the boundary value problem $y'' = (y'^2 + y)e^{y^2} + y^3 - 5y^2 + 1$, $y(0) = 0 = y(1)$, has a solution.

3.4.3 Show that the boundary value problem $y'' = e^y(1 + y'^2)$, $y(0) = y(1) = 0$, has a solution.

3.5 APPLICATION: NONLINEAR BVPs

We consider the following boundary value problem of a system of second-order differential equations:

$$(3.5.1) \qquad \begin{cases} y'' = f(t, y, y') \\ y(i) = g_i(y(0), y'(0), y(1), y'(1)), \qquad i = 0, 1 \end{cases}$$

where $f : [0, 1] \times \mathbb{R}^n \times \mathbb{R}^n \to \mathbb{R}^n$ is a continuous function satisfying the following conditions:

(H1) *There is $R > 0$ such that for all $(t, y, z) \in [0, 1] \times \mathbb{R}^n \times \mathbb{R}^n$ with $|y| > R$ and $y \cdot z = 0$, we have $y \cdot f(t, y, z) + |z|^2 > 0$.*

(H2) *There is a continuous function $\varphi : [0, \infty) \to (0, \infty)$ such that $\int_0^\infty \frac{s}{\varphi(s)}\, ds = \infty$ and $|f(t, y, z)| \le \varphi(|z|)$ for $(t, y, z) \in [0, 1] \times \mathbb{R}^n \times \mathbb{R}^n$ with $|y| \le R$.*

(H3) *There are constants K, $\alpha > 0$ such that if $|y| \le R$ and $z \in \mathbb{R}^n$, then $|f(t, y, z)| \le \alpha[y \cdot f(t, y, z) + |z|^2] + K$.*

We also assume that $g_i : \mathbb{R}^{4n} \to \mathbb{R}^n$, $i = 0, 1$, are two continuous functions satisfying the following condition:

(H4) *For all $(y_0, z_0, y_1, z_1) \in \mathbb{R}^n \times \mathbb{R}^n \times \mathbb{R}^n \times \mathbb{R}^n$ and $i = 0, 1$, we have $(-1)^i z_i \cdot g_1(y_0, z_0, y_1, z_1) \ge 0$.*

Then we have the following result.

Theorem 3.5.1

Suppose that $f : [0, 1] \times \mathbb{R}^n \times \mathbb{R}^n \to \mathbb{R}^n$ is a continuous mapping satisfying hypotheses (H1), (H2), and (H3) and that $g_0, g_1 : \mathbb{R}^{4n} \to \mathbb{R}^n$ are continuous and satisfy hypothesis (H4). Then problem (3.5.1) has at least one solution $y \in C^2([0, 1]; \mathbb{R}^n)$.

Proof. For $\lambda \in [0, 1]$, we consider the following family of boundary value problems:

$$(3.5.2_\lambda) \qquad \begin{cases} y'' = \lambda f(t, y, y') \\ y(i) = \lambda g_i(y(0), y'(0), y(1), y'(1)), \qquad i = 0, 1 \end{cases}$$

Put $X = C^2([0, 1]; \mathbb{R}^n)$, $Y = C([0, 1]; \mathbb{R}^n) \times \mathbb{R}^n \times \mathbb{R}^n$, and $Z = C^1([0, 1]; \mathbb{R}^n)$. We define the linear operator $L : X \to Y$ by $Lu = (u'', u(0), u(1)), u \in X$. Let $F : Z \to Y$ be defined by

$$F(u)(t) = (f(t, u(t), u'(t)), g_0(u(0), u'(0), u(1), u'(1)),$$
$$g_1(u(0), u'(0), u(1), u'(1)))$$

where $u \in Z$. Let $j : C^2([0, 1]; \mathbb{R}^n) \to C^1([0, 1]; \mathbb{R}^n)$ denote the natural inclusion. The operator j is a compact operator. We have the following diagram:

$$\begin{array}{ccc} C^2([0,1];\mathbb{R}^n) & \xrightarrow{\quad L \quad} & C([0,1];\mathbb{R}^n)\times\mathbb{R}^n\times\mathbb{R}^n \\ & {}_{j}\searrow \quad C^1([0,1];\mathbb{R}^n) \quad \nearrow_{F} & \end{array}$$

It is clear that system $(3.5.2_\lambda)$ is equivalent to the following family of problems:

$$(3.5.3_\lambda) \qquad Lu = \lambda F(j(u)), \qquad u \in X, \quad \lambda \in [0, 1]$$

Since L is an isomorphism, $(3.5.3_\lambda)$ can be written as the following fixed-point problem:

$$(3.5.4_\lambda) \qquad u = \lambda L^{-1} F(j(u)), \qquad u \in X, \quad \lambda \in [0, 1]$$

We define $\mathcal{F} : X \to X$ by $\mathcal{F}(u) = L^{-1}F(j(u))$, $u \in X$. The mapping $\mathcal{F}$ is a completely continuous map. As in the proof of Theorem 3.4.2, we consider the set $\Sigma(\mathcal{F}) := \{u \in C^2([0, 1]; \mathbb{R}^n);$ there exists $\lambda \in (0, 1)$ such that $u = \lambda\mathcal{F}(u)\}$. and want to show that $\Sigma(\mathcal{F})$ is bounded in X.

For this purpose, we need to establish a priori bounds on $\sup_{t\in[0,1]} |y^{(i)}(t)|$, $i = 0, 1, 2$, for all solutions y to $(3.5.2_\lambda)$.

We start with an estimation on $\sup_{t\in[0,1]} |y(t)|$ for a solution of $(3.5.2_\lambda)$. Assume $y(t)$ is a solution to $(3.5.2_\lambda)$ and put $r(t) = |y(t)|^2$. If $r(t)$ achieves its maximum at $t_0 = i$, where $i = 0$ or 1, then

$$\begin{aligned} 0 \geq (-1)^i r'(i) &= (-1)^i 2y(i) \cdot y'(i) \\ &= (-1)^i 2\lambda g_i(y(0), y'(0), y(1), y'(1)) \cdot y'(i) \geq 0 \end{aligned}$$

Thus, $r'(i) = 0$ and $y(t_0) \cdot y'(t_0) = 0$. Assume, for contradiction, that $|y(t_0)| > R$. Then, by (H1) we have

$$\begin{aligned} r''(t_0) &= 2y(t_0) \cdot y''(t_0) + 2|y'(t_0)|^2 = 2\lambda y(t_0) \cdot f(t_0, y(t_0), y'(t_0)) + 2|y'(t_0)|^2 \\ &\geq 2\lambda[y(t_0) \cdot f(t_0, y(t_0), y'(t_0)) + |y'(t_0)|^2] > 0 \end{aligned}$$

whenever $\lambda \in (0, 1)$. But this contradicts the maximum principle $r''(t_0) \leq 0$. Consequently, $|y(t_0)| \leq R$. If $r(t)$ achieves its maximum at $t_0 \in (0, 1)$, then $r'(t_0) = 0$ and $r''(t_0) \leq 0$. The above argument can also lead to $|y(t_0)| \leq R$. Therefore, we have $\sup_{t\in[0,1]} |y(t)| \leq R$.

To obtain an estimation for $\sup_{t\in[0,1]} |y'(t)|$, we need the following.

Lemma 3.5.2

Let $\varphi : [0, \infty) \to (0, \infty)$ be a continuous function such that $\int_0^\infty \frac{s\,ds}{\varphi(s)} = \infty$ and suppose that $x \in C^2([0, 1]; \mathbb{R}^n)$ is given so that for some constants $\alpha, K, R > 0$, $|x(t)| \leq R$, $|x''(t)| \leq \varphi(|x'(t)|)$ and $|x''(t)| \leq \alpha r''(t) + K$, $t \in [0, 1]$, where $r(t) = |x(t)|^2$. Then there is a constant $M = M(\varphi, \alpha, K, R)$ such that $\sup_{t\in[0,1]} |x'(t) \leq M$.

Proof. Suppose that $0 < \mu < 1$ and $0 \leq t \leq 1 - \mu$. Then

$$x(t+\mu) - x(t) - \mu x'(t) = \int_t^{t+\mu} (t + \mu - s)x''(s)\,ds \tag{3.5.5}$$

Since $t + \mu - s \geq 0$, it follows from the assumption that

$$\begin{aligned} \mu|x'(t)| &\leq 2R + \int_t^{t+\mu} (t + \mu - s)(\alpha r''(s) + K)\,ds \\ &\leq 2R - \mu\alpha r'(t) + \alpha[r(t+\mu) - r(t)] + \frac{\mu^2}{2}K \\ &= 2R + \alpha[r(t+\mu) - r(t) - \mu r'(t)] + \tfrac{1}{2}K\mu^2 \end{aligned}$$

and consequently,

$$\mu|x'(t)| \leq 2R(1 + \alpha R) + \tfrac{1}{2}K\mu^2 - \alpha\mu r'(t)\,, \qquad 0 \leq t \leq 1 - \mu \tag{3.5.6}$$

Similarly, for $\mu \leq t \leq 1$, we can use

$$x(t) - x(t-\mu) - \mu x'(t) = -\int_{t-\mu}^{t} (t - \mu - s)x''(s)\,ds$$

to obtain

$$\mu|x'(t)| \leq 2R(1 + \alpha R) + \tfrac{1}{2}K\mu^2 + \alpha\mu r'(t)\,, \qquad \mu \leq t \leq 1 \tag{3.5.7}$$

The function $\mu \mapsto \frac{2R(1+\alpha R)}{\mu} + \frac{1}{2}K\mu$, $0 < \mu$, attains its minimum $2[R(1 + \alpha R)K]^{1/2}$ at $\mu = 2\left[\frac{R(1+\alpha R)}{K}\right]^{1/2}$. We define the function $M_1(\mu)$ by

$$M_1(\mu) = \begin{cases} \dfrac{2R(1+\alpha R)}{\mu} + \dfrac{1}{2}K\mu\,, & \text{if } 0 < \mu \leq 2\left[\dfrac{R(1+\alpha R)}{K}\right]^{1/2} \\ 2[R(1+\alpha R)K]^{1/2}\,, & \text{if } \mu \geq 2\left[\dfrac{R(1+\alpha R)}{K}\right]^{1/2} \end{cases}$$

It is clear that $M_1(\mu)$ is a continuous and nonincreasing function for $\mu > 0$. If $\frac{1}{2} \le 2\left[\frac{R(1+\alpha R)}{K}\right]^{1/2}$, then by taking $\mu = \frac{1}{2}$, we obtain from (3.5.6) for $0 \le t \le \frac{1}{2} = 1 - \mu$

$$|x'(t)| \le M_1(\tfrac{1}{2}) - \alpha r'(t) \tag{3.5.8}$$

If $\frac{1}{2} \ge 2\left[\frac{R(1+\alpha R)}{K}\right]^{1/2}$ then (3.5.8) still holds for $0 \le t \le \frac{1}{2}$ by taking $\mu = 2\left[\frac{R(1+\alpha R)}{K}\right]^{1/2}$ in (3.5.6) and noting the fact that $M_1(\mu) = M_1(\frac{1}{2})$. Similarly, for $\frac{1}{2} \le t \le 1$, we can use (3.5.7) (for $\mu = \frac{1}{2}$ if $\frac{1}{2} \le 2\left[\frac{R(1+\alpha R)}{K}\right]^{1/2}$ and for $\mu = 2\left[\frac{R(1+\alpha R)}{K}\right]^{1/2}$ if $\frac{1}{2} \ge 2\left[\frac{R(1+\alpha R)}{K}\right]^{1/2}$ to obtain

$$|x'(t)| \le M_1(\tfrac{1}{2}) + \alpha r'(t) \tag{3.5.9}$$

For $t = \frac{1}{2}$, it follows from (3.5.8) and (3.5.9) that $|x'(\frac{1}{2})| \le M_1(\frac{1}{2})$. By assumption, we have

$$\frac{|x'(t) \cdot x''(t)|}{\varphi(|x'(t)|)} \le \frac{|x'(t) \cdot x''(t)|}{|x''(t)|} \le |x'(t)| \le M_1(\tfrac{1}{2}) \pm \alpha r'(t)$$

where the sign $\pm$ is taken to be $+$ for $t \ge \frac{1}{2}$ and $-$ for $t \le \frac{1}{2}$. We define the function $\Phi(s) = \int_0^s \frac{u\,du}{\varphi(u)}$. Then

$$\left|\int_{1/2}^{t} \frac{x'(t) \cdot x''(t)}{\varphi(|x'(t)|)}\,\mathrm{d}t\right| = \left|\int_{|x'(1/2)|}^{|x'(t)|} \frac{u\,du}{\varphi(u)}\right| = |\Phi(|x'(t)|) - \Phi(|x'(\tfrac{1}{2})|)|$$

Since $\frac{|x'(t) \cdot x''(t)|}{\varphi(|x'(t)|)} \le M_1(\frac{1}{2}) \pm \alpha r'(t)$, we obtain

$$|\Phi(|x'(t)|) - \Phi(|x'(\tfrac{1}{2})|)| \le \tfrac{1}{2} M_1(\tfrac{1}{2}) + \alpha |r(t) - r(\tfrac{1}{2})| \le \tfrac{1}{2} M_1(\tfrac{1}{2}) + 2\alpha R^2$$

Consequently, $\Phi(|x'(t)|) \le \Phi(|x'(\frac{1}{2})|) + \frac{1}{2} M_1(\frac{1}{2}) + 2\alpha R^2$. Since Φ is an increasing continuous unbounded function, we obtain

$$|x'(t)| \le M := \Phi^{-1}[\Phi(M_1(\tfrac{1}{2})) + \tfrac{1}{2} M_1(\tfrac{1}{2}) + 2\alpha R]$$

This completes the proof of Lemma 3.5.2. □

Let us now complete the proof of Theorem 3.5.1. Assume y is a solution of ($3.5.2_\lambda$) for some $\lambda \in (0, 1)$. Then by (H3), for $y = y(t)$, $y' = y'(t)$, and $y'' = y''(t)$, we have

$$\begin{aligned} |y''| = |\lambda f(t, y, y')| &\leq \alpha\lambda(y \cdot f(t, y, y') + |y'|^2) + K \\ &= \alpha(y \cdot y'' + \lambda|y'|^2) + K \leq \alpha r''(t) + K \end{aligned}$$

Therefore, by Lemma 3.5.2, there is a constant $M > 0$ such that $|y'(t)| \leq M$ for $t \in [0, 1]$.

Finally, we put $C = R + M + M_2$, where $M_2 = \max\{|f(t, y, z); t \in [0, 1], |y| \leq R, |z| \leq M\}$. Then for every $y \in \Sigma(\mathcal{F})$, we have $\sup_{t\in[0,1]} (|y(t)| + |y'(t)| + |y''(t)|) \leq C$. This shows that $\Sigma(\mathcal{F})$ is bounded in X. The remaining part of the proof is the same as that of Theorem 3.4.1. □

EXERCISES

Consider the following boundary value problem:

$$(*) \qquad \begin{cases} y'' = f(t, y), & t \in [0, 1] \\ y(0) = y(1), & y'(0) = y'(1) \end{cases}$$

where $f : [0, 1] \times \mathbb{R} \to \mathbb{R}$ is continuous. We say $\alpha : [0, 1] \to \mathbb{R}$ is a *lower solution* (respectively, $\beta : [0, 1] \to \mathbb{R}$ is an *upper solution*) of $y'' = f(t, y)$ if $-\alpha''(t) + f(t, \alpha(t)) \leq 0$ (respectively, $-\beta''(t) + f(t, \beta(t)) \geq 0$) for $t \in [0, 1]$.

Assume that the equation $y'' = f(t, y)$ has a lower solution α and an upper solution β such that:

(i) $\alpha(t) < \beta(t)$ for $t \in [0, 1]$.
(ii) $\alpha(0) < 0 < \beta(0)$ and $\alpha(1) < 0 < \beta(1)$.
(iii) $(\alpha(0), \beta(0), \alpha'(0), \beta'(0)) = (\alpha(1), \beta(1), \alpha'(1), \beta'(1))$.

Let $a > 0$ be a number such that $a > \max\left\{\dfrac{\beta''(t) - \alpha''(t)}{\beta(t) - \alpha(t)};\ t \in [0, 1]\right\}$ and consider the following family of BVPs:

$$(*_\lambda) \qquad \begin{cases} y'' = \lambda f(t, y) + (1 - \lambda)(ay - z), & \lambda \in [0, 1] \\ y(0) = y(1), \ y'(0) = y'(0) \end{cases}$$

where $z(t) = -\dfrac{(\beta(t) + \alpha(t))''}{2} + a\dfrac{\beta(t) + \alpha(t)}{2}$ for $t \in [0, 1]$.

3.5.1 Show that if y is a solution of ($*_\lambda$) such that $\alpha(t) \leq y(t) \leq \beta(t)$ for $t \in [0, 1]$, then there exist constants M_1 and M_2 such that $\max\{|y'(t)|; t \in [0, 1]\} < M_1$ and $\max\{|y''(t)|; t \in [0, 1]\} < M_2$.

3.5.2 Let $X := \{y \in C^2([0,1]; \mathbb{R});\ y(0) = y(1),\ y'(0) = y'(1)\}$, and $\Omega := \{y \in X;\ \alpha(t) < y(t) < \beta(t)$ and $|y'(t)| < M_1,\ |y''(t)| < M_2$ for $t \in [0,1]\}$. Show that there is no solution y of $(*_\lambda)$, with $\lambda \in [0,1)$, such that $y \in \partial\Omega$. *Hint:* Show that if $\alpha(s) = y(s)$ for some $s \in [0,1]$, then $y'(s) - \alpha'(s) = 0$ and $y''(s) - \alpha''(s) \geq 0$, which would imply that $\alpha''(s) \leq a\alpha(s) - z(s)$, which is impossible.

3.5.3 Use the results from (3.5.2) to prove that the system $(*)$, under the assumptions (i), (ii), and (iii), has a solution y such that $\alpha(t) \leq y(t) \leq \beta(t)$ for all $t \in [0,1]$.

3.6 A FIXED-POINT INDEX FOR ENRs AND ANRs

This section introduces the concept of a *fixed-point index* and extends it to the category of ANRs.

We begin with finite-dimensional spaces.

Definition 3.6.1 We denote by $\mathfrak{F}(\mathscr{F})$ the class of all triples $(U, F, \mathbb{R}^n)$ such that U is an open subset of $\mathbb{R}^n$, $F : U \to \mathbb{R}^n$ is a continuous mapping, and the *fixed-point set* $\mathrm{Fix}(F) := \{x \in U;\ F(x) = x\}$ is a compact subset of U. We will call a triple $(U, F, \mathbb{R}^n)$ from $\mathfrak{F}(\mathscr{F})$ a *fixed-point situation*.

For every $(U, F, \mathbb{R}^n) \in \mathfrak{F}(\mathscr{F})$, we define the *fixed-point index* $I(F) := I(U, F, \mathbb{R}^n)$ of F in U by $I(F) := \deg(\mathrm{Id} - F, U_0)$, where U_0 is a bounded, open subset of U such that $\mathrm{Fix}(F) \subset U_0 \subset \overline{U}_0 \subset U$. The following properties are immediate consequences of the definition of the fixed-point index:

(I1) *Normalization:* Let $x_0 \in \mathbb{R}^n$ and let x_0 also denote the constant mapping from U into $\mathbb{R}^n$ with the value x_0; then

$$I(U, x_0, \mathbb{R}^n) = \begin{cases} 1 & \text{if } x_0 \in U \\ 0 & \text{if } x_0 \notin U \end{cases}$$

(I2) *Additivity:* If $U_1 \cap U_2 = \emptyset$, then $I(U_1 \cup U_2, F, \mathbb{R}^n) = I(U_1, F, \mathbb{R}^n) + I(U_2, F, \mathbb{R}^n)$.

(I3) *Homotopy Invariance:* Let $H : [0,1] \times U \to \mathbb{R}^n$ be a continuous mapping such that $\mathrm{Fix}(H) = \{(t, x) \in [0,1] \times U;\ H(t, x) = x\}$ is a compact subset of $[0,1] \times U$; then $I(U, H_0, \mathbb{R}^n) = I(U, H_1, \mathbb{R}^n)$, where $H_t = H(t, x)$, $(t, x) \in [0,1] \times U$.

(I4) *Existence:* If $I(U, F, \mathbb{R}^n) \neq 0$, then the mapping F has a fixed point in U, that is, there is $x \in U$ such that $F(x) = x$.

(I5) *Excision:* If $U_0 \subseteq U_1$ and $F(x) \neq x$ for $x \in U_1 \backslash U_0$, then $I(U_0, F, \mathbb{R}^n) = I(U_1, F, \mathbb{R}^n)$.

(**I6**) *Commutativity:* Let $U \subset \mathbb{R}^n$ and $U' \subset \mathbb{R}^m$ be open and assume $F: U \to \mathbb{R}^m$ and $G: U' \to \mathbb{R}^n$. If $(F^{-1}(U'), G \circ F, \mathbb{R}^n)$ is a fixed-point situation, then so is $(G^{-1}(U), F \circ G, \mathbb{R}^m)$ and $I(F^{-1}(U'), G \circ F, \mathbb{R}^n) = I(G^{-1}(U), F \circ G, \mathbb{R}^m)$.

(**I7**) *Product Formula:* If $0 \in U_2 \subseteq \mathbb{R}^m$ and $(U_1, F. \mathbb{R}^n) \in \mathfrak{F}(\mathscr{F})$, then $I(U_1 \times U_2, F \times 0, \mathbb{R}^n \times \mathbb{R}^m) = I(U_1, F, \mathbb{R}^n)$.

Remark 3.6.2 In the above definition of a fixed-point situation, we did not assume that the open set U is bounded. Instead, we assumed that the fixed-point set Fix(F) is compact, so that we could use the extended definition of the degree, introduced in Section 2.1, for admissible maps in a larger sense.

Definition 3.6.3 Let $\mathfrak{U}(F)$ denote the family of all triples (U, F, X), where X is an ANR-space, U an open subset of X, and $F: U \to X$ a compact map such that $\text{Fix}(F) := \{x \in U; x = F(x)\}$ is a compact subset of U. We will call a triple $(U, F, X) \in \mathfrak{U}(\mathscr{F})$ a *fixed-point situation.*

Theorem 3.6.4

There exists a fixed-point index $I(U, F, X)$ defined for every fixed-point situation $(U, F, X) \in \mathfrak{U}(\mathscr{F})$ that satisfies the following properties:

(**I1**) *Normalization: Let $x_0 \in X$ and let x_0 also denote the constant mapping from U into X with the value x_0; then*

$$I(U, x_0, X) = \begin{cases} 1 & \text{if } x_0 \in U \\ 0 & \text{if } x_0 \notin U \end{cases}$$

(**I2**) *Additivity: If $U_1 \cap U_2 = \varnothing$; then $I(U_1 \cup U_2, F, X) = I(U_1, F, X) + I(U_2, F, X)$.*

(**I3**) *Homotopy Invariance: Let $H: [0, 1] \times U \to X$ be a continuous compact mapping such that* $\text{Fix}(H) = \{(t, x) \in [0, 1] \times U; H(t, x) = x\}$ *is a compact subset of $[0, 1] \times U$; then $I(U, H_0, X) = I(U, H_1, X)$, where $H_t = H(t, x)$, $(t, x) \in [0, 1] \times U$.*

(**I4**) *Existence: If $I(U, F, X) \neq 0$, then the mapping F has a fixed point in U, that is, there is $x \in U$ such that $F(x) = x$.*

(**I5**) *Excision: If $U_0 \subseteq U_1$ and $F(x) \neq x$ for $x \in U_1 \backslash U_0$, then $I(U_0, F, X) = I(U_1, F, X)$.*

(**I6**) *Commutativity: Let $U \subset X$ and $U' \subset X'$ be open and assume $F: U \to X'$ and $G: U' \to X$ are continuous such that one of them is compact. If $(F^{-1}(U'), G \circ F, X)$ are a fixed-point situation, then so is $(G^{-1}(U), F \circ G, X')$ and $I(F^{-1}(U'), G \circ F, X) = I(G^{-1}(U), F \circ G, X')$.*

(I7) *Product Formula: If* $0 \in U_2 \subseteq \mathbb{R}^m$ *and* $(U_1, F, X) \in \mathfrak{U}(\mathcal{F})$, *then* $I(U_1 \times U_2,\ F \times 0,\ X \times \mathbb{R}^m) = I(U_1, F, X)$.

Proof. Suppose that $(U, F, X) \in \mathfrak{U}(\mathcal{F})$ is a fixed-point situation. By Theorem 3.6.4, X is r-dominated by an open subset Ω in Banach space V. Put $\mathcal{U} := r^{-1}(U)$ and define $\tilde{F} := s \circ F \circ r : \mathcal{U} \to V$, where $s : X \to \Omega \subseteq V$ is given so that $rs = \mathrm{Id}_X$. Since $\overline{F(U)}$ is compact, $\tilde{F}$ is a compact map. Moreover, $\mathrm{Fix}(\tilde{F}) = s(\mathrm{Fix}(F))$. Let $\mathcal{U}_0$ be an open, bounded subset of $\mathcal{U}$ such that $\mathrm{Fix}(\tilde{F}) \subseteq \mathcal{U}_0 \subseteq \overline{\mathcal{U}_0} \subset \mathcal{U}$. Then we can define $I(U, F, X) := \deg(\mathrm{Id} - \tilde{F}, \mathcal{U}_0)$. It follows from the excision property that $I(U, F, X)$ does not depend on the choice of the set $\mathcal{U}_0$. We will show that the above definition does not depend on the choice of r and s. In fact, let $\Omega' \subset V'$ be another open set in a Banach space V' that r'-dominates X, that is, there are $s' : X \to \Omega'$, $r' : \Omega' \to X$ such that $r's' = \mathrm{Id}_X$. Then, since the second of the maps $sr' : \Omega' \to \Omega$, $s'Fr : r^{-1}(U) \to \Omega'$ is compact, we may apply the commutativity property for the Leray–Schauder degree to obtain

$$\begin{aligned}\deg(\mathrm{Id} - s'Fr', r'^{-1}(U)) &= \deg(\mathrm{Id} - (s'Fr) \circ (sr'), (sr')^{-1}(r^{-1}(U)))\\ &= \deg(\mathrm{Id} - (sr') \circ (s'Fr), r^{-1}(U))\\ &= \deg(\mathrm{Id} - sFr, r^{-1}(U))\end{aligned}$$

which proves that the above definition is independent of the choice of r and s.

The verification of the properties of the above fixed-point index is left to interested readers. □

Example 3.6.5 Let $X = \{(x, y) \in \mathbb{R}^2;\ xy = 0\}$ and $F : X \to X$ be given by $F(x, y) = (2x, 2y)$. It is clear that X is an ENR and we can easily construct a retraction $r : \mathbb{R}^2 \to X$, that is, a continuous mapping $r : \mathbb{R}^2 \to X$ such that $r(x, y) = (x, y)$ for all $(x, y) \in X$. Indeed, we define $r(x, y) = (x, y) - \min(|x|, |y|)(\operatorname{sign} x, \operatorname{sign} y)$, $(x, y) \in \mathbb{R}^2$. It is clear that r is continuous, $r_{|X} \equiv \mathrm{Id}_{|X}$, and that $r(x, y) \in X$ for all $(x, y) \in \mathbb{R}^2$. It is also clear that F has only one fixed point $(0, 0) \in X$. Thus, (X, F, X) is a fixed-point situation. We will compute the fixed-point index $I(X, F, X)$. By the definition $I(X, F, X) = \deg(\mathrm{Id} - F \circ r, U)$, where U denotes the unit ball in $\mathbb{R}^2$. We put $\tilde{f} := \mathrm{Id} - F \circ r$. Then

$$\tilde{f}(x, y) = (-x + 2\min(|x|, |y|) \operatorname{sign} x,\ -y + 2\min(|x|, |y|) \operatorname{sign} y)$$

Since $\tilde{f}(x, y) = (-x, -y)$ for all $(x, y) \in X$, the mapping $\tilde{f}$ may be illustrated as in Figure 3.6.1. In Figure 3.6.1 we have indicated by a dotted line the "folding lines," where $\tilde{f}$ acts as identity. The mapping $\tilde{f}$ transforms the plane in such a way that the dotted lines are mapped onto themselves, while the other lines are reversed. It is helpful to think that the mapping $\tilde{f}$ acts by folding the plane on the dotted lines. It is easy to see that any point (a, b) such that $a > b > 0$ is a regular value of $\tilde{f}$. Indeed, by direct computation the inverse image $\tilde{f}^{-1}(a, b)$ contains three points $(a, 2a - b)$, $(a, -2a - b)$, and $(-a - 2b, b)$, where the derivative $D\tilde{f}(x, y)$ has a negative determinant at each of these points. Since the derivative of $\tilde{f}$ at each of the points from $\tilde{f}^{-1}(a, b)$ has a negative determinant, we have $I(X, F, X) = -3$.

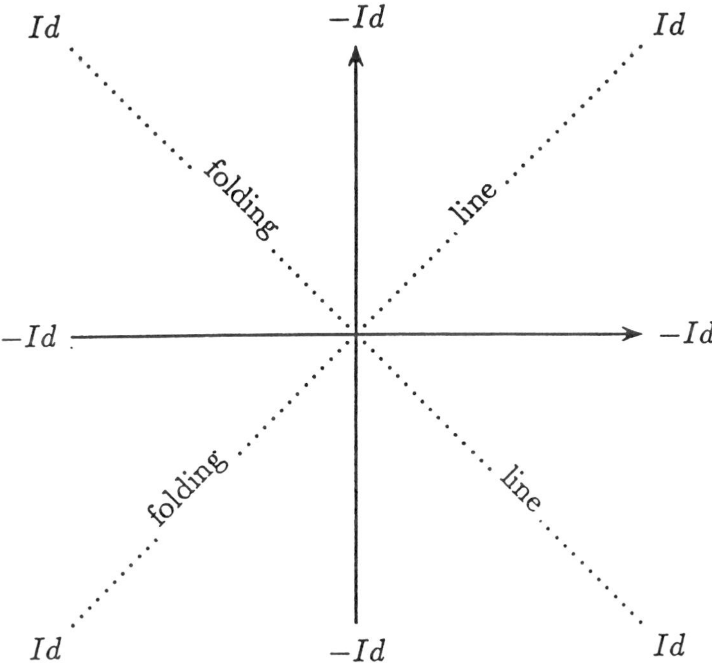

FIGURE 3.6.1. The map $\tilde{f}$ transforms the dotted lines onto themselves and reverses other lines.

EXERCISES

3.6.1 Prove all the properties of the fixed-point index, described in Theorem 3.6.4.

3.6.2 Let X be a subset of $\mathbb{R}^2$ composed of the origin, and $n > 0$ rays coming out from the origin. Let $F : X \to X$ be given by $F(p) = 2p$ for $p \in X$. Show that $I(X, F, X) = -(n - 1)$.

3.6.3 Let X be a subset of $\mathbb{R}^3$ composed of a plane passing through the origin and a straight line intersecting transversally the plane at the origin. Let $F : X \to X$ be given by $F(p) = 2p$ for $p \in X$. Compute $I(X, F, X)$.

3.6.4 Let $X = \{(x, y, z) \in \mathbb{R}^3;\ xyz = 0\}$ and $\Phi : X \to X$ be given by $\Phi(x, y, z) = (2x, 2y, 2z)$. Show that:

(i) The mapping $r : \mathbb{R}^3 \to \mathbb{R}^3$ defined by

$$r(x, y, z) = (x, y, z) - \min(|x|, |y|, |z|)(\operatorname{sign} x, \operatorname{sign} y, \operatorname{sign} z)$$

is a continuous retraction onto X.

(ii) The mapping $\tilde{\varphi} := \mathrm{Id} - \Phi \circ r$, restricted to each octant O_i of $\mathbb{R}^3$ is a 1–1 mapping, and that $\tilde{\varphi}(O_i)$ covers exactly seven octants, all except the octant $-O_i$, which is symmetric to O_i.

(iii) All the points (x, y, z) such that $x > y > z > 0$ are regular values of $\tilde{\varphi}$.

(iv) $I(X, \Phi, X) = 7$.

3.6.5 Let $X = \{x = (x_1, \dots, x_n) \in \mathbb{R}^n;\ \prod_{i=1}^{n} x_i = 0\}$ and let $\Phi : X \to X$ be given by $\Phi(x) = 2x$. Show that $I(X, \Phi, X) = (-1)^{n+1}(2^n - 1)$.

3.7 *BIBLIOGRAPHICAL NOTES*

The construction of the Leray–Schauder degree follows the standard approach and can be found in Deimling (1984), Dugundji and Granas (1982), Granas (1962a, 1962b, 1972), Krasnosiel'skii (1965, 1968), Krasnosiel'skii and Zabrejko (1984), Nirenberg (1974), and Zeidler (1986).

For an excellent account of the compact operators theory, various fixed-point theorems for compact mappings, the nonlinear alternative, and the invariance of domain, we refer the reader to Dugundji (1966), Dugundji and Granas (1982), and Granas (1962a, 1962b, 1972). We should mention that various properties of compact fields were established in a different approach in Dugundji (1966). We recommend Krasnosiel'skii (1965) for additional information on compact linear operators and refer the reader to Bott and Tu (1982), Brown (1967), Cronin (1964), Elworthy and Tromba (1970), Kolmogorov and Formin (1961), Kryszewski and Przeradzki (1986), Rothe (1986), and Zeidler (1986) for further details.

The application to Bernstein's theorem for second-order ODEs is based on the works of Granas et al. (1980, 1983). The application to nonlinear boundary value problems for second-order ODEs is based on a paper of Erbe et al. (1991) and some results from Hartman (1964). Related results can be found in Bebernes and Schmitt (1973), Bernfeld et al. (1974), Leray and Schauder (1934), and Schmitt (1973). For the material related to the fixed-point index for ENRs, we refer to Dold (1965, 1972, 1983). The extension of the fixed-point index to the category of ANRs was done by Granas in (1972). For more information on ANRs and ENRs, we refer the interested reader to Borsuk (1967).

Chapter Four

Nussbaum–Sadovskii Degree for Condensing Fields

This chapter's main purpose is to extend the Leray–Schauder degree to condensing fields.

We shall begin by introducing the measure of noncompactness and condensing fields and by establishing the bijection theorem that connects condensing mappings to compact mappings. We then, in Section 4.2, apply this bijection theorem to give a constructive definition of the Nussbaum–Sadovskii degree for condensing fields. Some useful fixed-point theorems and nonlinear alternative will also be established. In Section 4.3, some topological and analytical results, including the invariance of domain theorem and the Fredholm alternative theorem, will be presented. Section 4.4 discusses the homotopic properties of the class of linear condensing fields. These properties, together with the Sard–Smale theorem, enable us to develop regular approximations and to obtain the regular value formula for a C^1-condensing field. In Section 4.5, we present an application of the degree theory for condensing fields to boundary value problems of some neutral equations. In Section 4.6, we introduce the composite coincidence degree for some coincidence problems that will be used for boundary value problems of general neutral functional differential equations.

4.1 MEASURE OF NONCOMPACTNESS AND CONDENSING MAPPINGS

In what follows we assume that X is a real Banach space. We denote by $X^{n+\infty}$, $n = 0, 1, \ldots$, the space $\mathbb{R}^n \times X$ equipped with the norm $\|(p, x)\| = \max\{|p|, \|x\|\}$, $(p, x) \in \mathbb{R}^n \times X$, and we denote by $\pi : X^{n+\infty} \to X$ the natural projection onto X. For two bounded sets $A, B \subset X^{n+\infty}$, we define

$$d(A, B) = \inf\{\|x - y\|;\ x \in A,\ y \in B\}$$

$$d(x, B) = d(\{x\}, B)\ , \qquad x \in X^{n+\infty}$$

$$B_n(A, \varepsilon) = \{y \in X^{n+\infty};\ d(y, A) < \varepsilon\}$$

$$B(A, \varepsilon) = B_0(A, \varepsilon)\ , \qquad \text{that is, if } A \subset X$$

$$D^*(A, B) = \inf\{r > 0;\ A \subset B_n(B, r)\}$$

$$D(A, B) = \max\{D^*(A, B), D^*(B, A)\}$$

$$\|A\| = \sup\{\|x\|;\ x \in A\}$$

$$\mathrm{diam}(A) = \sup\{\|x - y\|;\ x, y \in A\}$$

The function D, defined on nonempty, bounded, and closed subsets of $X^{n+\infty}$, is a metric, called the Hausdorff metric. Finally, we denote by $\mathcal{M}_n$ the class of bounded subsets of $X^{n+\infty}$ and we put $\mathcal{M} = \bigcup_{n=0}^{\infty} \mathcal{M}_n$.

Definition 4.1.1 A function $\mu : \mathcal{M} \to [0, +\infty)$ is called a *measure of noncompactness* if the following conditions are satisfied:

(P1) $\mu(A) = 0 \Leftrightarrow \overline{A}$ is compact.
(P2) $\mu(A) = \mu(\overline{A})$.
(P3) $\mu(\mathrm{conv}(A)) = \mu(A)$.
(P4) $\mu(A \cup B) = \max\{\mu(A), \mu(B)\}$.
(P5) $\mu(rA) = |r| \cdot \mu(A)$, $r \in \mathbb{R}$.
(P6) $\mu(A + B) \le \mu(A) + \mu(B)$, where $A, B \in \mathcal{M}$.

Proposition 4.1.2

Let $\mu : \mathcal{M} \to [0, \infty)$ be a measure of noncompactness. Then we have the following:

(P7) *$A \subseteq B$ implies $\mu(A) \le \mu(B)$.*
(P8) $\mu(\pi(A)) = \mu(A)$.
(P9) *$\mu(B_n(A, r)) \le \mu(A) + r\mu(B_1(0))$, where $B_\varepsilon(x)$ denotes the open ball in X with center x and radius $\varepsilon > 0$.*
(P10) $|\mu(A) - \mu(B)| \le D(A, B) \cdot \mu(B_1(0))$.
(P11) *For two bounded sequences $\{x_k\}$, $\{y_k\}$ in $X^{n+\infty}$ and a compact set $K \subseteq X^{n+\infty}$, $\lim_{k\to\infty} d(x_k - y_k, K) = 0$ implies $\mu(\{x_k\}) = \mu(\{y_k\})$.*
(P12) *Let $A_1 \supseteq A_2 \supseteq \cdots$ be a sequence of closed, nonempty subsets of $\mathcal{M}_n$*

such that $\lim_{k\to\infty} \mu(A_k) = 0$. *Then*

$$A_\infty = \bigcap_{k=1}^{\infty} A_k \neq \emptyset$$

Proof. Since $B = A \cup (B \setminus A)$, by (P4) we have $\mu(B) = \mu(A \cup (B \setminus A)) = \max\{\mu(A), \mu(B\setminus A)\} \geq \mu(A)$, and this proves (P7).

For every $A \in \mathcal{M}_n$, there exists a compact convex $\Delta \subseteq \mathbb{R}^n$ such that $A \subseteq \pi(A) \times \Delta$. Since $A \subseteq \pi(A) + (\Delta \times \{0\})$ and $\pi(A) \subseteq A + ((-\Delta) \times \{0\})$, by (P6) and (P1) we get $\mu(A) \leq \mu(\pi(A)) + \mu(\Delta) = \mu(\pi(A))$, $\mu(\pi(A)) \leq \mu(A) + \mu(\Delta) = \mu(A)$ and, thus, $\mu(\pi(A)) = \mu(A)$.

By (P8), $\mu(B_n(0, 1)) = \mu(B_1(0))$, where $B_n(0, 1) := \{x \in X^{n+\infty};\ \|x\| < 1\}$. Therefore, by (P5) and (P6), $\mu(B_n(A, r)) = \mu(A + rB_n(0, 1)) \leq \mu(A) + r\mu(B_1(0))$.

Assume $D(A, B) = r$ and let $\varepsilon > 0$ be given. By the definition of the Hausdorff metric, $A \subseteq B_n(B, r + \varepsilon)$ and $B \subseteq B_n(A, r + \varepsilon)$. Therefore, by (P7) and (P9), $\mu(A) \leq \mu(B_n(B, r + \varepsilon)) \leq \mu(B) + (r + \varepsilon)\mu(B_1(0))$. Similarly, we have $\mu(B) \leq \mu(A) + (r + \varepsilon)\mu(B_1(0))$. Therefore, $|\mu(B) - \mu(A)| \leq (r + \varepsilon)\mu(B_1(0))$. Since $\varepsilon > 0$ is arbitrary, the property (P10) follows.

Now, for (P11), by assumption, for every $\varepsilon > 0$, there is an integer N such that for $k > N$, $(x_k - y_k) \in B_n(K, \varepsilon)$. That is, $x_k \in y_k + B_n(K, \varepsilon)$. Therefore,

$$\mu(\{x_k\}) = \mu(\{x_k; k \geq N\}) \leq \mu(\{y_k\}) + \varepsilon\mu(B_1(0))$$

Similarly, $\mu(\{y_k\}) \leq \mu(\{x_k\}) + \varepsilon\mu(B_1(0))$. Thus, $\mu(\{x_k\}) = \mu(\{y_k\})$.

Finally, in (P12), notice that the sets A_k are nonempty. Therefore, we can choose a sequence $\{x_k\}$ with $x_k \in A_k$. Since

$$\mu(\{x_k\}) = \mu(\{x_k; k \geq N\}) \leq \mu(A_N) \to 0 \qquad \text{as } N \to \infty$$

we have $\mu(\{x_k\}) = 0$. By (i), the sequence $\{x_k\}$ contains a convergent subsequence $\{x_{k_i}\}$. Let $x_0 = \lim_{i\to\infty} x_{k_i}$. Then $x_0 \in A_k$ for all $k = 1, 2, \ldots$, and thus $x_0 \in \bigcap_{k=1}^{\infty} A_k$. The proof is complete. □

It is easy to see that property (P8) can be used to extend a measure of noncompactness μ, defined only on $\mathcal{M}_0$ and satisfying the properties (P1–P6), to the class $\mathcal{M}_n$, for every $n \in \mathbb{N}$.

Examples 4.1.3

(i) *Hausdorff Measure of Noncompactness:* $\mathcal{X} : \mathcal{M} \to [0, \infty)$ is defined for

$A \in \mathcal{M}_0$ by

$$\mathscr{X}(A) = \inf\left\{ \varepsilon > 0;\ \text{there is } \{x_1, \ldots, x_N\} \subseteq X \text{ such that } A \subseteq \bigcup_{i=1}^{N} B_\varepsilon(x_i) \right\}$$

It is easy to see that the inclusion $A \subseteq \bigcup_{i=1}^{N} B_\varepsilon(x_i)$ implies that for any arbitrary measure of noncompactness μ,

$$(4.1.1)\quad \mu(A) \leq \mu\left(\bigcup_{i=1}^{N} B_\varepsilon(x_i)\right) = \max\{\mu(B_\varepsilon(x_i);\ i = 1, \ldots, N\}$$

Let $\varepsilon = r + \delta$, where $r = \mathscr{X}(A)$ and $\delta > 0$ is arbitrarily small. Then (4.1.1) implies

$$\mu(A) \leq (r + \delta)\mu(B_1(0))$$

and thus

$$(4.1.2)\qquad \mu(A) \leq \mathscr{X}(A)\mu(B_1(0))$$

(ii) *Kuratowski Measure of Noncompactness:* $\alpha : \mathcal{M} \to [0, \infty)$ is defined for $A \in \mathcal{M}_0$ by

$$\alpha(A) = \inf\{\varepsilon > 0;\ \text{there exists } A_1, \ldots, A_N \text{ such that } A = A_1 \cup \cdots \cup A_N \text{ and } \operatorname{diam}(A_i) < \varepsilon,\ i = 1, \ldots, N\}$$

In what follows, we assume that X is equipped with a measure of noncompactness μ.

Definition 4.1.4 Let $A \subseteq X^{n+\infty}$, $B \subseteq X^{m+\infty}$, and $F : A \to B$ be a continuous mapping sending bounded subsets of A into bounded subsets of B. We say that F is:

(i) A *μ-Lipschitzian mapping* with constant $k \geq 0$, if $\mu(F(D)) \leq k\mu(D)$ for every bounded subset $D \subseteq A$.

(ii) A *completely continuous mapping*, if it is μ-Lipschitzian with constant $k = 0$.

(iii) A *Darbó mapping* or *μ-set contraction*, if it is μ-Lipschitzian with constant $k < 1$.

(iv) A *condensing mapping*, if it is μ-Lipschitzian with constant $k = 1$ and $\mu(F(D)) < \mu(D)$ for every bounded subset $D \subseteq A$ satisfying $\mu(D) > 0$.

For the sake of convenience, we introduce the following notations:

$\mathscr{K}$: The class of *completely continuous mappings*.
$\mathscr{D}$: The class of *Darbó mappings*.
$\mathscr{C}$: The class of *condensing mappings*.

It is clear that $\mathscr{K} \subseteq \mathscr{D} \subseteq \mathscr{C}$. It is also easy to see that the composition of two μ-Lipschitzian mappings with constants k_1 and k_2 is a μ-Lipschitzian mapping with constant $k_1 k_2$. Moreover, if $k_1, k_2 \leq 1$ and if one of these mappings is condensing, then the composition is also condensing.

Proposition 4.1.5

Let $D \subseteq X^{n+\infty}$ be a bounded, closed subset and $F : D \to X^{m+\infty}$ be a condensing mapping. Then we have the following:

(i) *$(\pi - F)(A)$ is closed for every closed subset $A \subseteq D$.*
(ii) *$(\pi - F)^{-1}(K)$ is compact for every compact $K \subseteq X^{m+\infty}$.*

Proof. (i) Let $\{x_k\} \subseteq A$ be a sequence such that $y_k = \pi(x_k) - F(x_k) \to y \in X^{m+\infty}$ as $k \to \infty$. We need to show that $y \in (\pi - F)(A)$. Put $K = \{y\}$. Then by (P11) of Proposition 4.1.2, $\mu(\{x_k\}) = \mu(\{F(x_k)\})$. By the assumption, F is condensing, so $\mu(\{x_k\}) = 0$. Therefore, $\{x_k\}$ contains a subsequence convergent to $x \in A$. By the continuity of F, $\pi(x) - F(x) = y$.

(ii) Let $K \subseteq X^{m+\infty}$ be a compact set and $\{x_k\} \subset (\pi - F)^{-1}(K)$. Put $y_k = \pi(x_k) - F(x_k) \in K$. It follows again from (P11) and (P8) of Proposition 4.1.2 and the condensing property of F that

$$0 = \mu(\{\pi(x_k)\}) = \mu(\{x_k\})$$

Thus, $\{x_k\}$ is relatively compact and contains a convergent subsequence. Consequently, $(\pi - F)^{-1}(K)$ is compact. □

If F is a condensing mapping (respectively, Darbó mapping), then $f = \pi - F$ will be called a *condensing* (respectively, *Darbó*) *field*.

Definition 4.1.6 Let $\mathscr{A}$ denote one of the classes $\mathscr{K}$, $\mathscr{D}$, $\mathscr{C}$ and let (A, B) be a pair of closed bounded subsets in $X^{n+\infty}$ such that $A \supseteq B$. We denote by $\mathscr{A}(A, B)$ the class of all mappings $F : A \to X$ such that (i) $F \in \mathscr{A}$; (ii) $\pi(x) \neq F(x)$ for all $x \in B$, that is, F has no *fixed point* in B.

A mapping $H : [0, 1] \times A \to X$ is called a *homotopy* in $\mathscr{A}(A, B)$ if $H \in \mathscr{A}$ and $H(t, x) \neq \pi(x)$ for $(t, x) \in [0, 1] \times B$. We say F, $G \in \mathscr{A}(A, B)$ are *homotopic* in $\mathscr{A}(A, B)$ if there is a homotopy H in $\mathscr{A}(A, B)$ such that $H_0 = F$ and $H_1 = G$, where $H_t := H(t, \cdot)$ for $t \in [0, 1]$. The homotopy relation in $\mathscr{A}(A, B)$ is an equivalence relation. We denote by $\mathscr{A}[A, B]$ the set of homotopy classes in $\mathscr{A}(A, B)$.

Note that for every pair of closed bounded sets (A, B), we have the following inclusions:

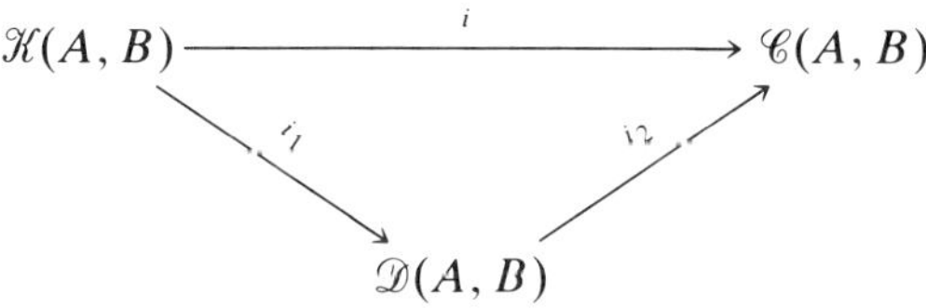

which induce the following mappings:

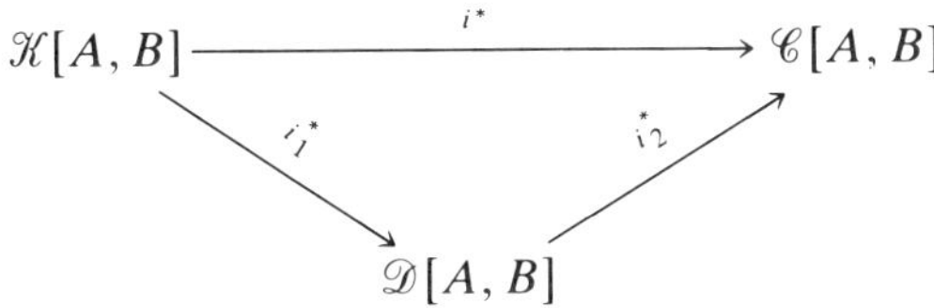

An important technical tool in dealing with condensing mappings is the following.

Theorem 4.1.7 (Bijection Theorem)

For every pair of closed, bounded sets (A, B), the induced mapping $i^* : \mathscr{K}[A, B] \to \mathscr{C}[A, B]$ is a bijection.

Proof. We divide the proof into two steps.

Step 1. *We show that $i_1^* : \mathscr{K}[A, B] \to \mathscr{D}[A, B]$ is a bijection.* In order to show that i_1^* is surjective, we assume $F \in \mathscr{D}[A, B]$, that is, there is $k \in [0, 1)$ such that F is a μ-Lipschitzian map with constant k. We want to show that F is homotopic in $\mathscr{D}(A, B)$ to a compact mapping.

We define by induction the convex sets $Q_i \subseteq X$, $i = 1, 2, \ldots,$

$$Q_1 = \overline{\operatorname{conv}(F(A) \cup \{0\})}\,, \qquad Q_{i+1} = \overline{\operatorname{conv}(F(A \cap \pi^{-1}(Q_i)) \cup \{0\})}$$

Notice that $Q_1 \supseteq Q_2 \supseteq \cdots$ is a sequence of nonempty, bounded, closed, and convex sets. Moreover,

$$\begin{aligned}\mu(Q_{i+1}) &= \mu(\operatorname{conv}(F(A \cap \pi^{-1}(Q_i)))) = \mu(F(A \cap \pi^{-1}(Q_i)))\\ &\leq k \cdot \mu(A \cap \pi^{-1}(Q_i)) \leq k \cdot \mu(Q_i)\end{aligned}$$

for $i = 1, 2, \ldots$. Therefore, $\mu(Q_i) \leq k^i \mu(A)$. Consequently, $\lim_{i\to\infty} \mu(Q_i) = 0$ and $Q = \bigcap_{i=1}^{\infty} Q_i$ is nonempty. Then $\mu(Q) = 0$ and $F(A \cap \pi^{-1}(Q)) \subseteq Q$. By

Corollary 1.5.5, there is a retraction $r: X \to Q$. Set

$$H(t, x) = (1-t)F(x) + tr(F(x)), \qquad x \in A, \qquad t \in [0, 1]$$

Suppose that there is $x_0 \in B$ and $t_0 \in [0, 1]$ such that $\pi(x_0) = H(t_0, x_0)$. Since $F(x_0) \in Q_1$ and $r(F(x_0)) \in Q \subseteq Q_1$, it follows that $\pi(x_0) \in Q_1$. By a similar argument, we can show that if $\pi(x_0) \in Q_i$, then $\pi(x_0) \in Q_{i+1}$. Consequently, we conclude that $\pi(x_0) \in Q$. Therefore, $F(x_0) \in Q$ and

$$\begin{aligned}\pi(x_0) &= (1-t_0)F(x_0) + t_0 r(F(x_0)) \\ &= (1-t_0)F(x_0) + t_0 F(x_0) = F(x_0)\end{aligned}$$

a contradiction of the assumption that $F \in \mathscr{D}(A, B)$. Therefore, $\pi(x) \neq H(t, x)$ for all $x \in B$ and $t \in [0, 1]$. It is clear that H is the required homotopy in $\mathscr{D}(A, B)$.

Now, we prove that $(i_1)^*$ is injective. Assume $F_0, F_1 \in \mathscr{K}(A, B)$ are such that there is a homotopy H in $\mathscr{D}(A, B)$ between F_0 and F_1. We will show that there is a homotopy H^* between F_0 and F_1 in $\mathscr{K}(A, B)$. By definition, $H \in \mathscr{D}([0, 1] \times A, [0, 1] \times B)$. Applying the surjectivity of i_1^* to the pair $[0, 1] \times B \subseteq [0, 1] \times A$, we can find a compact mapping $H' \in \mathscr{K}([0, 1] \times A, [0, 1] \times B)$ such that $\tilde{H}: [0, 1] \times [0, 1] \times A \to X$ given by $\tilde{H}_t = tH' + (1-t)H$ is a homotopy between H and H' in $\mathscr{D}([0, 1] \times A, [0, 1] \times B)$. Since $H|_{\{0\} \times A \cup \{1\} \times A}$ is compact, $\tilde{H}|_{[0,1] \times \{0\} \times A \cup [0,1] \times \{1\} \times A}$ is compact as well. Notice that $\tilde{H}|_{\{1\} \times [0,1] \times A}$ is also compact. Define

$$H^*(t, x) = \begin{cases} \tilde{H}(3t, 0, x) & \text{if } 0 \le t < \frac{1}{3} \\ \tilde{H}(1, 3t-1, x) & \text{if } \frac{1}{3} \le t < \frac{2}{3} \\ \tilde{H}(3-3t, 1, x) & \text{if } \frac{2}{3} \le t \le 1,\ x \in A \end{cases}$$

It is clear that H^* is the required homotopy in $\mathscr{K}(A, B)$ between F_0 and F_1.

STEP 2. *We show that i_2^*: $\mathscr{D}[A, B] \to \mathscr{C}[A, B]$ is a bijection.* We will show that i_2^* is surjective. The injectivity of i_2^* will follow from the surjectivity in a similar way as was done for i_1^*.

Suppose $F \in \mathscr{C}(A, B)$. Then we claim

$$\varepsilon = \inf_{x \in B} \|\pi(x) - F(x)\| > 0 \tag{4.1.3}$$

Indeed, if the claim is not true, then there exists a sequence $\{x_m\} \subseteq B$ such that $\pi(x_m) - F(x_m) \to 0$ as $m \to \infty$. Since $K = \overline{\{\pi(x_m) - F(x_m)\}_{m=1}^{\infty}}$ is compact, $(\pi - F)^{-1}(K) \supseteq \{x_m\}$ is a compact subset of $X^{n+\infty}$ and, hence, $\{x_m\}$ contains a subsequence convergent to $x_0 \in B$. By the continuity of F, $\pi(x_0) - F(x_0) = 0$, a contradiction of the assumption that $F \in \mathscr{C}(A, B)$.

Let k be a number such that

$$1-\frac{\varepsilon}{\|F(A)\|}<k<1$$

Since $k \cdot F$ is a μ-Lipschitzian mapping with constant k, and since for $x \in A$, $\|k \cdot F(x) \quad F(x)\| < \varepsilon$, it follows that the homotopy $H(t, x) = (1-t)F(x) + tkF(x)$, $(t, x) \in [0, 1] \times A$, is a condensing homotopy between F and kF in $\mathscr{C}(A, B)$. This completes the proof. □

EXERCISES

4.1.1 Show that the Hausdorff measure and the Kuratowski measure of noncompactness satisfy all the properties of Definition 4.1.1.

4.1.2 In a separable Hilbert space, find the Hausdorff measure and the Kuratowskii measure of noncompactness of the unit ball and the unit sphere.

4.1.3 Give an exampale of a condensing mapping that is not a Darbó mapping.

4.1.4 Let (K, d) be a compact metric space and let $C(K; \mathbb{R})$ denote the Banach space of all continuous functions φ from K into $\mathbb{R}$, equipped with the usual sup-norm $\|\varphi\| = \sup_{x \in K} |\varphi(x)|$. For a given $\varepsilon > 0$ and for a bounded subset $A \in C(K; \mathbb{R})$, we define $\omega(\varphi, \varepsilon) = \sup\{|\varphi(x) - \varphi(y)|;\ x, y \in K,\ d(x, y) < \varepsilon\}$ and $\omega(A, \varepsilon) = \sup\{\omega(x, \varepsilon);\ x \in A\}$. The measure of *nonequicontinuity* ω_0 is defined on $C(K; \mathbb{R})$ by $\omega_0(A) = \lim_{\varepsilon \to 0} \omega(A, \varepsilon)$, where A is a bounded subset of $C(K; \mathbb{R})$. Show that ω_0 is a measure of noncompactness.

4.1.5 Let Y be a Banach space and consider the Banach space $X := C([a, b]; Y)$ of all continuous mappings from the interval $[a, b]$ into F, equipped with the sup-norm. Assume that μ is a measure of noncompactness on the space Y. Then for every bounded subset $A \subseteq X$, we can define $\mu^*(A) = \sup\{\mu(A(t));\ t \in [a, b]\}$, where $A(t) = \{\varphi(t);\ \varphi \in A\}$. We put $\mu_c(A) = \omega_0(A) + \mu^*(A)$, where $\omega_0(A)$ is the measure of nonequicontinuity, defined in a similar way as in Exercise 4.1.4. Show that μ_c is a measure of noncompactness.

4.1.6 Let $X = L^p([a, b]; \mathbb{R})$ and $T_h : L^p([a, b]; \mathbb{R}) \to L^p(\mathbb{R})$, $h \in \mathbb{R}$, be the *shifting operator* $T_h\varphi(t) := \varphi(t+h)$, $\varphi \in X$. Given $\varepsilon > 0$, $\varphi \in X$, and a bounded subset $A \subseteq X$, we define $\nu(\varphi, \varepsilon) := \sup\{\|T_h\varphi - \varphi\|_p;\ |h| \leq \varepsilon\}$ and $\nu(A, \varepsilon) := \sup\{\nu(\varphi, \varepsilon);\ \varphi \in A\}$. Put $\nu_0 := \lim_{\varepsilon \to 0} \nu(A, \varepsilon)$. Show that ν_0 is a measure of noncompactness.

4.1.7 Consider again the space $X = L^p([a, b]; \mathbb{R})$. Let $S_h : L^p([a, b]; \mathbb{R}) \to L^p(\mathbb{R})$, $h \neq 0$, be defined by $S_h\varphi(t) = \dfrac{1}{2h} \displaystyle\int_{t-h}^{t+h} \varphi(s)\, ds$, $\varphi \in X$. The operator S_h is called a *Steklov operator*. We define $\eta(x, \varepsilon) := \sup\{\|S_h\varphi - \varphi\|_p;\ |h| \leq \varepsilon\}$ and $\eta(A, \varepsilon) := \sup\{\eta(\varphi, \varepsilon);\ \varphi \in A\}$. Put $\eta_0 := \lim_{\varepsilon \to 0} \eta(A, \varepsilon)$. Show that η_0 is a measure of noncompactness.

4.1.8 Let $X = L^p([a, b]; \mathbb{R})$. Show that the measures of noncompactness ν_0 (defined in Exercise 4.1.6) and η_0 (defined in Exercise 4.1.7) together with the Hausdorff measure of noncompactness $\mathscr{X}$ satisfy on X the following inequalities $\mathscr{X}(A) \leq \eta_0(A) \leq \nu_0(A)$, for all bounded subsets $A \subseteq X$.

4.1.9 Let X denote the space $C^n([a, b]; \mathbb{R})$ (respectively, the Sobolev space $W^{n,p}([a, b]; \mathbb{R})$), and let μ_0 be a measure of noncompactness on the space $C([a, b]; \mathbb{R})$ (respectively on $L^p([a, b]; \mathbb{R})$). For every bounded subset $A \in X$, we define $\mu(A) := \mu_0(A^{(n)})$, where $A^{(n)} := \{\varphi^{(n)};\ \varphi \in A\}$. Show that μ is a measure of noncompactness on the space X.

4.1.10 Let $X = C([0, 1]; \mathbb{R})$ (respectively, $X = L^p([0, 1]; \mathbb{R})$), and let $f : [0, 1] \times \mathbb{R} \to \mathbb{R}$ be a continuous function. Show that the *Nemytskii operator* $N_f : X \to X$, defined by $N_f(\varphi)(t) = f(t, \varphi(t))$, $\varphi \in C([0, 1]; \mathbb{R})$, is a $\mathscr{X}$-Lipschitzian mapping with constant k, if and only if $|f(t, x) - f(t, y)| \leq k|x - y|$ for all $t \in [0, 1]$ and $x, y \in \mathbb{R}$.

4.1.11 Let μ be a measure of noncompactness in a Banach space X. We denote by $B_1(0)$ the open unit ball in X and by $S = \partial B_1(0)$ the unit sphere in X. Prove that for any $\varepsilon > 0$ and any subset $A \subseteq \overline{B_1(0)}$, there exist subsets $A_i \subseteq S$ and real numbers $0 \leq s_i \leq t_i \leq 1$ for $1 \leq i \leq n$ such that $A \subseteq A^\varepsilon = \bigcup_{i=1}^{n} [s_i, t_i]A_i$ and $\mu(A^\varepsilon) \leq \mu(A) + \varepsilon$. *Hint*: For a fixed integer $n > 0$, let $t_i = i/n$, $0 \leq i \leq n$, and write $B_1(0) = \bigcup_{i=0}^{n-1} [t_i, t_{i+1}]S$. For $A \subseteq B_1(0)$, define $A_i' = A \cap [t_i, t_{i+1}]S$, $A_i = \left\{\dfrac{x}{\|x\|};\ x \in A_i'\right\}$ for $0 < i < n - 1$, and $A_0 = S$. Show that $A \subseteq A^\varepsilon = \bigcup_{i=0}^{n+1} [t_i, t_{i+1}]A_i$ and $D(A, A^\varepsilon)$ can be arbitrarily small if n is large. Then use (P10).

4.2 CONSTRUCTION AND APPLICATION OF NUSSBAUM–SADOVSKII DEGREE

We consider a Banach space X equipped with a measure of non-compactness μ. Let $\Omega \subseteq X$ be a bounded open set and let $F : \overline{\Omega} \to X$ be a continuous mapping such that:

(i) $F(x) \neq x$ for $x \in \partial\Omega$.
(ii) $F \in \mathscr{C}$, that is, F is condensing.

Put $f(x) = x - F(x)$, $x \in \overline{\Omega}$. We will call f an *Ω-admissible field*. We may define the *Nussbaum–Sadovskii degree* $\deg(f, \Omega)$ of the condensing field f in Ω as folows: Since $F \in \mathscr{C}(\overline{\Omega}, \partial\Omega)$, by the bijection theorem, there is a compact mapping $F_0 \in \mathscr{K}(\overline{\Omega}, \partial\Omega)$ such that $F_0 \in [F] \in \mathscr{C}[\overline{\Omega}, \partial\Omega]$. We define

$$\deg(f, \Omega) := \deg(\mathrm{Id} - F_0, \Omega) \tag{4.2.1}$$

where $\deg(\mathrm{Id} - F_0, \Omega)$ is the Leray–Schauder degree of the compact field $\mathrm{Id} - F_0$ on Ω.

The bijection theorem implies that the above definition does not depend on the choice of the compact mapping $F_0 \in [F] \in \mathscr{C}[\overline{\Omega}, \partial\Omega]$ and satisfies all the properties (P1–P10) in Section 3.1.

We may also apply this degree to obtain the following.

Theorem 4.2.1 (Sadovskii's Fixed-Point Theorem)

Let Ω be a nonempty, open, convex subset of X. Then every bounded condensing mapping $F : \overline{\Omega} \to \overline{\Omega}$ has at least one fixed point.

Proof. We assume, without loss of generality, that Ω is a bounded, convex neighborhood of zero. If $F(x) \neq x$ for $x \in \partial\Omega$, then $h(t, x) = x - tF(x)$, $t \in [0, 1]$, is a condensing Ω-admissible homotopy between $f(x) = x - F(x)$ and the identity map. Consequently, by the normalization property,

$$\deg(f, \Omega) = \deg(\mathrm{Id}, \Omega) = 1$$

The existence property then guarantees the existence of $x \in \Omega$ such that $F(x) = x$. □

Theorem 4.2.2 (Nonlinear Alternative)

Let Ω be an open, bounded neighborhood of zero in X. Then for every condensing mapping $F : \overline{\Omega} \to X$, we have the following alternative:

(i) *Either F has a fixed point in $\overline{\Omega}$.*
(ii) *Or there is $x \in \partial\Omega$ and $\lambda \in (0, 1)$ such that $x = \lambda F(x)$.*

Proof. We consider the map $H(t, x) = tF(x)$, $t \in [0, 1]$, $x \in \overline{\Omega}$. If $H(t, x) \neq x$ for $x \in \partial\Omega$ and $t \in (0, 1]$, then $h(t, x) = x - H(t, x)$ is a Ω-admissible homotopy and, thus, $\deg(\mathrm{Id} - F, \Omega) = 1$. Consequently, $F(x) = x$ for some $x \in \Omega$. If $H(1, x) = x$ for $x \in \partial\Omega$, then $F(x) = x$ and F has a fixed point. If $H(t, x) = x$ for some $x \in \partial\Omega$ and $t \in (0, 1)$, then $x = tF(x)$ and (ii) is satisfied. □

Corollary 4.2.3

Let $p : X \to \mathbb{R}_+$ be a function such that $p^{-1}(0) = 0$ and $p(\lambda x) = \lambda p(x)$ for $\lambda > 0$. Let Ω be an open, bounded neighborhood of zero in X such that one of the following conditions is satisfied:

Rothe's Condition: $p[F(x)] \le p(x)$ for all $x \in \partial\Omega$.
Altman's Condition: $[p(F(x))]^2 \le [p(F(x) - x)]^2 + [p(x)]^2$ for all $x \in \partial\Omega$.

Then F has a fixed point.

Proof. By Theorem 4.2.2, it is sufficient to show that either Rothe's condition or Altman's condition excludes the existence of $(\lambda, x) \in (0, 1) \times \partial\Omega$ such that $x = \lambda F(x)$. Suppose, for contradiction, that there is $(\lambda, x) \in (0, 1) \times \partial\Omega$ such that $x = \lambda F(x)$. Since $p(x) = \lambda p[F(x)]$, Rothe's condition may not be satisfied. On the other hand,

$$[p(x)^2] + [p(x - F(x))]^2 = (1 - 2\lambda(1 - \lambda))[p(F(x))]^2 < p[F(x)]^2$$

and, hence, Altman's condition also cannot be satisfied. □

Corollary 4.2.4 (Leray–Schauder Alternative)

Let $F : X \to X$ be a condensing mapping. We define

$$\Sigma(F) = \{x \in X; \text{there exists } \lambda \in (0, 1) \textit{ such that } x = \lambda F(x)\}$$

Then, either $\Sigma(F)$ is unbounded or F has a fixed point.

Proof. Suppose $\Sigma(F)$ is bounded; then there is $r > 0$ such that $\Sigma(F) \subseteq B_r(0) =: \Omega$. Since $\lambda F|_{\overline{\Omega}} : \overline{\Omega} \to X$, $\lambda \in (0, 1)$, is a condensing mapping without a fixed point on $\partial\Omega$, the condition (ii) of Theorem 4.2.2 is not satisfied. Therefore, $F|_{\overline{\Omega}}$ must have a fixed point. □

Definition 4.2.5 We say that a mapping $F : X$ is *quasibounded* if

$$|F|_\infty = \limsup_{\|x\| \to \infty} \frac{\|F(x)\|}{\|x\|} = \inf_{\rho > 0} \sup_{\|x\| \ge \rho} \frac{\|F(x)\|}{\|x\|} < \infty$$

The number $|F|_\infty$ is called the *quasinorm* of F.

Theorem 4.2.6

Let $F : X \to X$ be a quasibounded condensing mapping. Then for every $y \in X$ and every $\lambda \in [-1, 1]$ such that $|\lambda| < 1/|F|_\infty$, the equation $y = x - \lambda F(x)$ has a solution. In particular, the condensing mapping λF has a fixed point.

Proof. We consider the mapping $G(x) = y + \lambda F(x)$, $x \in X$. Since $|\lambda| \le 1$, the mapping $G : X \to X$ is condensing. It is easy to check that $|G|_\infty = |\lambda|\,|F|_\infty < 1$; thus, there is $r > 0$ such that

$$\frac{\|G(x)\|}{\|x\|} < 1 \qquad \text{for all } \|x\| \ge r$$

Applying the nonlinear alternative to $\Omega = \{x \in X;\ \|x\| < r\}$, we conclude that G has a fixed point. □

Definition 4.2.7 A mapping $F : X \to X$ is called *asymptotically linear* if there is a bounded linear operator $S : X \to X$ such that

$$\lim_{\|x\| \to \infty} \frac{\|Sx - F(x)\|}{\|x\|} = 0 \tag{4.2.2}$$

The operator S, which is uniquely determined by the condition (4.2.2), is called the *asymptotic derivative* of F and is denoted by $F'(\infty)$.

Corollary 4.2.8

Let $F : X \to X$ be a condensing asymptotically linear mapping. If $\|F'(\infty)\| < 1$, then for every $y \in X$ and for every $\lambda \in [-1, 1]$, the equation $y = x - \lambda F(x)$ has a solution. In particular, the condensing mapping λF has a fixed point.

Proof. Notice that F is a quasibounded mapping and $|F|_\infty \le \|F'(\infty)\|$. Indeed, for $x \ne 0$, we have

$$\frac{\|F(x)\|}{\|x\|} - \|F'(\infty)\| \le \frac{\|F(x)\|}{\|x\|} - \frac{\|F'(\infty)x\|}{\|x\|} \le \frac{\|F(x) - F'(\infty)x\|}{\|x\|}$$

Consequently,

$$\limsup_{\|x\| \to \infty} \frac{\|F(x)\|}{\|x\|} - \|F'(\infty)\| \le 0$$

Therefore, $|F|_\infty < 1$ and Theorem 4.2.6 applies. □

Theorem 4.2.9 (Borsuk's Fixed-Point Theorem)

Let Ω be a bounded, convex, symmetric, open neighborhood of zero in X. Let $f : \overline{\Omega} \to X$ be a condensing Ω-admissible field such that f is odd on $\partial\Omega$, that is, for $x \in \partial\Omega$, $f(-x) = -f(x)$. Then $\deg(f, \Omega)$ *is an odd integer. In particular,* $\deg(f, \Omega) \ne 0$ *and hence, there is $x \in \Omega$ such that $f(x) = 0$.*

Proof. Let $f(x) = x - F(x)$. We will show that the condensing mapping $F : \overline{\Omega} \to X$ is homotopic in $\mathscr{C}(\overline{\Omega}, \partial\Omega)$ to a compact mapping $F_0 : \overline{\Omega} \to X$ such that $F_0(-x) = -F_0(x)$ for $x \in \partial\Omega$.

Let $\varepsilon = \inf_{x \in \partial\Omega} \|x - F(x)\| > 0$ and let $k > 0$ be such that $1 - \varepsilon / \|F(\Omega)\| < k < 1$. Then kF is a μ-Lipschitzian mapping with constant $k < 1$. Moreover, the argument in Step 2 of the proof of the bijection theorem shows that there is a homotopy in $\mathscr{C}(\overline{\Omega}, \partial\Omega)$ between F and kF. Also note that kF is odd on $\partial\Omega$. Therefore, without loss of generality, we may assume that $F \in \mathscr{D}(\overline{\Omega}, \partial\Omega)$.

We define

$$Q_1 = \overline{\mathrm{conv}\{F(\overline{\Omega}) \cup (-F(\overline{\Omega}))\}}$$

$$Q_{i+1} = \overline{\mathrm{conv}\{F(\overline{\Omega} \cap Q_i) \cup (-F(\overline{\Omega} \cap Q_i))\}}\ , \qquad i \geq 1$$

Then $Q = \bigcap_{i=1}^{\infty} Q_i$ is a compact, convex, nonempty, symmetric subset of X. Let $\bar{r} : X \to Q$ be a retraction of X onto Q. Then $r_0(x) = \frac{1}{2}[\bar{r}(x) - \bar{r}(-x)]$ is an odd retraction onto Q. By an argument similar to that of Theorem 4.1.7, we show that $r_0 F$ and F are homotopic in $\mathscr{D}(\overline{\Omega}, \partial\Omega)$, and $F_0 = r_0 F$ is compact and symmetric on $\partial\Omega$. Therefore, by the definition, $\deg(f, \Omega) = \deg(\mathrm{Id} - F_0, \Omega)$. We can then apply a corresponding result for compact fields to obtain the required conclusion. □

EXERCISES

4.2.1 Use an example to show that the Birkhoff–Kellogg theorem is not true for Darbó maps.

4.2.2 Consider the boundary value problem

$$\begin{cases} x'' + f(x) = 0 & \text{in } J = [0, 1] \\ x(0) = x(1) = 0 \end{cases}$$

where $x \in \mathbb{R}^n$. Show that if $f(\mathbb{R}^n)$ is bounded, then the above boundary problem has at least one solution.

4.2.3 Assume that a continuous function $f : \mathbb{R}^n \to \mathbb{R}^n$ is asymptotically linear. Define the map $F : C([0, 1]; \mathbb{R}^n) \to C([0, 1]; \mathbb{R}^n)$ by $F(u)(t) = f(u(t))$.

(i) Show that F is asymptotically linear;

(ii) Define the map $\tilde{F} : C([0, 1]; \mathbb{R}^n) \to C([0, 1]: \mathbb{R}^n)$ by

$$\tilde{F}(v) = \begin{cases} 0 & \text{if } v = 0 \\ \|v\|^2 F\left(\dfrac{v}{\|v\|^2}\right) & \text{if } v \neq 0 \end{cases}$$

Show that $\tilde{F}$ is differentiable at zero and $D\tilde{F}(0) = F'(\infty)$.

4.2.4 Let $F: X \to X$ be a Darbó asymptotically linear mapping. Show that $F'(\infty)$ is a Darbó linear operator.

4.2.5 Give an example of a C^1-condensing mapping F such that $F'(x_0)$ is not a condensing linear operator for some x_0.

4.2.6 Give an example of an asymptotically linear condensing mapping F such that $F'(\infty)$ is not a condensing linear operator.

Let X denote a separable infinite-dimensional Hilbert space. Consider a sequence $\{X_n\}_{n=1}^{\infty}$ of finite-dimensional subspaces X_n of X satisfying the following conditions:

(i) $X_n \subset X_{n+1}$.

(ii) $\bigcup_{n=1}^{\infty} X_n$ is dense in X.

The sequence $\Gamma = \{X_n; P_n\}$, where $P_n : X \to X_n$ is the orthogonal projection onto X_n, is called a *projectionally complete scheme* for X. The following notion of A-proper mapping was introduced by Petryshyn (1975): A mapping $f: X \to X$ is said to be *A-proper* if every bounded sequence $\{x_{n_i}; x_{n_i} \in X_{n_i}\}$ such that $P_{n_i} f(x_{n_i}) \to y$ as $i \to \infty$ contains a convergent subsequence $x_{n_{i(k)}} \to x$ and $f(x) = y$. See Browder and Petryshyn (1969), Krawcewicz (1990) and Petryshyn (1975).

4.2.7 Show that every compact field is A-proper.

4.2.8 Show that every Darbó field is A-proper.

Let $\prod_{n=1}^{\infty} \mathbb{Z}$ be the group of all integer-valued sequences $\{x_i\}$ and $\bigoplus_{n=1}^{\infty} \mathbb{Z}$ denote the subgroup of all sequences $\{x_i\}$ such that $x_i = 0$ for all but finitely many $i \in \mathbb{N}$. We denote by $\mathbb{Z}^*$ the abelian group $\dfrac{\prod_{n=1}^{\infty} \mathbb{Z}}{\bigoplus_{n=1}^{\infty} \mathbb{Z}}$. Therefore, two sequences $\{x_i\}$ and $\{y_i\}$ determine the same element in $\mathbb{Z}^*$ if and only if $x_i = y_i$ for all but finitely many $i \in \mathbb{N}$.

Let Ω be an open, bounded subset of X and let $f: X \to X$ be a Ω-admissible A-proper map. Petryshyn defined $\deg_P(f, \Omega)$ as an element of $\mathbb{Z}^*$ that is the class of the sequence $\deg(f_n, \Omega_n)$, where $f_n := P_n \circ f$, $\Omega_n := X_n \cap \Omega$, and $\deg(f_n, \Omega_n)$ is the usual Brouwer degree of f_n on Ω_n.

4.2.9 Show that $\deg_P(f, \Omega)$ is well defined and satisfies additivity, excision, and existence properties.

4.2.10 An Ω-admissible homotopy $h:[0,1]\times X\to X$, continuous in t and uniformly continuous with respect to $x\in\partial\Omega$, is called an *A-proper homotopy*, if $h(t,\cdot):X\to X$ is A-proper for all $t\in[0,1]$. Show that the degree of A-proper mappings satisfies the homotopy invariance.

4.2.11 Let $f:X\to X$ be a compact Ω-admissible field. Show that $\deg_P(f,\Omega)$ is represented by the constant sequence $\{\deg(f,\Omega)\}$, where $\deg(f,\Omega)$ denotes the Leray–Schauder degree of f on Ω.

4.2.12 Let $f:X\to X$ be a Darbó Ω-admissible field. Show that $\deg_P(f,\Omega)$ is represented by the constant sequence $\{\deg(f,\Omega)\}$, where $\deg(f,\Omega)$ denotes the Nussbaum–Sadovskii degree of f on Ω.

4.3 OTHER PROPERTIES OF CONDENSING FIELDS

In this section we present other topological and analytical results such as the invariance of domain theorem and Fredholm alternative theorem. We begin with the following.

Proposition 4.3.1

Let Ω be an open, bounded, nonempty subset of X and let $f:\overline{\Omega}\to X$ be an Ω-admissible condensing field such that $\deg(f,\Omega)\neq 0$. *Then for every $\eta>0$ such that $f(\partial\Omega)\cap B_\eta(0)=\varnothing$, we have $B_\eta(0)\subseteq f(\Omega)$.*

Proof. Let $F=\mathrm{Id}-f$ and $x_0\in B_\eta(0)$. It is clear that $F_{x_0}(x)=F(x)+x_0$, $x\in\overline{\Omega}$, is homotopic to F in $\mathscr{C}(\overline{\Omega},\partial\Omega)$. Indeed, $\|F_{x_0}(x)-F(x)\|<\eta\leq\inf\{\|f(y)\|;\ y\in\partial\Omega\}$, $x\in\overline{\Omega}$. Thus, $H(t,x)=tF_{x_0}(x)+(1-t)F(x)$ is a homotopy between F_{x_0} and F in $\mathscr{C}(\overline{\Omega},\partial\Omega)$. For $f_{x_0}=\mathrm{Id}-F_{x_0}$, $\deg(f_{x_0},\Omega)=\deg(f,\Omega)\neq 0$. Therefore, F_{x_0} has a fixed point in Ω, that is, there is $x\in\Omega$ such that $x-F(x)=x_0$ and, thus, $B_\eta(0)\subseteq f(\Omega)$. □

Corollary 4.3.2

Let $f:X\to X$ be a condensing field such that for every bounded subset $A\subseteq X$, the inverse image $f^{-1}(A)$ is also bounded. Assume that there is a constant $C>0$ such that $f(-x)=-f(x)$ whenever $\|x\|\geq C$. Then f is surjective.

Definition 4.3.3 Let $D\subseteq X$ be a subset and $F:D\to X$ a condensing mapping. We say that F is a *regular condensing mapping* if for every bounded subset $A\subseteq D$ such that $\mu(A)>0$, we have $\lim\limits_{\varepsilon\to 0}\mu(F(B(A,\varepsilon)))<\mu(A)$. We say that a condensing field $f=\mathrm{Id}-F:D\to X$ is a *regular condensing field*, if F is a regular condensing mapping.

Proposition 4.3.4

Let $D \subseteq X$ be a subset. Every Darbó mapping $F: D \to X$ is a regular condensing mapping.

Proof. Let F be a μ-Lipschitzian mapping with constant $k < 1$. For every $\varepsilon > 0$ and a bounded subset $A \subseteq D$ with $\mu(A) > 0$, we have

$$\mu(F(B(A, \varepsilon))) \leq k \cdot \mu(B(A, \varepsilon)) \leq k \cdot [\mu(A) + \varepsilon \cdot \mu(B_1(0))]$$

Thus, $\lim_{\varepsilon \to 0} \mu(F(B(A, \varepsilon))) \leq k \cdot \mu(A) < \mu(A)$. □

Lemma 4.3.5

Let $F: \overline{B_r(0)} \to X$ be a regular condensing mapping. Then the mapping $H: [0, 1] \times B_r(0) \to X$ given by

$$H(t, x) = F\left(\frac{x}{1+t}\right) - F\left(\frac{-tx}{1+t}\right)$$

is a condensing mapping.

Proof. Let $A \subseteq [0, 1] \times \overline{B_r(0)}$ be a subset such that $\mu(A) > 0$. We may assume, without loss of generality, that A is closed. We define the following subsets:

$$A_{[s,t]} = \left\{\frac{1}{1+c} x; x \in \pi(A), s \leq c \leq t\right\}$$

$$B_{[s,t]} = \left\{\frac{-c}{1+c} x; x \in \pi(A), s \leq c \leq t\right\}$$

where $(s, t) \in C = \{(u, v) \in [0, 1] \times [0, 1]; u \leq v\}$. Note that

$$\mu(A_{[s,t]}) = \mu(\text{conv}(\{0\} \cup A_{[s,s]})) = \frac{1}{1+s} \mu(A) > 0$$

$$\mu(B_{[s,t]}) = \mu(\text{conv}(\{0\} \cup B_{[t,t]})) = \frac{t}{1+t} \mu(A)$$

$$\frac{1}{1+s} + \frac{t}{1+t} = 1 + o(|t-s|)$$

We now show that there exists $\varepsilon > 0$ such that

$$\mu(F(A_{[s,t]})) < \mu(A_{[s,t]}) - \varepsilon \tag{4.3.1}$$

for all $(s, t) \in C$. By way of contradiction, suppose that there exists a sequence $(s_i, t_i) \in C$ with the property

$$\mu(F(A_{[s_i,t_i]})) - \mu(A_{[s_i,t_i]}) \to 0 \qquad \text{as } i \to \infty$$

The set C is compact. Thus, we may assume, without loss of generality, that $(s_i, t_i) \to (s_0, t_0) \in C$ as $i \to \infty$. On the other hand, the mapping $(s, t) \to A_{[s,t]}$ is continuous with respect to the Hausdorff metric. Therefore, for $\eta > 0$, $A_{[s_i,t_i]} \subset B(A_{[s_0,t_0]}, \eta)$ for sufficiently large i. Consequently,

$$\mu(F(A_{[s_i,t_i]})) \leq \mu(F(B(a_{[s_0,t_0]}, \eta)))$$

$$\mu(F(B(A_{[s_0,t_0]}, \eta))) - \mu(A_{[s_i,t_i]}) \geq \mu(F(A_{[s_i,t_i]})) - \mu(A_{[s_i,t_i]}) \to 0$$

as $i \to \infty$. By (P10) of Proposition 4.1.2, $\mu(A_{[s_i,t_i]}) \to \mu(A_{[s_0,t_0]})$ as $i \to \infty$. Therefore, $\lim_{\eta \to 0} \mu(F(B(A_{[s_0,t_0]}, \eta))) - \mu(A_{[s_0,t_0]}) \geq 0$, a contradiction of the assumption that F is a regular condensing mapping.

Next, we consider a partition of the interval $[0, 1]$:

$$0 = t_0 \leq t_1 \leq \cdots \leq t_n = 1\,, \qquad t_i - t_{i-1} \leq \delta, \quad \delta > 0, \quad i = 1, 2, \ldots, n$$

By the properties of measure of noncompactness and (4.3.1), we have

$$\begin{aligned}
\mu(H([t_{i-1}, t_i] \times \pi(A)) &\leq \mu(F(A_{[t_{i-1},t_i]})) + \mu(F(B_{[t_{i-1},t_i]})) \\
&< \mu(A_{[t_{i-1},t_i]}) - \varepsilon + \mu(B_{[t_{i-1},t_i]}) \\
&= \frac{1}{1 + t_{i-1}} \mu(A) - \varepsilon + \frac{t_i}{1 + t_i} \mu(A) \\
&= \mu(A)(1 + o(\delta)) - \varepsilon
\end{aligned}$$

for all $i = 1, 2, \ldots, n$. Consequently,

$$\begin{aligned}
\mu(H(a)) &\leq \mu([0, 1] \times H(\pi(A))) \\
&\leq \max\{\mu(H([t_{i-1}, t_i] \times \pi(A))); i = 1, 2, \ldots, n\} \\
&< \mu(A)(1 + o(\delta)) - \varepsilon
\end{aligned}$$

Passing to the limit as $\delta \to 0$, we get $\mu(H(A)) \leq (A)) - \varepsilon$. $\square$

Definition 4.3.6 Let $f : X \to Y$ be a map of metric spaces. If there exist an $\varepsilon > 0$ and a $\delta \geq 0$ such that $\operatorname{diam} f^{-1}(\overline{B_\delta(y)}) < \varepsilon$ for all $y \in Y$, then we say that f is a *δ-based ε-map*. In the case where $\delta = 0$, the map f is called an *ε-map*. For $\delta > 0$, the map f is called an *ε-map in the narrow sense*.

Lemma 4.3.7

Let $f : \overline{B_\varepsilon(x_0)} \to X$ be a condensing field and $\varepsilon > 0$.

(i) *If f is a regular condensing field that is also an ε-map, then there is an $\eta > 0$ such that $B_\eta(f(x_0)) \subseteq f(B_\varepsilon(x_0))$.*
(ii) *If f is a δ-based $\varepsilon/3$-map with $\delta > 0$, then $B_\delta(f(x_0)) \subseteq f(B_\varepsilon(x_0))$.*

Proof. We may assume, without loss of generality, that $x_0 = f(x_0) = 0$. Since in both cases (1) and (2) the mapping f is an ε-map, $0 \in f(B_\varepsilon(0))$ implies $0 \notin f(\partial B_\varepsilon(0))$. Therefore, $F = \mathrm{Id} - f \in \mathscr{C}(\overline{B_\varepsilon(0)}, \partial B_\varepsilon(0))$. Moreover, if f is a δ-based $\varepsilon/3$-mapping, then $\delta \le d(0, f(\partial B_\varepsilon(0)))$. Consequently, by Proposition 4.3.1, it is sufficient to show that $\deg(f, B_\varepsilon(0)) \neq 0$.

CASE 1. Assume that f is a regular condensing field that is also an ε-map. Then by Lemma 4.3.5, the mapping

$$H(t, x) = F\left(\frac{x}{1+t}\right) - F\left(\frac{-tx}{1+t}\right)$$

is condensing. We claim that H has no fixed point in $[0, 1] \times \partial B_\varepsilon(0)$. Indeed, suppose there exists $(t, x) \in [0, 1] \times \partial B_\varepsilon(0)$ such that

$$x = \frac{x}{1+t} - \frac{-tx}{1+t} = F\left(\frac{x}{1+t}\right) - F\left(\frac{-tx}{1+t}\right)$$

that is, we have $f\left(\frac{x}{1+t}\right) = f\left(\frac{-tx}{1+t}\right)$, which contradicts the assumption that f is an ε-map.

We note that $H_0 \equiv F$ and H_1 is an odd mapping. Therefore, by Theorem 4.2.9, $\deg(f, B_\varepsilon(0))$ is an odd integer. In particular, $\deg(f, B_\varepsilon(0)) \neq 0$.

CASE 2. Let f be a δ-based $\varepsilon/3$-map. We choose a number k such that

$$0 < 1 - k < \frac{\delta}{\|F(B_\varepsilon(0))\|}$$

Since $g = \mathrm{Id} - k \cdot F$ is a Darbó field, by Proposition 4.3.4, it is a regular condensing field. Since $kF(x) \in B_\delta(F(x))$ for all $x \in B_\varepsilon(0)$, it follows that kF is homotopic to F in $\mathscr{C}(\overline{B_\varepsilon(0)}, \partial B_\varepsilon(0))$. Consequently, it is sufficient to prove that g is an ε-map. Let $y \in g(B_\varepsilon(0))$ and $x \in g^{-1}(y)$, that is, $y = x - kF(x)$. Let $y' = x - F(x)$. Then $\|y - y'\| = (1-k)\|F(x)\| < \delta$, so $y \in B_\delta(y')$. Therefore, $f^{-1}(y) \subseteq f^{-1}(B_\delta(y'))$. Since $x \in f^{-1}(B_\delta(y'))$ and $\mathrm{diam}\{f^{-1}(B_\delta(y'))\} < \varepsilon/3$, we have $d(x, f^{-1}(y)) < \varepsilon/3$. Consequently, for all $x_0, x \in g^{-1}(y)$, we have

$$\begin{aligned}\|x_0 - x\| &\le d(x_0, f^{-1}(y)) + d(x, f^{-1}(y)) + \mathrm{diam}\{f^{-1}(y)\} \\ &< \tfrac{2}{3}\varepsilon + \mathrm{diam}\{f^{-1}(y)\} < \varepsilon\end{aligned}$$

This implies that $\mathrm{diam}\{g^{-1}(y)\} < \varepsilon$ and the conclusion follows. □

Theorem 4.3.8

Let $f : X \to X$ be a condensing field.

(i) *If f is a regular condensing field that is also an ε-map, then $f(X)$ is open in X.*
(ii) *If f is an ε-map in the narrow sense, then $f(X) = X$.*

Proof. Statement (i) follows immediately from (i) of Lemma 4.3.7. In order to prove (ii), we notice, by Lemma 4.3.7, that for every $y \in f(X)$, $B_\delta(y) \subseteq f(X)$, where $\delta > 0$ is a constant such that f is a δ-based ε-map. As δ is independent of y, f has to be surjective. □

Theorem 4.3.9 (Invariance of Domain Theorem)

Let U be an open subset of X and $f : U \to X$ an injective regular condensing field. Then we have the following:

(i) *f is an open mapping. In particular, $f(U)$ is open in X.*
(ii) *$f(U)$ is homeomorphic to U.*

Moreover, if f is an ε-map in the narrow sense for $\varepsilon > 0$, then the assumption that f is regular can be omitted.

Proof. The proof follows directly from Theorem 4.3.8 and the fact that an injective map is an ε-map. □

Before we state the next result, we recall that a closed linear operator $L : \mathrm{Dom}(L) \subseteq X \to T$ between two Banach spaces is called a *Fredholm operator* if (i) $\mathrm{Ker}\, L$ is a finite-dimensional subspace of X; and (ii) $\mathrm{Im}\, L$ is closed and has finite codimension. The integer $\dim \mathrm{Ker}\, L - \mathrm{codim}\, \mathrm{Im}\, L$ is called the *index* of L.

Corollary 4.3.10 (Fredholm Alternative)

Let $T : X \to X$ be a linear condensing operator. Then the operator $\mathrm{Id} - T$ is a bounded Fredholm operator of index zero defined in X. In particular, we have the following alternative:

(i) *Either the equation $0 = x - Tx$ has a nonzero solution.*
(ii) *Or the equation $y = x - Tx$ has a unique solution for each $y \in X$.*

Proof. We first point out that if $f = \mathrm{Id} - T$ is injective, then f is a regular condensing field. Indeed, the mapping $A \to T(A)$, where A is a bounded, closed subset of X, is continuous with respect to the Hausdorff metric.

Consequently, $\lim_{\varepsilon\to 0} \mu(T(B(A,\varepsilon))) = \mu(T(A))$. Therefore, T is a regular condensing mapping.

We divide the remaining part of the proof into three steps.

STEP 1. $\operatorname{Ker}(\mathrm{Id}-T)$ *has a finite dimension.* We consider $B := \overline{B_1(0)}$ and let $T_0 := T|_B : B \to X$. By Proposition 4.1.5,

$$(\mathrm{Id}-T_0)^{-1}(0) = \operatorname{Ker}(\mathrm{Id}-T) \cap \overline{B_1(0)}$$

is a compact set. Since $\operatorname{Ker}(\mathrm{Id}-T)\cap\overline{B_1(0)}$ is a closed unit ball in $\operatorname{Ker}(\mathrm{Id}-T)$, we conclude $\dim \operatorname{Ker}(\mathrm{Id}-T) < \infty$.

STEP 2. $\operatorname{Im}(\mathrm{Id}-T) = (\mathrm{Id}-T)(X)$ *is a closed subspace.* Since $\operatorname{Ker}(\mathrm{Id}-T)$ is finite-dimensional, there exists a closed subspace X_1 of X such that $X = X_1 \oplus \operatorname{Ker}(\mathrm{Id}-T)$ and $A := (\mathrm{Id}-T)|_{X_1} : X_1 \to X$ is an injective operator. Let $\{x_n\} \subseteq X_1$ be a sequence such that $A(x_n) = y_n \to y_0 \in X$. We claim that the sequence $\{x_n\}$ is bounded. For, if $\|x_n\| \to \infty$ as $n \to \infty$, then $x_n' = x_n/\|x_n\|$ satisfies $\|x_n'\| = 1$ and $A(x_n') = y_n/\|x_n\| \to 0$. By (ii) of Proposition 4.1.5, there is a subsequence $\{x_{n_k}'\}$ convergent to $x_0' \in X_1$. By the continuity of A, we have $Ax_0' = 0$ and $\|x_0'\| = 1$, which is a contradiction of the injectivity of A. As the sequence $\{x_n\}$ is bounded, by using Proposition 4.1.5 again, we may find a subsequence $\{x_{n_k}\}$ convergent to $x_0 \in X_1$. By continuity of A, we have $A(x_0) = y_0$ and, thus, $\operatorname{Im}(\mathrm{Id}-T)$ is closed.

STEP 3. $X/\operatorname{Im}(\mathrm{Id}-T)$ *is finite-dimensional.* Suppose, by way of contradiction, that $X/\operatorname{Im}(\mathrm{Id}-T)$ is infinite-dimensional. Then there exists a subspace $Y_0 \subset X$ such that:

(i) $\dim(Y_0) = \dim(\operatorname{Ker}(\mathrm{Id}-T))$.

(ii) $Y_0 \cap \operatorname{Im}(\mathrm{Id}-T) = \{0\}$.

(iii) $Y_0 \oplus \operatorname{Im}(\mathrm{Id}-T) \neq X$.

Let P be a linear projection associated with the direct decomposition $X = X_1 \oplus \operatorname{Ker}(\mathrm{Id}-T)$, $\operatorname{Im}(P) = \operatorname{Ker}(\mathrm{Id}-T)$ and $\operatorname{Ker}(P) = X_1$, and assume that B: $\operatorname{Ker}(\mathrm{Id}-T) \to Y_0$ is an arbitrary isomorphism. Then it is clear that the operator $T + B\circ P : X \to X$ is an injective condensing linear operator. By Theorem 4.3.9, $\operatorname{Im}(\mathrm{Id}-(T+B\circ P)) = Y_0 \oplus \operatorname{Im}(\mathrm{Id}-T)$ is an open subset of X, but this is a contradiction of $Y_0 \oplus \operatorname{Im}(\mathrm{Id}-T) \neq X$.

STEP 4. *The index of* $(\mathrm{Id}-T)$ *is zero.* Let $L(X)$ denote the space of all bounded linear operators on X and let $L_{\mathscr{C}}(X)$ be a subspace of $L(X)$ of all operators $A = \mathrm{Id}-T$, where T is a linear condensing operator. Let $\Phi(X)$ denote the space of all Fredholm operators on X. Since index: $\Phi(X) \to \mathbb{Z}$ is a continuous function and $L_{\mathscr{C}}(X) \subset \Phi(X)$, it follows that the index of the operators $\mathrm{Id}-\lambda T$ is a constant independent of $\lambda \in [0,1]$ and, hence, is equal to zero. □

Corollary 4.3.11

Let $f: X \to X$ be a condensing field. Assume that there is a constant $M > 0$ such that

$$\|f(x) - f(y)\| \geq M\|x - y\| \qquad \textit{for all } x, y \in X$$

Then f is invertible.

Proof. Clearly, f is injective and f is an ε-map in the narrow sense for all $\varepsilon > 0$. By Theorem 4.3.9, f is invertible. □

Theorem 4.3.12

Let Ω be a bounded, convex, symmetric, open neighborhood of zero in X, and let $f: \overline{\Omega} \to X$ be a condensing field such that $f(x) \neq tf(-x)$ for all $(t, x) \in [0, 1] \times \partial\Omega$. Then f is Ω-admissible and $\deg(f, \Omega)$ *is an odd integer. In particular,* $\deg(f, \Omega) \neq 0$.

Proof. Let $f = \mathrm{Id} - F$ and define $H: [0, 1] \times \overline{\Omega} \to X$ by

$$H(t, x) = \frac{1}{1+t} F(x) - \frac{t}{1+t} F(-x), \qquad t \in [0, 1], \quad x \in \overline{\Omega}$$

It follows from the assumption that H has no fixed point in $[0, 1] \times \partial\Omega$. Moreover, for every $A \subseteq [0, 1] \times \overline{\Omega}$, $H(A) \subseteq \operatorname{conv}\{F(\pi(A)) \cup (-F(-\pi(A)))\}$; thus, H is a condensing mapping. Consequently, H is a homotopy in $\mathscr{C}(\overline{\Omega}, \partial\Omega)$ between F and $H_1(x) = \frac{1}{2}[F(x) - F(-x)]$ that is an odd mapping. The conclusion follows from Theorem 4.2.9. □

Theorem 4.3.13

Let Ω be an open, bounded neighborhood of zero in X and let $f: \overline{\Omega} \to X$ be a condensing field. For every $y \in \partial\Omega$, we define the ray from zero to y by $Ly := \{x \in X;\ x = \lambda y$ for some $\lambda > 0\}$. If f is Ω-admissible and $\deg(f, \Omega) \neq 0$, *then for every $y \in \partial\Omega$, we have $Ly \cap f(\partial\Omega) \neq \varnothing$.*

Proof. Suppose that there is $y \in \partial\Omega$ such that $x \neq F(x) + \lambda y$ for all $x \in \partial\Omega$ and $\lambda \in (0, \infty)$, where $f(x) = x - F(x)$, $x \in \overline{\Omega}$. Since $f(\overline{\Omega})$ is bounded, there is $k > 0$ such that $\|f(\overline{\Omega})\| < k \cdot \|y\|$. We put

$$H(t, x) = F(x) + t \cdot k \cdot y, \qquad x \in \overline{\Omega}, \quad t \in [0, 1]$$

H is a condensing mapping such that $H(t, x) \neq x$ for all $(t, x) \in [0, 1] \times \partial\Omega$. Moreover, $H_0 = F$ and H_1 has no fixed point in $\overline{\Omega}$. Indeed, if $x = H_1(x)$, then $k \cdot y = x - Fx = f(x)$ and, thus, $k \cdot \|y\| \leq \|f(\overline{\Omega})\|$, a contradiction of the choice of k. Therefore, $\deg(f, \Omega) = 0$, a contradiction of the assumption. □

Theorem 4.3.14

Let $S = \partial B_1(0)$ and let $X^{\infty-1}$denote a subspace of X of codimension 1. Let $f : \overline{B_1(0)} \to X^{\infty-1}$ be a condensing field. Then there exists a point $x \in S$ such that $f(x) = f(-x)$.

Proof. Suppose that $f(x) \neq f(-x)$ for all $x \in S$. Put $g(x) = \frac{1}{2}[f(x) - f(-x)]$, $x \in \overline{B_1(0)}$. Then $0 \notin g(S)$ and, hence, g is Ω-admissible. Since g is odd, $\deg(g, \Omega) \neq 0$. But for $y \in (X \backslash X^{\infty-1}) \cap S$, we have $Ly \cap g(S) = \emptyset$. This is a contradiction of Theorem 4.3.13. □

Lemma 4.3.15

Let $S = \partial B_1(0)$ and $F : S \to X$ ne a μ-Lipschitzian mapping with constant k. Then there exists a μ-Lipschitzian mapping $\tilde{F} : \overline{B_1(0)} \to X$ with the same constant k such that $F = \tilde{F}\|_S$.

Proof. Let $J \subseteq [0, 1]$, $A \subseteq S$. Put $J \cdot A = \{tx; t \in J, x \in A\}$ and $\|J\| = \max\{t; t \in J\}$. Then

$$\mu(JA) = \mu(\text{conv}(\{0\} \cup \|J\|A)) = \|J\|\mu(A)$$

We define $\tilde{F} : \overline{B_1(0)} \to X$ by

$$\tilde{F}(x) = \begin{cases} \|x\| \cdot F\left(\dfrac{x}{\|x\|}\right) & \text{if } x \neq 0 \\ 0 & \text{if } x = 0 \end{cases}$$

The mapping $\tilde{F}$ is continuous. We need only to show that $\tilde{F}$ is μ-Lipschitzian with constant k. Let $A \subseteq \overline{B_1(0)}$ and $\varepsilon > 0$. By (P10) of Proposition 4.1.2, there exists A^ε, $A \subseteq A^\varepsilon$, given by $A^\varepsilon = \bigcup_{i=1}^{n} [s_i, t_i] \cdot A_i$, $0 \leq s_i \leq t_i \leq 1$, $A_i \subseteq S$, satisfying the inequality (see Exercise 4.1.11)

$$\mu(A^\varepsilon) \leq \mu(A) + \varepsilon$$

Therefore,

$$\begin{aligned} \mu(\tilde{F}(A)) &\leq \mu(\tilde{F}(A^\varepsilon)) = \max_{1 \leq i \leq n} \{\mu(\tilde{F}([s_i, t_i]A_i))\} \\ &= \max_{1 \leq i \leq n} \{\mu([s_i, t_i]F(A_i))\} \leq k \cdot \max_{1 \leq i \leq n} \{t_i \mu(A_i)\} \\ &\leq k \cdot \mu(A^\varepsilon) \leq k \cdot \mu(A) + \varepsilon \cdot k \end{aligned}$$

Since $\varepsilon > 0$ is an arbitrary number, we have $\mu(\tilde{F}(A)) \leq k \cdot \mu(A)$. □

As an immediate consequence of Theorem 4.3.14 and Lemma 4.3.15, we get the following version of the antipodal theorem of Borsuk which, in the case of a sphere, asserts that every continuous mapping from a sphere into a plane has a pair of *antipodal points* (i.e., two diametrically opposite points on the sphere with the same image).

Corollary 4.3.16 (ANTIPODAL THEOREM)

For every Darbó field $f: S \to X^{\infty-1}$, *where* $S = \partial B_1(0)$ *and* $X^{\infty-1}$ *is a subspace of codimension* 1, *there exists a point* $x \in S$ *such that* $f(x) = f(-x)$.

Consequently, we have the following.

Corollary 4.3.17

There is no bijective Darbó field $f: S^{\infty-n} \to S^{\infty-k}$, $n < k$, *where* $S^{\infty-m}$ *denotes a unit sphere in a subspace* $X^{\infty-m}$ *of* X *of codimension* m.

EXERCISES

4.3.1 Give an example of a Banach space X such that the retraction r on the unit ball $\overline{B}$ defined by

$$r(x) = \begin{cases} x & \text{if } x \in \overline{B} \\ \dfrac{x}{\|x\|} & \text{if } x \notin \overline{B} \end{cases}$$

is not a $\mathscr{X}$-Lipschitzian map with constant 1.

4.3.2 Let U be an open, connected, bounded subset of a Banach space X and let $F: \overline{U} \to X$ be an U-admissible condensing mapping. The mapping F is said to be *essential* in $\mathscr{C}(\overline{U}, \partial U)$ if every mapping $G \in \mathscr{C}(\overline{U}, \partial U)$ such that $G_{|\partial U} \equiv F_{|\partial U}$ has a fixed point in U. Show that F is essential in $\mathscr{C}(\overline{U}, \partial U)$ if and only if $\deg(\mathrm{Id} - F, U) \neq 0$.

4.4 REGULAR VALUE FORMULA FOR CONDENSING FIELDS

We will use the following notations throughout the remainder of this book:

$L(X, Y)$	Space of all linear bounded operators from X to Y
$L(X)$	Space of all linear bounded operators in X
$\mathscr{K}(X)$	Subspace of $L(X)$ consisting of compact operators
$\mathscr{D}(X)$	Subset of $L(X)$ consisting of Darbó operators
$\mathscr{C}(X)$	Subset of $L(X)$ consisting of condensing operators

$\mathcal{F}(X)$	Subspace of $\mathcal{K}(X)$ consisting of finite-dimensional operators
$\Phi_0(X)$	Subset of $L(X)$ consisting of Fredholm operators of index zero
$Pr(X)$	Subset of $L(X)$ consisting of projections in X
$GL(X)$	Set of all linear automorphisms of X
$\mathrm{Iso}(X, Y)$	Set of all isomorphisms from X to Y, where X and Y are Banach spaces

Definition 4.4.1 A subset $\mathcal{P} \subseteq L(X)$ is called a *perturbation class* if the following properties are satisfied:

(i) $\mathcal{F}(X) + \mathcal{P} \subseteq \mathcal{P}$.
(ii) If $A \in \mathcal{P}$, then $t \cdot A \in \mathcal{P}$ for $t \in [0, 1]$.
(iii) If $A \in \mathcal{P}$, then $I + A \in \Phi_0(X)$.

Remark 4.4.2 By Corollary 4.3.10, the sets $\mathcal{F}(X)$, $\mathcal{K}(X)$, $\mathcal{D}(X)$, and $\mathcal{C}(X)$ are perturbation classes.

Let $\mathcal{P}$ denote a perturbation class. Define

$$L_{\mathcal{P}}(X) = \{\mathrm{Id} + T; T \in \mathcal{P}\}, \qquad GL_{\mathcal{P}}(X) = GL(X) \cap L_{\mathcal{P}}(X)$$

For $\mathcal{P} = \mathcal{F}(X)$, $\mathcal{K}(X)$, $\mathcal{D}(X)$, and $\mathcal{C}(X)$, we will denote $GL_{\mathcal{P}}(X)$ by $GL_{\mathcal{F}}(X)$ $GL_c(X) := GL_{\mathcal{K}}(X)$, $GL_{\mathcal{D}}(X)$, $GL_{\mathcal{C}}(X)$, respectively. It is clear that

$$GL_{\mathcal{F}}(X) \subseteq GL_c(X) \subseteq GL_{\mathcal{D}}(X) \subseteq GL_{\mathcal{C}}(X)$$

We are going to prove that all the above inclusions are homotopic equivalences.

Lemma 4.4.3

Let $\mathcal{P}$ be a perturbation class, $C_1 \subseteq C$ a pair of compact sets, and $a : C \to P_{\mathcal{P}}(X)$ a continuous mapping such that $a(C_1) \subseteq GL_{\mathcal{P}}(X)$. Then there exists a closed finite-codimensional subspace X^0 of X and a continuous mapping $p : C \to Pr(X)$ such that:

(i) $X^0 \cap \mathrm{Ker}\, a(\lambda) = \{0\}$ *for all* $\lambda \in C$.
(ii) $p(\lambda)X = a(\lambda)X^0$ *for all* $\lambda \in C$.
(iii) $p(\lambda) = a(\lambda) \circ Q \circ a(\lambda)^{-1}$ *for all* $\lambda \in C_1$, *where Q is a projection onto* X^0.
(iv) *The set $\xi(C) = \{(v, \lambda) \in X \times C; v \in \mathrm{Ker}\, p(\lambda)\}$ has a natural structure of a vector bundle over C.*

Proof. For every $\lambda_0 \in C$, $\mathrm{Ker}\, a(\lambda_0)$ is a closed subspace of finite dimension. Let $X(\lambda_0)$ be a closed subspace of X such that $X = X(\lambda_0) \oplus \mathrm{Ker}\, a(\lambda_0)$.

We consider the mapping $a_0 : C \to L(X(\lambda_0), X)$ defined by

$$a_0(\lambda) = a(\lambda)|_{X(\lambda_0} : X(\lambda_0) \to X$$

Clearly, $a_0(\lambda_0)$ is injective with a closed image. Since $GL(X)$ is an open subset of $L(X)$, there exists an open neighborhood V_{λ_0} of $\lambda_0 \in C$ such that $X(\lambda_0) \cap \operatorname{Ker} a(\lambda) = \{0\}$ for all $\lambda \in V_{\lambda_0}$. Since C is compact, there is a finite covering $\{V_{\lambda_i}\}_{i=1}^m$ of C and a sequence of finite-codimensional subspaces $X(\lambda_1), \ldots, X(\lambda_m)$ such that $X = X(\lambda_i) \oplus \operatorname{Ker} a(\lambda_i)$ and $X(\lambda_i) \cap \operatorname{Ker} a(\lambda) = \{0\}$ for all $\lambda \in V_{\lambda_i}$ and $i = 1, 2, \ldots, m$. Put $X^0 = \bigcap_{i=1}^m X(\lambda_i)$. It is clear that $X^0 \cap \operatorname{Ker} a(\lambda) = \{0\}$ for all $\lambda \in C$ and that X^0 is of finite codimension.

Since $a(\lambda_i)X(\lambda_i)$ is a closed finite-codimensional subspace, there is a finite-dimensional subspace $L(\lambda_i) \subseteq X$ such that $X = a(\lambda_i)X(\lambda_i) \oplus L(\lambda_i)$, $i = 1, 2, \ldots, m$. Since $GL(X)$ is open in $L(X)$ and a is continuous, we can assume, without loss of generality, that for all $\lambda \in V_{\lambda_i}$, $X = a(\lambda)X(\lambda_i) \oplus L(\lambda_i)$. We define the mappings $c_i : V_{\lambda_i} \to L_{\mathscr{P}}(X)$, $i = 1, \ldots, m$, by $c_i(\lambda) = a(\lambda)|_{X(\lambda_i)} \oplus A_i$, $\lambda \in V_{\lambda_i}$, where $A_i : \operatorname{Ker} a(\lambda_i) \to L(\lambda_i)$ is an isomorphism.

It is clear that $c_i(\lambda) \in GL_{\mathscr{P}}(X)$ and that $c_i(\lambda)$ is continuous for $\lambda \in V_{\lambda_i}$. Let P_i, $Q \in Pr(X)$, $i = 1, 2, \ldots, m$, be projections such that $P_i(X) = X(\lambda_i)$, $(\mathrm{Id} - P_i)(X) = \operatorname{Ker} a(\lambda_i)$, $i = 1, \ldots, m$, and $Q(X) = X^0$. Since $X^0 \subseteq X(\lambda_i)$ for all $i = 1, \ldots, m$, $Q \circ P_i \in Pr(X)$. We define $p_i : V_{\lambda_i} \to Pr(X)$, $i = 1, \ldots, , m$, by $p_i(\lambda) = c_i(\lambda) \circ Q \circ P_i \circ c_i(\lambda)^{-1}$. The mapping p_i is well defined, since

$$\begin{aligned} p_i(\lambda) \circ p_i(\lambda) &= [c_i(\lambda) \circ Q \circ P_i \circ c_i(\lambda)^{-1}] \circ [c_i(\lambda) \circ Q \circ P_i \circ c_i(\lambda)^{-1}] \\ &= c_i(\lambda) \circ Q \circ P_i \circ c_i(\lambda)^{-1} \circ c_i(\lambda) \circ Q \circ P_i \circ c_i(\lambda)^{-1} \\ &= c_i(\lambda) \circ Q \circ P_i \circ Q \circ P_i \circ c_i(\lambda)^{-1} \\ &= c_i(\lambda) \circ Q \circ P_i \circ c_i(\lambda)^{-1} = p_i(\lambda) \end{aligned}$$

We also have

$$\begin{aligned} p_i(\lambda)X &= [c_i(\lambda) \circ Q \circ P_i \circ c_i(\lambda)^{-1}](X) = [c_i(\lambda) \circ Q \circ P_i](X(\lambda_i) \oplus \operatorname{Ker} \alpha(\lambda_i)) \\ &= [c_i(\lambda) \circ Q]X(\lambda_i) = a(\lambda)X^0 \end{aligned}$$

Let $\{\alpha_i\}$, $i = 1, \ldots, m$, be a partition of unity subordinate to $\{V_{\lambda_i}\}$. Put

$$p(\lambda) = \sum_{i=1}^m \alpha_i(\lambda) p_i(\lambda)$$

Notice that $p(\lambda) \in Pr(X)$ for $\lambda \in C$. Indeed, if $\lambda \in V_{\lambda_{i_1}} \cap \cdots \cap V_{\lambda_{i_k}}$, then

$p_{i_1}(\lambda)X = \cdots = p_{i_k}(\lambda)X = a(\lambda)X^0$. Thus,

$$p(\lambda)X = \sum_{i=1}^{m} \alpha_i(\lambda)p_i(\lambda)X = a(\lambda)X^0$$

Now we only need to show that $\xi(C)$ has a natural structure of a vector bundle over C. Let $\lambda \in V_{\lambda_i}$. It is clear that $\varphi_i(\lambda) = (\mathrm{Id} - p_i(\lambda))|_{L(\lambda_i)} : L(\lambda_i) \to X$ is an injective operator satisfying $\mathrm{Im}(\varphi_i(\lambda)) = \mathrm{Ker}\, p_i(\lambda)$. Thus, we may define $\Phi_i : \xi(V_i) \to L(\lambda_i) \times V_i$ by $\Phi_i^{-1}(v, \lambda) = (\varphi_i(\lambda)v, \lambda)$. It is easy to verify that Φ_i is the required vector bundle chart of $\xi(C)$ over V_i. □

Lemma 4.4.4

Let $\mathcal{P}$ be a perturbation class, C a compact set, and $a : [0, 1] \times C \to L_{\mathcal{P}}(X)$ a continuous mapping such that $a(\{1\} \times C) \subseteq GL_{\mathcal{P}}(X)$. Then there exists a closed finite-codimensional subspace X^0 of X such that $X^0 \cap \mathrm{Ker}\, a(t, \lambda) = \{0\}$ for all $(t, \lambda) \in [0, 1] \times C$, and there is a continuous mapping $b : [0, 1] \times C \to GL_{\mathcal{P}}(X)$ such that:

(i) $b(t, \lambda)|_{X^0} = a(t, \lambda)|_{X^0}$ *for all* $(t, \lambda) \in [0, 1] \times C$.
(ii) $b|_{\{1\} \times C} = a|_{\{1\} \times C}$.

Proof. By Lemma 4.4.3, there is a finite-codimensional subspace X^0 of X and a continuous mapping $p : [0, 1] \times C \to Pr(X)$ such that $p(t, \lambda)X = a(t, \lambda)X^0$ for all $(t, \lambda) \in [0, 1] \times C$ and $p(1, \lambda) = a(1, \lambda) \circ Q \circ a(1, \lambda)^{-1}$ for $\lambda \in C$, where Q is a projection onto X^0. We consider the vector bundle

$$\xi([0, 1] \times C) = \{(t, \lambda, v) \in [0, 1] \times C \times X; v \in \mathrm{Ker}\, p(t, \lambda)\}$$

The restricted vector bundle $\xi(\{1\} \times C)$ over $\{1\} \times \mathrm{C}$ is trivial. Indeed, put $Y_0 = (\mathrm{Id} - Q)X$; then

$$\begin{aligned} a(1, \lambda)Y_0 &= a(1, \lambda)(\mathrm{Id} - Q)X = a(1, \lambda)(\mathrm{Id} - Q)a(1, \lambda)^{-1}X \\ &= [\mathrm{Id} - a(1, \lambda) \cdot Q \cdot a(1, \lambda)^{-1}]X = (\mathrm{Id} - p(1, \lambda))X = \mathrm{Ker}\, p(1, \lambda) \end{aligned}$$

for all $\lambda \in C$, and thus we can define the trivialization

$$\Psi : \xi(\{1\} \times \mathrm{C}) \to (\{1\} \times \mathrm{C}) \times \mathrm{V}$$

by $\Psi(1, \lambda, v) = (1, \lambda, a(1, \lambda)^{-1}v)$, where $v \in \mathrm{Ker}\, p(1, \lambda)$, $\lambda \in C$.

We have the following induced vector bundles:
where $i : \{1\} \times C \to [0, 1] \times C$ is the inclusion and $\rho : [0, 1] \times C \to \{1\}$ is the

$$\begin{array}{ccccc}
\xi([0,1]\times C) & \longleftarrow & i^*(\xi) & \longleftarrow & \rho^* i^*(\xi) \\
\Big\downarrow \pi & & \Big\downarrow \pi & & \Big\downarrow \pi \\
[0,1]\times C & \xleftarrow{\ i\ } & \{1\}\times C & \xleftarrow{\ \rho\ } & [0,1]\times C
\end{array}$$

projection. Since $i^*(\xi)$ is isomorphic to $\xi(\{1\}\times C)$, it is trivial. Consequently, $\rho^*i^*(\xi)$ is trivial. On the other hand, the map $i\circ\rho:[0,1]\times C\to[0,1]\times C$ is homotopic to the identity map of $[0,1]\times C$. Therefore, it follows from Proposition 1.4.10 that the vector bundle ξ is isomorphic to $\rho^*i^*(\xi)$. This shows that ξ is trivial.

Let $s_1,\dots,s_k:[0,1]\times C\to\xi([0,1]\times C)$ be linearly independent sections such that $a(1,\lambda)^{-1}s_i(1,\lambda)=v_i$, $\lambda\in C$, $i=1,\dots,k$, and $\{v_1,\dots,v_k\}\subseteq Y_0$ is a basis of Y_0. We define the mapping $f:[0,1]\times C\to L(Y_0,X)$ by

$$f(t,\lambda)\left(\sum_{i=1}^{k}\beta_i v_i\right)=\sum_{i=1}^{k}\beta_i s_i(t,\lambda)\,,\qquad (t,\lambda)\in[0,1]\times C$$

It is clear that f is continuous and $f(t,\lambda):Y_0\to\operatorname{Ker}p(t,\lambda)$ is an isomorphism for all $(t,\lambda)\in[0,1]\times C$. Therefore, the map $b:[0,1]\times C\to L_{\mathscr{P}}(X)$ defined by

$$b(t,\lambda)=a(t,\lambda)\circ Q+f(t,\lambda)\circ(I-Q)$$

is continuous and $b([0,1]\times C)\subseteq GL_{\mathscr{P}}(X)$. It is clear that $b(t,\lambda)|_{X^0}=a(t,\lambda)|_{X^0}$ and $b(1,\lambda)=a(1,\lambda)$ for $\lambda\in C$. □

Theorem 4.4.5

Let $\mathscr{P}$ be a perturbation class, C a compact set, and $a:C\to GL_{\mathscr{P}}(X)$ a continuous mapping. Then there exists a continuous map $h:[0,1]\times C\to GL_{\mathscr{P}}(X)$ and a direct sum decomposition of $X=X^0\oplus Y_0$, where X^0 is a closed subspace of finite codimension, such that for all $\lambda\in C$, we have:

(i) $h(1,\lambda)=a(\lambda)$.

(ii) $h(0,\lambda)|_{X^0}=\mathrm{Id}|_{X^0}$.

(iii) $h(0,\lambda)|_{Y_0}:Y_0\to Y_0$.

Proof. We define $b:[0,1]\times C\to L_{\mathscr{P}}(X)$ by $b(t,\lambda)x=x+tA(\lambda)x$, where $a(\lambda)=\mathrm{Id}+A(\lambda)$, $A(\lambda)\in\mathscr{P}$, $x\in X$, $\lambda\in C$. Applying Lemma 4.4.4 to $b:[0,1]\times C\to L_{\mathscr{P}}(X)$, we conclude that there is a finite-codimensional subspace X^0 and a continuous mapping $\tilde h:[0,1]\times C\to GL_{\mathscr{P}}(X)$ such that $\tilde h(1,\lambda)=a(\lambda)$ for all $\lambda\in C$, and $b(t,\lambda)|_{X^0}=\tilde h(t,\lambda)|_{X^0}$ for all $(t,\lambda)\in[0,1]\times C$. Therefore, $\tilde h(0,\lambda)|_{X^0}=\mathrm{Id}|_{X^0}$ and $\tilde h(0,\lambda):X^0\oplus Y_0\to X^0\oplus Y_0$ has the following block-matrix representation:

$$\begin{bmatrix}\mathrm{Id} & 0\\ M_1(\lambda) & M(\lambda)\end{bmatrix}$$

where $M(\lambda)$ is an automorphism of Y_0 and $M_1(\lambda): X^0 \to Y_0$. We define a continuous mapping $\hat{h}: [0, 1] \times C \to L_{\mathcal{P}}(X)$ by

$$\hat{h}(t, \lambda) = \begin{bmatrix} \mathrm{Id} & 0 \\ tM_1(\lambda) & M(\lambda) \end{bmatrix}$$

Combining $\tilde{h}$ with the above mapping $\hat{h}$, we obtain a continuous mapping $h: [0, 1] \times C \to GL_{\mathcal{P}}(X)$ satisfying the required conditions. □

Let $\{(X_n, X^n)\}$, $n = 1, 2, \ldots$ be a sequence of closed subspaces of X such that:

(i) $\dim X_n = n$.
(ii) $X_n \subseteq X_{n+1}$.
(iii) $X_n \oplus X^n = X$.
(iv) $X^{n+1} \subseteq X^n$.

If X is an infinite-dimensional space, then we can easily construct such a sequence $\{(X_n, X^n)\}$.

We define the inclusions $i_{n,n+k}: GL(n, \mathbb{R}) \to GL(n + k, \mathbb{R})$ and $i_n: GL(n, \mathbb{R}) \simeq GL(X_n) \to GL_{\mathcal{P}}(X)$ by $i_{n,n+k}(T): \mathbb{R}^n \times \mathbb{R}^k \to \mathbb{R}^n \times \mathbb{R}^k$, $i_{n,n+k}(T)(v_1, v_2) = (Tv_1, v_2)$, where $v_1 \in \mathbb{R}^n$, $v_2 = \mathbb{R}^k$, $T \in GL(n, \mathbb{R})$, and $i_n(T)(x_1, x_2) = (\tilde{T}x_1, x_2)$, with $\tilde{T} \in GL(X_n)$ being the operator corresponding to T under the identification $GL(n, \mathbb{R}) \simeq GL(X_n)$, and $x = x_1 + x_2 \in X_n \oplus X^n$. Then we obtain the mapping $i: \lim\limits_{n\to\infty} GL(n, \mathbb{R}) \to GL_{\mathcal{P}}(X)$, where $\lim\limits_{n\to\infty} GL(n, \mathbb{R})$ is the inductive limit of the direct system of spaces $\{GL(n, \mathbb{R}), i_{n,n+k}\}$.

Theorem 4.4.6

The mapping $i: \lim\limits_{n\to\infty} GL(n, \mathbb{R}) \to GL_{\mathcal{P}}(X)$ is a weak homotopy equivalence. That is, it induces the isomorphisms of all homotopy groups of $\lim\limits_{n\to\infty} GL(n, \mathbb{R})$ and $GL_{\mathcal{P}}(X)$.

Proof. We need the following result in order to prove Theorem 4.4.6.

Lemma 4.4.7

Let C be a compact set and $a: C \to GL(\mathbb{R}^n \times \mathbb{R}^k \times \mathbb{R}^n)$ a continuous map defined by the following block-matrix:

$$a(\lambda) = \begin{bmatrix} \mathrm{Id} & 0 & 0 \\ 0 & & \\ & A(\lambda) & \\ 0 & & \end{bmatrix}, \qquad \lambda \in C$$

Then there exists a continuous deformation $H:[0,1]\times C\to GL(\mathbb{R}^n\times\mathbb{R}^k\times\mathbb{R}^n)$ *such that:*

(i) $H(0,\lambda)=a(\lambda)$, $\lambda\in C$.
(ii) $H(1,\lambda)$ *has the following block-matrix representation:*

$$H(1,\lambda)=\begin{bmatrix} & & 0\\ & \tilde{A}(\lambda) & \\ & & 0\\ 0 & 0 & \mathrm{Id}\end{bmatrix},\qquad \lambda\in C$$

Proof. Let $\{e_1,\ldots,e_n\}$ be an orthonormal basis of $\mathbb{R}^n$ and $\{f_1,\ldots,f_n\}$ another orthonormal basis of $\mathbb{R}^n$. We define the vectors

$$e_i^t=\cos\left(\frac{\pi t}{2}\right)e_i-\sin\left(\frac{\pi t}{2}\right)f_i\,,\qquad i=1,2,\ldots,n$$

$$f_j^t=\sin\left(\frac{\pi t}{2}\right)e_j+\cos\left(\frac{\pi t}{2}\right)f_j\,,\qquad j=1,2,\ldots,n$$

where $t\in[0,1]$. It is clear that $\{e_1^t,\ldots,e_n^t,f_1^t,\ldots,f_n^t\}$ form an orthonormal basis of $\mathbb{R}^n\oplus\mathbb{R}^n$ for all $t\in[0,1]$. We define the mapping $A:[0,1]\to GL(\mathbb{R}^n\times\mathbb{R}^k\times\mathbb{R}^n)$ by $A(t)|_{\mathbb{R}^k}=\mathrm{Id}|_{\mathbb{R}^k}$, $A(t)e_i=e_i^t$, $A(t)f_j=f_j^t$, $t\in[0,1]$, $i,j=1,\ldots,n$. It is clear that $A(t)$ is an orthogonal matrix. We define $H:[0,1]\times C\to GL(\mathbb{R}^n\times\mathbb{R}^k\times\mathbb{R}^n)$ by $H(t,\lambda)=A(t)^{-1}\circ a(\lambda)\circ A(t)$, $(t,\lambda)\in[0,1]\times C$. Then $H(0,\lambda)=a(\lambda)$ and $A(1)^{-1}\circ a(\lambda)\circ A(1)f_i=A(1)^{-1}\circ a(\lambda)e_i=A(1)^{-1}e_i=f_i$, $i=1,\ldots,n$, $\lambda\in C$. Therefore, the homotopy $H(t,\lambda)$ satisfies the required conditions. □

Returning to the proof of Theorem 4.4.6, we need to show that the homomorphism

$$i_*:\lim_{n\to\infty}\pi_l(GL(n,\mathbb{R}))\to\pi_l(GL_{\mathscr{P}}(X))$$

is an isomorphism for all $l=1,2,\ldots$.

i_* *is injective:* Let $\sigma:S^l\to i(\lim_{n\to\infty}GL(n,\mathbb{R}))\subseteq GL_{\mathscr{P}}(X)$ be a continuous mapping and let $\tilde{h}:[0,1]\times S^l\to GL_{\mathscr{P}}(X)$ be a homotopy between σ and a constant mapping. By Theorem 4.4.5, there exists a homotopy $h:[0,1]\times S^l\to GL_{\mathscr{P}}(X)$ between σ and a constant mapping such that $h(t,\lambda)=\mathrm{Id}+T(t,\lambda)$, where $T(t,\lambda)$ is a finite-dimensional operator such that $X=X^0\oplus Y_0$ with $\dim Y_0=n+k<\infty$ and $h(t,\lambda)|_{X^0}=\mathrm{Id}|_{X_0}$ for $(t,\lambda)\in[0,1]\times S^l$. The map $h(t, \lambda) : X \to X$ has the following form:

$$
\begin{array}{c}
\qquad\quad \tilde{\mathbf{X}}^{\mathbf{0}} \quad \tilde{\mathbf{X}}_{\mathbf{n}} \quad \mathbf{Y}_{\mathbf{0}} \cap \mathbf{X}_{\mathbf{n+k}} \qquad\qquad \tilde{\mathbf{Y}}_{\mathbf{0}} \\
h(t, \lambda) = \left[\begin{array}{ccccc}
\mathrm{Id} & 0 & 0 & & 0 \\
0 & \mathrm{Id} & 0 & & 0 \\
0 & 0 & & & \\
 & & & \mathrm{Id} + T(t, \lambda) & \\
0 & 0 & & &
\end{array}\right]
\begin{array}{l}
\tilde{\mathbf{X}}^{\mathbf{0}} \\
\tilde{\mathbf{X}}_{\mathbf{n}} \\
\mathbf{Y}_{\mathbf{0}} \cap \mathbf{X}_{\mathbf{n+k}} \\
\\
\tilde{\mathbf{Y}}_{\mathbf{0}}
\end{array}
\end{array}
$$

where $\tilde{X}^0$, $\tilde{X}_n$, $\tilde{Y}_0$ are defined by the following decompositions:

$$(Y_0 \cap X_{n+k}) \oplus \tilde{Y}_0 = Y_0 , \qquad (Y_0 \cap X_{n+k} \oplus \tilde{X}_n = X_{n+k} , \qquad X^0 = \tilde{X}^0 \oplus X^{n+k}$$

Notice that $\dim(\tilde{X}_n) = \dim(\tilde{Y}_0) = n$. Therefore, by Lemma 4.4.7, there exists a deformation of the homotopy h to a homotopy of the type

$$
\left[\begin{array}{ccccc}
\mathrm{Id} & 0 & & 0 & 0 \\
0 & & & & 0 \\
 & & \mathrm{Id} + \tilde{T}(t, \lambda) & & \\
0 & & & & 0 \\
0 & 0 & & 0 & \mathrm{Id}
\end{array}\right]
$$

Therefore, σ is homotopic to a constant map in $i(\lim\limits_{n \to \infty} GL(n, \mathbb{R}))$ and the injectivity of i^* follows. Similarly, one can show that i^* is an epimorphism. □

Corollary 4.4.8

The following inclusions are all weak homotopic equivalences:

$$\lim_{n \to \infty} GL(n, \mathbb{R}) \subseteq GL_{\mathcal{F}}(X) \subseteq GL_c(X) \subseteq GL_{\mathcal{D}}(X) \subseteq GL_{\mathcal{C}}(X)$$

In particular, for $\mathcal{P} = \mathcal{F}$, $\mathcal{K}$, $\mathcal{D}$, or $\mathcal{C}$ the space $GL_{\mathcal{P}}(X)$ has two connected components $GL^+_{\mathcal{P}}(X)$ (which contains Id*) and $GL^-_{\mathcal{P}}(X)$.*

Note that Corollary 4.4.8 may be regarded as an infinite-dimensional analogue of Lemma 2.2.1.

In what follows, we will develop regular approximations and a regular value formula for C^1-Darbó fields that will permit us to give an analytic formula for the Nussbaum–Sadovskii degree.

Assume that X and Y are two Banach spaces and let $f : X \to Y$ be a mapping of class C^r $(r \geq 1)$. $x \in X$ is called a *critical point* of f if $Df(x) : X \to Y$ is not surjective. If $y = f(x)$ for some critical point of f, then y is called a *critical value* of f. A point $x \in X$ is called *regular* if it is not critical, and $y \in Y$ is called a *regular value* of f if it is not a critical value of f.

Definition 4.4.9 A C^1-mapping $f: X \to Y$ is called a *Fredholm map* if for every $x \in X$, the operator $Df(x): X \to Y$ is a Fredholm operator. Since f is of class C^1, index $(Df(x))$ does not depend on $x \in X$. This index is called the *index of the Fredholm map* f and is denoted by index (f).

Theorem 4.4.10 (SARD–SMALE THEOREM)

Let X and Y be two separable Banach spaces and $f: X \to Y$ a Fredholm map of class C^r, $1 \le r \le \infty$, such that $r >$ index(f). Then the set of regular values of f contains a dense G_δ-subset of Y.*

Proof. To establish the above result, we need the following two technical results.

Lemma 4.4.11

Let X and Y be two separable Banach spaces and $f: X \to Y$ a Fredholm map of class C^r such that $r >$ index(f). Then the set of critical points of f is closed in X.

Proof. It is sufficient to show that the set of regular points of f is open in X. Let $x \in X$ be a regular point of f. Let $W = \operatorname{Ker} Df(x)$. Since f is a Fredholm map, a $\dim W = \operatorname{index}(f) < \infty$. Let $P: X \to W$ be a projection onto W. The operator $I(x): X \to W \oplus Y$, $I(x)v = (Pv, Df(x)v)$, $v \in X$, is a bijective bounded operator. Hence, by the open mapping theorem, $I(x)$ is an isomorphism. Let $Q: W \oplus Y \to Y$ be the projection onto Y and define $\Lambda: L(X, W \oplus Y) \to L(X, Y)$ by $\Lambda(A) = Q \circ A$ for $A \in L(X, W \oplus Y)$. It is clear that Λ is a surjective linear operator. Therefore, it is an open mapping. Since $\operatorname{Iso}(X, W \oplus Y)$ is open in $L(X, W \oplus Y)$, $\mathcal{U} := \Lambda(\operatorname{Iso}(X, W \oplus Y))$ is open in $L(X, Y)$ and consists of only surjective operators. As $Df(x) = Q \circ I(x) \in \mathcal{U}$, $(Df)^{-1}(\mathcal{U})$ is an open neighborhood of x such that for all $x' \in (Df)^{-1}(\mathcal{U})$, $Df(x')$ is a surjective linear operator. The result then follows. □

Lemma 4.4.12

Let X and Y be two separable Banach spaces and $f: X \to Y$ a Fredholm map of class C^r such that $r > 0$. Then every $x \in X$ has a closed neighborhood such that f restricted to this neighborhood is a closed mapping.

Proof. Give $x \in X$. We denote by $W = \operatorname{Ker} Df(x)$, $V = \operatorname{Im} Df(x)$. Let $P: X \to W$ and $Q: Y \to V$ be two projections onto W and V, respectively. The operator $Av = (Pv, Df(x)v)$, $v \in X$, is an isomorphism from X onto

* Recall that a G_δ-set is the intersection of countably many open sets.

$W \oplus V$, and it is the derivative at x of the mapping $g(v) = (Pv, Qf(v))$, $v \in X$. It follows from the inverse function theorem that there exists a local coordinate system $w = \psi(v)$ in a neighborhood of x such that $Q \circ f(t_1, \ldots, t_n, w) = w$, $(t_1, \ldots, t_n) \in W$, $w \in V$. Consequently, f can be written as $f(t_1, \ldots, t_n, w) = (\varphi_1(t, w), \ldots, \varphi_p(t, w), w)$, $t = (t_1, \ldots, t_n)$, and $p - n = \text{index}(f)$. Let U be a closed neighborhood of x in which the coordinates $w = \psi(v)$ are defined. Let $\{x_n\}$ be a sequence in U such that $f(x_n) \to \bar{y} \in Y$ as $n \to \infty$. Put $x_n = (t^n, w^n)$; then $Q \circ f(x_n) = Q \circ f(t^n, w^n) = w^n \to Q(\bar{y})$ as $n \to \infty$. Since $\{t^n\}$ is a bounded sequence in W, it contains a subsequence $\{t^{n_k}\}$ convergent to $\bar{t}$. It is clear that $f(\bar{t}, Q(\bar{y})) = \bar{y}$. □

Returning to the proof of Theorem 4.4.10, we choose a local coordinate system $w = \psi(v)$ in a neighborhood U in the same way as in the proof of Lemma 4.4.12. That is, $f(t_1, \ldots, t_n, w) = (\varphi_1(t, w), \ldots, \varphi_p(t, w), w)$. Then $x = (t, w)$ is a critical point of f if and only if t is a critical point of $\varphi(\cdot, w) = (\varphi_1(\cdot, w), \ldots, \varphi_p(\cdot, w))$, $\varphi(\cdot, w) : Q(U) \subseteq \mathbb{R}^n \to \mathbb{R}^p$, and $y \in f(U)$ is a critical value of f if and only if $y - Q(y)$ is a critical value of $\varphi(\cdot, w)$, where $w = Q(y)$.

By Sard's theorem (Lemma 1.3.10), for every $w \in V$, the set of critical values of $\varphi(\cdot, w)$ has Lebesgue measure zero and, hence, is nowhere dense. That is, it has an empty interior. In other words, for every $w \in V$, the set of $y \in Q^{-1}(w)$, which are critical values of f in U, has an empty interior. Consequently, the set $f(\Sigma_U)$ of all critical values of f in U has an empty interior. Since Σ_U is closed in U, $f(\Sigma_U)$ is closed. On the other hand, X can be covered by a countable family of closed neighborhoods U satisfying the above properties; thus, the set of critical values of f is the complement of the intersection of a countable number of open dense subsets of Y. By Baire's theorem, we complete the proof. □

Theorem 4.4.13

Let $F : X \to X$ be a Darbó mapping of class C^1; then $f := \text{Id} - F$ is a Fredholm map of index zero.

Proof. Since $L_{\mathscr{C}}(X) \subseteq \Phi_0(X)$, it is sufficient to show that $DF(x)$ is a Darbó mapping. Assume that F is a μ-Lipschitzian map with constant $k < 1$. Let $A \subseteq X$ be a bounded set such that $A \subseteq B_r(0)$ for sufficiently large $r > 0$. Let $\varepsilon > 0$ be an arbitrary number. Then there is $\delta > 0$ such that if $\|v\| < \delta$, then

$$\|f(x + v) - f(x) - Df(x)v\| \le \varepsilon \|v\|$$

Consequently,

$$Df(x)\left(\frac{\delta}{r} A\right) \subseteq B\left(f\left(x + \frac{\delta}{r} A\right) - f(x), \delta\varepsilon\right)$$

This implies that

$$\begin{aligned}\frac{\delta}{r}\,\mu(Df(x)(A)) &\le \mu\Big(B\Big(f\Big(x+\frac{\delta}{r}A\Big)-f(x),\delta\varepsilon\Big)\Big)\\ &\le \mu\Big(f\Big(x+\frac{\delta}{r}A\Big)-f(x)\Big)+\delta\varepsilon\mu(B_1(0))\\ &\le k\mu\Big(\frac{\delta}{r}A\Big)+\delta\varepsilon\mu(B_1(0))\\ &=\frac{\delta}{r}k\mu(A)+\delta\epsilon\mu(B_1(0))\end{aligned}$$

That is, $\mu(Df(x)(A)) \le k\mu(A) + \varepsilon r\mu(B_1(0))$. Since $\varepsilon > 0$ is arbitrary, we get $\mu(Df(x)(A)) \le k\mu(A)$ and the conclusion follows. □

We can now present the following result that provides an analytic formula for the computation of the Nussbaum–Sadovskii degree for Darbó fields.

Theorem 4.4.14

Let Ω be an open, bounded, nonempty subset of X and let $f:\overline{\Omega}\to X$ be an *Ω-admissible Darbó field such that $f|_{\Omega}$ is of class C^1. Let $\varepsilon := \inf\{\|f(x)\|;\ x\in\partial\Omega\}$. Then there exists a regular value $y_0 \in B_\varepsilon(0)$ of $f|_\Omega$, and the Darbó field $f_0 := f - y_0$, which has zero as a regular value, satisfying*

$$\deg(f,\Omega) = \deg(f_0,\Omega)$$

Moreover, we have the following regular value formula:

$$\deg(f_0,\Omega) = \sum_{x\in f_0^{-1}(0)} \operatorname{sign} Df_0(x)$$

where sign $Df_0(x) = 1$ *if* $Df_0(x) \in GL^+_{\mathscr{D}}(X)$, *and* -1 *otherwise.*

Proof. This is an immediate consequence of Theorem 4.4.10, Corollary 4.4.13, and the additivity property of the degree. □

EXERCISES

4.4.1 Show that $GL_{\mathscr{F}}(X)$, $GL_{\mathscr{K}}(X)$, $GL_{\mathscr{D}}(X)$, and $GL_{\mathscr{C}}(X)$ are ANRs. *Hint:* Use the fact that $GL_{\mathscr{P}}(X)$ is an open subset of the convex set $L_{\mathscr{P}}(X)$, where $\mathscr{P} = \mathscr{F}, \mathscr{K}, \mathscr{D}$, or $\mathscr{C}$.

4.4.2 Show that all the following inclusions are homotopic equivalences:

$$\lim_{n\to\infty} GL(n,\mathbb{R}) \subseteq GL_{\mathscr{F}}(X) \subseteq GL_c(X) \subseteq GL_{\mathscr{D}}(X) \subseteq GL_{\mathscr{C}}(X)$$

Hint: Use Theorem 1.5.7 and the result from Exercise 4.4.1.

4.4.3 Give an example of a C^1-mapping $f : X \to X$, where X is a Banach space, such that the set of regular values of f is not dense in X.

4.4.4 Let X be a Banach space, Ω an open, bounded, nonempty subset of $\mathbb{R}^n \oplus X$, and $f : \Omega \to X$ a Darbó field of class C^1 such that $y \in f(\Omega) \subset X$ is a regular value of f. Show that $f^{-1}(y)$ has a natural structure of an orientable n-dimensional manifold. *Hint:* Follow the same idea as that in the proof of Theorem 1.1.8.

4.5 APPLICATION: BVPs OF NEUTRAL EQUATIONS

In what follows, we assume that T, $r > 0$ and $b \in \mathbb{R}$ are given constants and $\varphi \in C^1([-r, 0]; \mathbb{R})$. For any mapping $x : [-r, T] \to \mathbb{R}$ and $t \in [0, T]$, x_t denotes the mapping from $[-r, 0]$ to $\mathbb{R}$ defined by $x_t(s) = x(t + s)$ for $s \in [-r, 0]$. We denote by $C^i = C^i([0, T]; \mathbb{R})$ for $i = 0, 1, 2$, and $C_\varphi = \{x \in C^0;\ x(0) = \varphi(0)\}$ that is a closed affine subspace of C^0. We will fix a linear structure on the subspace C_φ. For any given $x \in C_\varphi$, let $\Phi x \in C([-r, T]; \mathbb{R})$ be the unique extension of x to $[-r, T]$ with $(\Phi x)_0 \equiv \varphi$. Evidently, $(\Phi x)_t \in C([-r, 0]; \mathbb{R})$ for any $x \in C_\varphi$ and $t \in [0, T]$. Moreover, for any $y \in C^0$, let Ψy be a mapping from $[-r, T]$ to $\mathbb{R}$ defined by

$$(\Psi y)(t) = \begin{cases} \dot{\varphi}(t) & t \in [-r, 0] \\ y(t) & t \in (0, T] \end{cases}$$

Obviously, $(\Psi y)_t$ belongs to $PC([-r, 0]; \mathbb{R})$, the set of all piecewise continuous functions from $[-r, 0]$ to $\mathbb{R}$.

Consider the following second-order neutral functional differential equation:

$$\frac{d^2}{dt^2}(Dx_t) = F(t, x_t, \dot{x}_t) \tag{4.5.1}$$

subject to the following two-point boundary value conditions:

$$x_0 = \varphi \tag{4.5.2}$$

$$Dx_T = b \tag{4.5.3}$$

where $D : C([-r, 0]; \mathbb{R}) \to \mathbb{R}$ is a bounded linear operator and $F : [0, T] \times C([-r, 0]; \mathbb{R}) \times PC([-r, 0]; \mathbb{R}) \to \mathbb{R}$ is a mapping satisfying the condition that for any $x \in C_\varphi$ and $y \in C^0$, the function $t \in [0, T] \to F(t, (\Phi x)_t, (\Phi y)_t) \in \mathbb{R}$ is continuous.

In order to reformulate the above boundary value problem as a fixed-point problem, we introduce the linear operator $\mathscr{L} : C^2 \to C^0 \times \mathbb{R} \times \mathbb{R}$ by $\mathscr{L}y = (\ddot{y} - \varepsilon y, y(0), y(T))$, which is an isomorphism for every sufficiently small $\varepsilon > 0$.

Let $X_{\varphi,b} := C^0 \times \{D\varphi\} \times \{b\}$ and $Y_{\varphi,b} := \mathscr{L}^{-1}(X_{\varphi,b})$. We define the following mappings:

$$\tilde{F} : C_\varphi \times C^0 \to C^0 , \qquad \text{by } \tilde{F}(u, v)(t) = F(t, (\Phi u)_t, (\Psi v)_t) - \varepsilon D(\Phi u)_t$$

$$\mathcal{F} : C_\varphi \times C^0 \to X_{\varphi,b} , \qquad \text{by } \mathcal{F}(u, v) = (\tilde{F}(u, v), D\varphi, b)$$

$$\mathcal{D} : C_\varphi \to C^0 , \qquad \text{by } \mathcal{D}u(t) = D(\Phi u)_t$$

where $t \in [0, T]$, $u \in C_\varphi$, and $v \in C^0$. Moreover, we put

$$C_D^{1,2} = \{x \in C_\varphi \cap C^1;\ D(\Phi x)_T = b \text{ and } \mathcal{D}x \in C^2\}$$

and define

$$j : C^1 \to C^0 \times C^0 \qquad \text{by} \quad j(u) = (u, \dot{u})$$

Then we have the following diagram:

$$\begin{array}{ccccc}
C_\varphi \times C^0 & \xrightarrow{\mathcal{F}} & X_{\varphi,b} & \xrightarrow{\subset} & C^0 \times \mathbb{R} \times \mathbb{R} \\
\Big\uparrow j & & \simeq \Big\downarrow \mathcal{L}^{-1} & & \simeq \Big\uparrow \mathcal{L} \\
C_D^{1,2} & \xrightarrow{\mathcal{D}} & Y_{\varphi,b} & \xrightarrow{\subset} & C^2
\end{array}$$

The boundary value problem (4.5.1–4.5.3) is equivalent to the following equation:

$$\mathcal{D}x = \mathcal{L}^{-1} \circ \mathcal{F} \circ j(x) , \qquad x \in C_D^{1,2} \tag{4.5.4}$$

In order to obtain a fixed-point reformation of (4.5.4), we notice that by the Riesz representation theorem there exists a constant α and a function $\eta : [-r, 0] \to \mathbb{R}$ of bounded variation such that

$$D\psi = \alpha\psi(0) - \int_{-r}^{0} d\eta(\theta)\psi(\theta) \qquad \text{for} \quad \psi \in C([-r, 0]; \mathbb{R})$$

We further assume:

(H1) The operator $D : C([-r, 0]; \mathbb{R}) \to \mathbb{R}$ is *atomic* at zero, that is, $\alpha \neq 0$ and $\mathrm{Var}_{[-s,0]}\eta \to 0$ as $s \to 0$.

Without loss of generality, we may assume that $\alpha = 1$. Therefore by introducing $\tilde{G} : C_\varphi \to C^0$, defined by $\tilde{G}(u)(t) = \int_{-r}^{0} d\eta(\theta)(\Phi u)(t + \theta)$, for $t \in$

$[0, T]$ and $u \in C_\varphi$, the coincidence equation (4.5.4) can be rewritten as the following fixed-point equation:

$$x = \tilde{G}x + \mathcal{L}^{-1} \circ \mathcal{F} \circ j(x)\,, \qquad x \in C_D^{1,2} \tag{4.5.5}$$

We will also assume the following:

(H2) If $x \in C^1$, then $\tilde{G}x \in C^1$ and

$$\frac{d}{dt}[(\tilde{G}x)(t)] = \int_{-r}^{0} d\eta(\theta)[\Psi x](t+\theta)$$

Consequently, by defining $G : X := C_\varphi \times C^0 \to C^0 \times C^0$ as

$$G(u, v)(t) = \left(\int_{-r}^{0} d\eta(\theta)(\Phi u)(t+\theta),\ \int_{-r}^{0} d\eta(\theta)(\Psi v)(t+\theta) \right) \tag{4.5.6}$$

we obtain from (4.5.5) the equivalent equation

$$j(x) = G(j(x)) + j \circ \mathcal{L}^{-1} \circ \mathcal{F}(j(x))\,, \qquad x \in C_D^{1,2} \tag{4.5.7}$$

This indicates that if $x \in C_D^{1,2}$ is a solution to the boundary value problem (4.5.1–4.5.3), then $w = j(x)$ solves the following fixed-point equation:

$$w = G(w) + j \circ \mathcal{L}^{-1} \circ \mathcal{F}(w)\,, \qquad w \in X := C_\varphi \times C^0 \tag{4.5.8}$$

The following result indicates the the converse is also true.

Theorem 4.5.1

Under the assumptions (H1) and (H2), the mapping $G + j \circ \mathcal{L}^{-1} \circ \mathcal{F} : C_\varphi \times C^0 \to C_\varphi \times C^0$ is well defined and the boundary value problem (4.5.1–4.5.3) has a solution if and only if the fixed-point equation (4.5.8) has a solution in $X := C_\varphi \times C^0$.

Proof. It suffices to prove that if $w \in X := C_\varphi \times C^0$ is a solution to Eq. (4.5.8), then $w = (u, \dot{u})$ for some $u \in C_D^{1,2}$ and Φu satisfies (4.5.1–4.5.3). In fact, $w - G(w) = j \circ \mathcal{L}^{-1} \circ \mathcal{F}(w)$ implies the existence of $\xi \in C^2$ such that $w - G(w) = (\xi, \dot{\xi})$. Let $w = (u, v)$. Then we have

$$u(t) - \int_{-r}^{0} d\eta(\theta)(\Phi u)(t+\theta) = \xi(t) \tag{4.5.9}$$

and

$$v(t) - \int_{-r}^{0} d\eta(\theta)(\Psi v)(t+\theta)) = \dot{\xi}(t) \tag{4.5.10}$$

Let

$$f(t) = \begin{cases} \varphi(t) & \text{for } t \in [-r, 0] \\ \varphi(0) + \int_0^t v(\theta)\, d\theta & \text{for } t \in [0, T] \end{cases}$$

Then $f \in C([-r, T]; \mathbb{R})$ and $f|_{[0,T]} \in C^1$.

It is easy to verify that

$$\frac{d}{dt}\left[f(t) - \int_{-r}^{0} d\eta(\theta) f(t+\theta) \right] = v(t) - \int_{-r}^{0} d\eta(\theta)(\Psi v)(t+\theta) \tag{4.5.11}$$

where $t \in [0, T]$. Therefore, by (4.5.9) and (4.5.11), for $m = f - \Phi u$, one has

$$\frac{d}{dt}\left[m(t) - \int_{-r}^{0} d\eta(\theta) m(t+\theta) \right] = 0\,, \qquad m_0 \equiv 0 \tag{4.5.12}$$

By (H1), D is atomic at zero. Thus, the fundamental theory of neutral equations [see, e.g., Hale (1977)] implies that the solution to (4.5.12) is unique. Therefore, $f = \Phi u$. This implies $v(t) = \dot{f}(t) = \dot{u}(t)$ on $[0, T]$, $w = j(u)$, and (4.5.8) can be reduced to

$$ju = G(j(u)) + j \circ \mathcal{L}^{-1} \circ \mathcal{F}(j(u))$$

which shows that u is a solution to the boundary value problem (4.5.1–4.5.3). □

In order to solve the fixed-point equation (4.5.8) by a homotopy argument, we consider the following family of fixed-point equations:

$$w = G(w) + \lambda j \circ \mathcal{L}^{-1} \circ \mathcal{F}(w) + (1-\lambda)(y_0, y_0') \tag{4.5.13$_\lambda$}$$

where $\lambda \in [0, 1]$ and y_0 is the unique solution to the following two-point boundary value problem:

$$\begin{cases} y'' - \varepsilon y = 0 \\ y(0) = D\varphi \\ y(T) = b \end{cases}$$

Using similar arguments as in the proof of Theorem 4.5.1, we can prove that the problem (4.5.13$_\lambda$) is equivalent to the following boundary value problem:

$$(4.5.14_\lambda)\quad \begin{cases} \dfrac{d^2}{dt^2}(Dy_t) - \varepsilon Dy_t = \lambda[F(t, y_t, \dot{y}_t) - \varepsilon Dy_t]\,, & t \in [0, T] \\ y_0 = \varphi \\ Dy_T = b \end{cases}$$

Our next goal is to provide some a priori bounds for solutions of the equation $(4.5.14_\lambda)$ for $\lambda \subset [0, 1]$.

Lemma 4.5.2

Suppose $\mathrm{Var}_{[-r,0]}\eta < 1$. *Then for any* $y \in C([-r, \tau]; \mathbb{R})$ *and* $M \geq 0$, *if*

$$\|y_0\| \leq \frac{M}{1 + \mathrm{Var}_{[-r,0]}\eta}$$

and $\max\{|Dy_s|; 0 \leq s \leq \tau\} \leq M$, *then*

$$\max_{0 \leq s \leq \tau} |y(s)| \leq \frac{M}{1 - \mathrm{Var}_{[-r,0]}\eta}$$

where $\tau > 0$ *is any given constant and* $\|\cdot\|$ *denotes the supremum norm in the space* $C([-r, 0]; \mathbb{R})$.

Proof. Let $C_0 = \{\varphi \in C((-\infty, 0]; \mathbb{R});\ \lim_{t \to -\infty} \varphi(t) = 0\}$ and $M_0 = \{v : [0, \infty) \to \mathbb{R};\ v$ is measurable and $\int_0^\infty d|v|(s) < \infty\}$, where $|v|$ denotes the total variation of v. For any $\varphi \in C_0$ and $v \in M_0$, we define $\|\varphi\|_0 = \sup_{t \leq 0} |\varphi(t)|$ and $\|v\|_0 = \int_0^\infty d(|v|(s)$. Evidently, the convolution operator $*$ defined by

$$v * \varphi(t) = \int_0^\infty \varphi(t - s)\, dv(s)\,, \qquad t \leq 0$$

maps C_0 into itself if $v \in M_0$.

For the given $\eta : [-r, 0] \to \mathbb{R}$, we define $\bar{\eta} : [0, \infty) \to \mathbb{R}$ by $\bar{\eta}(\theta) = \eta(-\theta)$ for $\theta \in [0, r]$ and $\bar{\eta}(\theta) = \eta(-r)$ for $\theta \geq r$. Obviously, $\bar{\eta} \in M_0$. If $\varphi \in C([-r, 0]; \mathbb{R})$ is extended to $(-\infty, 0]$ such that the extended function is in C_0, then $D\varphi = \mu * \varphi(0)$, where $\mu = \delta - \bar{\eta}$ and $\delta(0) = 0$, $\delta(t) = 1$ for $t > 0$.

Since $\mathrm{Var}_{[-r,0]}\eta < 1$, one can apply the contraction principle to show that $\mu *$ maps C_0 into itself continuously, it is invertible, and its inverse operator $\mu^{-1} *$ is continuous. For a general result, we refer to Staffans (1983).

For a given $y \in C([-r, \tau]; \mathbb{R})$, we now extend it to $\mathbb{R}$ such that $\lim_{t\to-\infty} y(t) = 0$ and $|y(t)| \le \|y_0\|$ for $t \in (-\infty, +\infty)$. Select a function $w \in C(\mathbb{R}; \mathbb{R})$ with $w(t) = y(t)$ for $t \in (-\infty, 0]$ and $|w(t)| \le \|y_0\|$ for $t \in [0, \infty)$. Then $z_t \in C_0$ for any $t \in \mathbb{R}$, where $z \in C(\mathbb{R}; \mathbb{R})$ is defined by $z(t) = y(t) - w(t)$ and $z_t(s) = z(t+s)$, $s \le 0$. Let

$$g(t) = \begin{cases} 0, & t \le 0 \\ Dy_t - \mu * w_t(0), & t > 0 \end{cases}$$

Then $\mu * z_t(0) = g(t)$ and $z(t) = \mu^{-1} * g_t(0)$, from which it follows that

$$\begin{aligned} |y(t) &\le |w(t) + z(t)| \\ &\le \left| w(t) + \int_0^t [d\mu^{-1}(s)] g(t-s) \right| \\ &\le \left| w(t) + \int_0^\infty [d\mu^{-1}(s)] \mu * w_{t-s}(0)\, ds \right| \\ &\quad + \left| \int_t^\infty [d\mu^{-1}(s)] \mu * w_{t-s}(0) \right| + \left| \int_0^t [d\mu^{-1}(s)] Dy_{t-s} \right| \\ &\le \left| \int_t^\infty [d\mu^{-1}(s)] \mu * w_{t-s}(0) \right| + \left| \int_0^t [d\mu^{-1}(s)] Dy_{t-s} \right| \\ &\le \|\mu\|_0 \|y_0\| \int_t^\infty d|\mu^{-1}|(s) + \max_{0\le s\le t} |Dy_s| \int_0^t d|\mu^{-1}|(s) \end{aligned}$$

Note that

$$\|\mu\|_0 \le 1 + \mathrm{Var}_{[-r,0]}\eta$$

$$\|\mu^{-1}\|_0 = \|(\delta - \bar{\eta})^{-1}\|_0 \le \sum_{k=0}^{\infty} \|\bar{\eta}\|_0^k = \frac{1}{1 - \mathrm{Var}_{[-r,0]}\eta}$$

Therefore, if $\|y_0\| \le \dfrac{M}{1 + \mathrm{Var}_{[-r,0]}\eta}$ and $\max_{0\le s\le r} |Dy_s| \le M$, then

$$\begin{aligned} |y(t)| &\le M \int_t^\infty d|\mu^{-1}|(s) + M \int_0^t d|\mu^{-1}|(s) \\ &\le M\|\mu^{-1}\|_0 \le \frac{M}{1 - \mathrm{Var}_{[-r,0]}\eta} \end{aligned}$$

This completes the proof. □

In what follows, we shall always assume:

(H3) $\text{Var}_{[-r,0]}\eta < 1$.

To obtain a priori bounds on the solution $y(t)$ of the boundary value problem $(4.5.14_\lambda)$, we assume that:

(H4) There exists a constant $R \geq (1 + Var_{[-r,0]}\eta)\|\varphi\|$ such that

$$F(t, (\Phi u)_t, (\Psi \dot{u})_t) \cdot D(\Phi_u)_t > 0$$

for $t \subset [0, T]$ and $u \in C_D^{1,2}$ satisfying $|D(\Phi u)_t| > R$, $D(\Psi \dot{u})_t = 0$ and $\|(\Phi u)_t\| \leq \dfrac{|D(\Phi u)_t|}{1 - \text{Var}_{[-r,0]}\eta}$.

Lemma 4.5.3

Assume that (H1–H4) hold and let y be a solution to the boundary value problem $(4.5.14_\lambda)$ for some $\lambda \in [0, 1]$. Then $|Dy_t| \leq \max\{R. |b|\}$ and

$$|y(t)| \leq \frac{\max\{R, |b|\}}{\text{Var}_{[-r,0]}\eta}$$

for $t \in [0, T]$.

Proof. It is clear that $|Dy_0| = |D\varphi| \leq (1 + \text{Var}_{[-r,0]}\eta)\|\varphi\| \leq R$ and $|Dy_T| = |b|$. Therefore, if $|Dy_t| \not\leq \max\{R, |b|\}$ on the interval $[0, T]$, then there exists $t_0 \in (0, T)$ such that $\max\{|Dy_s|;\ 0 \leq s \leq t_0\} = |Dy_{t_0}| > R$, $\dfrac{d}{dt}(Dy_t)|_{t=t_0} = 0$ and $\dfrac{d^2}{dt^2}(Dy_t)^2|_{t=t_0} \leq 0$. Since $\|y_0\| = \|\varphi\| \leq \dfrac{|Dy_{t_0}|}{1 + \text{Var}_{[-r,0]}\eta}$ and $\max\{|Dy_s|;$ $0 \leq s \leq t_0\} = |Dy_{t_0}|$, by Lemma 4.5.2, we obtain $\|y_{t_0}\| \leq \dfrac{|Dy_{t_0}|}{1 - \text{Var}_{[-r,0]}\eta}$. It follows from assumption (H4) that $F(t_0, y_{t_0}, \dot{y}_{t_0})Dy_{t_0} > 0$. On the other hand, at $t = t_0$, one has

$$\begin{aligned}
\frac{1}{2}\frac{d^2}{dt^2}(Dy_t)^2\bigg|_{t=t_0} &= \left(\frac{d}{dt}(Dy_t)\bigg|_{t=t_0}\right)^2 + (Dy_{t_0})\frac{d^2}{dt^2}(Dy_t)\bigg|_{t=t_0} \\
&= (Dy_{t_0})[\lambda F(t_0, y_{t_0}, \dot{y}_{t_0}) + (1-\lambda)\varepsilon(Dy_{t_0})] \\
&= \lambda F(t_0, y_{t_0}, \dot{y}_{t_0})Dy_{t_0} + (1-\lambda)\varepsilon(Dy_{t_0})^2
\end{aligned}$$

and this leads to the following inequalities:

$$\frac{d^2}{dt^2}(Dy_t)^2\Big|_{t=t_0} \geq \lambda F(t_0, y_{t_0}, \dot{y}_{t_0})Dy_{t_0} > 0 \qquad \text{for } \lambda \in (0, 1]$$

and

$$\frac{d^2}{dt^2}(Dy_t)^2\Big|_{t=t_0} = \varepsilon(Dy_{t_0})^2 > 0 \qquad \text{for } \lambda = 0$$

which is contrary to the assumption $\frac{d^2}{dt^2}(Dy_t)^2\Big|_{t=t_0} \leq 0$. Therefore, we observe that $|Dy_t| \leq \max\{R, |b|\}$ for all $t \in [0, T]$ and, hence, by Lemma 4.5.2, we get

$$\|y_t\| \leq \frac{\max\{R, |b|\}}{1 - \mathrm{Var}_{[-r,0]}\eta}$$

This completes the proof. □

To obtain an a priori bound for $|\dot{y}(t)|$, where $y(t)$ is a solution to $(4.5.14_\lambda)$, we assume the following Nagumo growth condition:

(H5) There is a nondecreasing, continuous, locally Lipschitz function $N : [0, \infty) \to (0, \infty)$ such that $\int_0^\infty \frac{s\,ds}{N(s)} = \infty$ and $\max\{R, |b|\} + |F(t, (\Phi u)_t), (\Psi v)_t)| \leq N(\|\Psi v)_t\|)$ for $t \in [0, T]$ and all $(u, v) \in C_\varphi \times C^0$ with $\sup_{t\in[0,T]} |u(t)| \leq \frac{\max\{R, |b|\}}{1 - \mathrm{Var}_{[-r,0]}\eta}$, where R is the same as in condition (H4).

Theorem 4.5.4

Assume that conditions (H1–H5) are satisfied. Then for any solution $y(t)$ to the boundary value problem $(4.5.14_\lambda)$ with $\lambda \in [0, 1]$, we have

$$|Dy_t| \leq \max\{R, |b|\} \qquad \textit{for } t \in [0, T]$$

$$|y(t)| \leq \frac{\max\{R, |b|\}}{1 - \mathrm{Var}_{[-r,0]}\eta} \qquad \textit{for } t \in [0, T]$$

$$|D\dot{y}_t| \leq M \qquad \textit{for } t \in [0, T]$$

and

$$|\dot{y}(t)| \leq \frac{M}{1 - \mathrm{Var}_{[-r,0]}\eta} \qquad \textit{for } t \in [0, T]$$

where $M > 0$ is a constant such that

$$M > \max\left\{\frac{1}{T}|b - D\varphi|, (1 + \mathrm{Var}_{[-r,0]}\eta)\|\dot{\varphi}\|\right\} \tag{4.5.15}$$

and

$$\int_{H}^{(1-\mathrm{Var}_{[-r,0]}\eta)^{-1}M} \frac{s}{N(s)}\,ds > \frac{2\max\{R, |b|\}}{(1 - \mathrm{Var}_{[-r,0]}\eta)^2} \tag{4.5.16}$$

with

$$H = (1 - \mathrm{Var}_{[-0r,0]}\eta)^{-1}\max\left\{\frac{1}{T}|b - D\varphi|, (1 + \mathrm{Var}_{[-r,0]}\eta)\|\dot{\varphi}\|\right\} \tag{4.5.17}$$

Proof. There exists a point $\tau \in (0, T)$ such that $TD\dot{y}_\tau = Dy_T - Dy_0 = b - D\varphi$, so $|D\dot{y}_\tau| = 1/T|b - D\varphi|$. Let $z(t) = \max\{|D\dot{y}_s|;\ s \in [0, t]\}$. Then $z(t)$ is nondecreasing on $[0, T]$. If $\max\{|D\dot{y}_s|;\ s \in [0, T]\} > M$, then there must exist t_1 and t_2 with $0 < t_1 < t_2 < T$ such that

$$z(t_1) = \max\left\{\frac{1}{T}|b - D\varphi|, (1 + \mathrm{Var}_{[-r,0]}\eta)\|\dot{\varphi}\|\right\}$$

$$z(t_2) = \max\{|D\dot{y}_t|; t \in [0, T]\} > M$$

and

$$z(t_1) < z(t) < z(t_2) \qquad \text{on} \quad (t_1 t_2)$$

For any given $t \in [t_1, t_2)$, we may distinguish two cases.

CASE 1. There exists $s \in [0, t)$ such that $z(t) = |D\dot{y}_s| > |D\dot{y}_t|$.

CASE 2. $|D\dot{y}_s| \le |D\dot{y}_t|$ for all $0 \le s \le t$.

In the first case, it is clear that

$$\frac{d^+}{dt}z(t) := \overline{\lim_{h \to 0^+}}\frac{z(t+h) - z(t)}{h} = 0$$

In the second case, we can find a sequence $h_n \to 0^+$ as $n \to \infty$ such that

$$\frac{d^+}{dt}z(t) = \lim_{n\to\infty}\frac{z(t+h_n) - z(t)}{h_n} = \lim_{n\to\infty}\frac{z(t+h_n) - |D\dot{y}_t|}{h_n}$$

If there are infinitely many n's such that $z(t+h_n) \le |D\dot{y}_t|$, then $\frac{d^+}{dt}z(t) \le 0$. If except for finitely many n's $z(t+h_n) > |D\dot{y}_t|$, then by the definition of $z(t)$, one can find $s_n \in (0, h_n]$ such that $z(t+h_n) = |D\dot{y}_{s_n+t}| > |D\dot{y}_t|$, and therefore,

$$\begin{aligned}\frac{d^+}{dt} z(t) &= \lim_{n\to\infty} \frac{|D\dot{y}_{s_n+t}| - |D\dot{y}_t|}{s_n} \cdot \frac{s_n}{h_n} \\ &\le \frac{d^+}{dt} |D\dot{y}_t| \le \left| \frac{d^2}{dt^2}(Dy_t) \right| \le N(\|\dot{y}_t\|)\end{aligned}$$

by the Nagumo growth condition (H5). Notice that in the second case, $|D\dot{y}_s| \le |D\dot{y}_t|$ for $0 \le s \le t$ and

$$\|\dot{\varphi}\| \le \frac{M}{1+\mathrm{Var}_{[-r,0]}\eta} \le \frac{|D\dot{y}_t|}{1+\mathrm{Var}_{[-r,0]}\eta}$$

Therefore, by Lemma 4.5.2, we have

$$\|\dot{y}_t\| \le \frac{|D\dot{y}_t|}{1+\mathrm{Var}_{[-r,0]}\eta}$$

and hence,

$$\frac{d^+}{dt} z(t) \le N\left(\frac{|D\dot{y}_t|}{1-\mathrm{Var}_{[-r,0]}\eta}\right) = N\left(\frac{z(t)}{1-\mathrm{Var}_{[-r,0]}\eta}\right)$$

Therefore, by the above arguments, we obtain

$$\dot{z}(t) \le \frac{1}{z(t)} N\left(\frac{z(t)}{1-\mathrm{Var}_{[-r,0}\eta}\right) g(t) \qquad \text{for } t \in [t_1, t_2] \tag{4.5.18}$$

where $g : [t_1, t_2] \to [0, \infty)$ is defined by

$$g(t) = \begin{cases} 0 & \text{if } |D\dot{y}_t| < z(t) \\ |D\dot{y}_t| & \text{if } |D\dot{y}_t| = z(t) \end{cases}$$

Let $r(t)$ denote the maximal solution of the following initial value problem:

$$\begin{cases} \dot{r}(t) = \dfrac{1}{r(t)} N\left(\dfrac{r(t)}{1-\mathrm{Var}_{[-r,0]}\eta}\right) g(t) \\ r(t_1) = z(t_1) \end{cases} \tag{4.5.19}$$

By the well-known comparison prinicple [see, e.g., Theorem 1.10.1 in Ladas and Lakshmikanthan (1972)], we have $z(t) \le r(t)$ for all $t \in [t_1, t_2]$ such that $r(t)$ exists. Therefore, there must be $t^* \in (t_1, t_2]$ such that

$$r(t^*) > M \tag{4.5.20}$$

On the other hand, from the equality (4.5.19) it follows that

$$\int_{r(t_1)}^{r(t')} \frac{r\,dr}{N\left(\frac{r}{1-\mathrm{Var}_{[-r,0]}\eta}\right)} = \int_E |D\dot{y}_t|\,dt$$

where

$$E = \{t \in [t_1, t^*]\,; \qquad |D\dot{y}_t| = z(t)\}$$

Therefore, we obtain

$$\int_{[1-\mathrm{Var}_{[-r,0]}\eta]^{-1}r(t_1)}^{[1-\mathrm{Var}_{[-r,0]}\eta]^{-1}r(t^*)} \frac{s\,ds}{N(s)}\cdot[1-\mathrm{Var}_{[-r,0]}\eta]^2 = \int_E |D\dot{y}_t|\,dt \tag{4.5.21}$$

Since $z(t)$ is continuous and nondecreasing on $[t_1, t^*]$, there exist at most a countable number of disjoint intervals $[\tau_i, \tau_i']$, $i = 1, 2, \ldots$, such that $z(t)$ is increasing only on each interval $[\tau_i, \tau_i']$ for $i = 1, 2, \ldots$ Therefore,

$$E = \bigcup_{i=1}^{\infty} [\tau_i, \tau_i']$$

and hence,

$$\int_E |D\dot{y}_t|\,dt = \sum_{i=1}^{\infty} \int_{\tau_i}^{\tau_i'} |D\dot{y}_t|\,dt = \sum_{i=1}^{\infty} \left| \int_{\tau_i}^{\tau_i'} D\dot{y}_t\,dt \right|$$

$$= \sum_{i=1}^{\infty} |Dy_{\tau_i'} - Dy_{\tau_i}| = \sum_{i=1}^{\infty} [|Dy_{\tau_i'}| - |Dy_{\tau_i}|]$$

We notice that if intervals $[\tau_i, \tau_i']$, $i = 1, 2, \ldots$, are in natural order, that is, $\tau_i' < \tau_{i+1}$ for $i = 1, 2, \ldots$, then for any $k \geq 2$, we get

$$\sum_{i=1}^{k} [|Dy_{\tau_i'}| - |Dy_{\tau_i}|] = |Dy_{\tau_k'}| - |Dy_{\tau_1}| \leq 2\max\{R, |b|\}$$

Therefore,

$$\int_E |D\dot{y}_t|\,dt \leq 2\max\{R, |b|\} \tag{4.5.22}$$

In the case where the intervals $[\tau_i, \tau_i']$, $i = 1, 2, \ldots$, are not in natural order, the series $\sum_{i=1}^{\infty} [|Dy_{\tau_i'}| - |Dy_{\tau_i}|]$ is a series of positive numbers (since $|Dy_{\tau_i'}| - |Dy_{\tau_i}| > 0$. So, one can rearrange the order of the intervals to get a new sequence of intervals in natural order and the sum of the series is not sequence of intervals in natural order and the sum of the series is not changed. Therefore, (4.5.22) holds. This implies, by equality (4.5.21), that

$$\int_{[1-\mathrm{Var}_{[-r,0]}\eta]^{-1}r(t_1)}^{[1-\mathrm{Var}_{[-r,0]}\eta]^{-1}r(t^*)} \frac{s\,ds}{N(s)} \le 2[1-\mathrm{Var}_{[-r,0]}\eta]^{-2}\max\{R,|b|\}$$

and then from (4.5.20), we obtain

$$\int_{H}^{(1-\mathrm{Var}_{[-r,0]}\eta)^{-1}M} \frac{s\,ds}{N(s)} < 2(1-\mathrm{Var}_{[-r,0]}\eta)^{-2}\max\{R,|b|\}$$

but this is a contradiction of assumption (4.5.16).

Therefore, $|D\dot{y}_t| \le M$ for all $t \in [0, T]$. Now, by using the assumption (4.5.15), we can again apply Lemma 4.5.2 to obtain

$$\|\dot{y}_t\| \le \frac{M}{1-\mathrm{Var}_{[-r,0]}\eta} \qquad \text{for } t \in [0, T]$$

This completes the proof. □

We are now in the position to state the main result of this section.

Theorem 4.5.5

Suppose that hypotheses (H1–H5) hold. Then there exists a solution to the boundary value problem (4.5.1–4.5.3).

Proof. Let E denote the Banach space $C^0 \times C^0$ with the supremum norm $\|\cdot\|$. Then $X := C_\varphi \times C^0$ is a closed affine subspace of E that, we assume, has a fixed vector space structure. Consequently, X is a Banach space. Using the same notation as above, we define

$$U = \left\{ (u, v) \in C_\varphi \times C^0;\ \|u\| < \frac{\max\{R, |b|\} + \max_{0\le s\le T}|y_0(s)|}{1-\mathrm{Var}_{[-r,0]}\eta} + 1 \right.$$

$$\left. \|v\| < \frac{M + \max_{0\le s\le T}|\dot{y}_0(s)|}{1-\mathrm{Var}_{[-r,0]}\eta} + 1 \right\}$$

Evidently, $U \subseteq X$ is an open subset. Let $A = \overline{U}$ and $B = \partial U$. Define $H : \overline{U} \times [0, 1] \to X$ by $H(w, \lambda) = G(w) + \lambda j \circ \mathcal{L}^{-1} \circ \mathcal{F}(w) + (1-\lambda)(y_0, \dot{y}_0)$, where $(w, \lambda) \in A \times]0, 1]$.

Since the mapping $G : C_\varphi \times C^0 \to C^0 \times C^0$ is an α-set contraction with constant $\mathrm{Var}_{[-r,0]}\eta < 1$ and $j \circ \mathcal{L}^{-1} \circ \mathcal{F} : C_\varphi \times C^0 \to C^0 \times C^0$ is completely continuous, H is an α-set contraction. Now, according to Theorem 4.5.1, if w is a fixed point of $H(\cdot, \lambda)$ for some $\lambda \in [0, 1]$, then $w = (u, \dot{u})$ and u solves the boundary problem (4.5.14$_\lambda$) and, hence, by Theorem 4.5.4, $(u, \dot{u}) \in A \backslash B$. This implies that H is a homotopy in $\mathscr{D}(A, B)$ between $G_0 := G + (y_0, \dot{y}_0)$ and $G + j \circ \mathscr{L}^{-1} \circ \mathscr{F}$.

On the other hand, we define $Q(w, \lambda) = \lambda G(w) + \lambda(y_0, \dot{y}_0) + (1 - \lambda)(\bar{\varphi}(0), 0)$ for any $(w, \lambda) \in A \times [0, 1]$, where $\bar{\varphi}(0)$ is a constant mapping from $[0, T]$ to $\mathbb{R}$ with the value $\varphi(0)$. Then Q maps $A \times [0, 1]$ into X and is an α-contraction. It is easy to verify that if w is a fixed point of $G(\cdot, \lambda)$ for some $\lambda \in [0, 1]$, then $w = (u, v)$ with

$$u(t) = \lambda \int_{-r}^{0} d\eta(\theta)(\Phi u)(t + \theta) + \lambda y_0(t) + (1 - \lambda)\varphi(0) , \qquad t \geq 0$$

and

$$v(t) = \lambda \int_{-r}^{0} d\eta(\theta)(\Psi v)(t + \theta) + \lambda \dot{y}_0(t) , \qquad t \geq 0$$

from which it follows that

$$\begin{aligned} |u(t) &\leq \max\left\{ \frac{\max_{0 \leq s \leq t} |y_0(s)| + |\varphi(0)|}{1 - \mathrm{Var}_{[-r,0]}\eta)}, \|\varphi\| \right\} \\ &< \frac{\max_{0 \leq s < t} |y_0(s)| + \max\{R, |b|\}}{1 - \mathrm{Var}_{[-r,0]}\eta} + 1 \end{aligned}$$

and

$$|v(t)| \leq \max\left\{ \frac{\max_{0 \leq s \leq t} |\dot{y}_0(s)}{1 - \mathrm{Var}_{[-r,0]}\eta}, \|\dot{\varphi}\| \right\} < \frac{M + \max_{0 \leq s \leq t} |\dot{y}_0(s)|}{1 - \mathrm{Var}_{[-r,0]}\eta} + 1$$

for $t \in [0, T]$. Therefore, $w \in A \backslash B$. This implies that Q is a homotopy in $\mathscr{D}(A, B)$ between G_0 and $G_2 : A \to X$ defined by $G_2(u, v) = (\overline{\varphi(0)}, 0)$. Clearly, $\deg(\mathrm{Id} - G_2, U) = 1$. So, $\deg(\mathrm{Id} - G - j \circ \mathscr{L}^{-1} \circ \mathscr{F}, U) = 1$. Therefore, there is a fixed point in U of $G + j \circ \mathscr{L}^{-1} \circ \mathscr{F}$, which implies that there is a solution to the boundary value problem (4.5.1–4.5.3) by Theorem 4.5.1. The proof is complete. □

EXERCISES

This set of exercises shows how a neutral equation is obtained from a nonlinear, mixed, initial boundary value problem of hyperbolic equations and how Theorem 4.5.5 can be applied to this neutral equation. We consider the current and voltage in a transmission line terminated at each end by linear or nonlinear circuit elements. We refer the interested reader to Brayton (1966, 1967), Brayton and Miranker (1964), and Cooke and Krumme (1968) for some discussion on the history and physical background of the subject.

Take an x-axis in the direction of the line, with the ends of the line at $x = 0$ and $x = 1$. Let $i(x, t)$ denote the current flowing in the line at time t and distance x down the line, and let $v(x, t)$ denote the voltage across the line at t and x. It is well known that the function v and i satisfy the partial differential equations

$$\begin{cases} \dfrac{\partial v}{\partial x} = -\left(Ri + L\dfrac{\partial i}{\partial t}\right) + e \\ \dfrac{\partial i}{\partial x} = -\left(Gv + C\dfrac{\partial v}{\partial t}\right) \end{cases}$$

where R, L, G, and C are the resistance, inductance, conductance, and capacitance per unit length of the line, and e is the voltage per unit length impressed along the line in series with it. We are interested in the lossless transmission line, that is, $R = G = 0$ and $e = 0$. Therefore, the basic equations are

$$\text{(4.5.23)} \qquad \begin{cases} \dfrac{\partial v}{\partial x} = -L\dfrac{\partial i}{\partial x} \\ \dfrac{\partial i}{\partial x} = -C\dfrac{\partial v}{\partial t} \end{cases}$$

The initial condition is

$$\text{(4.5.24)} \qquad v(x, 0) = u_0(x)\,, \qquad i(x, 0) = i_0(x)\,, \qquad 0 \le x \le 1$$

We further assume that the line is terminated by parallel and series circuits on the left and parallel circuits on the right; then we get the following boundary conditions:

$$\text{(4.5.25)} \qquad \begin{cases} v(0, t) + ri(0, t) = f_1(t) \\ \dfrac{d}{dt} v(1, t) = r_0(v(1, t)) + \displaystyle\int_0^t r_2(v(1, s))\, ds + \delta_0 i(1, t) + f_2(t) \end{cases}$$

where r and δ_0 are constants, and r_0, r_2, f_1, and f_2 are continuously differentiable functions. We are interested in the existence of solutions to the problem (4.5.23–4.5.25) subject to the conditions $v(1, \theta) = \varphi(\theta)$,

$-\frac{2}{\sqrt{LC}} \leq \theta \leq 0$, and $v(1, a) = b$ for any given $\varphi \in C^1\left(\left[-\frac{2}{\sqrt{LC}}, 0\right]; \mathbb{R}\right)$, $a \in \left[0, \frac{2}{\sqrt{LC}}\right]$, and $b \in \mathbb{R}$.

It is well known that each solution to Eq. (4.5.23) can be represented as a superposition of traveling waves moving to the left and right with velocity $s = \frac{1}{\sqrt{LC}}$, that is,

$$\begin{Bmatrix} v(x, t) = \psi_1(x - st) + \psi_2(x + st) \\ i(x, t) = \frac{1}{z}[\psi_1(x - st) - \psi_2(x + st)] \end{Bmatrix}$$

where $z = \sqrt{\frac{L}{C}}$, from which it follows:

$$\begin{cases} 2\psi_1(-st) = v\left(1, t + \frac{1}{s}\right) + zi\left(1, t + \frac{1}{s}\right) \\ 2\psi_2(st) = v\left(1, t - \frac{1}{s}\right) - zi\left(1, t - \frac{1}{s}\right) \end{cases}$$

4.5.1 Use these expressions in the general solution and the first boundary condition at $t - 1/s$ to obtain

$$i(1, t) - Ki\left(1, t - \frac{2}{s}\right) = \frac{2}{z + r} f_1\left(t - \frac{1}{s}\right) - \tfrac{1}{2}v(1, t) - \frac{K}{z} v\left(1, t - \frac{2}{s}\right)$$

where $K = \frac{z - r}{z + r}$. Use the second boundary condition and let $u(t) = v(1, t)$ to obtain the following boundary value problem:

$$(4.5.26) \qquad \begin{cases} \frac{d^2}{dt^2}\left[u(t) - Ku\left(t - \frac{2}{s}\right)\right] = H(t, u_t, \dot{u}_t) \\ u(\theta) = \varphi(\theta), \qquad -\frac{2}{s} \leq \theta \leq 0, \quad \varphi \in C^1\left(\left[-\frac{2}{s}, 0\right]; \mathbb{R}\right) \\ u(a) = b \in \mathbb{R} \end{cases}$$

where

$$F(t) = f_2'(t) - Kf_2'\left(t - \frac{2}{s}\right) + \frac{2\delta_0}{z + r} f_1'\left(t - \frac{1}{s}\right)$$

and $H : [0, a] \times (C([-2/s, 0]; \mathbb{R}))^2 \to \mathbb{R}$ is defined by

$$H(t, \varphi, \psi) = F(t) + r_0'(\varphi(0))\psi(0) - Kr_0'\left(\varphi\left(-\frac{2}{s}\right)\right)\psi\left(-\frac{2}{s}\right) + r_2(\varphi(0))$$
$$- Kr_2\left(\varphi\left(-\frac{2}{s}\right)\right) - \frac{\delta_0}{z}\psi(0) - \frac{K}{z}\delta_0\psi\left(-\frac{2}{s}\right)$$

where $(\varphi, \psi) \in \left(C\left(\left[-\frac{2}{s}, 0\right]; \mathbb{R}\right)\right)^2$.

4.5.2 Use Theorem 4.5.4 to show that if

(i) $|r_0'(x)| \leq M$ for a constant $M \geq 0$ and for $x \in \mathbb{R}$

(ii)
$$\lim_{|x-Ky|\to\infty} \frac{[r_2(x) - Kr_2(y)][x - Ky]}{|x - Ky|} = +\infty$$

then the boundary value problem (4.5.26) has a solution.

4.6 COMPOSITE COINCIDENCE DEGREE

This section obtains a variation of degrees for the study of a nonlinear problem of the type

$$L(\mathrm{Id} - B)(x) = G(x)\,, \qquad x \in \Omega \tag{4.6.1}$$

where X and Y are Banach spaces, $\Omega \subseteq X$ is an open and bounded subset, $L\colon \mathrm{Dom}(L) \subseteq X \to Y$ is a Fredholm operator of index zero, $B : \overline{\Omega} \to X$ is a condensing mapping, and $G : \overline{\Omega} \to Y$ is a mapping satisfying a certain compactness condition to be specified later.

Assume that $L : \mathrm{Dom}(L) \subseteq X \to Y$ is a Fredholm operator of index zero. A *compact resolvent* of L is a compact linear operator $K : X \to Y$ such that $L + K\colon \mathrm{Dom}(L) \to Y$ is a bijection. We will denote by $CR(L)$ the set of all compact resolvents of L. For any $K \in CR(L)$, we define $R_K := (L + K)^{-1} : Y \to X$. Since L is a closed operator, R_K is a bounded linear operator. One can easily verify that $CR(L) \neq \varnothing$. Also, if $K_1, K_2 \in CR(L)$ and $R_i := (L + K_i)^{-1}$, $i = 1, 2$, then

$$R_1(L + K_2) = \mathrm{Id}|_{\mathrm{Dom}(L)} + R_1(K_2 - K_1) \tag{4.6.2}$$

$$R_1 = R_2 + R_1(K_2 - K_1)R_2 \tag{4.6.3}$$

where $\mathrm{Id}|_{\mathrm{Dom}(L)} : \mathrm{Dom}(L) \to \mathrm{Dom}(L)$ denotes the identity map.

Let $\Omega \subseteq X$ be a bounded, open subset such that $\mathrm{Dom}(L) \cap \Omega \neq \varnothing$ and assume that $B : \overline{\Omega} \to X$ and $G : \overline{\Omega} \to Y$ are two continuous mappings. We are interested in the following *composite coincidence problem*:

$$(4.6.4) \qquad \begin{cases} \text{Find } x \in \overline{\Omega} \quad \text{such that } x - B(x) \in \text{Dom}(L) \text{ and} \\ L(x - B(x)) = G(x) \end{cases}$$

The problem (4.6.4) is equivalent to the following fixed-point problem:

$$(4.6.5) \qquad x = B(x) + R_K[G + K(\text{Id} - B)](x), \qquad x \in \overline{\Omega}$$

where $R_K = (L + K)^{-1}$, $K \in CR(L)$. Indeed, the problem (4.6.4) is equivalent to the equation

$$(4.6.6) \quad L(x - B(x)) + K(x - B(x)) = G(x) + K(x - B(x)), \qquad x \in \overline{\Omega}$$

whenever $x - B(x) \in \text{Dom}(L)$. Clearly, it can be transformed to (4.6.5). Conversely, $R_K[G + K(\text{Id} - B)](x) \in \text{Dom}(L)$. So, (4.6.5) implies that $x - B(x) \in \text{Dom}(L)$ and (4.6.4) holds.

Equation (4.6.5) motivates the introduction of the following transformation $\Theta_K(B, G)$ defined by

$$(4.6.7) \qquad \Theta_K(B, G) := B + R_K[G + K(\text{Id} - B)] : \overline{\Omega} \to X$$

With the use of this transformation, problem (4.6.4) may be written in the following form:

$$(4.6.8) \qquad \begin{cases} \text{Find a fixed point } x \in \overline{\Omega} \text{ of } \Theta_K(B, G), \text{ that is,} \\ x = \Theta_K(B, G)(x), \qquad x \in \overline{\Omega} \end{cases}$$

In what follows, we call Θ_K a *connection*.

Note that for $K_1, K_2 \in CR(L)$, $R_i = (L + K_i)^{-1}$, $i = 1, 2$, by using (4.6.2) and (4.6.3), we can get

$$\begin{aligned} \Theta_{K_1}(B, G) = \Theta_{K_2}(B, G) &+ R_1(K_2 - K_1)(\text{Id} - B) \\ &+ R_2(K_1 - K_2)R_1[G + K_2(\text{Id} - B)] \end{aligned}$$

Consequently,

$$(4.6.9) \quad \Theta_{K_1}(B, G) = \Theta_{K_2}(B, G) + R_1(K_2 - K_1)\Theta_{K_2}(B, G) + R_1(K_1 - K_2)$$

Definition 4.6.1 Let A be a closed subset of X, $B : A \to X$ and $G : A \to Y$ be given mappings. We say that the pair (B, G) is an *L-condensing pair* if the map $\Theta_K(B, G) : \overline{\Omega} \to X$ with some $K \in CR(L)$ is a condensing mapping.

It follows from (4.6.9) that the above definition does not depend on the choice of the resolvent $K \in CR(L)$.

Definition 4.6.2 Let Ω be a bounded, open subset of X and (B, G) be an L-condensing pair such that $x \neq \Theta_K(B, G)(x)$ for every $x \in \partial\Omega$. Then we say that the pair (B, G) is an *Ω-admissible pair*.

It is clear that if (B, G) is an Ω-admissible pair, then problem (4.6.4) has no solution $x \in \partial\Omega$.

Definition 4.6.3 Let (A, A_0) be a pair of closed, bounded subsets of X such that $\mathrm{Dom}(L) \cap A \neq \emptyset$. We denote by $C_L(A, A_0)$ the class of all L-condensing pairs (B, G), $B : A \to G : A \to Y$, such that $x \neq \Theta_K(B, G)(x)$ for every $x \in A_0$. We say that (B_t, G_t), $t \in [0, 1]$, is a *homotopy* in $C_L(A, A_0)$ if $B_t = B(t, \cdot\,)$, $G_t = G(t, \cdot\,)$, where $B : [0, 1] \times A \to X$, $G : [0, 1] \times A \to Y$ are continuous mappings such that $\Theta_K(B, G) : [0, 1] \times A \to X$ is a condensing map and $(B_t, G_t) \in C_L(\overline{A, A_0})$ for all $t \in [0, 1]$. For a bounded open set $\Omega \subseteq X$, a homotopy in $C_L(\overline{\Omega}, \partial\Omega)$ will be called an *Ω-admissible homotopy*.

We denote by $C_L[A, A_0]$ the set of all homotopy classes in $C_L(A, A_0)$. From the definition, $\Theta_K : C_L(A, A_0) \to \mathscr{C}(A, A_0)$ and, hence, it induces a map $\Theta_K^* : C_L[A, A_0] \to \mathscr{C}[A, A_0]$.

Now we are in a position to introduce the so-called composite coincidence degree for problem (4.6.4).

Let Ω be an open, bounded subset of X and let (B, G), $B : \overline{\Omega} \to X$, and $G : \overline{\Omega} \to Y$, be an Ω-admissible pair. We define the *composite coincidence degree* $\deg_L([B, G], \Omega)$ of the pair (B, G) and the operator L in the set Ω by

$$\deg_L([B, G], \Omega) := \deg(\mathrm{Id} - \Theta_K(B, G), \Omega) \tag{4.6.10}$$

where $K \in CR(L)$ is a fixed resolvent of L.

The definition (4.6.10) depends on the choice of the resolvent K. In fact, one can easily show that if K_1, $K_2 \in CR(L)$ and if $\mathrm{Id} + R_1 \circ (K_1 - K_2) \in GL_c^+(X)$, then $\deg(\mathrm{Id} - \Theta_{K_1}(B, G), \Omega) = \deg(\mathrm{Id} - \Theta_{K_2}(B, G), \Omega)$, but if $\mathrm{Id} + R_1 \circ (K_1 - K_2) \in GL_c^-(X)$, then $\deg(\mathrm{Id} - \Theta_{K_1}(B, G), \Omega) = -\deg(\mathrm{Id} - \Theta_{K_2}(B, G), \Omega)$. In other words, the definition (4.6.10) is unique up to sign for any choice of $K \in CR(L)$.

It follows from the above definition that the composite coincidence degree has all the standard properties of a degree. In particular, we have the following.

Theorem 4.6.4

The composite coincidence degree satisfies the following properties:

(i) *Additivity: If Ω_1 and Ω_2 are two disjoint subsets of Ω such that $(B, G) \in C_L(\overline{\Omega}, \overline{\Omega} \backslash (\Omega_1 \cup \Omega_2))$, then*

$$\deg_L([B, G], \Omega) = \deg_L([B, G], \Omega_1) + \deg_L([B, G], \Omega_2)$$

(ii) *Excision: If $\Omega_1 \subseteq \Omega$ is an open subset such that $(B, G) \in C_L(\overline{\Omega}, \overline{\Omega} \backslash \Omega_1)$, then $\deg_L([B, G], \Omega) = \deg_L([B, G], \Omega_1)$.*

(iii) *Existence: If $\deg_L([B, G], \Omega) \neq 0$, then there exists $x \in \Omega$ such that $x - B(x) \in \mathrm{Dom}(L)$ and $L(\mathrm{Id} - B)(x) = G(x)$.*

(iv) *Homotopy Invariance: If $(B_t, G_t) \in C_L(\overline{\Omega}, \partial\Omega)$, $t \in [0, 1]$, is an Ω-admissible homotopy, then $\deg_L([B_t, G_t], \Omega)$ is independent of t.*

The following result gives a class of condensing pairs.

Proposition 4.6.5

Suppose that $B : \overline{\Omega} \to X$ is a condensing map and $G : \overline{\Omega} \to Y$ is L-compact, that is, $R_K \circ G : \overline{\Omega} \to X$ is compact, then (B, G) is an L-condensing pair.

Proof. We have $Q_K(B, G) = B + R_K \circ [G + K(\mathrm{Id} - B)]$. Therefore, $Q_K(B, G)$ is a sum of the condensing map B and the compact map $R_K \circ G + R_K \circ K \circ (\mathrm{Id} - B)$ and, hence, is condensing. □

Remark 4.6.6 Let us point out some special situations contained in the definition of the composite coincidence degree:

(i) If $X = Y$ and $\dim X < \infty$, then $\deg_{\mathrm{Id}}([0, G], \Omega) = \deg(\mathrm{Id} - G, \Omega)$ is the *Brouwer degree* in finite-dimensional spaces.

(ii) If $X = Y$ and G is compact, then $\deg_{\mathrm{Id}}([0, G], \Omega) = \deg(\mathrm{Id} - G, \Omega)$ is exactly the *Leray–Schauder degree* of compact fields.

(iii) If $X = Y$ and G is a Darbó or a condensing mapping, then $\deg_{\mathrm{Id}}([0, G], \Omega) = \deg(\mathrm{Id} - G, \Omega)$ is the *Nussbaum–Sadovskii degree* for Darbó or condensing fields.

(iv) If G is an L-compact mapping, then $\deg_L(G, \Omega) := \deg_L([0, G], \Omega)$ is called the *Mawhin coincidence degree* for the coincidence problem $Lx = Gx$, see Mawhin (1979), and Gaines and Mawhin 1977a,b) for details.

(v) If $B(\overline{\Omega}) \subseteq \mathrm{Dom}(L)$ and G is L-compact, then the coincidence degree $\deg_L([B, G], \Omega) =: \deg_L(LB + G, \Omega)$ is the so-called *Hetzer degree for the coincidence problem* $x = LB(x) + G(x)$; see Hetzer (1975a,b,c,d) for details.

We have the following result that will be used later to reduce the solvability of boundary value problems of neutral equations to that of a corresponding retarded of ordinary differential equation.

Corollary 4.6.7

If $G:[0,1]\times\overline{\Omega}\to Y$ is L-compact and $B:\overline{\Omega}\to X$ is condensing such that (tB, G_t), $t\in[0,1]$, is an Ω-admissible homotopy, then $\deg_L([B, G_1],\Omega)=\deg_L(G_0,\Omega)$, *where the latter is the Mawhin coincidence degree of (L, G_0) with respect to Ω.*

Proof. Since $\Theta(tB, G_t)=t\Theta(B, G_t)+(1-t)\Theta(0, G_t)$, by Proposition 4.6.5, (tB, G_t), $t\in[0,1]$, is a homotopy in $C_L(\overline{\Omega},\partial\Omega)$. Consequently, from the homotopy property, $\deg_L([B, G_1],\Omega)=\deg_L([0, G_0],\Omega)=\deg_L(G_0,\Omega)$. □

Theorem 4.6.8 (Borsuk Antipodal Theorem)

Let Ω be an open, bounded, symmetric, convex subset of X such that $\Omega\cap\mathrm{dom}(L)\neq\emptyset$. Let $(B, G)\in C_L(\overline{\Omega},\partial\Omega)$ be such that $G(-x)=-G(x)$, and $B(-x)=-B(x)$ for $x\in\partial\Omega$; then $\deg_L([B, G],\Omega)$ *is an odd integer.*

Proof. It follows immediately from the definition of $\Theta(B, G)$ that if both B and G are odd on $\partial\Omega$, then $\Theta(B, G)(-x)=-\Theta(B, G)(x)$, $x\in\partial\Omega$. Therefore, the result follows from Theorem 4.3.12. □

The following result gives an extension of the *Krasnosiel'skii fixed-point theorem* [see Krasnosiel'skii (1965)] for compact fields.

Theorem 4.6.9

Suppose that Ω is an open, bounded, convex, and symmetric neighborhood of $0\in X$. Assume that $(B, G)\in C_L(\overline{\Omega},\partial\Omega)$ and there is no point $x\in\partial\Omega$ such that $x-Bx+\lambda(x+B(-x))\in\mathrm{Dom}(L)$ and

$$L(x-B(x)+\lambda(x+B(-x))]=G(x)-\lambda G(-x)$$

for $\lambda\in[0,1]$. Then $\deg_L([B, G],\Omega)\neq 0$ *and the composite coincidence problem*

$$L(\mathrm{Id}-B)(x)=G(x)\,,\qquad x\in\overline{\Omega}$$

has a solution in Ω.

Proof. Note that $\Theta(-B,-G)(-x)=-\Theta(B, G)(-x)$ for every $x\in\overline{\Omega}$. Therefore, if we define $B'(x):=-B(-x)$ and $G'(x):=-G(-x)$, then (B', G') is also an L-condensing pair. For $t\in[0,1]$, we define

$$\tilde{B}(t,x)=\frac{1}{1+t}[B(x)-tB(-x)]\,,\qquad \tilde{G}(t,x)=\frac{1}{1+t}[G(x)-tG(-x)]$$

where $x \in \overline{\Omega}$. Since $(\tilde{B}_t, \tilde{G}_t) = \frac{1}{1+t}(B, G) + \frac{t}{1+t}(B', G')$, where $\tilde{B}_t := \tilde{B}(t, \cdot)$, $\tilde{G}_t := \tilde{G}(t, \cdot)$, $(\tilde{B}_t, \tilde{G}_t)$ is an L-condensing pair for every $t \in [0, 1]$. We need to show that $(\tilde{B}_t, \tilde{G}_t)$ is a homotopy in $C_L(\overline{\Omega}, \partial\Omega)$. Assume, by way of contradiction, that $\Theta(\tilde{B}_t, \tilde{G}_t)(x) = x$ for some $(t, x) \in [0, 1] \times \partial\Omega$; then $L[x - \tilde{B}(t, x)] = \tilde{G}(t, x)$, that is, $L[(1+t)x - B(x) + tB(-x)] = G(x) - tG(-x)$. Then $L[x - B(x) + t(x + B(-x))] = G(x) - tG(-x)$, which is a contradiction of the assumption. By the homotopy invariance property, we get $\deg_L([\tilde{B}_1, \tilde{G}_1], \Omega) = \deg_L([B, G], \Omega)$. On the other hand, $\tilde{B}(1, x) - \frac{1}{2}[B(x) - B(-x)]$ and $\tilde{G}(1, x) = \frac{1}{2}[G(x) - G(-x)]$ are odd mappings. Therefore, by Theorem 4.6.8, $\deg_L([\tilde{B}_1, \tilde{G}_1], \Omega) \equiv 1 \pmod 2$. Consequently, $\deg_L([B, G], \Omega) \neq 0$. □

Proposition 4.6.10

Suppose that $P : X \to X$ and $Q : Y \to Y$ are bounded projections onto $\operatorname{Ker} L$ *and* $\operatorname{Im} L$, *respectively, where* $\operatorname{Im} L$ *denotes the range of L. Let $T_P : \operatorname{Im} L \to \operatorname{Dom} L \cap \operatorname{Ker} P$ denote the right inverse of L. Moreover, let $J : \operatorname{Im} Q \to \operatorname{Ker} L$ be an isomorphism. Then $K := J^{-1} \circ P$ is a compact resolvent of L and*

$$\Theta_K(B, G) = B + P(\mathrm{Id} - B) + JQG + T_P(\mathrm{Id} - Q)G$$

Proof. Evidently, $K := J^{-1} \circ P$ is a linear operator with a finite-dimensional range; thus, K is compact. It is an easy observation that $L + K : \operatorname{Dom}(L) \to Y$ is a bijection. Moreover, $R_K Qy = Jy$ and $R_K(\mathrm{Id} - Q)y = T_P y$. Thus, we have

$$\begin{aligned} Q_K(B, G) &= B + R_K \circ [G + J^{-1}P(\mathrm{Id} - B)] \\ &= B + R_K[QG + (\mathrm{Id} - Q)G + J^{-1}P(\mathrm{Id} - B)] \\ &= B + JQG + T_P(\mathrm{Id} - Q)G + P(\mathrm{Id} - B) \quad \square \end{aligned}$$

In what follows, we assume that $P : X \to X$ and $Q : Y \to Y$ are two projections such that $\operatorname{Im} P = \operatorname{Ker} L$ and $\operatorname{Ker} Q = \operatorname{Im} L$. We also fix an isomorphism $J : \operatorname{Im} Q \to \operatorname{Ker} L$. Then, by the above proposition, $K = J^{-1} \circ P$ is a compact resolvent of L. We fix this resolvent and we will consider the composite coincidence degree with respect to the operator K. Moreover, by $T_P : \operatorname{Im} L \to \operatorname{Dom}(L) \cap \operatorname{Ker} P$, we denote the right inverse of L.

The following is a continuation theorem for composite coincidence degrees.

Theorem 4.6.11

Suppose that $y \in \operatorname{Im} L$ is given. Let $B^* : [0, 1] \times \overline{\Omega} \to X$ be condensing and $G^* : [0, 1] \times \overline{\Omega} \to Y$ be L-compact such that:

(**i**) $B^*(1,\cdot)=B$, $G^*(1,\cdot)=G$.

(**ii**) *For every* $\lambda\in(0,1)$, *the equation* $L[\mathrm{Id}-\lambda B^*(\lambda,x)]=\lambda G^*(\lambda,x)+y$ *has no solution in* $\partial\Omega$.

(**iii**) $QG^*(0,x)\neq 0$ *for every* $x\in L^{-1}\{y\}\cup\partial\Omega$.

(**iv**) *The Brouwer degree* $\deg(JQG^*(0,\cdot)|_{L^{-1}\{y\}\cap\overline{\Omega}},\ \Omega\cap L^{-1}\{y\})\neq 0$, *where* $JQG^*(0,\cdot):\overline{\Omega}\to X$ *and we identify* $L^{-1}\{y\}$ *with* $\operatorname{Ker}L$ *by a shifting* $h(x)=x+x_0$, $x_0=R_Ky$.

Then the equation $L(\mathrm{Id}-B)(x)=G(x)+y$ *has at least one solution in* $\overline{\Omega}$.

Proof. Let $\tilde{B}:[0,1]\times\overline{\Omega}\to X$ and $\tilde{G}:[0,1]\times\overline{\Omega}\to Y$ be defined as

$$\tilde{B}(t,x)=tB^*(t,x)\,,\quad \tilde{G}(t,x)=QG^*(t,x)+t(\mathrm{Id}-Q)G^*(t,x)+y$$

for $x\in\overline{\Omega}$ and $t\in[0,1]$. Evidently, $\tilde{B}$ is condensing and $\tilde{G}$ is L-compact.

If $L(x-\tilde{B}(t,x))=\tilde{G}(t,x)$ for some $(t,x)\in(0,1)\times\overline{\Omega}$, then

$$L[x-tB^*(t,x)]=QG^*(t,x)+t(\mathrm{Id}-Q)G^*(t,x)+y$$

Therefore, by assumption (ii), $x\notin\partial\Omega$.

If $t=0$ and $x\in\overline{\Omega}$ are given so that $L[x-\tilde{B}(t,x)]=\tilde{G}(t,x)$, that is, $Lx=QG^*(0,x)+y$, then $QG^*(0,x)=0$ and $Lx=y$. Therefore, by assumption (ii), $x\notin\partial\Omega$.

If $t=1$ and $x\in\partial\Omega$ are given so that $L[x-\tilde{B}(t,x)]=\tilde{G}(t,x)$, that is, $L[x-B(x)]=QG(x)+(\mathrm{Id}-Q)G(x)+y$, then $L(\mathrm{Id}-B)(x)=G(x)+y$ has at least one solution in $\overline{\Omega}$.

Therefore, if there exists no $x\in\partial\Omega$ such that $L(\mathrm{Id}-B)(x)=G(x)+y$, then the above argument indicates that

$$L[\mathrm{Id}-\tilde{B}(t,x)]\neq\tilde{G}(t,x)\qquad\text{for } t\in[0,1] \text{ and } x\in\partial\Omega$$

So, $(\tilde{B}_t,\tilde{G}_t)$ is a homotopy in $C_L(\overline{\Omega},\partial\Omega)$ and, by the homotopy invariance and Proposition 4.6.10, we obtain

$$\begin{aligned}
&\deg_L([B,G],\overline{\Omega})\\
&\quad=\deg_L([0,\tilde{G}_0],\Omega)=\deg_L([0,QG^*(0,\cdot)+y],\Omega)\\
&\quad=\deg_L(QG^*(0,\cdot)+y,\Omega)\quad\text{(Mawhin coincidence degree)}\\
&\quad=\deg(\mathrm{Id}-JQG^*(0,\cdot)-P-x_0,\Omega)\quad\text{(Leray–Schauder degree)}
\end{aligned}$$

where $x_0=R_Ky$. Put $f(x)=x-JQG^*(0,\cdot)-Px-x_0$ and note that

$$\deg(f,\Omega)=\deg(f\circ h,\Omega-\{x_0\})$$

where $h(x) = x + x_0$. It is easy to check that

$$(f \circ h)(x) = x + x_0 - JQG^*(0, x + x_0) - Px - x_0$$
$$= (\mathrm{Id} - P)x - JQG^*(0, x + x_0)$$

where $x \in \overline{\Omega}$. Therefore, by the reduction formula, we get

$$\deg(f \circ h, \Omega - \{x_0\} = \deg(-JQG^*(0, \cdot\,) \circ h|_{\mathrm{Ker}\, L}, (\Omega - \{x_0\}) \cap \mathrm{Ker}\, L)$$
$$= \pm \deg(JQG^*(0, \cdot\,)|_{L^{-1}\{y\}}, \Omega \cap L^{-1}\{y\})$$

Consequently, by assumption (iv),

$$|\deg_L([B, G], \Omega)| = |\deg(JQG^*(0, \cdot\,)|_{L^{-1}\{y\}}, \Omega \cap L^{-1}\{y\})| \neq 0$$

and the conclusion follows. □

EXERCISES

4.6.1 Let Ω be a bounded, open subset in X and $(\overline{B}, \overline{G})$ be an L-condensing Ω-admissible pair. The pair $(\overline{B}, \overline{G})$ is called *L-universal* on Ω, if Problem 4.6.4 has a solution for every L-condensing pair (B, G) such that $G_{|\partial\Omega} \equiv \overline{G}_{|\partial\Omega}$ and $B_{|\partial\Omega} \equiv \overline{B}_{|\partial\Omega}$. Show the following:

- **(i)** If $(B_t, G_t) \in C_L(\overline{\Omega}, \partial\Omega)$ is an Ω-admissible homotopy between (B_0, G_0) and (B_1, G_1), then (B_0, G_0) is L-universal if and only if (B_1, G_1) is L-universal.
- **(ii)** The pair $(\overline{B}, \overline{G})$ is L-universal on a connected, open, bounded set Ω if and only if $\deg_L([\overline{B}, \overline{G}], \Omega) \neq 0$.

4.6.2 Prove the following version of a Fredholm alternative: Let (B, G) be an L-condensing pair of linear operators; then we have:

- **(i)** Either the equation $L(x - Bx) = Gx$ has a nonzero solution.
- **(ii)** Or the equation $L(z - Bx) = Gx + y$ has a unique solution for every $y \in Y$.

4.7 *APPLICATION: PERIODIC SOLUTIONS OF NEUTRAL EQUATIONS*

For a fixed $r \geq 0$, let $C := C([-r, 0]; \mathbb{R}^n)$. If $x \in C([\sigma - r, \sigma + \delta]; \mathbb{R}^n)$ for some $\delta > 0$ and $\sigma \in \mathbb{R}$, then $x_t \in C$ for $t \in [\sigma, \sigma + \delta]$ is defined by $x_t(\theta) = x(t + \theta)$ for $\theta \in [-r, 0]$. The supremum norm in C is denoted by $\|\cdot\|$.

We consider the following neutral functional differential equation:

$$\frac{d}{dt}[a(t) - b(t, x_t)] = f(t, x) \tag{4.7.1}$$

where $f : \mathbb{R} \times C \to \mathbb{R}^n$ is completely continuous and $b : \mathbb{R} \times C \to \mathbb{R}^n$ is continuous. Moreover, we assume:

(H1) *There exists a constant $k < 1$ such that $|b(t, \varphi) - b(t, \psi)| \leq k\|\varphi - \psi\|$ for $t \in \mathbb{R}$ and $\varphi, \psi \in C$.*

(H2) *There exists $\omega > 0$ such that for every $(t, \varphi) \in \mathbb{R} \times C$, we have $b(t + \omega, \varphi) = b(t, \varphi)$ and $f(t + \omega, \varphi) = f(t, \varphi)$.*

We want to find an ω-periodic solution to the problem (4.7.1), that is, we want to find an ω-periodic continuous function $x(t)$ such that $x(t) - b(t, x_t)$ is continuously differentiable and (4.7.1) holds for all $t \in \mathbb{R}$.

Theorem 4.7.1

Suppose that there exists a constant $M > 0$ such that:

(i) *For any $\lambda \in (0, 1)$ and any ω-periodic solution x of the system*

$$\frac{d}{dt}[x(t) - \lambda b(t, x_t)] = \lambda f(t, x_t) \tag{4.7.2$_\lambda$}$$

we have $|x(t)| < M$ for $t \in \mathbb{R}$;

(ii) $g(u) := \int_0^\omega f(s, \hat{u})\, ds \neq 0$ *for any $u \in \partial B_M(\mathbb{R}^n)$, where $B_M(\mathbb{R}^n) = \{u \in \mathbb{R}^n;\ |u| < M\}$, and $\hat{u}$ denotes the constant mapping from $[-r, 0]$ to $\mathbb{R}^n$ with the value $u \in \mathbb{R}^n$.*

(iii) $\deg(g, B_M(\mathbb{R}^n)) \neq 0$.

Then there exists at least one ω-periodic solution of system (4.7.1) that satisfies $\sup_{t \in \mathbb{R}} |x(t)| < M$.

Proof. We define the following spaces:

$$X_\omega := \{x \in C(\mathbb{R}, \mathbb{R}^n);\ x(t + \omega) = x(t) \text{ for } t \in \mathbb{R}\}$$

$$Y_\omega := \{y \in C(\mathbb{R}, \mathbb{R}^n);\ y(0) = 0,\ y(t) = \alpha t + h(t) \text{ for some } \alpha \in \mathbb{R}^n \text{ and } h \in X_\omega\}$$

For any $x \in X_\omega$, let $\|x\| = \sup_{t \in [0,\omega]} |x(t)|$, and for any $y \in Y_\omega$, $y(t) = \alpha t + h(t)$, $\alpha \in \mathbb{R}^n$, $h \in X_\omega$, let $\|y\| = |\alpha| + \|h\|$. It is clear that X_ω and Y_ω are Banach spaces. We define the following mappings:

$$L : X_\omega \to Y_\omega\,, \qquad Lx(t) = x(t) - x(0)$$

$$G : X_\omega \to Y_\omega\,, \qquad Gx(t) = \int_0^t f(s, x_s)\, ds$$

$$B : X_\omega \to X_\omega\,, \qquad Bx(t) = b(t, x_t)\,, \quad x \in X_\omega\,, \quad t \in \mathbb{R}$$

Since $\int_0^t f(s, x_s)\, ds - t/\omega \int_0^\omega f(s, x_s)\, ds$ is an ω-periodic function of $t \in \mathbb{R}$ for any $x \in X_\omega$, G is a well-defined operator from X_ω into Y_ω. It is also easy to prove that L is a continuous Fredholm operator of index zero, B is a Darbó mapping with constant k, and G is a completely continuous mapping. Therefore, finding an ω-periodic solution of Eq. (4.7.1) is equivalent to solving the following composite coincidence problem:

$$L(\mathrm{Id} - B)(x) = G(x)\,, \qquad x \in X_\omega$$

On the other hand, direct verification leads to $\mathrm{Ker}\, L = \{x \in X_\omega;\, x(t) = x(0)$ for $t \in \mathbb{R}\}$ and $\mathrm{Im}\, L = Y_\omega \cap X_\omega$. We define $P : X_\omega \to X_\omega$ and $Q : Y_\omega \to Y_\omega$ by $Px(t) = x(0)$, $x \in X_\omega$, $Qy(t) = \alpha t$, $y(t) = \alpha t + h(t)$, where $\alpha \in \mathbb{R}^n$, $h \in X_\omega$, and $t \in \mathbb{R}$. The operators P and Q are projections and $X_\omega = \mathrm{Ker}\, P \oplus \mathrm{Ker}\, L$, $Y_\omega = \mathrm{Im}\, L \oplus \mathrm{Im}\, Q$. Let $\Omega = \{x \in X_\omega;\, |x(t)| < M$ for $t \in \mathbb{R}\}$. Evidently, the mapping $B^* : [0, 1] \times \overline{\Omega} \to X_\omega$, defined by $B^*(\lambda, x)(t) = b(t, x_t)$ for $x \in X_\omega$, $t \in \mathbb{R}$, and $\lambda \in [0, 1]$, is a Darbó mapping. The mapping $G^* : [0, 1] \times \overline{\Omega} \to Y_\omega$, defined by $G^*(\lambda, x)(t) = \int_0^t f(s, x_s)\, ds$ for $x \in \overline{\Omega}$, $t \in \mathbb{R}$, and $\lambda \in [0, 1]$, is L-compact.

By assumption (i), $L[\mathrm{Id} - \lambda B^*(\lambda, x)] \neq \lambda G^*(\lambda, x)$ for $(\lambda, x) \in (0, 1) \times \partial\Omega$. Since $QG^*(0, x)(t) = t/\omega \int_0^\omega f(s, x_s)\, ds$, $x \in \overline{\Omega}$, $t \in \mathbb{R}$, assumption (ii) implies that $QG^*(0, x) \neq 0$ for $x \in L^{-1}(0) \cap \partial\Omega$. The isomorphism $J : \mathrm{Im}\, Q \to \mathrm{Ker}\, L$ is defined by $J(\alpha t) = \alpha$, $\alpha \in \mathbb{R}^n$. Therefore, from assumption (iii), it follows that $\deg(JQG^*(0, \cdot)|_{\mathrm{Ker}\, L},\ \Omega \cap \mathrm{Ker}\, L) \neq 0$, where $JQG^*(0, x) = 1/\omega \int_0^\omega f(s, \hat{x})\, ds$, $x \in \mathbb{R}^n$. Therefore, the conclusion follows from Theorem 4.6.11. □

In order to apply Theorem 4.7.1, it is crucial to establish a priori bounds for ω-periodic solutions of $(4.7.2_\lambda)$. In what follows, we show that the idea of guiding functions and the Liapunov–Razumikhin technique provide a useful tool to establish a priori bounds.

For $\lambda \in [0, 1]$, $\varphi \in C$, and $t \subset \mathbb{R}$, define $D_\lambda(t, \varphi) = \varphi(0) - \lambda b(t, \varphi)$. The following definition is a generalization to neutral equations, introduced by Erbe et al. (1990), of the guiding functions developed by Krasnosiel'skii and others for ordinary differential equations, and by Mawhin (1979), Hetzer (1975c), Gustafson and Schmitt (1974), Schmitt (1973), and others for retarded equations.

Definition 4.7.2 A continuously differentiable function $V : \mathbb{R}^n \to \mathbb{R}$ is said to be a *guiding function* for the ω-periodic boundary value problem of Eq. (4.7.1), if there exists a constant $\rho > 0$ such that for any $x \in X_\omega$ and $\lambda \in [0, 1]$, at any t such that $|D_\lambda(t, x_t)| \geq \rho$ and $|V(D_\lambda(t, x_t))| \geq |V(D_\lambda(s, x_s))|$ for $s \in \mathbb{R}$, we have

$$\langle \nabla V(D_\lambda(t, x_t)), f(t, x_t) \rangle > 0$$

The following result provides a simple criterion for a function to be a guiding function.

Lemma 4.7.3

A continuous differentiable function $V : \mathbb{R}^n \to \mathbb{R}$ is a guiding function for the ω-periodic boundary value problem of neutral equation (4.7.1) if there exists a constant $\rho > 0$ and a continuous function $p : [0, \infty) \to [0, \infty)$ such that:

(i) *For any given $h > 0$, $\lambda \in [0, 1]$, and any $x \in X_\omega$, if $|V(D_\lambda(t, x_t))| \leq h$ for $t \in \mathbb{R}$, then $|x(t)| \leq p(h)$ for $t \in \mathbb{R}$.*

(ii) *$\langle \nabla V(D_\lambda(t, x_t)), f(t, x_t) \rangle > 0$ for all $x \in X_\omega$, $t \in \mathbb{R}$ with $|D_\lambda(t, x_t)| \geq \rho$ and $\|x_t\| \leq p(V(D_\lambda(t, x_t)))$.*

Proof. Assume $x \in X_\omega$, $\lambda \in [0, 1]$, and $t \in \mathbb{R}$ are given so that $|D_\lambda(t, x_t)| \geq \rho$ and $|V(D_\lambda(s, x_s))| \leq |V(D_\lambda(t, x_t))| =: h$ for all $s \in \mathbb{R}$. Then (i) implies $\|x_s\| \leq p(h)$ for all $t \in \mathbb{R}$. Therefore, by (ii), we have $\langle \nabla V(D_\lambda(t, x_t)), f(t, x_t) \rangle > 0$. □

Example 4.7.4 We consider the following equation:

$$\frac{d}{dt}[x(t) - B(t)x(t - r)] = f(t, x_t) \tag{4.7.3}$$

where $B : \mathbb{R} \to \mathbb{R}^{n \times n}$ is an ω-periodic and continuous map such that $\sup_{t \in \mathbb{R}} |B(t)| \leq k < 1$. We claim that the function $V : \mathbb{R}^n \to \mathbb{R}$ defined by $V(x) = \varepsilon \langle x, x \rangle$, where $\varepsilon = 1$ or $\varepsilon = -1$, is a guiding function for the ω-periodic boundary value problem of Eq. (4.7.3), provided that there is a constant $\rho > 0$ such that

$$\varepsilon\langle x(t) - \lambda B(t)x(t-r), f(t, x_t)\rangle > 0$$

for every $x \in X_\omega$, $\lambda \in [0, 1]$, $t \in \mathbb{R}$ with $|x(t) - \lambda B(t)x(t-r)| \geq \rho$ and

$$\|x_t\| \leq \frac{|x(t) - \lambda B(t)x(t-r)|}{1-k}$$

Indeed, it is easy to verify that for any ω-periodic function $g : \mathbb{R} \to \mathbb{R}^n$, the functional equation

$$x(t) - \lambda B(t)x(t-r) = g(t)$$

has an ω-periodic solution

$$x(t) = \sum_{i=1}^{\infty} \lambda^i \left(\prod_{j=0}^{i-1} B(t-jr) \right) g(t-ir) + g(t)$$

Therefore, if $|g(t)| \leq h$ for $t \in \mathbb{R}$, then $|x(t)| \leq \sum_{i=0}^{\infty} \lambda^i k^i h \leq \dfrac{h}{1-k}$, and our conclusion follows from Lemma 4.7.3.

Theorem 4.7.5

If there exists a guiding function $V : \mathbb{R}^n \to \mathbb{R}$ for the ω-periodic boundary value problem of Eq. (4.7.1) such that

$$\lim_{\|x\| \to \infty} \sup_{t \in \mathbb{R}} \inf_{\lambda \in [0,1]} |V(D_\lambda(t, x_t))| = \infty, \qquad x \in X_\omega$$

then there exists a constant $\rho^* > 0$ such that any ω-periodic solution to Eq. (4.7.2$_\lambda$), $\lambda \in [0, 1]$, satisfies $|x(t)| < \rho^*$ for $t \in \mathbb{R}$.

Proof. Let $x(t)$ be an ω-periodic solution to Eq. (4.7.2$_\lambda$) for some $\lambda \in [0, 1]$. Then $\Phi(t) := |V(D_\lambda(t, x_t))|$ is also an ω-periodic function. Thus, there exists $\tau > 0$ such that $\Phi(\tau) = \max_{t \in [0,\omega]} \Phi(t)$ and

$$\begin{aligned} 0 = \Phi'(\tau) &= \pm \left\langle \nabla V(D_\lambda(\tau, x_\tau)), \frac{d}{dt} \frac{d}{dt} D_\lambda(t, x_t)|_{t=\tau} \right\rangle \\ &= \pm \langle \nabla V(D_\lambda(\tau, x_\tau)), f(\tau, x_\tau)\rangle \end{aligned}$$

Therefore, by Definition 4.7.2, $|D_\lambda(\tau, x_\tau)| < \rho$. This implies that

$$\Phi(t) = |V(D_\lambda(t, x_t))| \leq \max_{|z| \leq \rho} |V(z)| \qquad \text{for all } t \in \mathbb{R}$$

Therefore, we obtain $|x(t)| < \rho^*$, where $\rho^* > 0$ is a given constant such that for any $x \in X_\omega$ and $\lambda \in [0, 1]$, if $\|x\| \geq \rho^*$, then $|V(D_\lambda(t, x_t))| > \max_{|z| \leq \rho} |V(z)|$ for some $t \in [0, \omega]$. This completes the proof. □

Theorem 4.7.6

Suppose that there exists a guiding function $V : \mathbb{R}^n \to \mathbb{R}$ for the periodic boundary value problem (4.7.1) such that

$$\lim_{\|x\| \to \infty} \sup_{t \in \mathbb{R}} \inf_{\lambda \in [0,1]} |V(D_\lambda(t, x_t))| = \infty, \qquad x \in X_\omega$$

If the Brouwer degree $\deg(\nabla V, G) \neq 0$, *where* $G := \{u \in \mathbb{R}^n;\ |V(u)| < \nu\}$ *with* $\nu > \max_{|u| \leq \rho} V(u)$, *then there exists at least one ω-periodic solution to Eq. (4.7.1).*

Proof. By the definition of a guiding function, for $\lambda = 0$, we get

$$\langle \nabla V(x(t)), f(t, x_t) \rangle > 0, \qquad x \in X_\omega, \quad t \in \mathbb{R}$$

providing $|x(t)| \geq \rho$ and $|V(x(s))| \leq |V(x(t))|$ for all $s \in \mathbb{R}$. In particular, for all $u \in \mathbb{R}^n$ with $|u| \geq \rho$, we have $\langle \nabla V(u), f(t, \hat{u}) \rangle > 0$ for $t \in \mathbb{R}$. Consequently, $\left\langle \nabla V(u), 1/\omega \int_0^\omega f(s, \hat{u})\, ds \right\rangle > 0$. Let $M = \rho^*$, where the number ρ^* is given in Theorem 4.7.5. We consider $\Omega = \{x \in X_\omega;\ \|x\|_\infty < M\}$. It is clear that the assumptions imply that conditions (i) and (ii) of Theorem 4.7.1 are satisfied. In order to show that condition (iii) of Theorem 4.7.1 is also satisfied, we put $g(u) := 1/\omega \int_0^\omega f(s, \hat{u})\, ds$, $u \in \mathbb{R}^n$, and let $h(\tau, u) = \tau \nabla V(u) + (1 - \tau) g(u)$, $\tau \in [0, 1]$, $u \in \mathbb{R}^n$. Clearly, $\langle h(\tau, u), g(u) \rangle = \tau \langle \nabla V(u), g(u) \rangle + (1 - \tau)|g(u)|^2 > 0$ if $|u| \geq \rho$. Consequently, $\deg(g, B_M(\mathbb{R}^n)) = \deg(\nabla V, B_M(\mathbb{R}^n)) = \deg(\nabla V, G) \neq 0$. So the conclusion follows from Theorem 4.7.2. □

Corollary 4.7.7

Let $B : \mathbb{R} \to \mathbb{R}^{n \times n}$ be an ω-periodic and continuous map such that $|B(t)| \leq k < 1$ for $t \in \mathbb{R}$, and assume that for $\varepsilon = 1$ or $\varepsilon = -1$, there is a constant $\rho > 0$ such that

$$\varepsilon \langle x(t) - \lambda B(t) x(t - r), f(t, x_t) \rangle > 0$$

for every $x \in X_\omega$, $\lambda \in [0, 1]$, $t \in \mathbb{R}$ *with* $|x(t) - \lambda B(t)x(t-r)| \geq \rho$ *and* $\|x_t\| \leq \dfrac{|x(t) - \lambda B(t)x(t-r)|}{1-k}$, *where* $f : \mathbb{R} \times C \to \mathbb{R}^n$ *is a completely continuous mapping* ω*-periodic in the first argument. Then the neutral equation*

$$\frac{d}{dt}[x(t) - B(t)x(t-\tau)] = f(t, x_t)$$

has at least one ω*-periodic solution.*

Corollary 4.7.8

Let $|a| < \frac{1}{2}$ *and* $f : \mathbb{R} \times \mathbb{R} \to \mathbb{R}$ *be a continuous and* 1*-periodic function with respect to the first variable, that is,* $f(t+1, x) = f(t, x)$ *for* $(t, x) \in \mathbb{R} \times \mathbb{R}$. *Assume that there is a constant* $M > 0$ *such that one of the following two conditions is satisfied:*

(**i**) $xf(t, x) > 0$ *for* $|x| \geq M$.
(**ii**) *Or* $xf(t, x) < 0$ *for* $|x| \geq M$.

Then the neutral equation

$$\frac{d}{dt}[x(t) - ax(t-r)] = f(t, x(t)) \tag{4.7.4}$$

has at least one 1*-periodic solution.*

Proof. Suppose $x \in X_1 = \{x \in C(\mathbb{R}; \mathbb{R});\ x(t+1) = x(t)\}$. Put $g(t) = x(t) - \lambda ax(t-r)$ for $\lambda \in [0, 1]$. Suppose $t \in [0, 1]$ is such that $\|g\| = |g(t)| = \max_{s \in [0,1]} |g(s)| \geq \rho = \dfrac{1-|a|}{1-2|a|} M$. Note that $x(t) = \sum_{i=0}^{\infty} \lambda^i a^i g(t - ir)$. Without loss of generality, we assume $g(t) \geq \rho$. Then

$$\begin{aligned} x(t) &\geq g(t) - \sum_{i=1}^{\infty} (\lambda|a|)^i |g(t - ir)| \\ &\geq g(t) - \frac{|\lambda a|}{1 - \lambda|a|} g(t) = \frac{1 - 2\lambda|a|}{1 - \lambda|a|} g(t) \\ &\geq \frac{1 - 2|a|}{1 - |a|} g(t) \geq M \end{aligned}$$

Therefore, $[x(t) - \lambda ax(r - r)]f(x(t)) > 0$. Consequently, $V(x) = x^2$ is a guiding function for (4.7.4). On the other hand, for any $x \in X_1$ and $t \in \mathbb{R}$

with $|x(t)| = \sup_{s\in\mathbb{R}} |x(s)|$, we have $|x(t) - \lambda ax(t-r)| \geq |x(t)| - \lambda|a||x(t-r| \geq [1-|a|]|x(t)|$. Therefore, $\lim_{\|x\|\to\infty} \sup_{t\in\mathbb{R}} \inf_{\lambda\in[0,1]} |V(x(t)-\lambda ax(t-r))| = \infty$. Therefore, the conclusion follows from Theorem 4.7.6. □

EXERCISES

4.7.1 Formulate (4.7.4) as a coincidence problem $Lx = Gx$ and apply Theorem 4.6.11 to establish the existence of 1-periodic solutions under the conditions:

(i) $|a| < 1$ (not necessarily $|a| < \frac{1}{2}$ as required by Corollary 4.7.8).

(ii) $\varepsilon xf(t, x) > 0$ for $t \in \mathbb{R}$ and $|x| \geq M$, where $M > 0$ is a constant and $\varepsilon = 1$ or -1.

4.7.2 Extend Corollary 4.7.8 to the neutral equation $\frac{d}{dt}[x(t) - g(t, x(t-r))] = f(t, x(t))$, where $g, f : \mathbb{R}^2 \to \mathbb{R}$ are continuous and 1-periodic in the first argument. Note that the technique in Exercise 4.7.1 cannot be applied if g is nonlinear in the second argument.

4.8 BIBLIOGRAPHICAL NOTES

A detailed discussion on the measures of noncompactness has been given by Banaś and Goebel (1980) and Akhmerov et al. (1992).

The bijection theorem is mostly due to Nussbaum (1969a, 1969b, 1971, 1972), but we should also mention the contribution of Borisovich and Sapronov (1968), Ewert (1980), and Webb (1971). The version of this theorem presented here is taken from Krawcewicz (1988). The construction of the degree for condensing fields and its applications to various properties of condensing fields, as well as the homotopic properties of $GL_{\mathscr{C}}(X)$, also follow the approach of Krawcewicz (1988). The properties of the condensing fields, established in Sections 4.2 and 4.3, can be found in Borisovich and Sapronov (1968), Ewert (1980), Nussbaum (1969a, 1969b, 1971, 1972), Petryshyn (1975), Sadovskii (1968, 1970), and Webb (1971). More materials on condensing fields and their degrees are contained in Appell (1981), and Appell and Zabrejko (1983). The regular value formula is based on the results of Smale (1965). See also Quinn (1970).

The results on boundary value problems of neutral equations presented in Section 4.5 are due to Erbe et al. (1992). Hale (1977) and Hale and Verduyn Lunel (1993) provide an excellent treatment of functional differential equations. More motivations and discussions of neutral equations can be found in Brayton (1966, 1967), Brayton and Miranker (1954), Cooke and

Krumme (1968), Haddock et al. (1994), Hale (1971, 1974), Hale and Meyer (1967), Muhamadiev and Sadovskii (1973), Staffans (1983), and Wu (1986). The composite coincidence degree and its applications to general neutral functional differential equations are taken from Erbe et al. (1993). An excellent account of earlier results related to coincidence degree theories has been provided by Gaines and Mawhin (1977a,b), Hetzer (1975a, 1975b, 1975c, 1975d), and Mawhin (1971, 1972, 1979, 1981). Further details can be found in Erbe and Palamides (1987), Gopalsamy and Zhang (1988), Gustafson and Schmitt (1977), Hale and Mawhin (1974), Kuang (1991), Lopes (1975), Muhamadiev and Sadovskii (1973), Schmitt (1973), Staffans (1983), and Waltman and Wong (1972).

Chapter Five

Applications to Bifurcation Theory

In this chapter we apply the established degree theory to investigate the structure of the solution set for a large class of nonlinear eigenvalue problems in a Banach space. A local bifurcation theorem due to Krasnosiel'skii will be presented that relates an eigenvalue of odd algebraic multiplicity to the occurrence of a bifurcation point, and a global bifurcation theorem of Rabinowitz will be discussed which shows that bifurcation from eigenvalues of odd multiplicity is a global rather than a local phenomenon. An application to Sturm–Liouville problems for second-order ordinary differential equations will be provided to illustrate the general theory.

The bifurcation problem discussed in this chapter usually arises from the steady-state solutions of dynamical systems. Some special dynamic bifurcations, in particular, Hopf bifurcations, will be the central subject of Chapter 7.

5.1 LOCAL BIFURCATION THEORY

Let X be a Banach space, $\Omega \subseteq X$ an open subset containing 0, $J = (\lambda_0 - \delta, \lambda_0 + \delta)$ for some $\lambda_0 \in \mathbb{R}$ and $\delta > 0$, and $F : J \times \Omega \to X$ a condensing mapping satisfying $F(\lambda, 0) = 0$ for all $\lambda \in J$. Consider the following operator equation:

$$x = F(\lambda, x) \tag{5.1.1}$$

A pair $(\lambda, x) \in J \times \Omega$ is called a *solution* of Eq. (5.1.1) if it satisfies the equality (5.1.1). Clearly, for each $\lambda \in J$, $(\lambda, 0)$ is a solution of (5.1.1). This solution is called a *trivial solution*. Other solutions are called *nontrivial solutions*.

Definition 5.1.1 A point $(\lambda_0, 0) \in J \times \Omega$ is called a *bifurcation point* of (5.1.1) if every neighborhood of $(\lambda_0, 0)$ contains a nontrivial solution of (5.1.1), that is,

$$(\lambda_0, 0) \in \overline{\{(\lambda, x) \in J \times \Omega; x = F(\lambda, x), x \neq 0\}}$$

Clearly, $(\lambda_0, 0)$ is a bifurcation point of (5.1.1) if and only if there exists a sequence $\{(\lambda_n, x_n)\}_{n=1}^{\infty} \subseteq J \times \Omega$ such that $(\lambda_n, x_n) \to (\lambda_0, 0)$ as $n \to \infty$ and $x_n = F(\lambda_n, x_n)$ but $x_n \neq 0$ for all $n = 1, 2, \ldots$.

Let $\mathscr{C}(X)$ denote the set of all bounded condensing linear operators defined in X. We will concentrate on the following important class of mappings $F : J \times \Omega \to X$ that may usually be obtained from (5.1.1) via linearization.

$$F(\lambda, x) = K(\lambda)x + G(\lambda, x) \tag{5.1.2}$$

where $K : J \to \mathscr{C}(X)$ is a continuous mapping and $G(\lambda, x) = o(|x|)$ as $x \to 0$, uniformly for $\lambda \in J$. As a particular example of such a mapping F, we consider the mapping $F(\lambda, x) = \lambda Kx + G(\lambda, x)$, where $K \in \mathscr{C}(X)$, for which Eq. (5.1.1) is usually called a *nonlinear eigenvalue problem*. This type of problem arises in many contexts in mathematical physics. The following result gives a necessary condition for $(\lambda_0, 0)$ to be a bifurcation point of (5.1.1) with F given by (5.1.2).

Theorem 5.1.2

Suppose F is given by (5.1.2). If $(\lambda_0, 0)$ is a bifurcation point of (5.1.1), then 1 is an eigenvalue of $K(\lambda_0)$.

Proof. If 1 is not an eigenvalue of $K(\lambda_0)$, then the operator $\mathrm{Id} - K(\lambda_0)$ has a bounded inverse. Therefore, there exists $k := (\|(\mathrm{Id} - K(\lambda_0))^{-1}\|)^{-1} > 0$ such that $\|(\mathrm{Id} - K(\lambda_0))x\| \geq k\|x\|$ for all $x \in X$. Consequently,

$$\begin{aligned}\|x - F(\lambda, x)\| &\geq \|(\mathrm{Id} - K(\lambda_0))x\| - \|K(\lambda) - K(\lambda_0)\| \, \|x\| - \|G(\lambda, x)\| \\ &\geq k\|x\| - \|K(\lambda) - K(\lambda_0\| \, \|x\| - o(\|x\|) > 0\end{aligned}$$

if $|\lambda - \lambda_0|$ and $\|x\|$ are sufficiently small and $\|x\| \neq 0$. Consequently, if $|\lambda - \lambda_0|$ and $\|x\|$ are sufficiently small, then (5.1.1) has only the trivial solution. So $(\lambda_0, 0)$ is not a bifurcation point. □

Remark 5.1.3 Clearly, the fact that 1 is an eigenvalue of $K(\lambda_0)$ does not necessarily imply that $(\lambda_0, 0)$ is a bifurcation point. For example, let $X = \mathbb{R}^2$, $J = \mathbb{R}$, and

$$F(\lambda, x_1, x_2) = -\lambda(x_1, x_2) + (x_2^3, -x_1^3)$$

Then for $\lambda_0 = -1$, the operator $K(\lambda_0) = \mathrm{Id} : \mathbb{R}^2 \to \mathbb{R}^2$ has an eigenvalue equal to 1, but $(-1, 0)$ is not a bifurcation point. In fact, $F(\lambda, x_1, x_2) = (x_1, x_2)$ implies that

$$(1 + \lambda)x_1 x_2 = -x_1^4 = x_2^4$$

and, hence, $x_1 = x_2 = 0$.

Theorem 5.1.2 implies that the necessary condition for a point $(\lambda_0, 0)$ to be a bifurcation point is that the linear operator $\mathrm{Id} - K(\lambda_0)$ is not invertible. Since the operator $K(\lambda_0)$ is a condensing operator, Corollary 4.3.10 implies that $\mathrm{Id} - K(\lambda_0)$ is a Fredholm operator of index zero. Therefore, if the point $(\lambda_0, 0)$ is a bifurcation point of (5.1.1), then $\mathrm{Id} - K(\lambda_0) \notin GL(X)$. We introduce the following definition.

Definition 5.1.4 The set $\Lambda := \{(\lambda, 0) \in J \times \Omega;\ \mathrm{Id} - K(\lambda) \notin GL(X)\}$ is called the set of *singular points* of (5.1.1).

We now focus our attention on sufficient conditions for $(\lambda_0, 0)$ to be a bifurcation point of (5.1.1). Assume that $(\lambda_0, 0)$ is an isolated singular point. The point $(\lambda_0, 0)$ is a bifurcation point of (5.1.1) if and only if in every neighborhood of $(\lambda_0, 0)$, there is a nontrivial solution of (5.1.1). Consequently, the problem of finding bifurcation points reduces to the problem of finding nontrivial solutions of (5.1.1). In order to be able to distinguish nontrivial solutions from trivial solutions, we introduce a continuous function $\theta : J \times \Omega \to \mathbb{R}$ that is negative on the set of trivial solutions $J \times \{0\}$. Such a function θ is called an *auxiliary function* for the bifurcation problem (5.1.1). It is clear that every solution of the system

$$\begin{cases} x = F(\lambda, x) \\ \theta(\lambda, x) = 0 \end{cases} \tag{5.1.3}$$

is a nontrivial solution of Eq. (5.1.1). Therefore, the problem of finding nontrivial solutions of (5.1.1) is reduced to the problem of the existence of solutions of (5.1.3). Motivated by the above remark, we define a mapping $f_\theta : J \times \Omega \to \mathbb{R} \times X$ by $f_\theta(\lambda, x) = (\theta(\lambda, x), x - F(\lambda, x))$ for $(\lambda, x) \in J \times \Omega$. Let U be a small neighborhood of $(\lambda_0, 0)$. If the mapping f_θ is U-admissible, then the degree $\deg(f, U)$ is well defined, and the existence of a nontrivial solution in U could be deducted from the existence property of the degree. On the other hand, other properties, such as the excision property and homotopy invariance, can be applied to conclude that $\deg(f_\theta, U) \neq 0$ does not depend on the size of the neighborhood U. Consequently, this reduction technique coupled with the application of degrees could lead to a bifurcation result.

In order to pursue the above idea, we introduce the notion of a complementing function that was first introduced by Ize (1974, 1976).

Definition 5.1.5 Let $(\lambda_0, 0)$ be an isolated singular point of (5.1.1), and let $\varepsilon > 0$, $r > 0$ be such that:

(i) $0 < |\lambda - \lambda_0| \leq \varepsilon$ implies $\lambda \notin \Lambda$.
(ii) $F(\lambda, x) \neq x$ for $|\lambda - \lambda_0| = \varepsilon$ and $0 < \|x\| \leq r$.

Then the neighborhood $U(\varepsilon, r) := (\lambda_0 - \varepsilon,\ \lambda_0 + \varepsilon) \times B_r(0)$ is called a *special neighborhood* of $(\lambda_0, 0)$. A function $\varphi : \overline{U(\varepsilon, r)} \times \mathbb{R}$, satisfying the following properties:

(a) $\varphi(\lambda, 0) = -|\lambda - \lambda_0|$
(b) $\varphi(\lambda, x) = r$ for $\|x\| = r$

is called a *complementing function* (or *Ize's function*) for $(\lambda_0, 0)$.

The existence of a special neighborhood of an isolated singular point follows from the assumption that the mapping F is of the form (5.1.2). Indeed, let $\kappa = \min\{\|(\mathrm{Id} - K(\lambda))^{-1}\|^{-1};\ |\lambda - \lambda_0| = \varepsilon\}$; then it is sufficient to choose $r > 0$ such that $\kappa > \dfrac{\|G(\lambda, x)\|}{\|x\|}$ for $0 < \|x\| \leq r$. We can assume, without loss of generality, that this inequality is always satisfied by a special neighborhood of $(\lambda_0, 0)$. It is also clear that a complementing function can be defined simply by the formula $\varphi(\lambda, x) = |\lambda - \lambda_0|\left(\dfrac{\|x\| - r}{r}\right) + \|x\|$. We should also note that a complementing function φ is almost an auxiliary function, in the sense that the function $\theta_\rho(\lambda, x) := \varphi(\lambda, x) - \rho$, for every $\rho > 0$, is an auxiliary function. Consequently, it follows from the homotopy property of the degree that $\deg(f_{\theta_\rho}, U(\varepsilon, r)) = \deg(f_\varphi, U(\varepsilon, r))$ for sufficiently small $\rho > 0$, so the existence of nontrivial solutions in the neighborhood will follow from the inequality $\deg(f_\varphi, U(\varepsilon, r)) \neq 0$.

Proposition 5.1.6

Suppose that F is given by (5.1.2), $(\lambda_0, 0)$ is an isolated singular point for (5.1.1), $U(\varepsilon, r)$ is a special neighborhood of $(\lambda_0, 0)$. Then for every special neighborhood $U(\varepsilon', r')$ such that $0 < \varepsilon' < \varepsilon$ and $0 < r' < r$, and for every complementing function φ' defined on $U(\varepsilon', r')$, we have

$$\deg(f_\varphi, U(\varepsilon, r)) = \deg(f_{\varphi'}, U(\varepsilon', r'))$$

Proof. Let $k(\lambda, x) := x - K(\lambda)x$ and $f(\lambda, x) = x - F(\lambda, x)$. Define $h : [0, 1] \times \overline{U(\varepsilon, r)} \to X$ by $h(t, \lambda, x) := x - K(\lambda)x - tG(\lambda, x)$ and $h_\varphi : [0, 1] \times$

$\overline{U(\varepsilon, r)} \to \mathbb{R} \times X$ by $h_\varphi(t, \lambda, x) = (\varphi(\lambda, x),\ h(t, \lambda, x))$. Since $\kappa > \frac{\|G(\lambda, x)\|}{\|x\|}$, where $\kappa = \min\{\|(\mathrm{Id} - K(\lambda))^{-1}\|^{-1};\ |\lambda - \lambda_0| = \varepsilon\}$, h_φ is an $U(\varepsilon, r)$-admissible homotopy (in the class of condensing fields) between f_φ and k_φ. Consequently, $\deg(f_\varphi, U(\varepsilon, r)) = \deg(k_\varphi, U(\varepsilon, r))$, where $k_\varphi(\lambda, x) = (\varphi(\lambda, x), k(\lambda))$. Similarly, we can show that $\deg(f_{\varphi'},\ U(\varepsilon', r')) = \deg(k_{\varphi'},\ U(\varepsilon', r'))$. Since $k_\varphi^{-1}(0) = (\lambda_0, 0)$, the conclusion follows from the excision property. □

Theorem 5.1.7

Suppose that F is given by (5.1.2), $(\lambda_0, 0)$ is an isolated singular point for (5.1.1), $U(\varepsilon, r)$ is a special neighborhood of $(\lambda_0, 0)$, and $\varphi : \overline{U(\varepsilon, r)} \to \mathbb{R}$ is a complementing function. Then

$$\begin{aligned}
&\deg(f_\varphi, U(\varepsilon, r)) \\
&\quad = \deg(\mathrm{Id} - K(\lambda_0 - \varepsilon), B_r(0)) - \deg(\mathrm{Id} - K(\lambda_0 + \varepsilon), B_r(0)) \\
&\quad = \mathrm{sign}(\mathrm{Id} - K(\lambda_0 - \varepsilon)) - \mathrm{sign}(\mathrm{Id} - K(\lambda_0 + \varepsilon))
\end{aligned}$$

Consequently, if $\mathrm{sign}(\mathrm{Id} - K(\lambda_0 - \varepsilon)) \neq \mathrm{sign}(\mathrm{Id} - K(\lambda_0 + \varepsilon))$, *then the point $(\lambda_0, 0)$ is a bifurcation point of (5.1.1).*

Proof. We notice, by using the same argument as in the proof of Proposition 5.1.6, that $\deg(f_\varphi,\ U(\varepsilon, r)) = \deg(k_\varphi,\ U(\varepsilon, r))$, where $k_\varphi(\lambda, x) = (\varphi(\lambda, x),\ x - K(\lambda)x)$ for $(\lambda, x) \in \overline{U(\varepsilon, r)}$. By the homotopy invariance we may assume, without loss of generality, that the mapping $K : J \to \mathscr{C}(X)$ is (locally) constant on some neighborhood $I_\pm$ of the point $\lambda_0 \pm \varepsilon$. We define the following homotopy, k_{θ_t}, where $\theta_t(\lambda, x) := \varphi(\lambda, x) + t\rho$ and $0 < \rho < \varepsilon$ is chosen so that $\lambda_0 \pm \rho \in I_\pm$. Then the mapping k_{θ_1} has exactly two zeros on $U(\varepsilon, r)$, namely, at the points $(\lambda_0 \pm \rho, 0)$, and it follows from Theorem 4.4.14 that

$$\begin{aligned}
&\deg(k_{\theta_1}, U(\varepsilon, r)) \\
&\quad = \deg(\mathrm{Id} - K(\lambda_0 - \varepsilon), B_r(0)) - \deg(\mathrm{Id} - K(\lambda_0 + \varepsilon), B_r(0)) \\
&\quad = \mathrm{sign}(\mathrm{Id} - K(\lambda_0 - \varepsilon)) - \mathrm{sign}(\mathrm{Id} - K(\lambda_0 + \varepsilon))
\end{aligned}$$

The last statement follows from Proposition 5.1.6. □

Definition 5.1.8 Let F be given by (5.1.2), $(\lambda_0, 0)$ an isolated singular point for (5.1.1), $U(\varepsilon, r)$ a special neighborhood of $(\lambda_0, 0)$, and $\varphi : \overline{U(\varepsilon, r)} \to \mathbb{R}$ a complementing function. We will call the number $\gamma(\lambda_0) := \mathrm{sign}(\mathrm{Id} - K(\lambda_0 - \varepsilon)) - \mathrm{sign}(\mathrm{Id} - K(\lambda_0 + \varepsilon))$ a *crossing number* at $(\lambda_0, 0)$.

Assume now that $F(\lambda, x) = \lambda Kx + G(\lambda, x)$, where $K: X \to X$ is a compact linear operator. Recall that the algebraic multiplicity of an eigenvalue σ_0 of K is defined as $\dim X_{\sigma_0}$, where $X_{\sigma_0} := \bigcup_{j=1}^{\infty} \operatorname{Ker}(\sigma_0 \mathrm{Id} - K)^j$ is the generalized eigenspace of K corresponding to σ_0.

Theorem 5.1.9

Assume that $F: J \times \Omega \to X$ is a condensing mapping given by $F(\lambda, x) = \lambda Kx + G(\lambda, x)$, where $K: X \to X$ is a compact linear operator. If λ_0^{-1} is an eigenvalue of K of odd algebraic multiplicity m, then $\deg(f_\varphi, U(\varepsilon, r)) \neq 0$ and, hence, $(\lambda_0, 0)$ is a bifurcation point of (5.1.1).

Proof. Since λ^{-1} is not an eigenvalue of K if $\lambda \in [\lambda_0 - \varepsilon, \lambda_0 + \varepsilon] \setminus \{\lambda_0\}$, by using the computational formula of the Leray–Schauder degree for linear compact fields, we know that as λ increases from $\lambda_0 - \varepsilon$ to $\lambda_0 + \varepsilon$, the degree $\deg(\mathrm{Id} - \lambda K,\ B_r(0)) = \operatorname{sign}(I - \lambda K)$ changes by a multiplicative factor $(-1)^m$. □

EXERCISES

5.1.1 Assume that X is a complex finite-dimensional vector space, $J = (\lambda_0 - \delta, \lambda_0 + \delta)$ for some $\lambda_0 \in \mathbb{R}$ and $\delta > 0$, and $F: J \times X \to X$ is a C^1-map, analytic with respect to $x \in X$ such that $f(\lambda, 0) = 0$ for all $\lambda \in J$. Suppose that $(\lambda_0, 0)$ is an isolated singular point for $x = F(\lambda, x)$. Show that $\deg(f_\varphi, U(\varepsilon, r)) = 0$, where $f(\lambda, x) = x - F(\lambda, x)$, $U(\varepsilon, r)$ is a special neighborhood of $(\lambda_0, 0)$ and $\varphi: \overline{U(\varepsilon, r)} \to \mathbb{R}$ a complementing function.

5.1.2 Using the Nussbaum–Sadovskii degree for condensing fields to extend Theorem 5.1.9 to the case where K is a Darbó map with μ-Lipschitz constant $k < \dfrac{1}{|\lambda_0|}$ and G is a Darbó map with sufficiently small μ-Lipschitz constant.

In the next two exercises, we let V and W be two Banach spaces and $L: \operatorname{Dom}(L) \subseteq V \to W$ a closed Fredholm operator of index zero. Assume that $N: \mathbb{R} \times V \to W$ is a continuous mapping such that: (i) N is *L-completely continuous*, that is, $R_K \circ N$ is a completely continuous mapping for some $K \in CR(L)$, where $R_K = (L + K)^{-1}$ (see Section 4.6 for more details), and $N(\lambda, 0) = 0$ for all λ; (ii) N is *asymptotically linear* with respect to V, that is, there is a continuous map $N_\infty: \mathbb{R} \to L(V, W)$ such that

$$\lim_{\|v\| \to \infty} \frac{N(\lambda, v) - N_\infty(\lambda) v}{\|v\|} = 0$$

uniformly for λ in bounded intervals of $\mathbb{R}$. We consider the following nonlinear coincidence problem:

$$(*) \qquad Lv = N(\lambda, v), \qquad v \in \mathrm{Dom}(L)$$

We say that $(*)$ has a *bifurcation from infinity* at $\lambda_0 \in \mathbb{R}$ if for every $R > 0$ and $\varepsilon > 0$, there exists a solution (λ, v) of $(*)$ such that $\|v\| > R$ and $|\lambda - \lambda_0| < \varepsilon$. We will call a point $\lambda_0 \in \mathbb{R}$ *a V-singular point at infinity* if $\mathrm{Id}_V - \Theta_K(N_\infty)(\lambda_0, \cdot): V \to V$, where $\Theta_K(\tilde{N})(\lambda, v) := R_K(\tilde{N}(\lambda, v) + Kv)$, is not an isomorphism. A V-singular point at infinity λ_0 is called *isolated*, if λ_0 has a neighborhood containing no other V-singular points at infinity.

5.1.3 Show that if $(*)$ has a bifurcation from infinity at λ_0, then λ_0 is a V-singular point at infinity.

5.1.4 Define the mapping $\hat{N}: \mathbb{R} \times V \to W$ by

$$\hat{N}(\lambda, v) = \begin{cases} 0 & \text{if } v = 0 \\ \|v\|^2 N\left(\lambda, \dfrac{v}{\|v\|^2}\right) & \text{if } v \neq 0 \end{cases}$$

and assume that λ_0 is an isolated V-singular point at infinity. Show that:

(i) $(*)$ has a bifurcation from infinity at λ_0 if and only if the equation

$$(**) \qquad v = \Theta_K(\hat{N})(\lambda, v)$$

has a bifurcation at $(\lambda_0, 0)$, where we assume that $\{0\} \times \mathbb{R}$ are trivial solutions of $(**)$.

(ii) The map $\hat{N}$ is differentiable with respect to the v-variable at all points $(\lambda, 0)$, $\lambda \in \mathbb{R}$, and $D_v\hat{N}(\lambda, 0) = N_\infty(\lambda)$.

(iii) If $\mathrm{sign}(\mathrm{Id} - \Theta_K(N_\infty)(\lambda_0 - \varepsilon, 0) \neq \mathrm{sign}(\mathrm{Id} - \Theta_K(N_\infty)(\lambda_0 + \varepsilon, 0)$ for a sufficiently small $\varepsilon > 0$, then $(*)$ has a bifurcation from infinity at λ_0.

5.2 GLOBAL BIFURCATION THEORY

Let us start with the following result.

Lemma 5.2.1

Let (M, d) be a compact metric space, $A \subseteq M$ a component, and $B \subseteq M$ a closed subset such that $A \cap B = \emptyset$. Then there exist compact neighborhoods $M_1 \supseteq A$ and $M_2 \supseteq B$ such that $M = M_1 \cup M_2$ and $M_1 \cap M_2 = \emptyset$.

Proof. First of all, let us recall that for a given $\varepsilon > 0$, two points $a \in M$ and $b \in M$ are said to be ε-chainable if there are finitely many points $x_1, x_2, \ldots, x_n \in M$ such that $x_1 = a$, $x_n = b$, and $d(x_{i+1}, x_i) \in \varepsilon$ for $i = 1, \ldots, n-1$. In this case, $x_1, \ldots, x_n$ is called an ε-chain joining a and b. Let

$$A_\varepsilon = \{x \in M; \text{ there exists } a \in A \text{ such that } x \text{ and } a \text{ are } \varepsilon\text{-chainable}\}$$

Clearly, $A \subseteq A_\varepsilon$ and A_ε is both open and closed in M. Consequently, it suffices th show that $B \cap A_\varepsilon = \emptyset$ for some $\varepsilon > 0$.

By way of contradiction, if $B \cap A_\varepsilon \neq \emptyset$ for every $\varepsilon > 0$, then there exists a sequence of positive numbers $\{\varepsilon_n\}$ and sequences $\{a_n\} \subseteq A$, $\{b_n\} \subseteq B$ such that $\varepsilon_n \to 0$ as $n \to \infty$, and b_n are ε_n-chainable for every $n \geq 1$. Since both A and B are compact, without loss of generality, we may assume that $a_n \to a_0 \in A$ and $b_n \to b_0 \in B$. Therefore, for each $n \geq 1$, there exists an ε_n-chain M_n joining a_0 and b_0. Consider now the set

$$M_0 = \{y \in M; \, y = \lim_{k \to \infty} x_{n_k} \text{ with } x_{n_k} \in M_{n_k}\}$$

Evidently, we have:

(i) M_0 is closed and, hence, compact.
(ii) $a_0, b_0 \in M_0$.
(iii) For any y, $z \in M_0$ and any $n \geq 1$, y and z are ε_n-chainable.

Suppose that M_0 is not connected. Then there exist nonempty compact sets C_1 and C_2 and a positive number $\delta > 0$ such that $M_0 = C_1 \cup C_2$ and $\text{dist}(y, z) > \delta$ for all $y, z \in M$ with $\text{dist}(y, C_1) < \delta$ and $\text{dist}(z, C_2) < \delta$. Choose $\varepsilon_n < \delta$, $c_1 \in C_1$, and $c_2 \in C_2$. Then c_1 and c_2 are ε_n-chainable. This contradicts the fact that $\text{dist}(y, z) > \delta$ for $y, z \in M$ with $\text{dist}(y, C_1) < \delta$ and $\text{dist}(z, C_2) < \delta$. Hence, M_0 is connected. As M_0 contains $a_0 \in A$ and A is a component, we have $M_0 \subseteq A$. This leads to the contradiction of $b_0 \in M_0$, as $A \cap B = \emptyset$. The proof is then complete. □

We now are in the position to state the following version of the Rabinowitz-type global bifurcation theorem.

Theorem 5.2.2

Let X be a real Banach space, $\Omega \subseteq \mathbb{R} \times X$ a bounded neighborhood of a given point $(\lambda_0, 0)$, $F : \overline{\Omega} \to X$ a condensing mapping such that $F(\lambda, x) = K(\lambda)x + G(\lambda, x)$, $(\lambda, x) \in \overline{\Omega}$, where $\lambda \in \mathbb{R} \to K(\lambda) \in \mathscr{C}(X)$ is a continuous mapping and $G(\lambda, x) = o(\|x\|)$ as $x \to 0$ uniformly in λ. Assume that every singular point of the set $\Lambda = \{(\lambda, 0) \in \overline{\Omega}; \, \text{Id} - K(\lambda) \notin GL(X))$ is isolated. Let

$$M = \{(\lambda, x) \in \overline{\Omega}; \, F(\lambda, x) = x, x \neq 0\}$$

Then the component C *of* $\overline{M}$ *containing* $(\lambda_0, 0)$ *has at least one of the following properties:*

(**i**) $C \cap \partial\Omega \neq \emptyset$.

(**ii**) *C contains a finite number of trivial solutions* $(\lambda_i, 0)$, $i = 0, 1, \ldots, k$, *and* $\sum_{i=0}^{k} \gamma(\lambda_i) = 0$ *holds, where* $\gamma(\lambda_i)$ *denotes the crossing number of* $(\lambda_i, 0)$.

Proof. Assume that $C \cap \partial\Omega = \emptyset$. In view of the fact that F is a condensing mapping, we know that $\overline{M}$ is compact. We divide the remaining part of the proof into two steps.

Step 1. We show that there exists an open and bounded subset Ω_0 such that

$$C \subseteq \Omega_0 \subseteq \overline{\Omega} \qquad \text{and} \qquad \overline{M} \cap \partial\Omega_0 = \emptyset$$

Let $U_\delta = \{(\lambda, x) \in \Omega;\ \mathrm{dist}((\lambda, x), C) < \delta\}$. Evidently, $\overline{U_\delta} \cap \overline{M}$ is compact and $C \cap \partial U_\delta = \emptyset$. We now distinguish two cases: (i) $\overline{U_\delta} \cap \overline{M}$ is connected. In this case, $C = \overline{U_\delta} \cap \overline{M}$, as C is a maximal connected subset of $\overline{M}$. Consequently, we can choose $\Omega_0 = U_\delta$ with sufficiently small $\delta > 0$. (ii) $\overline{U_\delta} \cap \overline{M}$ is not connected. By Lemma 5.2.1, there exist compact sets $C_1 \supseteq C$ and $C_2 \supseteq \partial U_\delta \cap \overline{M}$ such that $C_1 \cap C_2 = \emptyset$ and $\overline{U_\delta} \cap \overline{M} = C_1 \cup C_2$. Let $\beta = \mathrm{dist}(C_1, C_2)$. Clearly, $\beta > 0$. Define $\Omega_0 = U_\delta \cap B(C_1, \beta/2)$, where $B(C_1, \beta/2) = \{(\lambda, x) \in \mathbb{R};\ \mathrm{dist}((\lambda, x), C_1) < \beta/2\}$. Then for sufficiently small $\delta > 0$, $C \subseteq \Omega_0 \subseteq \overline{\Omega_0} \subseteq \Omega$ and $\overline{M} \cap \partial\Omega_0 = \emptyset$.

Step 2. Since every singular point of the set Λ is isolated and $C \cap \{(\lambda, 0);\ (\lambda, 0) \in \Omega\} \subset \Lambda$, it follows from the compactness of C that C contains only a finite number of trivial solutions that we denote by $\{(\lambda_i, 0)\}$, $i = 0, 1, \ldots, k\}$. We may assume that the set $\overline{\Omega_0} \cap \{(\lambda, 0);\ (\lambda, 0) \in \overline{\Omega}\}$ is composed of a union of disjoint intervals $[\lambda_i - \varepsilon_i, \lambda_i + \varepsilon_i] \times \{0\}$, where $i = 0, 1, \ldots, k$, and without loss of generality we can assume that there is an open subset U_0 of Ω_0 such that $\Omega_0 \backslash \overline{U_0}$ is a union of special neighborhoods $\mathcal{U}_i := U(\varepsilon_i, r_i)$ of isolated singular points $(\lambda_i, 0)$, $i = 0, 1, \ldots, k$.

By taking $\varepsilon = \min \varepsilon_i$ and $r = \min r_i$, we may assume that each special neighborhood $\mathcal{U}_i$ of $(\lambda_i, 0)$ is exactly $\mathcal{U}_i = (\lambda_i - \varepsilon, \lambda_i + \varepsilon) \times B_r(0)$. Moreover, let $\varphi_i : \overline{\mathcal{U}_i} \to \mathbb{R}$ be a complementing function for $(\lambda_i, 0)$, that is, $\varphi_i(\lambda, 0) = -|\lambda - \lambda_i|$ and $\varphi_i(\lambda, x) = r$ for $\|x\| = r$. Then we can define a function $\varphi : \overline{\Omega_0} \to \mathbb{R}$ by

$$\varphi(\lambda, x) = \begin{cases} \varphi_i(\lambda, x), & \text{if } (\lambda, x) \in \overline{\mathcal{U}_i} \\ r, & \text{if } (\lambda, x) \in U_0 \end{cases}$$

It is clear that φ is a continuous function. Since $f_\varphi(\lambda, x) \neq 0$ for $(\lambda, x) \in \partial\Omega_0$, the degree $\deg(f_\varphi, \Omega_0)$ is well defined.

Let $R > \varepsilon > 0$. We define the following homotopy $h : [0, 1] \times \overline{\Omega}_0 \to \mathbb{R} \times X$ by $h(t, \lambda, x) = ((1-t)\varphi(\lambda, x) - tR,\ f(\lambda, x))$, $(\lambda, x) \in \overline{\Omega}_0$, $t \in [0, 1]$. It is easy to check that h is an Ω_0-admissible homotopy. Indeed, if $h(t, \lambda, x) = 0$ for some $t \in [0, 1]$ and $(\lambda, x) \in \partial\Omega_0$, then the fact that $f(\lambda, x) \neq 0$ for $(\lambda, x) \in \partial\Omega_0 \backslash \overline{U}_0$ implies $(\lambda, x) \in \partial\Omega_0 \cap \overline{\mathcal{U}}_i$. In this case, $f(\lambda, x) = 0$ implies that $\|x\| = 0$, and consequently $\varphi(\lambda, x) = \varphi_i(\lambda, x) = -|\lambda - \lambda_i| = -\varepsilon$. Therefore, $(1-t)\varphi(\lambda, x) - tR = -(1-t)\varepsilon - tR < 0$, which is a contradiction of the assumption that $h(t, \lambda, x) = 0$.

Clearly, $h_0 = f_\varphi$. Therefore, by the homotopy property, $\deg(f_\varphi, \Omega_0) = \deg(h_1, \Omega_0)$. However, since $h_1(\lambda, x) = (-R, f(\lambda, x)) \neq 0$ for all $(\lambda, 0) \in \Omega_0$, it follows from the existence property that $\deg(h_1, \Omega_0) = 0$. Consequently, $\deg(f_\varphi, \Omega_0) = 0$. On the other hand, since $f_\varphi(\lambda, x) \neq 0$ for $(\lambda, x) \in U_0$, by the excision and additivity properties,

$$0 = \deg(f_\varphi, \Omega_0) = \deg\left(f_\varphi, \bigcup_{i=0}^{k} \mathcal{U}_i\right) = \sum_{i=0}^{k} \deg(f_\varphi, \mathcal{U}_i)$$

By Theorem 5.1.7 and Definition 5.1.8, we obtain $\deg(f_\varphi, \mathcal{U}_i) = \gamma(\lambda_i)$. Thus, $\sum_{i=0}^{k} \gamma(\lambda_i) = 0$ and the conclusion follows. □

As an immediate consequence of Theorem 5.2.2, we have the following global bifurcation theorem due to Rabinowitz (1971).

Theorem 5.2.3

Let X be a real Banach space, $\Omega \subseteq \mathbb{R} \times X$ be a bounded neighborhood of $(\lambda_0, 0)$, $G : \overline{\Omega} \to X$ be compact, and $G(\lambda, x) = o(|x|)$ as $x \to 0$ uniformly in λ. Assume also that K is a compact linear operator defined on X and λ_0^{-1} is its eigenvalue of odd algebraic multiplicity. Let

$$F(\lambda, x) = \lambda Kx + G(\lambda, x)\,, \qquad (\lambda, x) \in \overline{\Omega}$$

$$M = \{(\lambda, x) \in \Omega;\ F(\lambda, x) = x,\ x \neq 0\}$$

Then the component C of $\overline{M}$, containing $(\lambda_0, 0)$, has at least one of the following properties:

(i) $C \cap \partial\Omega \neq \varnothing$.

(ii) *C contains an odd number of trivial solutions $(\lambda_i, 0) \neq (\lambda_0, 0)$, where λ_i^{-1} is an eigenvalue of K of odd algebraic multiplicity.*

Remark 5.2.4 In the case where $\Omega = \mathbb{R} \times X$, $G : \mathbb{R} \times X \to X$ is completely continuous, $G(\lambda, x) = o(|x|)$ as $x \to 0$ uniformly for λ in any bounded interval, by applying Theorem 5.2.3, we claim that either: (i) C is unbounded, or (ii) C contains an odd number of trivial solutions $(\lambda_i, 0) \neq (\lambda_0, 0)$, where λ_i^{-1} is an eigenvalue of K of odd algebraic multiplicity.

Remark 5.2.5 Both alternatives of Theorem 5.2.3 and Remark 5.2.4 are possible. The simplest example of (i) is the linear case $G \equiv 0$. Examples of (ii) are more complicated. A sufficient condition for (ii) to occur is that C is bounded (in Remark 5.2.4). We give an example of this nature due to Crandall and Rabinowitz (1974). Let $X = \mathbb{R}^2$ and $x = (x_1, x_2) \in X$, $\|x\| = \sqrt{x_1^2 + x_2^2}$. Consider the equation

$$Ax = \lambda[x - B(x)x] \tag{5.2.1}$$

where

$$A = \begin{bmatrix} 1 & 0 \\ 0 & 2 \end{bmatrix}, \qquad B(x) = \begin{bmatrix} 4x_1^2 + 6x_2^2 & -2x_1x_2 \\ -2x_1x_2 & 6x_1^2 + 4x_2^2 \end{bmatrix}$$

By inverting A, Eq. (5.2.1) may be put in the form (5.1.2) with $K = A^{-1}$ that has characteristic values $\frac{1}{2}$ and 1 of algebraic multiplicity 1. Taking the inner product of Eq. (5.2.1) with x leads to the following equality:

$$x_1^2 + 2x_2^2 = \lambda[x_1^2 + x_2^2 - 4(x_1^2 + x_2^2)^2]$$

from which it follows that all solutions in the connected component $C_{1/2}$ and C_1 through $(\frac{1}{2}, 0) \in \mathbb{R} \times \mathbb{R}^2$ and $(1, 0) \in \mathbb{R} \times \mathbb{R}^2$, respectively, satisfy the a priori bound

$$x_1^2 + x_2^2 \leq \tfrac{1}{4}$$

Consequently, if the first alternative of Remark 5.2.4 were to hold, there would exist a sequence (λ_n, x_n) of solutions of (5.2.1) with $\lambda_n \geq n$. Dividing Eq. (5.2.1) by λ_n, letting $n \to \infty$, and using the bounds for $\|x_n\|$, we can find a subsequence of $\{x_n\}$ that converges to a solution v of the limit equation

$$B(v)v = v \tag{5.2.2}$$

Moreover, $v \neq 0$ for otherwise, dividing (5.2.1) by $\|x_n\|$ and letting $n \to \infty$ give a contradiction. The matrix $B(x)$ can be written as

$$B(x) = T^{-1}(\theta)D(r)T(\theta)$$

where $x_1 = r\cos\theta$, $x_2 = r\sin\theta$ and

$$D(r) = \begin{bmatrix} 4r^2 & 0 \\ 0 & 6r^2 \end{bmatrix}, \qquad T(\theta) = \begin{bmatrix} \cos\left(\theta + \frac{\pi}{4}\right) & \sin\left(\theta + \frac{\pi}{4}\right) \\ -\sin\left(\theta + \frac{\pi}{4}\right) & \cos\left(\theta + \frac{\pi}{4}\right) \end{bmatrix}$$

Rewriting (5.2.2) as $D(r)T(\theta)x = T(\theta)x$ leads to $1 = 4r^2 = 6r^2$, which is impossible. Thus, alternative (i) of Remark 5.2.4 cannot occur here, (ii) must.

EXERCISES

5.2.1 State and prove an analog of Theorem 5.2.3, where K is a condensing operator.

In the following three exercises, we consider the nonlinear coincidence problem

$$(*) \qquad Lv = N(\lambda, v), \qquad v \in \mathrm{Dom}(L)$$

where $L : \mathrm{Dom}(L) \subseteq V \to W$ is a closed Fredholm operator of index zero, $N : \mathbb{R} \times V \to W$ is a completely continuous asymptotically linear map with respect to V (see the exercises in Section 5.1) such that $N(0, \lambda) = 0$ for all $\lambda \in \mathbb{R}$. So, all the points $(0, \lambda)$ are solutions, called *trivial* solutions, of Eq. (*). In addition, we make the following assumption: All V-singular points $(\lambda_0, 0)$ and V-singular points at infinity of Eq. (*) are isolated. We will use all notations from the exercises in Section 5.1.

5.2.2 Consider the topological space $\hat{V} := V \cup \{\infty\}$, where a basis of neighborhoods of ∞ is composed of sets $\{\infty\} \cup \{v \in V;\ \|v\| > R\}$. Show that the stereographic embedding $h : V \to V \times \mathbb{R}$ given by

$$h(v) = \begin{cases} (v, \|v\| - 1) & \text{if } \|v\| \le 1 \\ \left(\dfrac{v}{\|v\|^2}, 1 - \dfrac{1}{\|v\|}\right) & \text{if } \|v\| > 1 \end{cases}$$

extends to a homeomorphism $\hat{h} : \hat{V} \to S$, where S is the unit sphere in $V \times \mathbb{R}$. Show that a closed subset $A \subset \hat{V}$ is compact in $\hat{V}$ if and only if $A \cap B_R(0)$, $B_R(0) = \{v \in V;\ \|v\| \le R\}$, is compact in V for all $R > 0$.

5.2.3 We define the following transformation:

$$\tilde{N}(v, \lambda) := \begin{cases} 0 & \text{if } v = \hat{v} \\ \|v - \hat{v}\|^2 N\left(\lambda, \dfrac{v - \hat{v}}{\|v - \hat{v}\|^2}\right) & \text{if } v \ne \hat{v} \end{cases}$$

where $\hat{v} \in V$ is a nonzero vector such that $(\lambda, \hat{v})$ is not a solution of (*) for all λ. Consider the following coincidence problem:

$$(**) \qquad Lv = \tilde{N}(\lambda, v)\,, \qquad v \in \mathrm{Dom}(L)$$

Show that (λ, v), $v \neq \hat{v}$, is a solution of (**) if and only if $\left(\lambda, \dfrac{v - \hat{v}}{\|v - \hat{v}\|^2}\right)$ is a solution of (*).

5.2.4 Use Theorem 5.2.2 to formulate a global bifurcation result for Eq. (**) and apply it to prove that if $\mathscr{C}$ is a bounded, connected component of the closure of the set of nontrivial solutions of (*) in the space $\mathbb{R} \times \hat{V}$, then $\mathscr{C}$ is compact in $\mathbb{R} \times \hat{V}$ and it contains only a finite number of bifurcation points, say, $\{(\lambda_{01}, 0), \ldots, (\lambda_{0k}, 0)\}$, and a finite number of bifurcation points at infinity $\{\lambda_{\infty 1}, \ldots, \lambda_{\infty r}\}$. Moreover, the following is satisfied:

$$\sum_{i=1}^{k} \gamma(\lambda_{0i}) - \sum_{j=1}^{r} \gamma(\lambda_{\infty j}) = 0$$

where we put $\gamma(\lambda_{0i}) = \mathrm{sign}(\mathrm{Id} - \Theta_K(D_v N(\lambda_{0i} - \varepsilon, 0)) - \mathrm{sign}(\mathrm{Id} - \Theta_K(D_v N(\lambda_{0i} + \varepsilon, 0))$ and $\gamma(\lambda_{\infty j}) = \mathrm{sign}(\mathrm{Id} - \Theta_K(N_\infty(\lambda_{1j} - \varepsilon, 0)) - \mathrm{sign}(\mathrm{Id} - \Theta_K(N_\infty(\lambda_{\infty j} + \varepsilon, 0))$.

5.3 APPLICATION: STURM–LIOUVILLE PROBLEMS

Let $u : [0, \pi] \to \mathbb{R}$ be a function of a second-order continuous derivative. Define

$$Lu = -(pu')' + qu$$
$$B_1 u = a_0 u(0) + b_0 u'(0)$$
$$B_2 u = a_1 u(\pi) + b_1 u'(\pi)$$
$$Bu = (B_1 u, B_2 u)^T$$

where $p : [0, \pi] \to (0, \infty)$ is continuously differentiable, $q : [0, \pi] \to \mathbb{R}$ is continuous, a_i and b_i, $i = 0, 1$, are constants such that

$$(a_0^2 + b_0^2)(a_1^2 + b_1^2) \neq 0$$

We will need the following result.

Lemma 5.3.1

Assume that the boundary value problem

(5.3.1) $$\begin{cases} Lu = 0\,, & 0 < x < \pi \\ Bu = 0 \end{cases}$$

has no nontrivial solution. Then there exists a unique continuous function $g = g(x, y)$ *defined for all* $0 \le x, y \le \pi$ *such that:*

(i) $\frac{\partial}{\partial x} g(x, y)$ *is continuous for* (x, y) *on each of the triangles* $0 \le x \le y \le \pi$ *and* $0 \le y \le x \le \pi$.

(ii) $\frac{\partial}{\partial x} g(y+0, y) - \frac{\partial}{\partial x} g(y-0, y) = -\frac{1}{p(y)}$, $0 \le y \le \pi$.

(iii) *As a function of* x, g *satisfies* $Lu = 0$ *for* $y \ne x$.

(iv) *As a function of* x, g *satisfies* $Bu = 0$ *for* $0 \le y \le \pi$.

(v) *For any continuous function* $f : [0, \pi] \to \mathbb{R}$, *the solution of the nonhomogeneous problem*

$$\begin{cases} Lu = f\,, & 0 < x < \pi \\ Bu = 0 \end{cases}$$

is given by

$$u(x) = \int_0^\pi g(x, y) f(y)\, dy$$

This result can be easily proved by using the fundamental matrix of $Lu = 0$, $0 < x < \pi$, and the variation-of-constant formula for the nonhomogeneous equation $Lu = f$, $0 < x < \pi$. For details, we refer to Coddington and Levinson (1955).

We now consider the following *linear eigenvalue problem*:

(5.3.2) $$\begin{cases} Lu = \lambda a u\,, & 0 < x < \pi \\ Bu = 0 \end{cases}$$

where λ is a real number, $a : [0, \pi] \to (0, \infty)$ is a given continuous function. The values of λ for which the linear eigenvalue problem (4.3.3) has a nontrivial solution are called *eigenvalues* and the corresponding nontrivial solutions are called *eigenfunctions*.

The following result can be found in Coddington and Levinson (1955).

Lemma 5.3.2

The linear eigenvalue problem (5.3.2) has a countable number of eigenvalues $\lambda_1 < \lambda_2 < \cdots < \lambda_n < \lambda_{n+1} < \cdots \to \infty$. *Moreover, every eigenfunction corresponding to* λ_n *has exactly* $(n-1)$ *simple zeros in* $(0, \pi)$.

The purpose of this section is to use the above information about the linear eigenvalue problem (5.3.2) and to apply the global bifurcation theorem (Theorem 5.2.3) to investigate the global continuation of the solution set of the following nonlinear eigenvalue problem:

$$\begin{cases} Lu = \lambda F(x, u, u') \,, & 0<x<\pi \\ Bu = 0 \end{cases} \tag{5.3.3}$$

where $F : [0, \pi] \times \mathbb{R}^2 \to \mathbb{R}$ is locally Lipschitz continuous and

$$F(x, y, z) = a(x)y + o(|y| + |z|) \qquad \text{as } |y| + |z| \to 0$$

We first formulate problem (5.3.3) as an operator equation. Assume that (5.3.1) has no nontrivial solution. By Lemma 5.3.1, u is a solution of (5.3.3) if and only if it solves the following integral equation:

$$\begin{aligned} u(x) &= \lambda \int_0^\pi g(x, y)F(y, u(y), u'(y))\, dy \\ &= [\lambda Ku + G(\lambda, u)](x) \,, \qquad 0<x<\pi \end{aligned} \tag{5.3.4}$$

where

$$Ku(x) = \int_0^\pi g(x, y)a(y)u(y)\, dy \,, \qquad 0<x<\pi$$

Let

$$X = \{u \in C^1([0, \pi]; \mathbb{R});\, a_0u(0) + b_0u'(0) = 0,\, a_1u(\pi) + b_1u'(\pi) = 0\}$$

Then X is a Banach space under the usual maximal norm

$$\|u\|_1 = \max_{x\in[0,\pi]} |u(x)| + \max_{x\in[0,\pi]} |u'(x)|$$

Clearly, $K : X \to X$ is compact and the mapping $(\lambda, u) \in \mathbb{R} \times X \to \lambda Ku + G(\lambda, u) \in X$ is compact. Moreover, the assumption on $F(x, u, u')$ near $(u, u') = (0, 0)$ implies that

$$G(\lambda, u) = o(\|u\|_1) \qquad \text{as } \|u\|_1 \to 0$$

uniformly on any bounded interval of λ.

Eigenvalues of K correspond to the inverse of eigenvalues of the linear eigenvalue problem (5.3.3). In particular, each λ_n^{-1} is an eigenvalue of K. It is easy to employ an argument similar to that of Example 3.3.4 to show that the algebraic multiplicity of each λ_n^{-1} is 1. Hence, by Theorem 5.2.3, there exists a component C_k of

$$M = \overline{\{(\lambda, u);\, u = \lambda Ku + G(\lambda, u),\, u \neq 0\}}$$

that contains $(\lambda_k, 0)$ and is either unbounded in $\mathbb{R} \times X$ or meets $(\lambda_j, 0)$ with $j \neq k$. Actually, only the first alternative is possible as shall be shown next.

Let S_k^+ denote the set $\varphi \in X$ such that φ has exactly $k-1$ simple zeros in $(0, \pi)$ and φ is positive for $x \neq 0$ near 0. Let $S_k^- = -S_k^+$ and $S_k = S_k^+ \cup S_k^-$. Then the sets $S_k^\pm$ and S_k are open in X.

Theorem 5.3.3

C_k is unbounded and is contained in $\mathbb{R} \times S_k \cup \{(\lambda_k, 0)\}$.

Proof. To justify the above results, we need two technical lemmas.

Lemma 5.3.4

If (λ, u) is a solution of (5.3.3) and u has a double zero (i.e., $u(\tau) = u'(\tau) = 0$ for some $\tau \in [0, \pi]$), then $u \equiv 0$.

Proof. The result follows from the basic uniqueness theorem for initial value problems of ordinary differential equations. □

Lemma 5.3.5

For each $j > 0$, there exists a neighborhood N_j of $(\lambda_j, 0)$ such that $(\lambda, u) \in N_j \cap M$ and $u \neq 0$ imply $u \in S_j$.

Proof. If not, there exists a sequence $(\mu_n, u_n) \in M$ such that $0 \neq u_n \notin S_j$ and $(\mu_n, u_n) \to (\lambda_j, 0)$. If we write (5.3.4) as

$$w_n = \mu_n K w_n + G(\mu_m, u_n)/\|u_n\|_1 \tag{5.3.5}$$

where $w_n = u_n/\|u_n\|_1$, it follows that the second term on the right converges to zero as $n \to \infty$. Since K is compact, a subsequence of Kw_n converges. Hence the left-hand side of (5.3.5) has a convergent subsequence $w_{n_i} \to w$ with $\|w\|_1 = 1$ and satisfying

$$w = \lambda_j K w \tag{5.3.6}$$

Consequently, by Lemma 5.3.2, $w \in S_j$. Since this set is open, w_{n_i} and, therefore, $u_{n_i} \in S_j$ for sufficiently large i. This is contrary to the assumption and completes the proof. □

We now are in a position to finish a proof of Theorem 5.3.3: If $C_k \subseteq (\mathbb{R} \times S_k) \cup \{(\lambda_k, 0)\}$, then as $S_j \cap S_k = \emptyset$ for $j \neq k$, it follows from Remark 5.2.4 and Lemma 5.3.5 that C_k must be unbounded in $\mathbb{R} \times S_k$. Hence, the claim will be established once we show that $C_k \not\subseteq (\mathbb{R} \times S_k) \cup \{(\lambda_k, 0)\}$ is impossible. By way of contradiction, if $C_k \not\subseteq (\mathbb{R} \times S_k) \cup \{(\lambda_k, 0)\}$, then there exists $(\lambda, u) \in C_k \cap (\mathbb{R} \times \partial S_k)$ with $(\lambda, u) \neq (\lambda_k, 0)$ and

$(\lambda, u) = \lim_{n\to\infty} (\mu_n, u_n)$, $u_n \in S_k$. If $u \in \partial S_k$, by Lemma 5.3.4, $u \equiv 0$. Hence, $\lambda = \lambda_j$ for some $j \neq k$. But then, by Lemma 5.3.5, $(\mu_n, u_n) \in N_j \cap (\mathbb{R} \times S_k)$ for sufficiently large n, which is impossible. This completes the proof. □

EXERCISES

5.3.1 Consider the following nonlinear boundary value problem:

$$\begin{cases} 0 = \dfrac{d^2}{dx^2} u(x) + \alpha u(x)(1 - u(x)) , & x \in (0, \pi) \\ u(0) = u(\pi) = 0 \end{cases}$$

where α is a constant. Prove that if $\alpha > 1$, then the above problem has a unique solution that is positive in $(0, \pi)$. *Hint:* Use an argument similar to that of Theorem 5.3.3.

5.3.2 Consider the integral equation

$$(*) \qquad x(t) = \mu \int_a^b A(t,s)(x(s) + f(x(s)))\, ds , \qquad t \in [a, b]$$

where $A : [a, b] \times [a, b] \to \mathbb{R}$ and $f : \mathbb{R} \to \mathbb{R}$ are continuous and $f(x) = o(|x|)$ as $x \to 0$ in $\mathbb{R}$. The associated linearized equation is $x(t) = \mu \int_a^b A(t, s)x(s)\, ds$ and can be formulated as the operator equation $x = \mu Lx$ in $X = C([a, b], \mathbb{R})$ by setting $(Lx)(t) = \int_a^b A(t, s)x(s)\, ds$ for $x \in X$. A characteristic number μ_0 of $x = \mu Lx$ is the reciprocal of an eigenvalue of L. Prove that to every characteristic number μ_0 of $x = \mu Lx$ of odd algebraic multiplicity, there corresponds a bifurcation point $(\mu_0, 0)$ of (*). In addition, the solution component $C(\mu)$ in $\mathbb{R} \times X$ containing the point $(\mu_0, 0)$ is either (i) unbounded or (ii) compact and contains exactly an even number of points $(\mu_j, 0)$, where μ_j is a characteristic number of odd algebraic multiplicity.

5.4 BIBLIOGRAPHICAL NOTES

For the class of compact fields, the local bifurcation theorem is due to Krasnosiel'skii (1965) and its global version has been proved by Rabinowitz (1971). We present a slightly more general version of those results, following the ideas of Ize (1974, 1976). Our presentation of the application of the global bifurcation theory to Sturm–Liouville problems follows Deimling (1984), and some details on the related ordinary differential equations can be found in Coddington and Levinson (1955). We refer the interested reader to Appell and Zabrejko (1983), Chiappinelli (1989), Chow and Hale (1982), Crandall and Rabinowitz (1971, 1974), Ize (1993), Krasnosiel'skii and Zabrejko (1984), Makhamudov and Aliev (1989), Nirenberg (1974), Sattinger (1973), Stuart (1973), and Thomas (1973) for further applications and developments.

Chapter Six

S^1-Equivariant Degree

In this chapter we develop the S^1-equivariant degree theory for a class of nonlinear mappings preserving the symmetry related to the compact Lie group S^1.

Section 6.1 provides a brief introduction to the representation theory that makes our presentation self-contained. Then, in Section 6.2, we establish an equivariant generic approximation theorem. This important result is applied, in Section 6.3, to give a construction of the S^1-degree, denoted by $S^1\text{-deg}(f, \Omega)$, for an S^1-equivariant mapping $f : \overline{\Omega} \to V$ defined in the closure of an open invariant and bounded subset $\Omega \subseteq V \oplus \mathbb{R}$, where V is a finite-dimensional representation of S^1. This degree is a sequence of integers, with each term measuring the number of zeros of a certain type. It is shown that such a degree has all the standard properties of a topological degree as discussed in the previous chapters. In the final section we establish an equivariant bijection theorem so that the S^1-degree can be extended to equivariant condensing fields and coincidence problems.

6.1 ELEMENTS OF REPRESENTATION THEORY

In what follows, let V denote a finite-dimensional real (respectively, complex) vector space and let G denote a *compact* Lie group.

Definition 6.1.1 A *representation* of the Lie group G on the vector space V is a continuous mapping $\cdot : G \times V \to V$ such that (i) for each $g \in G$, the translation $T_g : V \to V$ given by $T_g(v) = g \cdot v$, $v \in V$, is a linear map; (ii) $T_e = \text{Id}$, where e is the identity element of G; (iii) $T_{gh} = T_g T_h$ for $g, h \in G$. We call the pair $(V, \cdot\,)$ a *real* (respectively *complex*) representation and V the *representation space*. The dimension of V is called the *dimension* of the representation. We will also write gv for $g \cdot v$ for the sake of simplicity.

Suppose that $(V, \cdot\,)$ is a representation of the Lie group G. We define

$T: G \to GL(V)$ by $T(g) := T_g : V \to V$. The map T is a continuous homomorphism of topological groups. In fact, the map T must be analytic [see Montgomery and Zioppin (1955)] and, therefore, is a homomorphism of Lie groups. In what follows, we will not distinguish a *representation of the Lie group G* from the continuous *homomorphism* $T: G \to GL(V)$.

Definition 6.1.2 A *real* (respectively, *complex*) *matrix representation* of the Lie group G is a continuous homomorphism $T: G \to GL(n, \mathbb{R})$ (respectively, $T: G \to GL(n, \mathbb{C})$).

Definition 6.1.3 Let $(V, \cdot)$ and $(W, \cdot)$ be two finite-dimensional real (respectively, complex) representations of the Lie group G. A *morphism* $A: V \to W$ between the representations $(V, \cdot)$ and $(W, \cdot)$ is a linear map that is *equivariant*, that is, $A(gv) = gA(v)$ for $g \in G$, $v \in V$. We denote by $L_G(V, W)$ the set of all such morphisms. A morphism $A \in L_G(V, W)$ is called an *isomorphism* between the representations $(V, \cdot)$ and $(W, \cdot)$ if $A: V \to W$ is a linear isomorphism. We say that $(V, \cdot)$ and $(W, \cdot)$ are *equivalent representations* of G, if there is an isomorphism between $(V, \cdot)$ and $(W, \cdot)$. We will denote by $V \cong W$ two equivalent representations.

Definition 6.1.4 Let V be a real (respectively, complex) representation space of G. An inner product (respectively, Hermitian inner product), $\langle \cdot, \cdot \rangle : V \times V \to \mathbb{R}$ (respectively, $\langle \cdot, \cdot \rangle : V \times V \to \mathbb{C}$) is called *G-invariant* if $\langle gu, gv \rangle = \langle u, v \rangle$ for all $g \in G$, $u, v \in V$. A representation together with a G-invariant inner product is called an *orthogonal* (respectively, *unitary*) *representation*.

It is well known that every real (respectively, complex) representation is equivalent to an orthogonal (respectively, unitary) representation.

By choosing an orthonormal basis for the space V of an orthogonal (respectively, unitary) representation, we obtain an associated matrix representation that is clearly a homomorphism $G \to O(n)$ [respectively, $G \to U(n)$], where $n = \dim V$ and $O(n)$ [respectively, $U(n)$] is the group of all orthogonal (respectively, unitary) matrices.

Definition 6.1.5 Let V be a finite-dimensional representation of a compact Lie group G. A subspace $U \subseteq V$ is called a *subrepresentation* if U is G-invariant, that is, $gu \in U$ for $g \in G$ and $u \in U$. A nonzero representation V is called *irreducible* if it has no subrepresentations other than $\{0\}$ and V. A representation that is not irreducible is called *reducible*.

Proposition 6.1.6

Let U be a subrepresentation of a finite-dimensional orthogonal (respectively, *unitary) representation V of the Lie group G. Then there is a complementary subrepresentation W of V such that $V = U \oplus W$.*

Proof. Let $W = U^{\perp} = \{x \in V;\ x \perp U\}$. Then W is G-invariant and $V = U \oplus W$. □

Corollary 6.1.7 (Complete Reducibility Theorem)

Let G be a compact Lie group and V a finite-dimensional representation of G. Then there exists irreducible subrepresentations $V_1, \ldots, V_m$ of V such that $V = V^G \oplus V_1 \oplus \cdots \oplus V_m$, where $V^G = \{x \in V;\ gx = x$ for all $g \in G\}$ is the fixed-point subspace.

Theorem 6.1.8 (Schur's Lemma)

Let G be a compact Lie group and let V and W be two complex irreducible representations of G. Then we have the following:

(i) *A morphism $A: V \to W$ is either zero or an isomorphism.*
(ii) *Let $A: V \to V$ be a morphism. Then there exists $\lambda \in \mathbb{C}$ such that $Av = \lambda v$ for every $v \in V$.*
(iii) $\dim_{\mathbb{C}} L_G(V, W) = 1$ *if V and W are equivalent, and* $\dim_{\mathbb{C}} L_G(V, W) = 0$ *otherwise.*

Proof. (i) Since V is irreducible, the kernel of A is either $\{0\}$ or V. Therefore, A is either injective or zero. If A is injective, its image is a nonzero subrepresentation of W; hence by the irreducibility of W, $A(V) = W$ and (i) follows.

(ii) Assume that $A: V \to V$ is a nontrivial morphism and let $\lambda \in \mathbb{C}$ be an eigenvalue of A. We denote by W the corresponding eigenspace of A. Since $W = \{w \in V;\ Aw = \lambda w\}$ is a nonzero subrepresentation of V, $W = V$ and (ii) follows.

(iii) The proof of (iii) follows immediately from (i) and (ii). □

Definition 6.1.9 A representation V of a compact Lie group G is called *absolutely irreducible* if the only G-equivariant linear mappings on V are scalar multiples of the identity.

Proposition 6.1.10

Any absolutely irreducible representation V of a compact Lie group G is irreducible.

Proof. Suppose that the representation V is reducible. Then there is a proper subrepresentation $W \neq \{0\}$. We can assume, without loss of generality, that V is an orthogonal (or unitary) representation. Then $V = W \oplus W^{\perp}$, where $W^{\perp}$ is the orthogonal complement of W with respect to the G-invariant inner product on V. Let $A: W \oplus W^{\perp} \to V$ be the orthogonal projection onto W. It is easy to see that A is equivariant but it is not a scalar multiple of the identity. Therefore, V is not absolutely irreducible. □

Corollary 6.1.11

A complex representation of a compact Lie group G is absolutely irreducible if and only if it is irreducible.

Proof. This is an immediate consequence of Proposition 6.1.10 and Schur's lemma. □

Corollary 6.1.12

Any complex irreducible representation of an abelian compact Lie group G is one-dimensional.

Proof. Suppose that $(V, \cdot)$ is a complex irreducible representation of the Lie group G. We consider the translation operator $T_g : V \to V$. Since G is abelian, T_g is an equivariant isomorphism. Therefore, it follows from Schur's lemma that there exists $\lambda(g) \in \mathbb{C}$ such that $T_g v = \lambda(g)v$. But this implies that any subspace of V is G-invariant. This completes the proof. □

We should point out that it is not true, in general, that a real irreducible representation is absolutely irreducible, which may be highlighted by the following example.

Example 6.1.13 Let $G = SO(2)$ and let $V = \mathbb{R}^2$. The group $SO(2)$ consists precisely of the planar rotations:

$$R_\theta = \begin{bmatrix} \cos\theta & -\sin\theta \\ \sin\theta & \cos\theta \end{bmatrix}, \qquad \theta \in [0, 2\pi]$$

Define the action of $SO(2)$ on $\mathbb{R}^2$ by matrix multiplication $(R_\theta, v) = R_\theta v$, where $v = \begin{bmatrix} x \\ y \end{bmatrix} \in \mathbb{R}^2$. It is easy to see that all linear mappings $A : \mathbb{R}^2 \to \mathbb{R}^2$ of the form $A = cR_\theta$, $c \in \mathbb{R}$, are $SO(2)$-equivariant. Hence, V is not absolutely irreducible. But clearly this representation is irreducible.

To characterize equivalent representations and irreducibility, we introduce the following notions.

Definition 6.1.14 Let V be a finite-dimensional real (respectively, complex) representation of a compact Lie group G and $T : G \to GL(V)$ the corresponding homomorphism of Lie groups. The function $\chi_T : G \to \mathbb{R}$ (respectively, $\chi_T : G \to \mathbb{C}$) defined by $\chi_T(g) := Tr\, T(g)$, where Tr denotes the trace, is called the *character* of the representation T of G. We will also use $\chi_V(g)$ to denote the character $\chi_T(g)$, if no confusion should arise.

Proposition 6.1.15

Suppose that $T: G \to GL(V)$ and $T': G \to GL(V')$ are two representations of G. Then we have the following:

(i) $\chi_{T\oplus T'}(g) = \chi_T(g) + \chi_{T'}(g),\ g \in G$
(ii) $\chi_T(ghg^{-1}) = \chi_T(h),\ g,\ h \in G$

where $T \oplus T'$ denotes the direct sum of the representations T and T' defined by $(T \oplus T')(g) = T(g) \oplus T'(g),\ g \in G$.

Proof. The properties (i) and (ii) follow immediately from the properties of the trace: $\text{Tr}(A \oplus B) = \text{Tr}\,A + \text{Tr}\,B$ and $\text{Tr}(ABA^{-1}) = \text{Tr}\,B$ for A, $B \in GL(n, \mathbb{R})$, or $GL(n, \mathbb{C})$. □

For $h \in G$, the map $L_h : G \to G$ defined by $L_h(g) = gh$ is called a *left translation map* and the map $R_h : G \to G$ defined by $R_h(g) = hg$ is called a *right translation map*.

We summarize some facts about Haar integrals on a compact Lie group. Let G be a compact Lie group. For $h \in C(G; \mathbb{R})$, where $C(G; \mathbb{R})$ denotes the space of continuous functions $f: G \to \mathbb{R}$, define three maps $R_h : C(G; \mathbb{R}) \to C(G; \mathbb{R})$, $L_h : C(G; \mathbb{R}) \to C(G; \mathbb{R})$, and $B : C(G; \mathbb{R}) \to C(G; \mathbb{R})$ by $(R_h f)(g) = f(gh)$, $(L_h f)(g) = f(h^{-1}g)$, and $(Bf)(g) = f(g^{-1})$ for $g \in G$. The *Haar integral* on G is a continuous map $I : C(G; \mathbb{R}) \to \mathbb{R}$ satisfying the following properties:

(i) $I(f_1 + f_2) = I(f_1) + I(f_2)$ for $f_1, f_2 \in C(G; \mathbb{R})$.
(ii) $I(cf) = cI(f)$ for $c \in \mathbb{R}$ and $f \in C(G; \mathbb{R})$.
(iii) If $f \in C(G; \mathbb{R})$ satisfies $f(g) \geq 0$ for all $g \in G$, then $I(f) \geq 0$.
(iv) If $f(g) = 1$ for all $g \in G$, then $I(f) = 1$.
(v) $I(R_h f) = I(L_h f) = I(f)$ for all $f \in C(G; \mathbb{R})$ and $h \in G$.
(vi) $I(Bf) = I(f)$ for $f \in C(G; \mathbb{R})$.

There exists a unique Haar integral on every compact Lie group. We will use the following notation for $I(f)$:

$$I(f) = \int_G f(g)\, d\mu(g)$$

The Haar integral induces a regular σ-additive measure μ on the Borel sets $\mathcal{B}(G)$ of G such that (i) $\mu(A) = \mu(R_h A) = \mu(L_h A)$ for $h \in G$ and $A \in \mathcal{B}(G)$ and (ii) $\mu(G) = 1$.

The measure μ is called the *Haar measure* on G. The associated integral $\int_G f(g)\,d\mu(g)$ is called the *Haar integral* [see Pontriagin (1946), Hochschild (1965), Kirillov (1976), or Kawakubo (1991)].

The Haar measure and integral will often be employed in the so-called *averaging technique*, illustrated by the proof of the following theorem.

Theorem 6.1.16 (Schur Orthogonality Relations)

Let G be a compact Lie group and $T: G \to U(n)$, $T': G \to U(m)$ two nonequivalent, irreducible, complex matrix representations of G. Let $T(g) = [T_{ij}(g)]$ and $T'(g) = [T'_{k\ell}(g)]$; then we have the following:

(i) $\displaystyle\int_G T_{ij}(g)\overline{T'_{k\ell}(g)}\,d\mu(g) = 0$

(ii) $\displaystyle\int_G T_{ij}(g)\overline{T_{k\ell}(g)}\,d\mu(g) = 1/n\;\delta_{ik}\,\delta_{j\ell}$

Proof. Let $V = \mathbb{C}^m$ and $W = \mathbb{C}^n$. The space $L(V, W)$ of all linear mappings from V into W is a representation of G with the action of G defined by

$$gA := T(g)A[T'(g)]^{-1} = T(g)A\overline{[T'(g)]}^t, \qquad A \in L(V, W)$$

Clearly, $\tilde{A} = \int_G (gA)\,d\mu(g)$ is an invariant element, that is, $g\tilde{A} = \tilde{A}$ for all $g \in G$. Since $\tilde{A} = T(g)\tilde{A}\overline{[T'(g)]}^t$ for $g \in G$, by Schur's lemma, $\tilde{A}$ is either zero or an isomorphism. But T and T' are nonequivalent; thus, $\tilde{A}$ must be zero for all $A \in L(V, W)$. Let $\{E^{jk}\}$ denote the usual basis of $L(V, W)$, that is, E^{jk} is the matrix whose entries are all zero except the (j, k)-entry that is 1. Then, $\tilde{E}^{jk}$ is the zero matrix. But

$$\tilde{E}^{jk} = \int_G T(g)E^{jk}\overline{[T'(g)]}^t\,d\mu(g) = \left[\int_G T_{ij}(g)\overline{T'_{k\ell}(g)}\,d\mu(g)\right]_{i\ell}$$

Consequently, (i) follows.

Similarly, it follows again from Schur's lemma that for any $B \in L(W, W)$,

$$\tilde{B} = \int_G T(g)B[T(g)]^{-1}\,d\mu(g) = \lambda_B \cdot \mathrm{Id}_W$$

holds for a suitable $\lambda_B \in \mathbb{C}$. On the other hand, the invariance of trace under conjugation enables us to compute λ_B as follows:

$$n \cdot \lambda_B = \operatorname{Tr} \tilde{B} = \int_G \operatorname{Tr}[T(g)B(T(g))^{-1}]\, d\mu(g) = \int_G \operatorname{Tr} B\, d\mu(g) = \operatorname{Tr} B$$

Hence, we have $\operatorname{Tr} \tilde{E}^{ik} = \operatorname{Tr} E^{ik} = \delta_{ik}$ and, thus,

$$\tilde{E}^{ik} = \int_G T(g)E^{ik}[T(g)]^{-1}\, d\mu(g) = \frac{\delta_{ik}}{n} \operatorname{Id}$$

This implies that $\displaystyle\int_G T_{ij}(g)\overline{T_{k\ell}(g)}\, d\mu(g) = \frac{\delta_{ik} \cdot \delta_{j\ell}}{n}$. □

Corollary 6.1.17

We have the following characterizations of irreducible or equivalent complex representations in terms of characters:

(i) *The characters of the nonequivalent, irreducible, complex, unitary representations T and T' are orthogonal to each other in $L^2(G, \mu)$, that is,*

$$\langle \chi_T, \chi_{T'} \rangle_{L^2} = \int_G \chi_T(g)\overline{\chi_{T'}(g)}\, d\mu(g) = 0$$

(ii) *A complex unitary representation T is irreducible if and only if the norm of χ_T in $L^2(G, \mu)$ is 1, that is,*

$$\|\chi_T\|_{L^2}^2 = \langle \chi_T, \chi_T \rangle_{L^2} = \int_G \chi_T(g)\overline{\chi_T(g)}\, d\mu(g) = 1$$

(iii) *Two complex unitary representations T_1 and T_2 (not necessarily irreducible) are equivalent if and only if*

$$\chi_{T_1}(g) = \chi_{T_2}(g) \qquad \text{for all } g \in G$$

Proof. The proof follows immediately from Theorem 6.1.16, Proposition 6.1.10, and Corollary 6.1.7. □

For $\varphi, \psi : G \to \mathbb{C}$, we denote by $\varphi * \psi$ *the convolution* of φ and ψ that is given by $\varphi * \psi(g) = \displaystyle\int_G \varphi(gh^{-1})\psi(h)\, d\mu(h)$, $g \in G$.

Proposition 6.1.18

Let G be a compact Lie group and $T : G \to U(n)$, $T' : G \to U(m)$ two *nonequivalent irreducible unitary matrix representations of G with characters* χ_T *and* $\chi_{T'}$*, and dimensions* d_T *and* $d_{T'}$*, respectively. Then we have the following:*

(i) $T_{ij} * T_{k\ell} = \frac{1}{d_T}\delta_{jk}T_{i\ell}\,, \quad T_{ij} * T'_{k\ell} = 0$

(ii) $\chi_{T'} * T'_{ij} = T'_{ij} * \chi_{T'} = \frac{1}{d_{T'}} T'_{ij}\,, \quad \chi_T * T'_{k\ell} = T'_{k\ell} * \chi_T = 0$

(iii) $\chi_T * \chi_T = \frac{1}{d_T}\chi_T\,, \quad \chi_T * \chi_{T'} = 0$

Proof. (i) Substituting the relation

$$T_{ij}(gh^{-1}) = \sum_t T_{it}(g)T_{tj}(h^{-1}) = \sum_t T_{it}(g)\overline{T_{jt}}(h)$$

in the definition of the convolution

$$T_{ij} * T_{k\ell}(g) = \int_G T_{ij}(gh^{-1})T_{k\ell}(h)\, d\mu(h)$$

we obtain

$$T_{ij} * T_{k\ell}(g) = \sum_t T_{it}(g) \int_G T_{k\ell}(h)\overline{T_{jt}}(h)\, d\mu(h)$$

Now, the Schur orthogonality relations lead to the first identity of (i). Similar arguments apply to the second identity of (ii). Moreover, (ii) and (iii) follow from (i) and the definition $\chi_T = \sum_t T_{tt}$. □

If V is a complex vector space, we may define the conjugate space $\overline{V}$ that has the same additive structure as V, but the scalar multiplication $\mathbb{C} \times V \to V$ on $\overline{V}$ is defined by $(z, v) \to \overline{z}v$. If V is a complex representation of G, then $\overline{V}$ is also a complex representation of G, called the *conjugate representation* of V.

Proposition 6.1.19

Let G be a compact Lie group and $T : G \to U(n)$ *a unitary matrix representation of G. Then the conjugate representation* $\overline{T} : G \to U(n)$ *satisfies* $\chi_{\overline{T}}(g) = \overline{\chi_T(g)} = \chi_T(g^{-1})$, $g \in G$.

We now briefly discuss the interaction between real and complex representations. Given a real irreducible representation U of a compact Lie group G, there exists a natural complexification of U, denoted by $e(U) =$

$\mathbb{C}\underset{\mathbb{R}}{\otimes} U$, such that $e(U)$ is a complex representation of G. $e(U)$ may be irreducible or reducible. In the first case, we call U an irreducible representation of the *real type*. In the second case, it is known that $e(U) \simeq V \oplus \overline{V}$ for some irreducible complex representation V of G, where $\overline{V}$ is the conjugate representation of V [see, e.g., Bröcker and tom Dieck [1985)]. If V is not equivalent to $\overline{V}$, we call U an irreducible representation of the *complex type*. Otherwise, we call U an irreducible representation of the *quaternionic type*. Let $\mathrm{Hom}_G(U, U)$ denote the set of all morphisms $A: V \to V$. It is clear that every nonzero $A \in \mathrm{Hom}_G(U, U)$ is invertible and, hence, $\mathrm{Hom}_G(U, U)$ is a division algebra over $\mathbb{R}$. This division algebra is isomorphic to either $\mathbb{R}$, $\mathbb{C}$, or $\mathbb{H}$, and U is of the real (respectively, complex or quaternionic) type if $\mathrm{Hom}_G(U, U)$ is isomorphic to $\mathbb{R}$ (respectively, $\mathbb{C}$ or $\mathbb{H}$).

Clearly, if U is a real irreducible representation, then $\chi_U(g) = \chi_{\mathbb{C}\otimes_{\mathbb{R}} U}(g)$.

Proposition 6.1.20

Let U be a real irreducible representation of a compact Lie group G. Let $d_U = \dim_{\mathbb{R}} U$. Then

$$\chi_U * \chi_U = \begin{cases} \dfrac{1}{d_U}\chi_U & \text{if } U \text{ is of the real type} \\ \dfrac{2}{d_U}\chi_U & \text{if } U \text{ is of the complex type} \\ \dfrac{4}{d_U}\chi_U & \text{if } U \text{ is of the quaternionic type} \end{cases}$$

Moreover, if U' is another irreducible real representation of G that is not equivalent to U, then $\chi_U * \chi_{U'} = 0$.

Proof. If U is of the real type, then the conclusion follows from (iii) of Proposition 6.1.18.

If U is not of the real type, then $\mathbb{C}\underset{\mathbb{R}}{\otimes} U \cong V_1 \oplus \overline{V_1}$, $\chi_U(g) = \chi_{\mathbb{C}\otimes_{\mathbb{R}} U}(g) = \chi_{V_1 \oplus \overline{V_1}}(g) = \chi_{V_1}(g) + \chi_{\overline{V_1}}(g)$, $\dim_{\mathbb{R}} U = \dim_{\mathbb{C}} \mathbb{C}\underset{\mathbb{R}}{\otimes} U = 2\dim_{\mathbb{C}} V_1 = 2\dim_{\mathbb{C}} \overline{V_1}$. For the complex type U, (iii) of Proposition 6.1.18 implies

$$\begin{aligned} \chi_U * \chi_U &= (\chi_{V_1} + \chi_{\overline{V_1}}) * (\chi_{V_1} + \chi_{\overline{V_1}}) \\ &= \chi_{V_1} * \chi_{V_1} + \chi_{\overline{V_1}} * \chi_{V_1} + \chi_{V_1} * \chi_{\overline{V_1}} + \chi_{\overline{V_1}} * \chi_{\overline{V_1}} \\ &= \frac{1}{d_{V_1}}(\chi_{V_1} + \chi_{\overline{V_1}}) = \frac{2}{d_U}\chi_U \end{aligned}$$

For the quaternionic type U, we use (iii) of Corollary 6.1.17 to get

$$\chi_U * \chi_U = (2\chi_{V_1}) * (2\chi_{V_1}) = 4\chi_{V_1} * \chi_{V_1} = \frac{4}{d_{V_1}} \chi_{V_1} = \frac{4}{d_U} \chi_U$$

The last statement can be proved in a similar way. □

Definition 6.1.21 Let U be a real irreducible representation. We define

$$n(U) = \begin{cases} \dim_{\mathbb{R}} U & \text{if } U \text{ is of the real type} \\ \dfrac{\dim_{\mathbb{R}} U}{2} & \text{if } U \text{ is of the complex type} \\ \dfrac{\dim_{\mathbb{R}} U}{4} & \text{if } U \text{ is of the quaternionic type} \end{cases}$$

and we will call $n(U)$ the *intristic dimension of U*.

Therefore, the results in Proposition 6.1.20 can be expressed, in terms of intristic dimensions, as $\chi_U * \chi_U = \dfrac{1}{n(U)} \chi_U$.

Proposition 6.1.22

Let W be a real representation of G and U a real irreducible matrix representation of G. Then the linear mapping $P_U : W \to W$ defined by

$$P_U x = n(U) \int_G \chi_U(g) gx \, d\mu(g), \qquad x \in W$$

is a G-equivariant projection of W that satisfies the following:

(i) *If $x \in W$ belongs to a representation space of an irreducible subrepresentation of W that is equivalent to U, then $P_U x = x$.*

(ii) *If $x \in W$ belongs to a representation space of an irreducible subrepresentation of W that is not equivalent to U, then $P_U x = 0$.*

Proof. We first show that P_U is a projection, that is, $P_U \circ P_U = P_U$. Indeed, for $x \in W$, we have

$$\begin{aligned} P_U \circ P_U(x) &= (n(U))^2 \int_G \chi_U(h) \int_G \chi_U(g) hgx \, d\mu(g) \, d\mu(h) \\ &= (n(U))^2 \int_G \chi_U(h) \int_G \chi_U(gh^{-1}) gx \, d\mu(g) \, d\mu(h) \\ &= (n(U))^2 \int_G \int_G \chi_U(gh^{-1}) \chi_U(h) \, d\mu(h) gx \, d\mu(g) \end{aligned}$$

$$= (n(U))^2 \int_G (\chi_U * \chi_U)(g)gx\, d\mu(g)$$

$$= (n(U))^2 \int_G \frac{\chi_U(g)}{n(U)}\, gx\, d\mu(g)$$

$$= n(U) \int_G \chi_U(g)gx\, d\mu(g) = P_U x$$

Next, assume that x belongs to an irreducible subrepresentation U' of W. Identifying U' with a matrix representation $[T'_{ij}]$, we get

$$P_U(x) = n(U) \int_G \chi_U(g)[T'_{ij}(g)]x\, d\mu(g)$$

$$= n(U) \int_G \sum_t T_{tt}(g)[T'_{ij}(g)]x\, d\mu(g)$$

$$= n(U) \sum_t \int_G [T_{tt}(g)T'_{ij}(g)]x\, d\mu(g)$$

Consequently, (i) and (ii) follow from Theorem 6.1.16 if U and U' are of the real type. The cases where U or U' is of either the complex or quaternionic type can be dealt with by using an argument similar to those of Proposition 6.1.20. □

We now turn to infinite-dimensional representations.

Definition 6.1.23 A (*Banach*) *representation* (T_g) of the Lie group G on the real (respectively, complex) Banach space V is a continuous map $\cdot : G \times V \to V$ such that:

(i) For each $g \in G$, the translation $T_g : V \to V$ given by $T_g(v) := gv$, $v \in V$, is a (bounded) linear operator.
(ii) $T_e = \mathrm{Id}$, where e is the identity in G.
(iii) $T_{gh} = T_g T_h$ for $g, h \in G$.

If (T_g) is a Banach representation of G on V, then its *dual representation* $(\hat{T}_g)$ of G on the dual space V^* is given by $(\hat{T}_g f)(v) := f(T_g v)$ for $f \in V^*$ and $v \in V$. If V is a Hilbert space, the representation (T_g) is called a *Hilbert representation* of G.

Similarly to the finite-dimensional case, we can define G-equivariant maps, representation spaces, etc.

Let V be a Banach representation of the compact Lie group G. Recall that a function $f : G \to V$ is called *weakly integrable* if for every $y \in V^*$, the function $g \mapsto \langle f(g), y \rangle$ is integrable. Define $\int_G f(g)\, d\mu(g) \in V^{**}$, where V^{**} denotes the second dual of V, by

$$\left\langle \int_G f(g)\, d\mu(g), y \right\rangle = \int_G \langle f(g), y \rangle \, d\mu(g)$$

The function f is called *integrable* if $\int_G f(g)\, d\mu(g) \in V \subseteq V^{**}$. It can be verified that a continuous function $f : G \to V$ is integrable [see, e.g., Dinculeanu (1974), or Kirillov (1976)].

Remark 6.1.24 Let V be a Banach representation of the Lie group G and $V_0 \subseteq V$ a G-invariant finite-dimensional subspace. Let $P : V \to V$ be a continuous linear projection on V_0. It is clear that the mapping $g \mapsto gPg^{-1}(x)$ is continuous for every $x \in V$. Therefore, this map is integrable. Consequently, we can use the averaging technique and define the linear projection by $P_0 x := \int_G gPg^{-1}(x)\, d\mu(g)$, $x \in V$, which is evidently a G-equivariant projection. Let us remark that this also implies that the subspace V_0 has an invariant complementing subspace $V^0 := (I - P_0)V$, that is, $V = V_0 \oplus V^0$.

Example 6.1.25

(i) Let $C(G; \mathbb{K})$ denote the Banach space of all continuous functions $x \in G \to \mathbb{K}$, where $\mathbb{K} = \mathbb{R}$ or $\mathbb{C}$. Left and right translations in G induce actions of G on $C(G; \mathbb{K})$ as follows:

$$L : G \times C(G; \mathbb{K}) \to C(G; \mathbb{K}); \qquad L(g, x)(h) = x(g^{-1}h)$$

$$R : G \times C(G; \mathbb{K}) \to C(G; \mathbb{K}); \qquad R(g, x)(h) = x(hg)$$

The spaces $C(G; \mathbb{K})$ with any of these two actions is a well-defined Banach representation of G. Let us notice that, contrary to the finite-dimensional case, the homomorphisms T_L or T_R from G into $GL(V)$, $V = C(G; \mathbb{K})$, defined by $T_L(g)x = L(g, x)$, $T_R(g)x = R(g, x)$, are generally not *continuous*. For example, if $G = S^1$, then for $T = T_R$ and $x_n : S^1 \to \mathbb{C}$ given by $x_n(u) = u^n$, $n \in \mathbb{N}$, we have

$$\|T(1)x_n - T(z)x_n\| = \sup_{u \in S^1} |u^n - z^n u^n| = |1 - z^n|$$

Therefore, if we take $z \in S^1$ such that $\{z^n\}$ is dense in S^1, then $\|T(1) - T(z)\| = 2$ and, thus, T is not continuous.

(ii) Similarly, we can define the Hilbert representations $(L^2(G, \mu; \mathbb{K}), L)$ and $(L^2(G, \mu; \mathbb{K}), R)$. The inclusion $C(G; \mathbb{K}) \hookrightarrow L^2(G, \mu; \mathbb{K})$ is a G-equivariant, bounded, linear operator with respect to the action L or R on both spaces.

Remark 6.1.26 Let V be a Banach space and let $\cdot : G \times V \to V$ be an action of G on V such that:

(i) The translation operators $T_g : V \to V$ are continuous for every $g \in G$.

(ii) The map $x_v : G \to V$ defined by $x_v(g) = gv$, $g \in G$, is continuous for every $v \in V$.

Then, it follows from the Banach–Steinhaus theorem [see, e.g., Rudin (1976)] that V is a Banach representation of G.

A Banach representation $(V, \cdot\,)$ is called *isometric* if for each $g \in G$, the translation operator $T_g : V \to V$ is an isometry, that is, $\|T_g v\| = \|v\|$ for $g \in G$, $v \in V$. It is clear that the representations $C(G; \mathbb{K})$ and $L^2(G; \mu; \mathbb{K})$, $\mathbb{K} = \mathbb{R}$ or $\mathbb{C}$, are isometric for both actions L and R.

Proposition 6.1.27

Let V be a Banach representation of a compact Lie group G. Then there exists an equivalent G-invariant norm on V. Therefore, the space V equipped with this G-invariant norm is an isometric representation of G.

Proof. Let $\|\cdot\| : V \to \mathbb{R}$ be the norm on V. We define $\|v\|_0 := \int_G \|gv\| \, d\mu(g)$, $v \in V$. It is clear that $\|\cdot\|_0 : V \to \mathbb{R}$ is a G-invariant norm V. Since $\sup_{g \in G} \|T_g v\| < \infty$ for all $v \in V$, it follows from the Banach–Steinhaus theorem that there exists a constant $M > 0$ such that $\|T_g\| < M$ for all $g \in G$. Therefore, $\|v\|_0 \leq M\|v\|$. Moreover, $\|v\| = \|g^{-1}gv\| \leq \|T_{g^{-1}}\| \, \|gv\| \leq M\|gv\|$ for $g \in G$. Thus, $\|v\| \leq M \int_G \|gv\| \, d\mu(g) = M\|v\|_0$ and $\|v\|_0 \leq M \int_G \|v\| \, d\mu(g) = M\|v\|$. □

Let X be a Hausdorff topological space. By a *topological transformation group*, we mean a triple $(G, X, \cdot\,)$, where $\cdot : G \times X \to X$ is a continuous map such that (i) $g \cdot (h \cdot x) = (gh) \cdot x$; (ii) $e \cdot x = x$, where $g, h \in G$, $x \in X$, and $e \in G$ is the identity of G. The map $\cdot$ is called an *action* of G on X, and the space X is called a *G-space*. In what follows, we will simply write gx instead of $g \cdot x$. A representation V (finite- or infinite-dimensional) is a G-space. A subset $A \subseteq X$ is called *G-invariant* if $gx \in A$ for all $g \in G$ and $x \in A$.

We refer to Murayama (1983) for the proof of the following equivariant version of Dugundji's extension theorem.

Theorem 6.1.28

Let C be a convex invariant subset in a Banach representation V of the Lie group G such that $C^G \neq \varnothing$ (i.e., there exists $x \in C$ such that $gx = x$ for every $g \in G$), X a metrizable G-space, and A a closed G-invariant subspace of X. Assume that $f : A \to C$ is a continuous G-equivariant map [i.e., $f(gx) = gf(x)$ for all $g \in G$ and $x \in A$]. Then f has a continuous G-equivariant extension $\tilde{f} : X \to C$.*

We denote by $\mathrm{Irr}(G; \mathbb{C})$ the set of all equivalence classes of complex, finite-dimensional, irreducible representations of the Lie group G. It is convenient to identify the set $\mathrm{Irr}(G; \mathbb{C})$ with the set of unitary matrix representations $T = [T_{ij}]$ representing these equivalence classes.

Theorem 6.1.29 (Completeness Theorem of Peter–Weyl)

Let G be a compact Lie group. Then we have the following:

- **(i)** *The linear combinations of $\{T_{ij};\ T \in \mathrm{Irr}(G; \mathbb{C})\}$ span a subalgebra B of $C(G; \mathbb{C})$ that is dense in $C(G; \mathbb{C})$.*
- **(ii)** *The functions $\{\sqrt{d_T} T_{ij};\ T \in \mathrm{Irr}(G; \mathbb{C})\}$, where d_T denotes the dimension of T, form a Hilbert basis in $L^2(G; \mu; \mathbb{C})$.*

We refer interested readers to Bröcker and tom Dieck (1985) for a proof of the above theorem.

Proposition 6.1.30

Every irreducible Banach representation (real or complex) of a compact Lie group G is finite-dimensional.

Proof.

Step 1. Assume first that (τ_g) is a complex, irreducible Banach representation of G on V and let $T \in \mathrm{Irr}(G, \mathbb{C})$ be an arbitrary unitary matrix representation of G on the space $E = \mathbb{C}^n$. Suppose that $A : V \to E$ is a bounded, G-equivariant linear operator [i.e., $A(\tau_g v) = T_g(Av)$ for $g \in G$ and

* It is possible, even for the group $G = S^1$, that a convex set C may not contain a G-fixed point [Murayama (1983)].

$v \in V$]. There exist $f_i \in V^*$, $i = 1, 2, \ldots, n$, such that $Av = \sum_{i=1}^{n} f_i(v)e_i$, where $(e_1, e_2, \ldots, e_n)$ is the standard basis in $E = \mathbb{C}^n$. Since A is G-equivariant and linear, we have

$$\sum_{i=1}^{n} f_i(\tau_g v)e_i = A(\tau_g v) = T_g Av = \sum_{i=1}^{n} f_i(v)T_g e_i$$

$$= \sum_{j=1}^{n} \left[\sum_{i=1}^{n} f_i(v)T_{ij}(g) \right] e_j$$

Therefore, we obtain

$$f_i(\tau_g v) = \sum_{j=1}^{n} f_j(v)T_{ij}(g) \in \text{span}\{f_1, \ldots, f_n\}$$

for all $g \in G$ and $v \in V$. This implies that the subspace $\text{span}\{f_1, \ldots, f_n\}$ is G-invariant.

Suppose now that V is infinite-dimensional. By the irreducibility of V, it follows that $f_1 \equiv f_2 \equiv \cdots \equiv f_n \equiv 0$. Let $\xi \in E$ be an arbitrary vector and $f \in V^*$. Define $A_0 v = f(v)\xi$ and let $A : V \to E$ be its average over G, that is,

$$Av := \int_G T_g^{-1} A\tau_g v \, d\mu(g) = \int_G T_g^* \xi \cdot f(\tau_g v) \, d\mu(g)$$

Since A is G-equivariant, $Av = 0$ for all $v \in V$. On the other hand, the vector $\xi \in E$ was chosen arbitrarily; thus, it implies that the function $\varphi : G \to \mathbb{C}$, $\varphi(g) = f(\tau_g v)$ (for $v \in V$ fixed), is orthogonal in $L^2(G; \mu; \mathbb{C})$ to the functions T_{ij}. On the other hand, the unitary irreducible representation (T_g) of G on E was also chosen arbitrarily; thus, we see that φ is orthogonal to all the functions T_{ij}, $T \in \text{Irr}(G; \mathbb{C})$. By Theorem 6.1.29, $\varphi \equiv 0$; thus, $f(\tau_g v) = 0$ for all $f \in V^*$. Hence, $\tau_g v = 0$, but this is a contradiction.

Step 2. In order to prove that this result also holds for a real Banach irreducible representation U, we consider the complexification $e(U)$ of the representation U, which is the tensor product of U with the complex numbers. The representation space of $e(U)$ is denoted by $V = \mathbb{C} \otimes_{\mathbb{R}} U$. The operator $J : V \to V$ defined by $J(z \otimes u) = \bar{z} \otimes u$ is a complex-conjugate linear G-equivariant map with the spectrum $\{1, -1\}$. Let V_+ and V_- be the eigenspaces of J associated with $+1$ and -1, respectively. Note that

$$v = \frac{v - Jv}{2} + \frac{v - Jv}{2}, \qquad v \in V$$

So, $V = V_+ \oplus V_-$. Note also that multiplication by $i \in \mathbb{C}$ induces isomorphisms $V_+ \to V_-$, $V_- \to V_+$. The complex representation V can be regarded as a real representation, denoted by $r(V)$, which is isomorphic to $U \oplus U$. Then, we have

$$r(V) = r(V_+ \oplus V_-) = r(V_+) \oplus r(V_-) \cong U \oplus U$$

Since U is a real Banach irreducible representation, it may happen that $e(U)$ is also irreducible and the result will follow from the first step. However, if $e(U)$ is reducible, then $e(U)$ may contain Banach subrepresentations $V_1, V_2, \ldots, V_m$. Since $r(V) \cong U \oplus U$ and $r(V_i)$ is a subrepresentation of $U \oplus U$, we conclude that $V = V_1 \oplus V_2$, where $r(V_1) \cong r(V_2) \cong U$. We claim that $V \cong V_1 \oplus \overline{V_1}$. Indeed, the map $\alpha : V_1 \oplus \overline{V_1} \to \mathbb{C} \otimes_{\mathbb{R}} V_1$ defined by

$$\alpha(v, w) = \tfrac{1}{2}[1 \otimes v - i \otimes v] + \tfrac{1}{2}[1 \otimes w + i \otimes w]$$

is $\mathbb{R}$-linear and satisfies $\alpha(iv, -iw) = i\alpha(v, w)$. Therefore, we can regard α as a $\mathbb{C}$-linear map $V_1 \oplus \overline{V_1} \to \mathbb{C} \otimes_{\mathbb{R}} V_1$ with the inverse given by $\beta : \mathbb{C} \otimes_{\mathbb{R}} V_1 \to V_1 \oplus \overline{V_1}$, $\beta(z \otimes v) = (zv, \bar{z}v)$. Therefore, $e(U) = e(r(V_1)) = V_1 \oplus \overline{V_1}$.

Consequently, if U is a real irreducible and $e(U)$ is complex reducible, then $V = V_1 \oplus \overline{V_1}$, where V_1 is irreducible. Therefore, by step 1, V_1 is finite-dimensional. □

Example 6.1.31

(i) We consider the group $S^1 := \{z \in \mathbb{C}; |z| = 1\}$. Since the functions $\xi_n : S^1 \to \mathbb{C}$, $\xi_n(z) = z^n$, $n = \cdots -2, -1, 0, 1, 2, \ldots$, form an orthogonal basis in $L^2(S^1; \mathbb{C})$, it follows from Theorem 6.1.29 that the sequence of the *homomorphisms* $\xi_n : S^1 \to \mathbb{C}^* = \{z \in \mathbb{C};\ z \neq 0\} = GL(1, \mathbb{C})$, $n \in \mathbb{Z} \setminus \{0\}$, determines all complex irreducible representations of S^1.

(ii) Since for an abelian compact group G, an irreducible real representation U of G has its complexification $V = \mathbb{C} \otimes_{\mathbb{R}} U$ as either irreducible (thus of complex dimension 1), or isomorphic to $V_1 \oplus \overline{V_1}$ (thus of complex dimension 2), it follows that U is either one-dimensional and of the real type or two-dimensional and of the complex type. This implies that real irreducible representations of S^1 are exactly the representations ξ_n, $n = 1, 2, \ldots$.

We conclude this section with some facts about the topological properties of the group $GL(n, \mathbb{C})$, $n \geq 1$.

Proposition 6.1.32

Let $M \in GL(n, \mathbb{C})$, $n \geq 1$. Then M can be represented uniquely as a product

$M = TS$, where T is a unitary matrix and S a symmetric, positive definite matrix. This representation is called the polar decomposition of M.

Proof. We first show the existence of the representation. Since the matrix $\overline{M^t}M$ is a symmetric, positive definite matrix, it can be diagonalized with respect to an orthonormal basis, that is,

$$\overline{M^t}M = U_1 \begin{bmatrix} \lambda_1 & & 0 \\ & \ddots & \\ 0 & & \lambda_n \end{bmatrix} U_1^{-1}, \qquad U_1 \in U(n)$$

where all the eigenvalues are positive. We put

$$S := U_1 \begin{bmatrix} \sqrt{\lambda_1} & & 0 \\ & \ddots & \\ 0 & & \sqrt{\lambda_n} \end{bmatrix} U_1^{-1}$$

The matrix S is symmetric and positive definite. Clearly, $T = MS^{-1} \in U(n)$ and $M = TS$. This shows the existence of the required representation.

For uniqueness of the representation, we remark that if $M = TS$ and $M = \tilde{T}\tilde{S}$ are two representations, where T, $\tilde{T} \in U(n)$ and S, $\tilde{S}$ are symmetric positive definite, then $\overline{M^t}M = S^2 = \tilde{S}^2$. This implies $S = \tilde{S}$, since there is only one positive definite square root of $\overline{M^t}M$. Consequently, $T = \tilde{T}$ and uniqueness follows. $\square$

Let us denote by $SPD(n)$ the set of all complex, *symmetric, positive definite* $(n \times n)$ matrices. We have the following.

Proposition 6.1.33

The polar decomposition $M = TS$ defines a surjective homomorphism Φ: $U(n) \times SPD(n) \to GL(n; \mathbb{C})$, $(T, S) \mapsto TS$. The homomorphism Φ is a homeomorphism.

Proof. We define $\Psi : GL(n, \mathbb{C}) \to U(n) \times SPD(n)$ by $\Psi(M) = \Psi(TS) = (T, S)$, where $T \in U(n)$, $S \in SPD(n)$. Since $\Phi \circ \Psi = \mathrm{Id}|_{GL(n,\mathbb{C})}$ and $\Psi \circ \Phi = \mathrm{Id}|_{U(n)\times SPD(n)}$, the statement follows. $\square$

Corollary 6.1.34

The inclusion $i : U(n) \hookrightarrow GL(n, \mathbb{C})$ is a homotopy equivalence.

Proof. This follows from Proposition 6.1.33, since $SPD(n)$ is contractible to $\{\mathrm{Id}\}$. $\square$

Proposition 6.1.35

The groups $GL(n, \mathbb{C})$ and $U(n)$ are arc-connected.

Proof. Let $A, B \in GL(n, \mathbb{C})$. The set $\{z \in \mathbb{C};\ \det_{\mathbb{C}}(zA + (1-z)B) = 0\}$ is finite and, therefore, the set $U := \{z \in \mathbb{C};\ \det_{\mathbb{C}}(zA + (1-z)B) \neq 0\}$ is connected by an arc. Since $1, 0 \in U$, we can find a continuous path $\sigma : [0, 1] \rightarrow U$ such that $\sigma(0) = 0$ and $\sigma(1) = 1$. Then $\Sigma : [0, 1] \rightarrow GL(n, \mathbb{C})$, defined by $\Sigma(t) = \sigma(t)A + (1 - \sigma(t))B$, is a path from B to A in $GL(n, \mathbb{C})$. This shows that $GL(n, \mathbb{C})$ is connected. The connectedness of $U(n)$ follows from Corollary 6.1.34. □

We set $SU(n) := \{A \in U(n); \det_{\mathbb{C}} A = 1\}$. It is clear that $SU(n)$ is a closed subgroup of $U(n)$.

Proposition 6.1.36

The group $U(n)$ is homeomorphic to $SU(n) \times S^1$.

Proof. The mapping $\Phi : SU(n) \times S^1 \rightarrow U(n)$ defined by

$$\Phi(U, \lambda) = \begin{bmatrix} \lambda & 0 & & 0 \\ 0 & 1 & & 0 \\ & & \ddots & \\ 0 & 0 & & 1 \end{bmatrix} \cdot U$$

is a bijection from the compact set $SU(n) \times S^1$ onto $U(n)$. Therefore, it is a homeomorphism. □

Proposition 6.1.37

$SU(2)$ is homeomorphic to S^3 and $SU(n+1)/SU(n)$ is homeomorphic to S^{2n+1}.

Proof. The mapping $\Phi : S^3 \rightarrow SU(2)$ defined by

$$\Phi(x_1, x_2, x_3, x_4) = \begin{bmatrix} x_1 + ix_2 & -x_3 + ix_4 \\ x_3 + ix_4 & x_1 - ix_2 \end{bmatrix}$$

is the required homeomorphism in the first part of the proposition.

In order to show the second part of the statement, we need some additional notations. Let G be a compact Lie group and V a finite-dimensional representation of G. For $x \in V$, we denote by G_x the subgroup of G defined by $G_x := \{g \in G;\ gx = x\}$. The subgroup G_x is a closed subgroup of G and is called the *isotropy group of* x. The set $Gx := \{gx;\ g \in G\}$ is called the *orbit* of x. It is a simple exercise to show that the orbit Gx is homeomorphic to G/G_x.

It is clear that $G := SU(n+1)$ acts on $\mathbb{C}^{n+1}$ and $G(e_{n+1}) = S^{2n+1}$, where $x = e_{n+1} = (0, \ldots, 0, 1)^T$. On the other hand, $Ax = x$ for $A \in SU(n+1)$ if and only if

$$A = \begin{bmatrix} A_0 & 0 \\ 0 & 1 \end{bmatrix}, \qquad A_0 \in SU(n)$$

Therefore, $G_x = SU(n)$ and the statement follows from the above remark. □

Theorem 6.1.38

The complex determinant $\det_{\mathbb{C}} : GL(n, \mathbb{C}) \to \mathbb{C}^* = \{z \in \mathbb{C};\ z \neq 0\}$ *induces the isomorphism* $\deg : \pi_1(GL(n, \mathbb{C})) \to \pi_1(\mathbb{C}^*) = \mathbb{Z}$ *that is given by* $\deg[\sigma] = \deg(\det_{\mathbb{C}}(\sigma(\cdot)),\ B_1(0))$, *where* $[\sigma]$ *denotes the homotopy class of* $\sigma : S^1 \to GL(n, \mathbb{C})$ *and* $B_1(0) = \{z \in \mathbb{C};\ |z| < 1\}$.

Proof. It follows from the Weierstrass theorem and transversality theorem that every continuous mapping $\sigma : S^1 \to S^{2n+1}$ is homotopic to a smooth mapping $\sigma_0 : S^1 \to S^{2n+1}$ that is not surjective, hence, homotopic to a constant mapping. Consequently, $\pi_1(S^{2n+1}) = 0$ for $n > 0$. Therefore, the fibration

$$SU(2) \overset{i}{\hookrightarrow} SU(3) \overset{p}{\longrightarrow} SU(3)/SU(2) \cong S^5$$

induces the following exact sequence of fundamental groups:

$$0 = \pi_1(SU(2)) \overset{i_*}{\longrightarrow} \pi_1(SU(3)) \overset{p_*}{\longrightarrow} \pi_1(S^5) = 0$$

Therefore, $\pi_1(SU(3)) = 0$. By induction, we see that if $\pi_1(SU(n)) = 0$, then the fibration

$$SU(n) \overset{i}{\hookrightarrow} SU(n+1) \overset{p}{\longrightarrow} SU(n+1)/SU(n) \cong S^{2n+1}$$

induces the exact sequence

$$0 = \pi_1(SU(n)) \overset{i_*}{\longrightarrow} \pi_1(SU(n+1)) \overset{p_*}{\longrightarrow} \pi_1(S^{2n+1}) = 0$$

and thus, $\pi_1(SU(n+1)) = 0$. Consequently, we have $\pi_1(SU(n)) = 0$ for all $n > 0$. This implies that

$$\pi_1(U(n)) \cong \pi_1(SU(n) \times S^1) = \pi_1(SU(n)) \oplus \pi_1(S^1) \cong \pi_1(S^1) = \mathbb{Z}$$

Note that the isomorphism $\pi_1(U(n)) \cong \pi_1(S^1)$ is induced by $\det_{\mathbb{C}} : U(n) \to S^1$. So, the conclusion follows from Corollary 6.1.34 and the properties of the Brouwer degree. □

Proposition 6.1.39

Every homotopy class α from $\pi_1(GL(n;\mathbb{C}))$ contains a representation $\sigma : S^1 \to GL(n,\mathbb{C})$ such that

$$\sigma(z) = \begin{bmatrix} z^d & & & 0 \\ & 1 & & \\ & & \ddots & \\ 0 & & & 1 \end{bmatrix} : \mathbb{C}^n \to \mathbb{C}^n$$

where $d = \deg \alpha$, $z \in S^1$.

Proof. Since for $n=1$, $GL(1,\mathbb{C}) = \mathbb{C}^*$ the statement thus follows from the definition of $\pi_1(S^1)$. For $n>1$, the inclusion $GL(1,\mathbb{C}) \hookrightarrow GL(n,\mathbb{C})$ defined by

$$\sigma(z) \mapsto \begin{bmatrix} \sigma(z) & & & 0 \\ & 1 & & \\ & & \ddots & \\ 0 & & & 1 \end{bmatrix}, \qquad \sigma : S^1 \to \mathbb{C}^*$$

induces, by Theorem 6.1.38, an isomorphism $i_* : \pi_1(GL(1,\mathbb{C}) \to \pi_1(GL(n,\mathbb{C}))$. Thus, the conclusion follows. □

EXERCISES

6.1.1 Let $G = S^1$. Give an example of an L^2-function $x : G \to \mathbb{R}$ such that the orbit Gx spans an infinite-dimensional subspace of $L^2(G,\mathbb{R})$.

6.1.2 Let $G = S^1$. Give an example of a continuous function $x : G \to \mathbb{R}$ such that the orbit Gx spans an infinite-dimensional subspace of $C(G,\mathbb{R})$.

6.1.3 Let $V = \mathbb{C}^n$ be a complex, irreducible representation of a compact Lie group G and A be a complex $n \times n$ matrix such that $(gx)^t A(gy) = x^t A y$ for all $g \in G$ and $x, y \in V$. Use the Schur lemma to show that:

(i) The matrix A is nonsingular.

(ii) The matrix A is symmetric or antisymmetric.

(iii) The matrix A can be uniquely determined up to a scalar coefficient.

6.1.4 Let G be a compact Lie group and H a closed subgroup of G. Show that if every complex irreducible representation of G is also an irreducible representation of H, where the G-action is reduced to H, then $G = H$. *Hint:* Use the fact that the average value of a function $f : G \to \mathbb{C}$, given by $[f] = \int_G f(g)\, d\mu(g)$, coincides with the average value of the function $f_{|H}: H \to \mathbb{C}$ on H. In particular, if $f(g)$ denotes the distance from $g \in G$ to the subgroup H, then $f(g)=0$ for all $g \in G$, thus; $H=G$.

6.2 EQUIVARIANT NORMAL APPROXIMATIONS

Let $G=S^1$. Denote by $O(G)$ the set of all closed subgroups of G. That is,

$$O(G)=\{\mathbb{Z}_k; k \in \mathbb{N} \cup \{\infty\}\}$$

where $\mathbb{Z}_k=\{k \in S^1; z^k=1\}$ for $k=1,2,\ldots$, and $\mathbb{Z}_\infty=S^1$. We introduce an *order relation* in $O(G)$ by $\mathbb{Z}_k < \mathbb{Z}_m$ if and only if $m<k$. Clearly, $\mathbb{Z}_\infty$ is the minimal element in $O(G)$.

Assume that V is a real, finite-dimensional, orthogonal representation of G. Consider the product space $V \oplus \mathbb{R}^n$, where G acts on the second argument trivially. As mentioned in the previous section, the mapping $\cdot: G \times (V \oplus \mathbb{R}^n) \to V \oplus \mathbb{R}^n$ is smooth. Therefore, the orbit Gx is a smooth submanifold diffeomorphic to the quotient group G/G_x. This implies that the isotropy group G_x describes the orbit Gx.

In what follows, for a given $x \in V \oplus \mathbb{R}^n$, we will call the subgroup $G_x=\mathbb{Z}_k$ of S^1 the *orbit type* of x. It is easy to verify that $V^G:=\{x \in V;\ gx=x$ for all $g \in G\}$ is a linear subspace of V, called the subspace of *G-fixed points*, and the orbit type of $x \in V^G \oplus \mathbb{R}^n$ is exactly $\mathbb{Z}_\infty=S^1$.

Assume that $\Omega \subseteq V \oplus \mathbb{R}^n$ is a given G-invariant, bounded, open set. Recall that a continuous mapping $f: \overline{\Omega} \to V$ is *G-equivariant* if $f(gx)=gf(x)$ for all $x \in \overline{\Omega}$ and $g \in G$. A G-equivariant mapping will also be called a *G-map*. A G-map $f: \overline{\Omega} \to V$ is said to be *Ω-admissible* if $f(x) \neq 0$ for $x \in \partial\Omega$.

For a G-invariant set $X \subseteq V \oplus \mathbb{R}^n$ and a closed subgroup H of G, we put

$$X^H := \{x \in X; hx=x \text{ for all } h \in H\}$$

$$X_H := \{x \in X; G_x=H\}$$

$$J(X) := \{G_x; x \in X\} \subseteq O(G)$$

The set $J(X)$ is called the set of *orbit types* in X, and a subgroup $H \subseteq G$ such that $H=G_x$ for some $x \in X$ is also called an *orbit type in X*. It is known that X_H is open and dense in X^H. Moreover, $J(X)$ is a finite set, partially ordered by the relation $<$ defined for the set $O(G)$. For more details, see Kawakubo (1991).

Definition 6.2.1 Let $\Omega \subseteq V \oplus \mathbb{R}^n$ be an open, bounded, G-invariant set and $f: \overline{\Omega} \to V$ a G-equivariant map. We say that f satisfies the *normality* condition at $x \in \Omega$ if there exists $\delta_x > 0$ such that for all $h \in (V^H \oplus \mathbb{R}^n)^\perp$, where $H = G_x$, with $|h| < \delta_x$ and $x + h \in \Omega$, we have

$$f(x+h) = f(x) + h$$

Definition 6.2.2 Let $\Omega \subseteq V \oplus \mathbb{R}^n$ be an open, bounded, G-invariant set and $f : \overline{\Omega} \to V$ an Ω-admissible G-map. We say that f is *normal* if it satisfies the normality condition at every $x \in f^{-1}(0)$.

Theorem 6.2.3 (NORMAL APPROXIMATION THEOREM)

Let $\Omega \subseteq V \oplus \mathbb{R}^n$ be an open, bounded, G-invariant set and let $f : \overline{\Omega} \to V$ be an Ω-admissible G-map. Then for every $\eta > 0$, there exists a normal G-map $f_0 : \overline{\Omega} \to V$ such that $\sup\limits_{x \in \overline{\Omega}} |f_0(x) - f(x)| < \eta$.

Proof. We will construct a G-map $f_0 : \overline{\Omega} \to V$ such that

$$\sup_{x \in \overline{\Omega}} |f_0(x) - f(x)| < \eta$$

and for every $\mathbb{Z}_k \in J(f_0^{-1}(0))$, there exists an open G-invariant set D_k such that $\overline{D}_k \subseteq \Omega$ and the collection $\{D_k\}$ satisfies:

(i) $\overline{D}_k \cap \overline{D}_l = \emptyset$ for $k \neq \ell$, where $\mathbb{Z}_k, \mathbb{Z}_\ell \in J(f_0^{-1}(0))$.
(ii) $\bigcup\limits_k \overline{D}_k$ is a compact neighborhood of $f^{-1}(0)$, where the summation is over all k such that $\mathbb{Z}_k \in J(f_0^{-1}(0))$.
(iii) For every k such that $\mathbb{Z}_k \in J(f_0^{-1}(0))$, $f_0|_{\overline{D}_k} : \overline{D}_k \to V$ is $\mathbb{Z}_k$-*normal*, that is, for $H = \mathbb{Z}_k$, we have

(a) $\overline{D}_k = \overline{D}_k^H \times B_{\varepsilon_k}$; $B_{\varepsilon_k} = \{y \in V;\ y \perp (V^H),\ |y| \leq \varepsilon_k\}$ for some $\varepsilon_k > 0$.
(b) H is the minimal orbit type in $J(D_k)$.
(c) $f_0(x+h) = f_0(x) + h$ for $x \in \overline{D_k}^H$, $h \in B_{\varepsilon_k}$.

It is clear that f_0 is normal. We will call such a G-map f_0 a *strongly normal G-map*, and we will say that $\{\overline{D}_k\}_k$ is the *normal covering* of $f_0^{-1}(0)$ *associated* with f_0.

We may assume, without loss of generality, that f is defined on the whole space $W := V \oplus \mathbb{R}^n$. Indeed, let $\bar{f}$ be a continuous extension of f to $V \oplus \mathbb{R}^n$; then $x \mapsto \int_G g\bar{f}(g^{-1}x)\, d\mu(g)$ is the required G-equivariant extension of f to $V \oplus \mathbb{R}^n$.

Let $\mathbb{Z}_{r_1} < \mathbb{Z}_{r_2} < \cdots < \mathbb{Z}_{r_m}$ denote the ordered sequence of all elements of $J(\overline{\Omega})$, that is, $r_1 > r_2 > \cdots > r_m$, $r_i \in \mathbb{N} \cup \{\infty\}$, $i = 1, \ldots, m$. We consider the *minimal orbit type* in $J(\overline{\Omega})$, that is, $H = \mathbb{Z}_{r_1}$, and the space $W = W^H \oplus W_1$, where W_1 is the orthogonal complement of W^H. It is clear that both subspaces W^H and W_1 are G-invariant. Assume $f^{-1}(0) \cap \Omega^H \neq \emptyset$. Then there exists a *compact neighborhood* K_1 of $f^{-1}(0) \cap \Omega^H$ such that for sufficiently small $\varepsilon_1 > 0$ and $\varepsilon > 0$, $B(K_1, \varepsilon_1) \times B_\varepsilon(W_1) \subseteq \Omega$, where $B(K_1, \varepsilon_1) = K_1 + B_{\varepsilon_1}(W^H)$, $B_{\varepsilon_1}(W^H) = \{v \in W^H;\ |v| \le \varepsilon_1\}$, $B_\varepsilon(W_1) = \{y \in W_1;\ |y| \le \varepsilon\}$. We put $\overline{D}_1 = \overline{K_1 \times B_{\varepsilon/2}(W_1)}$ and let $\gamma_1 : W \to [0, 1]$ be a continuous G-invariant function such that $\gamma_1(x) = 1$ for $x \in \overline{D}_1$ and $\gamma_1(x) = 0$ for $x \in W \setminus (B(K_1, \varepsilon_1) \times B_\varepsilon(W_1))$. Next we define the following G-map $f_1 : W \to V$ by

$$f_1(x) := \begin{cases} f(x)\,, & x \in W \setminus (B(K_1, \varepsilon_1) \times B_\varepsilon(W_1)) \\ \gamma_1(x)[f(v) + y] + (1 - \gamma_1(x))f(x)\,, & x = v + y \in N_1 \end{cases}$$

where $N_1 = B(K_1, \varepsilon_1) \times B_\varepsilon(W_1)$, $v \in B(K_1, \varepsilon_1)$, $y \in B_\varepsilon(W_1)$. Then $f_1|_{\overline{D}_1}$ is $\mathbb{Z}_{r_1}$-normal and

$$\begin{aligned} \sup_{x \in N_1} |f_1(x) - f(x)| &= \sup_{v + y \in N_1} |\gamma(v + y)[f(v) + y - f(v + y)]| \\ &\le \sup_{v + y \in N_1} |f(v) - f(v + y)| + \varepsilon \end{aligned}$$

Without loss of generality, we may assume that $\varepsilon > 0$ and $\varepsilon_1 > 0$ are sufficiently small so that $\overline{\Omega} \cap f_1^{-1}(0) \setminus \overline{D_1}$ is compact and its smallest orbit type is greater than $\mathbb{Z}_{r_1}$, and that

$$\sup_{v + y \in N_1} |f(v) - f(v + y)| < \frac{\eta}{m} - \varepsilon$$

Consequently,

$$\sup_{x \in N_1} |f_1(x) - f(x)| < \frac{\eta}{m}$$

For the purpose of induction, assume that for $k \ge 1$, we have already constructed a map $f_k : W \to V$ satisfying the following conditions:

(**a**) For every $\ell \in \{1, \ldots, k\}$, there is a compact invariant set $\overline{D}_\ell$ such that $\overline{D}_\ell \subseteq \Omega$ and $f_\ell|_{\overline{D}_\ell}$ is $\mathbb{Z}_{r_\ell}$-normal.

(**b**) For every $\ell, s \in \{1, \ldots, k\}$ such that $\ell \ne s$, the sets $\overline{D}_\ell$ and $\overline{D}_s$ are disjoint.

(**c**) $(\overline{\Omega} \cap f_k^{-1}(0)) \setminus \bigcup_{i=1}^{k} \overline{D}_i$ is compact and its smallest orbit type is greater than $\mathbb{Z}_{r_k}$.

(**d**) $\sup_{x \in \overline{\Omega}} |f_k(x) - f(x)| < k \dfrac{\eta}{m}$.

Let us denote by A the set $(\overline{\Omega} \cap f_k^{-1}(0)) \setminus \bigcup_{i=1}^{k} \overline{D}_i$. A is a compact subset of $\Omega_k := \Omega \setminus \bigcup_{i=1}^{k} \overline{D}_i$. Let $H = \mathbb{Z}_{r_{k+1}}$. Then $A \cap (\Omega_k)_H = A \cap \Omega_k^H$ and there exists a compact neighborhood K_{k+1} of $A \cap \Omega_k^H$ in $(\Omega_k)_H$ such that for sufficiently small numbers $\delta_1 > 0$ and $\delta > 0$, we have $B(K_{k+1}, \delta_1) \times B_\delta(W_{k+1}) \subset \Omega_k$, where $B(K_{k+1}, \delta_1) = K_{k+1} + B_{\delta_1}(W^H) \subseteq (\Omega_k)_H$, $W = W^H \oplus W_{k+1}$ is the orthogonal decomposition of W and $B_\delta(W_{k+1}) = \{y \in W_{k+1};\ |y| \le \delta\}$, $B_{\delta_1}(W^H) = \{v \in W^H;\ |v| \le \delta_1\}$. We put $\overline{D}_{k+1} := K_{k+1} \times \overline{B_{\delta/2}(W_{k+1})}$ and let $\gamma_{k+1} : W \to [0, 1]$ be a continuous G-invariant function such that $\gamma_{k+1}(x) = 1$ for $x \in \overline{D}_{k+1}$ and $\gamma_{k+1}(x) = 0$ for $x \in W \setminus (B(K_{k+1}, \delta_1) \times B_\delta(W_{k+1}))$. Let $N_{k+1} = B(K_{k+1}, \delta_1) \times B_\delta(W_{k+1})$. Then we define the following G-map $f_{k+1} : \Omega \to V$ by

$$f_{k+1}(x) = \begin{cases} f_k(x)\,, & x \in W \setminus N_{k+1} \\ \gamma_{k+1}(x)[f_k(v) + y] + (1 - \gamma_{k+1}(x)) f_k(x)\,, & x = v + y \in N_{k+1} \end{cases}$$

Without loss of generality, we may assume that

$$\sup_{v+y \in N_{k+1}} |f_k(v) - f_k(v+y)| < \frac{\eta}{m} - \delta$$

and, hence,

$$\begin{aligned} \sup_{x \in \overline{\Omega}} |f_{k+1}(x) - f(x)| &\le \sup_{x \in \overline{\Omega}} |f_{k+1}(x) - f_k(x)| + \sup_{x \in \overline{\Omega}} |f_k(x) - f(x)| \\ &\le \frac{\eta}{m} + k\frac{\eta}{m} = (k+1)\frac{\eta}{m} \end{aligned}$$

It follows from the construction that f_{k+1} satisfies the conditions formulated in (a–d). Therefore, by the induction principle, we obtain the normal approximation $f_0 := f_m$. □

Definition 6.2.4 Let $\Omega \subseteq V \oplus \mathbb{R}^n$ be an open, bounded, G-invariant set and let $f : \overline{\Omega} \to V$ be an Ω-admissible G-map. We say that f is a *regular normal* map if:

- **(i)** $f|_\Omega$ is of class C^1.
- **(ii)** f is normal.
- **(iii)** Zero is a regular value of f.

Remark 6.2.5 Let $\Omega \subseteq V \oplus \mathbb{R}^n$ be an open, bounded, G-invariant set and let $f : \overline{\Omega} \to V$ be an Ω-admissible continuous G-map. Then for every $\varepsilon > 0$, there exists an Ω-admissible G-map $f_0 : \overline{\Omega} \to V$ of class C^1 such that

$$\sup_{x \in \overline{\Omega}} |f(x) - f_0(x)| < \varepsilon$$

Indeed, let $\overline{f}_0 : \overline{\Omega} \to V$ be a map of class C^1 (not necessarily G-equivariant) such that

$$\sup_{x\in\overline{\Omega}}|f(x)-\overline{f}_0(x)|<\varepsilon$$

Put $f_0(x):=\int_G g\overline{f}_0(g^{-1}x)\,d\mu(g)$, $x\in\overline{\Omega}$. It is clear that f_0 is G-equivariant, of class C^1, and

$$|f(x)-f_0(x)|\leq\int_G|f(g^{-1}x)-f_0(g^{-1}x)|\,d\mu(g)<\varepsilon$$

Thus, (6.2.1) is satisfied.

Remark 6.2.6 Let $\Omega\subseteq V\oplus\mathbb{R}^n$ be an open, bounded, G-invariant set and let $f:\overline{\Omega}\to V$ be an Ω-admissible G-map of class C^1. Then for every $\eta>0$, there exists a normal G-map $f_0:\overline{\Omega}\to V$ of class C^1 such that

$$\sup_{x\in\overline{\Omega}}|f_0(x)-f(x)|<\eta$$

Indeed, it is sufficient to notice that in the proof of Theorem 6.2.3, it is possible to construct the G-invariant functions γ_1 and γ_{k+1} that are of class C^1. By Corollary 1.2.7, there exists a C^1-function $\tilde{\gamma}_1$ (not necessarily G-invariant) separating the sets $\overline{D_1}$ and $W\setminus(B(K_1,\varepsilon_1)\times B_\varepsilon(W_1))$. It is clear that $\gamma_1(x):=\int_G\tilde{\gamma}_1(g^{-1}x)\,d\mu(x)$, $x\in W$, is a C^1-function separating $\overline{D_1}$ and $W\setminus(B(K_1,\varepsilon_1)\times B_\varepsilon(W_1))$. A C^1-function γ_{k+1} may be constructed in a similar way.

Now we can formulate the main result of this section.

Theorem 6.2.7 (Regular Normal Approximation Theorem)

Let $\Omega\subseteq V\oplus\mathbb{R}^n$ be an open, bounded, G-invariant set and let $f:\overline{\Omega}\to V$ be an Ω-admissible G-map. Then for every $\eta>0$, there exists a regular normal G-map $f_0:\overline{\Omega}\to V$ such that $\sup_{x\in\overline{\Omega}}|f_0(x)-f(x)|<\eta$.

Proof. Before we can go forward with the proof, we need the following lemma.

Lemma 6.2.8

Assume that $\Omega\subseteq V\oplus\mathbb{R}^n$ is an open, bounded, G-invariant set such that for all $x\in\overline{\Omega}$, $G_x=\mathbb{Z}_k$, that is, $\overline{\Omega}$ contains only one orbit type $\mathbb{Z}_k$. Assume that $f:\overline{\Omega}\to V$ is an Ω-admissible G-map of class C^1 and let $A\subseteq\overline{\Omega}$ be a closed set such that all the points $x\in f^{-1}(0)\cap A$ are regular. Then for every $\varepsilon>0$, there *exists an Ω-admissible map $f_0 : \overline{\Omega} \to V$ such that:*

(i) *f_0 is of class C^1.*
(ii) $f|_A \equiv f_0|_A$.
(iii) *Zero is a regular value of f_0.*
(iv) $\sup_{x \in \overline{\Omega}} |f(x) - f_0(x)| < \varepsilon$.

Proof. We may assume, without loss of generality, that $\mathbb{Z}_k \neq \mathbb{Z}_\infty$ and f is defined on the whole space $V \oplus \mathbb{R}^n$. Since $\overline{\Omega} \subseteq V \oplus \mathbb{R}^n =: W$ contains only one orbit type, $G_0 := G/H$ with $H = \mathbb{Z}_k$ acts freely on $\overline{\Omega}_H$. Let $x \in \overline{\Omega}$ and consider the orbit $G_0 x$ that is a one-dimensional, smooth submanifold of W_H diffeomorphic to S^1. Let S be the orthogonal complement to $T_x(G_0 x)$. We denote by $B_\delta := B_\delta(S)$ the closed δ-ball of S centered at zero and define, for sufficiently small δ, $\varphi : G_0 \times B_\delta \to W_H$ by

$$\varphi(g, v) = g(x + v)\ , \qquad v \in B_\delta$$

It is clear that φ is a diffeomorphism that is also G_0-equivariant, where G_0 acts on $G_0 \times B_\delta$ by translation on G_0 and by trivial action on B_δ. We may regard φ as a local coordinate system on W_H. The set $\varphi(G_0 \times B_\delta) =: U$ is called a *G_0-tube* or *G_0-tubular neighborhood* around the orbit $G_0 x$, and S is called a *slice* at x to the orbit $G_0 x$. Consider the mapping $f|_U \circ \varphi$. For $(g, v) \in G_0 \times B_\delta$, we have $(f|_U \circ \varphi)(g, v) = f(g(x+v)) = gf(x+v) = g\tilde{f}(v)$, where $\tilde{f} : B_\delta \to V$ is defined by $\tilde{f}(v) = f(x + v)$. Revising the above operation, we can see that in order to define a G_0-equivariant mapping on U, it is enough to define it on the subset $x + B_\delta$ of the slice.

Suppose that $\{U_i\}_{i=1}^N$ is a finite covering of $\overline{\Omega}$ such that:

(a) $\varphi_i(G_0 \times B^i_\delta) = U_i$ is a G_0-tube at x_i for $i = 1, \ldots, N$.
(b) $\varphi_i(G_0 \times B^i_{2\delta}) = V_i$ is also a G_0-tube at x_i for $i = 1, \ldots, N$.

We now present the construction of the mapping f_0. We consider the set $V_1 = \varphi_1(G_0 \times B^1_{2\delta})$. We put $A_1 = A$ and define $\tilde{f}_1 : B^1_{2\delta} \to V$ by $\tilde{f}_1(v) = f(x_1 + v)$, $v \in B^1_{2\delta}$. By Proposition 1.3.14, there exists $\hat{f}_1 : B^1_{2\delta} \to V$ such that:

(i) $\hat{f}_1$ is of class C^1.
(ii) $\hat{f}_1|_{B^1_\delta}$ has zero as a regular value.
(iii) $\hat{f}_1|_{B_1} = \tilde{f}_1|_{B_1}$, where $B_1 = (B^1_{2\delta} \backslash B^1_{3/2\delta}) \cup (A_1 \cap B^1_{2\delta})$.
(iv) $\sup_{v \in B^1_{2\delta}} |\hat{f}_1(v) - \tilde{f}_1(v)| < \dfrac{\varepsilon}{N}$.

As remarked above, $\hat{f}_1$ induces a G_0-equivariant mapping $\bar{f}_1$ on U_1, which is

ε/N-close to $f|_{U_1}$. On the other hand, the map $f_1 : \overline{\Omega} \to V$ defined by

$$f_1(x) = \begin{cases} \bar{f}_1(x) & \text{if } x \in U_1 \\ f(x) & \text{if } x \in \overline{\Omega} \setminus U_1 \end{cases}$$

satisfies $\sup_{x \in \overline{\Omega}} |f_1(x) - f(x)| < \varepsilon/N$.

Assume, for the purpose of induction, that for $k \geq 1$, we have constructed a G_0-equivariant mapping $f_k : \overline{\Omega} \to V$ satisfying the following conditions:

(i) f_k is of class C^1.
(ii) $f_k|_{(U_1 \cup \cdots U_k)}$ has zero as a regular value.
(iii) $f_k|_{A_{k+1}} \equiv f|_{A_{k+1}}$, $A_{k+1} = (U_1 \cup \cdots \cup U_k) \cap A$.
(iv) $\sup_{x \in \overline{\Omega}} |f_k(x) - f(x)| < k \dfrac{\varepsilon}{N}$.

We consider $\tilde{f}_{k+1} : B_{2\delta}^{k+1} \to V$, $\tilde{f}_{k+1}(v) = f_k(x_{k+1} + v)$, $v \in B_{2\delta}^{k+1}$. We first use Proposition 1.3.14 to find $\hat{f}_{k+1} : B_{2\delta}^{k+1} \to V$ such that:

(i) $\hat{f}_{k+1}$ is of class C^1.
(ii) $\hat{f}_{k+1}|_{B_\delta^{k+1}}$ has zero as a regular value.
(iii) $\hat{f}_{k+1}|_{B_{k+1}} = \tilde{f}_{k+1}|_{B_{k+1}}$, where $B_{k+1} = (B_{2\delta}^{k+1} \setminus B_{3/2\delta}^{k+1}) \cup (A_{k+1} \cap B_{2\delta}^{k+1})$.
(iv) $\sup_{v \in B_{2\delta}^{k+1}} |\hat{f}_{k+1}(v) - \tilde{f}_{k+1}(v)| < \dfrac{\varepsilon}{N}$.

Next, we use $\hat{f}_{k+1}$ to induce an G_0-equivariant mapping f_{k+1} on U_{k+1}, which is ε/N-closed to $f_k|_{U_1}$. Finally, we extend it to a G_0-equivariant mapping $f_{k+1} : W \to V$ by

$$f_{k+1}(x) = \begin{cases} f_{k+1}(x), & x \in U_{k+1} \\ f_k(x), & x \in \overline{\Omega} \setminus U_{k+1} \end{cases}$$

By the induction assumption that $\sup_{x \in \overline{\Omega}} |f_k(x) - f(x)| < k \dfrac{\varepsilon}{N}$ and the properly (iv) of $\hat{f}_{k+1}$, we have

$$\begin{aligned} \sup_{x \in \overline{\Omega}} |f(x) - f_{k+1}(x)| &\leq \sup_{x \in \overline{\Omega}} |f(x) - f_k(x)| + \sup_{x \in \overline{\Omega}} |f_k(x) - f_{k+1}(x)| \\ &< k \frac{\varepsilon}{N} + \frac{\varepsilon}{N} = (k+1) \frac{\varepsilon}{N} \end{aligned}$$

Therefore, the conclusion follows by the induction principle. □

Let us return to the proof of Theorem 6.2.7. Using the same argument as in the proof of Theorem 6.2.3 and by Remark 6.2.6, we may assume, without loss of generality, that $f : \overline{\Omega} \to V$ is an Ω-admissible, strongly normal

G-map of class C^1. Let $\{\overline{D}_k\}_k$ be the associated normal covering of $f^{-1}(0)$. We may also assume that there exists a refinement $\{\overline{D}'_k\}_k$ of the covering $\{\overline{D}_k\}_k$ such that $\overline{D}'_k \subseteq \operatorname{Int} \overline{D}_k$ and $\{\overline{D}'_k\}_k$ is also a normal covering of $f^{-1}(0)$. Let $\gamma : W \to [0, 1]$ be a G-invariant C^1-function such that $\gamma(x) = 1$ if $x \in \bigcup_k \overline{D}'_k$ and $\gamma(x) = 0$ if $x \in W \setminus \bigcup_k \overline{D}_k$. We put

$$\delta := \inf_{x \in \overline{\Omega} \setminus \cup_k \overline{D}'_k} |f(x)| > 0$$

Fix k. Since $H = \mathbb{Z}_k$ is the minimal orbit type in $\overline{D}_k$, the set $(\overline{D}_k)^H = (\overline{D}_k)_H \subseteq W^H$, $W := V \oplus \mathbb{R}^n$, has only one orbit type. By Lemma 6.2.8, there is an admissible G-map of class C^1 $f_k : (\overline{D}_k)_H \to V^H$ such that

$$\sup_{x \in (\overline{D}_k)_H} |f(x) - f_k(x)| < \min(\delta, \eta) = \eta_0$$

and zero is a regular value of f_k. We define $f_0 : \overline{\Omega} \to V$ by

$$f_0(x) = \begin{cases} f(x), & x \in W \setminus \bigcup_k \overline{D}_k \\ \gamma(x)(f_k(v) + y) + (1 - \gamma(x))f(x), & x = v + y \in \overline{D}_k \end{cases}$$

We have

$$\begin{aligned} &\sup_{x \in \cup_k \overline{D}_k} |f(x) - \gamma(x)(f_k(v) + y) - (1 - \gamma(x))f(x)| \\ &\quad \leq \sup_{x \in \cup_k \overline{D}_k} |f(x) - [f_k(v) + y]| \\ &\quad = \sup_{x \in \cup_k \overline{D}_k} |f(v) - f_k(v)| < \delta \end{aligned}$$

where $x = v + y \in \overline{D}_k$. Therefore, for $x \in \overline{\Omega} \setminus \bigcup_k \overline{D}'_k$,

$$|f_0(x)| \geq |f_0(x)| - |f_0(x) - f(x)| > \delta - \delta = 0$$

that is, $f_0(x) \neq 0$ for $x \in \overline{\Omega} \setminus \bigcup_k \overline{D}'_k$. Consequently, $f_0^{-1}(0) \subseteq \operatorname{Int}(\bigcup_k \overline{D}'_k)$. Hence, it is clear that f_0 has zero as a regular value, and $\sup_{x \in \overline{\Omega}} |f_0(x) - f(x)| < \eta$. $\square$

Remark 6.2.9 Suppose that $f : \overline{\Omega} \to V$, $\Omega \subset V \oplus \mathbb{R}^n$, is a regular normal G-map. Consider $H = \mathbb{Z}_k$ such that $H \in J(f^{-1}(0))$. Since H is the minimal orbit type in some open neighborhood $\mathcal{U}_k$ of $(f^{-1}(0))_H$ such that $\mathcal{U}_k \cap f^{-1}(0) = (f^{-1}(0))_H$ and zero is a regular value of $f_k := f|_{\mathcal{U}_k}$, $f_k^{-1}(0) \subseteq (\mathcal{U}_k)_H$

is a submanifold of dimension n. It is convenient to consider $f_k^{-1}(0)$ as zeros of f of the orbit type $H = \mathbb{Z}_k$. It is clear that if $k < \infty$ and $n = 0$, then f has no zeros of orbit type $\mathbb{Z}_k$. Moreover, if $k < \infty$ and $n = 1$, then every zero of f of the orbit type $\mathbb{Z}_k$ belongs to an isolated orbit of zeros (with respect to the action of S^1). The idea of a regular normal map is based on the desire for separation of zeros being "essentially" of different orbit types. It is now clear that, in the case $n = 1$, there are many possible orbit types $\mathbb{Z}_k$, $k < \infty$. The equivariant S^1-degree, which we will introduce in the following section, is a topological invariant characterizing (in a similar way as Brouwer degree) the existence of zeros of the normal map f for every orbit type $\mathbb{Z}_k$, $k < \infty$, in Ω.

EXERCISES

6.2.1 Let V be an orthogonal representation of the group $G = S^1$ and $\Omega \subseteq V \oplus \mathbb{R}$ be an open, bounded, G-invariant subset such that $\Omega^G = \emptyset$. Show the following version of the transversality theorem: For every Ω-admissible G-equivariant continuous mapping $f : V \oplus \mathbb{R} \to V$ and for every $\varepsilon > 0$, there exists a G-equivariant C^1-mapping $\bar{f} : V \oplus \mathbb{R} \to V$ such that:

(i) zero is a regular value of $\bar{f}_{|\Omega}$.

(ii) $\bar{f}^{-1}(0) \cap \Omega$ is composed of a finite number of G-orbits.

(iii) $\max\{|f(x) - \bar{f}(x)|; x \in \overline{\Omega}\} < \varepsilon$.

6.2.2 Let G be a finite abelian group, V an orthogonal representation of G, and $\Omega \subseteq V$ an open, bounded, G-invariant subset. Use the same argument, as that of the regular normal approximation theorem, to prove the following version of the transversality theorem for G-equivariant mappings: For every Ω-admissible G-equivariant continuous mapping $f : V \to V$ and for every $\varepsilon > 0$, there exists a G-equivariant C^1-mapping $\bar{f} : V \to V$ such that:

(i) zero is a regular value of $\bar{f}_{|\Omega}$.

(ii) $\bar{f}^{-1}(0) \cap \Omega$ is composed of a finite number of G-orbits.

(iii) $\max\{|f(x) - \bar{f}(x)|; x \in \overline{\Omega}\} < \varepsilon$.

6.2.3 Show, by giving an example, that in the case of an arbitrary compact Lie group G, it is impossible in general to approximate a G-equivariant mapping $f : V \to V$, where V is an orthogonal representation of G by a G-equivariant C^1-mapping $\bar{f}$ such that $\bar{f}$ restricted to a bounded invariant set has zero as a regular value. *Hint:* Consider an irreducible representation of $O(2)$.

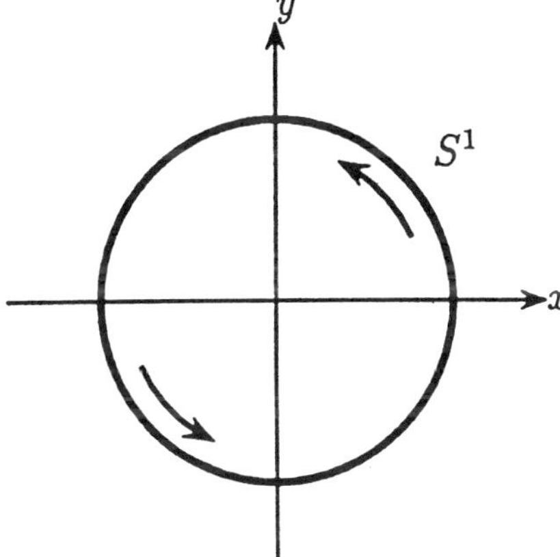

FIGURE 6.3.1. Standard orientation of S^1.

6.3 CONSTRUCTION OF S^1-EQUIVARIANT DEGREE

Throughout this section we assume that we have chosen the orientation of S^1 that determines the standard orientation of $\mathbb{R}^2$; see Figure 6.3.1.

For every $k \in \mathbb{N}$, the quotient group $S^1/\mathbb{Z}_k$ is diffeomorphic to S^1. We choose an orientation of $S^1/\mathbb{Z}_k$ such that the quotient map $\pi : S^1 \to S^1/\mathbb{Z}_k$, which is a local diffeomorphism, preserves the orientations. Consequently, for every $k \in \mathbb{N}$, we have *fixed* the orientation of $S^1/\mathbb{Z}_k$, which we call the *preferred orientation*.

Let $G = S^1$, $\Omega \subseteq V \oplus \mathbb{R}$ be an open, bounded, G-invariant set and let $f : \overline{\Omega} \to V$ be an Ω-admissible G-map. By Theorem 6.2.7, there exists a regular normal G-map $f_0 : \overline{\Omega} \to V$ such that

$$\sup_{x \in \overline{\Omega}} |f_0(x) - f(x)| < \frac{\delta}{2}$$

where

$$\delta = \inf_{x \in \partial\Omega} |f(x)|$$

Assume that $H = \mathbb{Z}_k \neq \mathbb{Z}_\infty$ and $H \in J(f_0^{-1}(0))$. We will associate with f an integer $\deg_k(f, \Omega)$, called the k*th component of the S^1-degree* of f in Ω, to characterize the zeros of f "essentially of the orbit type" $H = \mathbb{Z}_k$. In order to present the definition, we denote by U_k the set Ω_H that is an open set in $W^H = V^H \oplus \mathbb{R}$. We choose and fix an orientation of V^H that determines also an orientation of $W^H = V^H \oplus \mathbb{R}$. The map $f_k := f_0|_{\overline{U}_k} : \overline{U}_k \to V^H$ is U_k-admissible, G-equivariant and has zero as its regular value. Thus, $f_k^{-1}(0) = Gx_1 \cup \cdots \cup Gx_m$ is a finite union of disjoint orbits Gx_j, $j = 1, \ldots, m$. The action of S^1 on W^H induces a natural orientation of each of the orbits Gx_j as follows: Define $\varphi_j : S^1/\mathbb{Z}_k \to Gx_j$ by $\varphi_j(\gamma\mathbb{Z}_k) = \gamma x_j$, $\gamma \in S^1$. The map φ_j is a diffeomorphism and, hence, determines an orientation on Gx_j such that it preserves the chosen orientations.

Let S_j be the subspace of W^H orthogonal to $T_{x_j}Gx_j$, that is, S_j is the slice to Gx_j at x_j. We choose an ordered orthonormal basis in S_j such that the oriented vector of Gx_j followed by the chosen basis in S_j gives us a basis of W^H that determines the previously chosen orientation of W^H. We denote by $Df_k(x_j)$ the Jacobian of f_k at x_j with respect to chosen bases and let $D_{\hat{j}}f_k(x_j)$ denote the part of the Jacobian matrix $Df_k(x_j)$ composed of the partial derivatives corresponding to basis vectors from S_j. The matrix $D_{\hat{j}}f_k(x_j)$ is square and, therefore, we can write

$$\operatorname{sign} D_{\hat{j}}f_k(x_j) = \operatorname{sign} \det_{\mathbb{R}} D_{\hat{j}}f_k(x_j)$$

See Figure 6.3.2.

We define the integer $\deg_k(f, \Omega)$ by

$$\deg_k(f, \Omega) := \sum_{j=1}^{m} \operatorname{sign} D_{\hat{j}}f_k(x_j) \tag{6.3.1}$$

and

$$S^1\text{-}\deg(f, \Omega) = \{\deg_k(f, \Omega)\}_{k=1}^{\infty}$$

as an element of the group $\bigoplus_{k=1}^{\infty} \mathbb{Z}$, where $\deg_k(f, \Omega) = 0$ if $\mathbb{Z}_k \notin J(f_0^{-1}(0))$.

We will present here a sketch of the proof that the above construction does not depend on the choice of a regular normal approximation f_0.

Suppose that $\tilde{f}_0$ is another regular normal G-map such that $\sup_{x\in\bar{\Omega}} |\tilde{f}_0(x) - f(x)| < \delta/4$. Then $h(t, x) = tf_0(x) + (1-t)\tilde{f}_0(x)$ is an Ω-admissible G-homotopy between f_0 and $\tilde{f}_0$. Without loss of generality, we may assume

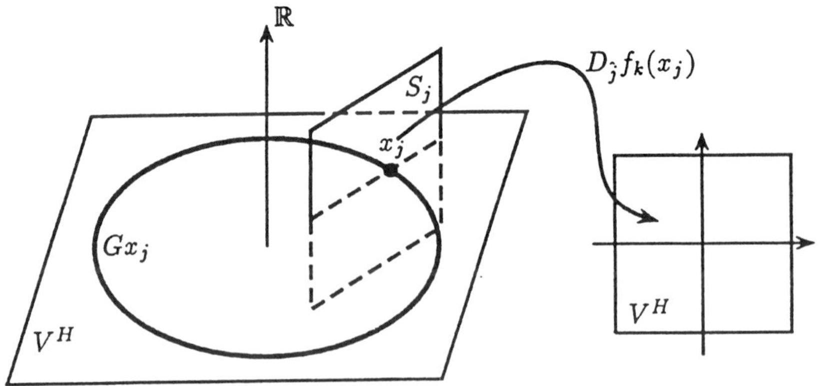

FIGURE 6.3.2. Linearization of f_k on the slice S_j.

$h : \mathbb{R} \oplus V \oplus \mathbb{R} \to V$. Let $a : \mathbb{R} \to [0, 1]$ be a function of class C^∞ such that

$$\alpha(t) = \begin{cases} 0, & t \leq \frac{1}{3} \\ 1, & t > \frac{2}{3} \end{cases}$$

Define a new Ω-admissible G-homotopy h^* between $\tilde{f}_0$ and f_0 by $h^*(t, x) = h(\alpha(t), x)$, where $t \in \mathbb{R}$, and $x \in V \oplus \mathbb{R}$. By the regular normal approximation theorem, there is a regular normal G-homotopy $\tilde{h}$ such that

$$\sup_{(t,x) \in [0,1] \times \Omega} |h^*(t, x) - \tilde{h}(t, x)| < \frac{\delta}{4}$$

where $\tilde{h}_t = \tilde{f}_0$ for $t \leq \frac{1}{3}$, and $\tilde{h}_t = f_0$ for $t \geq \frac{2}{3}$. Let $\mathcal{U}$ be an invariant neighborhood of $[0, 1] \times \overline{\Omega}^H$ in $[0, 1] \times V^H \oplus \mathbb{R}$. Then $M := \tilde{h}_H^{-1}(0) \cap \mathcal{U}_H$ is a two-dimensional submanifold of $\mathcal{U}_H$. The submanifold M, as the inverse image of a regular value, is naturally oriented. Since $\tilde{h}$ is an Ω-admissible and regular normal homotopy, the only points belonging to $M \cap \partial([0, 1] \times U_k) \subset \mathbb{R} \times V^H \oplus \mathbb{R}$ are the orbits $(0, Gx_i)$, where $x_i \in \tilde{f}_k^{-1}(0)$, and $(1, Gx_j)$, where $x_j \in f_k^{-1}(0)$. Let us consider an oriented, connected component $\tilde{M}$ of M that contains $(0, Gx_i)$ or $(1, Gx_j)$. Since $\tilde{M}$ is composed of G/H-orbits, which are naturally oriented, we may construct on $\tilde{M}$ a unit vector field composed of tangent (to $\tilde{M}$) vectors v orthogonal to orbits Gx and such that the natural orientation of $Gx \simeq G/H$ followed by v gives the orientation of $\tilde{M}$. By the existence result for ODEs, there exists a family of integral curves for this vector field. For each of the orbits $(0, Gx_i)$ or $(1, Gx_j)$, we may choose an oriented integral curve C, which starts or ends at any point of these orbits. The assumption that $\tilde{h}$ is regular normal prevents the curve C from leaving the region $[0, 1] \times U_k$ and passing through a point in $[0, 1] \times \Omega^H$ having a smaller orbit type. In addition, if C joins two orbits of points $\tilde{h}^{-1}(0) \cap U_k$, then these two orbits will remain the same independent of the choice of a starting point, for the curve C. Consequently, any such oriented curve C contained in $[0, 1] \times U_k$ must be one of the following types:

(i) It contains points $(0, x_i)$ and $(1, x_j)$, and

$$\text{sign } D\tilde{f}_0(x_i)|_{\tilde{S}_i} = \text{sign } Df_0(x_j)|_{S_j}$$

(ii) It contains points $(0, x_i)$ and $(0, x_i^*)$ with x_i, $x_i^* \in \tilde{f}_k^{-1}(0)$, where $x_i^* \notin Gx_i$, and

$$\text{sign } D\tilde{f}_0(x_i)|_{\tilde{S}_i} = -\text{sign } D\tilde{f}_0(x_i^*)|_{\tilde{S}_i^*}$$

(iii) It contains points $(1, x_j)$ and $(1, x_j^*)$ with x_j, $x_j^* \in f_k^{-1}(0)$, where

$x_j^* \notin Gx_j$, and

$$\operatorname{sign} Df_0(x_j)|_{S_j} = -\operatorname{sign} Df_0(x_j^*)|_{S_j^*}$$

Consequently, it follows that the formula (6.3.1) applied to $\tilde{f}_0$ and f_0 will give the same number.

The following properties of the S^1-degree can be verified in a standard way. See Dylawerski et al. (1991), or Gęba et al. (1994).

Theorem 6.3.1

Let V be an orthogonal representation of $G = S^1$ and $\Omega \subseteq V \oplus \mathbb{R}$ an open, bounded, invariant subset. For every Ω-admissible G-map $f : \overline{\Omega} \to V$, we can assign an element $S^1\text{-deg}(f, \Omega) \in \bigoplus_{k=1}^{\infty} \mathbb{Z}$ such that the following properties are satisfied:

(P1) *Existence: If $S^1\text{-deg}(f, \Omega) = \{\deg_k(f, \Omega)\} \neq 0$, that is, there is $k \in \{1, 2, \ldots\}$ such that $\deg_k(f, \Omega) \neq 0$, then there exists $x \in f^{-1}(0)$ with $\mathbb{Z}_k \subseteq G_x$.*

(P2) *Homotopy Invariance: If $h : [0, 1] \times \overline{\Omega} \to V$ is an Ω-admissible G-equivariant homotopy, then $S^1\text{-deg}(h_t, \Omega)$ does not depend on $t \in [0, 1]$, where G acts trivially on $[0, 1]$.*

(P3) *Excision: If $\Omega_0 \subseteq \Omega$ is an open and invariant subset such that $f^{-1}(0) \subseteq \Omega_0$, then*

$$S^1\text{-deg}(f, \Omega) = S^1\text{-deg}(f, \Omega_0)$$

(P4) *Additivity: If Ω_1 and Ω_2 are two open, invariant subsets of Ω such that $\Omega_1 \cap \Omega_2 = \emptyset$ and $f^{-1}(0) \subseteq \Omega_1 \cup \Omega_2$, then*

$$S^1\text{-deg}(f, \Omega) = S^1\text{-deg}(f, \Omega_1) + S^1\text{-deg}(f, \Omega_2)$$

(P5) *Product: Let U be another orthogonal representation of $G = S^1$, $\Omega \subseteq U \times V \oplus \mathbb{R}$ an open, bounded, invariant subset, and $\tilde{f} : \overline{\Omega} \to U \times V$ an Ω-admissible map such that $\tilde{f}(u, v, \lambda) = (u, f(v, \lambda)) \in U \times V$ for $(u, v, \lambda) \in U \times V \oplus \mathbb{R}$. Then*

$$S^1\text{-deg}(\tilde{f}, \Omega) = S^1\text{-deg}(f|_{\overline{\Omega}_0}, \Omega_0)$$

where $\Omega_0 = \Omega \cap \mathrm{V} \oplus \mathbb{R}$.

Example 6.3.2 We now demonstrate the definition of S^1-degree by an example. Assume that the action of S^1 on $V = \mathbb{C}$ is given by $\gamma \cdot z = \gamma^n z$, where $\gamma \in S^1$, $z \in \mathbb{C}$, and $n \in \{1, 2, \ldots\}$ is a fixed integer. Let $\Omega \subseteq W :=$

$V \oplus \mathbb{R}$ be the set

$$\Omega := \{(z,t) \in \mathbb{C} \times \mathbb{R};\ \tfrac{1}{2} < |z| < 2;\ -1 < t < 1\}$$

and we define $f : \overline{\Omega} \to V$ by

$$f(z,t) = \frac{z}{|z|}(1 - |z| + it)\,, \qquad (z,t) \in \overline{\Omega}$$

We want to compute the S^1-degree of f in Ω. First, we note that Ω contains points of only one orbit type $\mathbb{Z}_n$ and $f^{-1}(0) = \{(z,0);\ |z| = 1\}$. We choose the point $(1,0) \in f^{-1}(0)$ and consider the slice S to the orbit $f^{-1}(0)$ at $(1,0)$, that is, $S = \{(x,0,t) \in \mathbb{R}^3;\ x,t \in \mathbb{R}\} \subseteq V \oplus \mathbb{R}$. See Figure 6.3.3.

The chosen basis of S is such that the oriented vector to $f^{-1}(0)$ at $(1,0,0)$, that is, the vector $(0,1,0)$, followed by this basis of S, is a basis of $\mathbb{R}^3$ giving the same orientation as the standard basis $e_1 = (1,0,0)$, $e_2 = (0,1,0)$, $e_3 = (0,0,1)$. For example, we can choose the basis $\{(0,0,1), (1,0,0)\}$. Next, we consider the restriction f_0 of f to a neighborhood of $(1,0,0)$ in S that can be expressed by

$$f_0(t,x) = \frac{x}{x}(1 - x + it) = (1 - x, t)$$

Since f is a regular normal mapping, we can compute the degree as

$$\det{}_{\mathbb{R}} Df_0(0,1) = \det{}_{\mathbb{R}} \begin{bmatrix} 0 & 1 \\ -1 & 0 \end{bmatrix} = 1 > 0$$

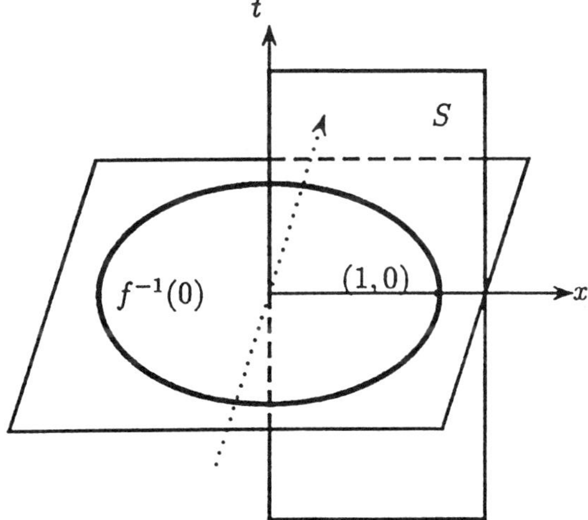

FIGURE 6.3.3. The orbit $f^{-1}(0)$ and the slice S to the orbit $f^{-1}(0)$ at $(1,0)$.

Therefore,

$$\deg_k(f,\Omega)=\begin{cases}1 & \text{if } k=n\\ 0 & \text{if } k\neq n\end{cases}$$

In the remaining part of this section, we develop several computational formulae for the S^1-degree.

We first consider an open invariant subset $\Omega\subseteq V\oplus\mathbb{R}$ and an S^1-equivariant Ω-admissible C^1-mapping $f:\overline{\Omega}\rightarrow V$ such that:

(i) 0 is a regular value of f.
(ii) $f^{-1}(0)\cap\Omega=Gx_0$ for some $x_0\in\Omega$.
(iii) $G_{x_0}=\mathbb{Z}_n=:H$

Denote by S the linear orthogonal slice of the orbit Gx_0 and x_0, that is, $S=\{x\in V\oplus\mathbb{R};\ x-x_0\perp T_{x_0}(Gx_0)\}$. It is easy to verify that S is an orthogonal representation of H and $A:=Df(x_0)|_S:S\rightarrow V$ is an H-equivariant linear mapping. As before, we fix a basis in S such that the oriented vector to Gx_0 at x_0, followed by the basis of S, determines the same orientation in $V\oplus\mathbb{R}$. We consider the subspaces S^H and $S^{\perp}:=(S^H)^{\perp}$; thus, $S=S^H\oplus S^{\perp}$. The subspaces S^H and $S^{\perp}$ are H-invariant. Let $B:S\rightarrow V$ be an S^1-equivariant isomorphism preserving the orientations in S and V. We can identify the spaces S and V via this isomorphism. Since $B(S^H)=V^H$ and $B(S^{\perp})=(V^H)^{\perp}$, under the above identification, we note that S^H can be identified with V^H and $S^{\perp}$ with $(V^H)^{\perp}=:V^{\perp}$.

We denote by $A^H:=A|_{S^H}$ and $A^{\perp}:=A|_{S^{\perp}}$. Then we have the following block-matrix representation of $A:S^H\oplus S^{\perp}\rightarrow V^H\oplus V^{\perp}$:

$$A:=\begin{bmatrix}A^H & 0\\ 0 & A^{\perp}\end{bmatrix}$$

Theorem 6.3.3 (Regular Value Formula)

Under the above assumptions, S^1-deg$(f,\Omega)=\{\deg_k(f,\Omega)\}$ is given by

$$\deg_k(f,\Omega)=\begin{cases}\operatorname{sign} A^H & \textit{if } k=n\\ -\operatorname{sign} A^H & \text{if } k=\dfrac{n}{2} \textit{ and } \det A^{\perp}<0\\ 0 & \textit{otherwise}\end{cases}$$

Proof. We consider a G-tube U around the orbit Gx_0, that is, $U=\varphi(G\times D)$, where $D=\{v\in S;\ |v|<\delta\}$ and $\varphi(g,v)=g(x_0+v)$ for $v\in D$ and $g\in G$. It follows from the excision property that we may assume, without loss of generality, that $\Omega=U$ is a tube with the slice D.

Note that f is not necessarily a regular normal mapping. Our next goal is to approximate it, up to an equivariant homotopy, by a regular normal mapping. Since it is enough to deform the mapping $f_0 := f|_{\overline{D}} : \overline{D} \to V$ to a regular normal H-equivariant mapping, we will apply several H-equivariant deformations to the mapping f_0. First, we remark that f_0 can be replaced by A (by assuming that $\delta > 0$ is a sufficiently small number). $A : S \to V$ is an H-equivariant isomorphism. By Corollary 6.1.7 (the complete reducibility theorem), $S = S^H \oplus S_{\xi_1} \oplus \cdots \oplus S_{\xi_m}$, where S_{ξ_j} are irreducible subrepresentations of $H = \mathbb{Z}_n$. By Example 6.1.31, an irreducible subrepresentation S_{ξ_j} is equivalent to either the (complex) representation $\xi_k : \mathbb{Z}_n \to S^1$; $\xi(\gamma) = \gamma^k$, $k \in \{1, 2, \ldots, n-1\}$, $\gamma \in \mathbb{Z}_n \subset S^1$, and $\xi(\mathbb{Z}_n) \neq \mathbb{Z}_2$ (i.e., $k \neq n/2$), or to the one-dimensional (real) representation $\xi : \mathbb{Z}_n \to \mathbb{Z}_2$; $\xi(\gamma) = \gamma^{n/2}$, for even n. We denote by S_k the direct sum of all subrepresentations S_{ξ_j} equivalent to the representation $\xi_k(\gamma) = \gamma^k$, $k = \{1, 2, \ldots\}$. S_k is called the *isotypical* component of S^1 corresponding to ξ_k. Consequently, we obtain the following isotypical decomposition of S:

$$S = S^H \oplus S_1 \oplus \cdots \oplus S_{n-1}$$

(where we do not exclude that $S_j = \{0\}$). It follows from Schur's lemma (Theorem 6.1.10) that if $j \neq n/2$, then S_j has a natural complex structure (induced by the action of $\mathbb{Z}_n = H$ on S_j) and a linear operator $T : S_j \to S_j$ is H-equivariant if and only if it is $\mathbb{C}$-linear with respect to this complex structure.

If $j = n/2$, then $\mathbb{Z}_n$ acts on S_j as $\mathbb{Z}_2 = \mathbb{Z}_n / \mathbb{Z}_j$. Therefore, every linear operator $T : S_j \to S_j$ is H-equivariant. On the other hand, A is H-equivariant, so $A(S_j) = S_j$ for all j and $A(S^H) = S^H$ (under the H-equivariant identification B of S with V). We denote by A_j the restriction of A to the subspace S_j. Consequently, the operator A can be represented by the following matrix:

$$A = \begin{bmatrix} A^H & 0 & \cdots & 0 \\ 0 & A_1 & \cdots & 0 \\ & & \ddots & \\ 0 & 0 & \cdots & A_{n-1} \end{bmatrix}$$

Since for $j \neq n/2$, S_j can be identified with $\mathbb{C}^{n_j}$ and $A_j \in GL(n_j, \mathbb{C})$, by Proposition 6.1.35, the isomorphism A_j can be connected by a path in $GL(n_j; \mathbb{C})$ to Id. Consequently, A is H-equivariantly homotopic to the following H-equivariant operator with the matrix representation:

$$\tilde{A} := \begin{bmatrix} A^H & 0 & 0 \\ 0 & A_{n/2} & 0 \\ 0 & 0 & \mathrm{Id} \end{bmatrix}$$

where we can assume, without loss of generality, that $n = 2k$. It is clear that $\operatorname{sign}\det_{\mathbb{R}}(A^{\perp}) = \operatorname{sign}\det_{\mathbb{R}}(A_k)$. If $\operatorname{sign}\det_{\mathbb{R}}(A_k) > 0$, then A_k can be connected by a path in $GL(S_k)$ to Id. In this case, $\tilde{A}$ can be replaced by the following operator $\hat{A}$ with the matrix representation

$$\hat{A} := \begin{bmatrix} A^H & 0 \\ 0 & \mathrm{Id} \end{bmatrix}$$

This matrix $\hat{A}$ defines a regular normal mapping $\hat{f}:\overline{\Omega}\to V$ and by the definition

$$\deg_j(f,\Omega) = \begin{cases} \operatorname{sign}\det_{\mathbb{R}}(A^H) & \text{if } j=n \\ 0 & \text{otherwise} \end{cases}$$

If $\det_{\mathbb{R}}(A_k)<0$, then A_k can abe connected by a continuous path in $GL(S_k)$ to an operator $\tilde{A}_k$ that has the following representation:

$$\tilde{A}_k(t_1, t_2, \ldots, t_N) = (-t_1, t_2, \ldots, t_N), \qquad (t_1, \ldots, t_N) \in S_k$$

with respect to a certain basis in S_k. Let $C_k: S_k \to S_k$ be given by

$$C_k(t_1, t_2, \ldots, t_N) = \left(-t_1\left(t_1 - \frac{\delta}{2}\right)\left(t_1 + \frac{\delta}{2}\right), t_2, \ldots, t_N\right)$$

and define $C: S\to V$ by

$$C(v) = \begin{cases} A^H v & \text{if } v\in S^H \\ v & \text{if } v\in S_j,\ j\neq k \\ C_k(v) & \text{if } v\in S_k \end{cases}$$

The mapping $\tilde{f}:\overline{\Omega}\to V$ determined by C is a regular normal mapping Ω-homotopic to f. Since $C^{-1}(0)$ is composed of exactly two H-orbits: $\{0\}$ and $\left(\pm\frac{\delta}{2}, 0, \ldots, 0\right)$ the S^1-degree may be computed now from the definition. Indeed, $\tilde{f}^{-1}(0) = Gx_0 \cup Gx_1$, where x_1 is the point $x_0 + \left(\frac{\delta}{2}, 0, \ldots, 0\right)$, $G_{x_0} = \mathbb{Z}_n$, $G_{x_1} = \mathbb{Z}_k$, and $2k=n$. Therefore,

$$\deg_j(\tilde{f},\Omega) = \begin{cases} \operatorname{sign}\deg_{\mathbb{R}}(A^H) & \text{if } j=n \\ -\operatorname{sign}\det_{\mathbb{R}}(A^H) & \text{if } j=k \\ 0 & \text{otherwise} \end{cases}$$

This completes the proof. □

We illustrate the regular value formula with the following example.

Example 6.3.4 Let $V := \mathbb{C}\oplus\mathbb{C}$ and assume that S^1 acts on V by

$$\gamma\circ(z_1, z_2) = (\gamma^n z_1, \gamma^{2n} z_2), \qquad (z_1, z_2)\in V,\ \gamma\in S^1$$

where $n \in \mathbb{N}$. Let $\Omega = \{(z_1, z_2, t) \in V \oplus \mathbb{R};\ |z_1| < 1,\ \frac{1}{2} < |z_2| < 2,\ -1 < t < 1\}$, and we define $f : \overline{\Omega} \to V$ by

$$f(z_1, z_2, t) = \left(\bar{z}_1 z_2, \frac{z_2}{|z_2|}(1 - |z_2| + it)\right), \qquad (z_1, z_2, t) \in \overline{\Omega}$$

The mapping f is S^1-equivariant. Indeed, we have

$$\begin{aligned} f(\gamma \circ z_1, \gamma \circ z_2, t) &= f(\gamma^n z_1, \gamma^{2n} z_2, t) \\ &= \left(\gamma^{-n} \bar{z}_1 \gamma^{2n} z_2, \frac{\gamma^{2n} z_2}{|z_2|}(1 - |z_2| + it)\right) \\ &= \left(\gamma^n (\bar{z}_1 z_2), \frac{\gamma^{2n} z_2}{|z_2|}(1 - |z_2| + it)\right) \\ &= \left(\gamma \circ (\bar{z}_1 z_2), \frac{\gamma \circ z_2}{|z_2|}(1 - |z_2| + it)\right) \\ &= \gamma \circ \left(\bar{z}_1 z_2, \frac{z_2}{|z_2|}(1 - |z_2| + it)\right) \\ &= \gamma \circ f(z_1, z_2, t) \end{aligned}$$

Since $f^{-1}(0) = \{(0, z_2, 0);\ |z_2| = 1\}$, $f^{-1}(0) = G \circ (0, 1, 0)$ is exactly the orbit of the element $x_0 = (0, 1, 0) \in \mathbb{C} \times \mathbb{C} \times \mathbb{R}$. The slice S to the orbit $f^{-1}(0)$ at $(0, 1, 0)$ is exactly

$$\begin{aligned} S &= \{(z_1, z_2, t) \in \mathbb{C} \times \mathbb{C} \times \mathbb{R};\ \operatorname{Im} z_2 = 0\} \\ &= \mathbb{C} \times \mathbb{R} \times \mathbb{R} \subseteq \mathbb{C} \times \mathbb{C} \times \mathbb{R} \end{aligned}$$

The restriction f_0 of f to the slice $S \cap \overline{\Omega}$ can be expressed by

$$f_0(z_1, t, x_2) = (\bar{z}_1, 1 - x_2 + it), \qquad (z_1, t, x_2) \in S \cap \overline{\Omega}$$

where the ordered basis

$$\begin{aligned} \{&e_1 = (1, 0, 0, 0, 0),\ e_2 = (0, 1, 0, 0, 0),\ e_3 = (0, 0, 0, 0, 1), \\ &e_4 = (0, 0, 1, 0, 0)\} \end{aligned}$$

is chosen according to the convention explained in the definition of the degree. The isotypical decomposition of S is exactly

$$S = S_n \oplus S_{2n}$$

where $S_n \cong \mathbb{C}$ is exactly the first $\mathbb{C}$-component of V and $S_{2n} = \{(t, x_2) \in \mathbb{R}^2;\ t, x_2 \in \mathbb{R}\}$ is the space of the stationary points of the action of $H = \mathbb{Z}_{2n}$, that is, $S_{2n} = S^H$. Consequently, $(S^H)^\perp = S_n = S^\perp$. The derivative $A = Df_0(0,1,0)$ has the following matrix:

$$
\begin{array}{c c}
 & \begin{array}{cccc} x_1 & y_1 & t & x_2 \end{array} \\
A = & \left[\begin{array}{cc|cc} 1 & 0 & & \\ 0 & -1 & \multicolumn{2}{c}{\mathbf{0}} \\ & & 0 & -1 \\ \multicolumn{2}{c}{\mathbf{0}} & 1 & 0 \end{array}\right] \begin{array}{c} x_1 \\ y_1 \\ x_2 \\ y_2 \end{array}
\end{array}
$$

Thus,

$$A^H = \begin{bmatrix} 0 & 1 \\ 1 & 0 \end{bmatrix} \quad \text{and} \quad A^\perp = \begin{bmatrix} 1 & 0 \\ 0 & -1 \end{bmatrix}$$

According to the regular value formula,

$$S^1\text{-deg}(f, \Omega) = \{\deg_k(f, \Omega)\}$$

where

$$\deg_k(f, \Omega) = \begin{cases} 1 & \text{if } k = 2n \\ -1 & \text{if } k = n \\ 0 & \text{otherwise} \end{cases}$$

We now establish a computational formula that will play an important role in Hopf bifurcation theory.

Let V denote a finite-dimensional, real, orthogonal representation of $G = S^1$ and let

$$V = V_1 \oplus V_2 \oplus \cdots \oplus V_k \oplus V_\infty$$

be the *isotypical* decomposition of V with respect to the S^1-action, that is, for $x \in V_j \backslash \{0\}$, we have $G_x = \mathbb{Z}_j$, $j = 1, 2, \ldots, k, \infty$. The subspaces V_j can be defined by the following formula:

$$V_j = P_j(V), \qquad P_j x = 2 \int_G \cos jt\, gx\, d\mu(g), \qquad j \neq \infty$$

$$\text{(6.3.2)} \qquad P_\infty x := \int_G gx\, d\mu(g), \qquad j = \infty$$

where $g = \cos t + i \sin t$. It is easy to verify that $P_j^2 = P_j$ and the adjoint operator P_j^* of P_j is exactly P_j. Therefore, P_j is an orthogonal projection. Since $\cos jt = \frac{1}{2}\chi_{\rho_j}(g)$, where χ_{ρ_j} is the character of the complex irreducible representation $\rho_j(g) = g^j$, $g \in S^1$, P_j is an G-equivariant linear operator. Moreover, if x belongs to a representation space of an irreducible subrepresentation of V that is equivalent to ρ_j, then $P_j(x) = x$, and if x belongs to a representation space of an irreducible subrepresentation of V that is not equivalent to ρ_j, then $P_j x = 0$.

For every $j \neq \infty$, the subspace V_j has the natural complex structure induced by the action of S^1 on V_j, namely,

$$(a + ib)x := ax + b \exp\left(i \frac{\pi}{2j}\right)x , \qquad a + ib \in \mathbb{C},\ x \in V_j$$

It can be easily shown that an $\mathbb{R}$-linear operator $A : V_j \to V_j$ is S^1-equivariant if and only if it is $\mathbb{C}$-linear.

Assume that $\omega : S^1 \to GL^G\left(\bigoplus_{i\neq\infty} V_i\right)$ is a continuous mapping, where for any orthogonal representation W of $G = S^1$, we denote by $GL^G(W)$ the subspace of all operators in $GL(W)$ that are G-equivariant. Since $A \in GL^G\left(\bigoplus_{i\neq\infty} V_i\right)$ implies $A(V_i) = V_i$ for all $i = 1, \dots, k$, the following restrictions are well defined:

$$\omega_i : S^1 \to GL^G(V_i)\,, \qquad \omega_i(\lambda) = \omega(\lambda)|_{V_i}\,, \qquad \lambda \in S^1$$

If $i \neq \infty$, then $GL^G(V_i) = GL_{\mathbb{C}}(V_i) = GL(d_i, \mathbb{C})$, $d_i = \dim_{\mathbb{C}} V_i$, where the last identification is made by choosing a complex basis in V_i. Consequently, we may define an integer μ_i by setting

$$\mu_i := \deg[\omega_i] = \deg(\det_{\mathbb{C}}(\omega_i(\cdot)), B_1(0))$$

In what follows, we call μ_i the *winding number* of ω_i. For every continuous map $\omega : S^1 \to GL^G\left(\bigoplus_{i\neq\infty} V_i\right)$, we define an element $\mu = \mu(\omega) = \{\mu_i\} \in \bigoplus_{i\neq\infty}^{\infty} \mathbb{Z}$, where μ_i is the winding number of ω_i if $1 \leq i \leq k$, and $\mu_i = 0$ if $i > k$. We call $\mu(\omega)$ the *winding element* of ω.

Assume $f : V \oplus \mathbb{R} \to V$ is an S^1-equivariant C^1-map satisfying the following hypothesis:

(A) There is an open, bounded invariant set $\Omega \subseteq V \oplus \mathbb{R}$ such that f is Ω-admissible, zero is a regular value for $f|_\Omega$, and $\Sigma := f^{-1}(0) \cap \Omega \subseteq V^G \oplus \mathbb{R}$ is diffeomorphic to S^1.

It is clear from the assumption (A) that $V^G \neq \{0\}$. We choose an orientation of V^G and we orient $V^G \oplus \mathbb{R}$ with the product orientation. Put $f_0 := f|_{V^G \oplus \mathbb{R}} : V^G \oplus \mathbb{R} \to V^G$. Since zero is a regular value of $f|_\Omega$, for $x \in \Sigma$, the derivative $Df_0(x)$ maps $N_x := (T_x\Sigma)^{\perp} \cap V^G \oplus \mathbb{R}$ S^1-isomorphically onto $V^G = V_\infty$, and thus it induces an orientation of N_x. Let $\eta : S^1 \to \Sigma$ be a diffeomorphism such that the chosen orientation of $T_x\Sigma$ followed by the orientation of N_x induced by η gives the chosen orientation of $V^G \oplus \mathbb{R}$. We define $\omega : S^1 \to GL^G\Big(\bigoplus_{i\neq\infty} V_i\Big)$ by

$$\omega(\lambda) = Df(\eta(\lambda))|_{\oplus_{i\neq\infty} V_i} \in GL^G\Big(\bigoplus_{i\neq\infty} V_i\Big)$$

Consequently, the winding element $\mu(\omega) \in \bigoplus_{j=1}^{\infty} \mathbb{Z}$ is well defined. We have the following.

Theorem 6.3.5

Suppose $f : V \oplus \mathbb{R} \to V$ satisfies assumption (A). Then S^1-deg$(f, \Omega) = \mu(\omega)$.

Proof. We begin with some simplification by using the excision and homotopy properties of the S^1-equivariant degree. We can assume, without loss of generality, that:

(i) $\Omega_0 = \Omega \cap V^G \oplus \mathbb{R}$ is a small G-invariant neighborhood of Σ in $V^G \oplus \mathbb{R}$, where every element $x \in \Omega_0$ can be written uniquely as $x = \sigma + w$, $\sigma \in \Sigma$, and w is normal to Σ at σ. We will denote by $\pi : \overline{\Omega_0} \to \Sigma$ the natural projection $\pi(x) = \sigma$.

(ii) Ω is the product $\prod_{i\neq\infty} B(V_i) \times \Omega_0$, where $B(V_i)$ denotes the unit ball in the isotypical component V_i. We will also denote

$$V^* = \bigoplus_{i\neq\infty} V_i\,, \qquad \Omega^* = \prod_{i\neq\infty} B(V_i)$$

(iii) $f(v, x) = f_0(x) + A(\pi(x))v$, where $x \in \Omega_0$, $v \in \overline{\Omega^*}$ and $A(\pi(x)) = Df(\pi(x))|_{V^*} : V^* \to V^*$.

We will use the notation $\lambda = \eta^{-1}(\sigma) \in S^1$, $\sigma \in \Sigma$, to denote the correspondence between $\lambda \in S^1$ and $\sigma \in \Sigma$.

We write the isotypical decomposition of V as $V = \bigoplus V_i$. For a vector $v \in V = V_1 \oplus V_2 \oplus \cdots \oplus V_k \oplus V_\infty$, we denote by v_i the V_i-component of v and we put $v^j = v_1 + v_2 + \cdots + v_j$. For all $i = 1, 2, \ldots, k$, we can choose a complex basis for V_i and identify it with $\mathbb{C}^{d_i}$, where $d_i = \dim_{\mathbb{C}} V_i$. By Proposition 6.1.39, the homotopy class of $(A \circ \eta)_i : S^1 \to GL(d_i, \mathbb{C})$ contains a representative $b_i : S^1 \to GL(d_i, \mathbb{C})$ that has the following matrix form:

$$b_i(\lambda) = \begin{bmatrix} \lambda^{\mu_i} & 0 & \cdots & 0 \\ 0 & 1 & \cdots & 0 \\ \vdots & \vdots & \ddots & \vdots \\ 0 & 0 & \cdots & 1 \end{bmatrix} : \mathbb{C}^{d_i} \to \mathbb{C}^{d_i}$$

where $\mu_i := \deg([\omega_i])$.

For $i = \infty$, we can choose a (real) basis for $V_\infty = V^G$ and identify it with $\mathbb{R}^{d_\infty}$, where $d_\infty = \dim_{\mathbb{R}} V_\infty$. If the homotopy class of $\omega_\infty := (A \circ \eta)_\infty$ is trivial, then it contains a representative

$$b_\infty(\lambda) = \begin{bmatrix} (-1)^\nu & 0 & \cdots & 0 \\ 0 & 1 & \cdots & 0 \\ \vdots & \vdots & \ddots & \vdots \\ 0 & 0 & \cdots & 1 \end{bmatrix} : \mathbb{R}^{d_\infty} \to \mathbb{R}^{d_\infty}$$

where $\nu = 0$ if $\det A_\infty(x) > 0$ and $\nu = 1$ if $\det A_\infty(x) < 1$, for some $x \in \Sigma$. Otherwise, it contains a representative

$$b_\infty(\lambda) = \begin{bmatrix} (-1)^\nu \cos(l\delta) & -\sin(l\delta) & \cdots & 0 \\ (-1)^\nu \sin(l\delta) & \cos(l\delta) & \cdots & 0 \\ \vdots & \vdots & \ddots & \vdots \\ 0 & 0 & \cdots & 1 \end{bmatrix} : \mathbb{R}^{d_\infty} \to \mathbb{R}^{d_\infty}$$

where $\lambda = \cos\delta + i \sin\delta$, and l is an integer.

Consequently, we can define a mapping $g : \overline{\Omega} \to V$ by

$$g(v, x) = f_0(x) + a(\pi(x))v \,, \qquad v \in V^* \,, \quad x \in \overline{\Omega^0}$$

where for $a(\sigma) = \sum_i a_i(\sigma)$, $a_i(\sigma) : V_i \to V_i$, $a_i(\sigma) = (b_i \circ \eta^{-1})(\sigma)$ for $j = 1, 2, \ldots, \infty$ and $\sigma \in \Sigma$.

It follows from the homotopy invariance that S^1-$\deg(f, \Omega) = S^1$-$\deg(g, \Omega)$.

We will show, without loss of generality, that we can always assume the homotopy class of $\omega_\infty : S^1 \to GL(V_\infty)$ is trivial. To achieve this, suppose that the homotopy class of ω_∞ is not trivial. We may assume for simplicity that $V_\infty \simeq \mathbb{C}$ and $\nu = 0$. So, $b_\infty(\lambda)z = \lambda^l z$, $z \in \mathbb{C}$. We need the following.

Lemma 6.3.6

Let $U = \{z \in \mathbb{C};\ \frac{1}{4} < |z| < 1\}$ *and* $\varphi : \overline{U} \to \mathbb{C}$ *be given by* $\varphi(z) = 4(1 - 2|z|)z$, $z \in \overline{U}$. *Then* $\deg(\varphi, U) = 0$.

Proof. By identifying $\mathbb{C}$ with $\mathbb{R}^2$, we can define $\varphi_\varepsilon(x, y) = (4x(1 - 2\sqrt{x^2 + \varepsilon y^2}),\ 4y(1 - 2\sqrt{\varepsilon x^2 + y^2}))$, where ε is a number sufficiently close to

1. Then by the homotopy property, $\deg(\varphi, U) = \deg(\varphi_\varepsilon, U)$. On the other hand, it is easy to verify that the equation $\varphi_\varepsilon(x, y) = 0$ has the following solutions $(\pm\frac{1}{2}, 0)$; $(0, \pm\frac{1}{2})$; $\left(\pm\frac{1}{2\sqrt{1+\varepsilon}}, \pm\frac{1}{2\sqrt{1+\varepsilon}}\right)$; $\left(\pm\frac{1}{2\sqrt{1+\varepsilon}}, \mp\frac{1}{2\sqrt{1+\varepsilon}}\right)$. Direct computation gives $\deg(\varphi_\varepsilon, U) = 0$. □

Returning to the proof of Theorem 6.3.5, we introduce the following piecewise linear functions

$$q_j(t) = \begin{cases} 1 & \text{if } 0 \le t < s_j \\ -\dfrac{1}{\varepsilon_j}(t - t_j) & \text{if } s_j \le t < t_j \\ 0 & \text{if } t_j \le t \end{cases}$$

where

$$\begin{cases} s_j = \dfrac{j}{j+1} - \dfrac{1}{2(j+2)^2} \\ t_j = \dfrac{j}{j+1} + \dfrac{1}{2(j+2)^2} \\ \varepsilon_j = t_j - s_j = \dfrac{1}{(j+1)^2}, \qquad j = 1, 2, \ldots \end{cases}$$

Clearly, $\dfrac{s_j + t_j}{2} = \dfrac{j}{j+1}$, $q_j\left(\dfrac{j}{j+1}\right) = \frac{1}{2}$, and $q_j(t) \neq 0, 1$ if $t \in (s_j, t_j)$. See Figure 6.3.4.

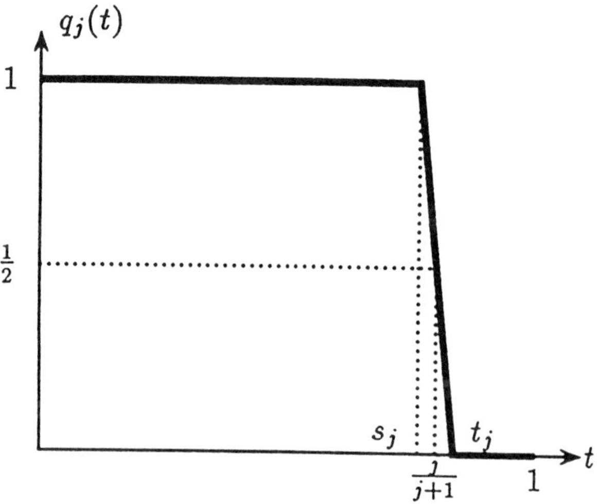

FIGURE 6.3.4. The piecewise linear function q_j.

Define the map $c_\infty : S^1 \times \mathbb{C} \to \mathbb{C}$ by $c_\infty(\lambda, z) = q_1(|z|)z + (1 - q_1(|z|)b_\infty(\lambda)z$. It is easy to verify that $c_\infty(\lambda, z) = 0$ if and only if $z = 0$ or $|z| = \frac{1}{2}$ and $\lambda^l = -1$. Define $\tilde{f} : \overline{\Omega} \to V$ by $\tilde{f}(v, x) = (c_\infty(\eta^{-1}(\pi(x)), z), f(v, x))$, where $x \in \overline{\Omega}_0$, $v \in \overline{\Omega}^*$, and $x = (z, t) \in V_\infty \times \mathbb{R}$. By the homotopy invariance, we have S^1-deg$(f, \Omega) = S^1$-deg$(\tilde{f}, \Omega)$. The set $\tilde{f}^{-1}(0) \cap \Omega = \Sigma \cup \bigcup_{j=1}^{k} \Sigma_j$, where $\Sigma_j = \{(v, z, t) \in V^* \times \mathbb{C} \times \mathbb{R};\ |z| = \frac{1}{2},\ (z, t) = x \in \Sigma,\ \eta^{-1}(x) = \lambda_j,\ v = 0\}$ for $\lambda_j = e^{i(2j-1/l)\pi}$, and $j = 1, \dots, l$. Since in a small neighborhood Ω_j of Σ_j, $j = 1, \dots, l$, the mapping $\tilde{f}$ is S^1-homotopic to the mapping $\hat{f}(v, x) = (4(1-2)|z|)z,\ f(v, x)) = (\varphi(z),\ f(v, x))$, it follows from Lemma 6.3.6 that $\hat{f}$ is S^1-homotopic to a mapping $g : \Omega_j \to V$ such that $g(v, x)$ has no solution in Ω_j. This implies, by excision and additivity properties, that S^1-deg$(f, \Omega) = S^1$-deg$(\tilde{f}, \tilde{\Omega})$, where $\tilde{\Omega}$ is a small neighborhood of Σ, and the mapping ω_∞, defined for $\tilde{f} : \tilde{\Omega} \to V$, has a trivial homotopy class. Consequently, we can assume, without loss of generality, that the homotopy class of ω_∞ is trivial. In the next step, we further deform g to a new mapping h that will allow us to use the regular value formula. We define

$$c_i : V^* \times \overline{\Omega_0} \to V_i\,, \qquad i = 1, 2, \dots, k, \infty$$

by the following formula:

$$\begin{cases} c_\infty(v, x) = a_\infty(\pi(x))v_\infty \\ c_i(v, x) = q_{N-i+1}(|v^i|)v_j + (1 - q_{N-i+1}(|v^i|)a_i(\pi(x))v_i\,, \qquad i = 1, \dots, k \end{cases}$$

where $v \in V$, $x \in \overline{\Omega_0}$.

Let $h : \Omega \to V \oplus \mathbb{R}$ be defined by

$$h(v, x) = f_0(x) + c(v, x)\,, \qquad x \in \overline{\Omega_0}\,, \quad v \in V^*$$

where $c(x, v) = \bigoplus c_i(x, v)$, $i = 1, \dots, k, \infty$. Then again, by the homotopy invariance

$$S^1\text{-deg}(f, \Omega) = S^1\text{-deg}(h, \Omega)$$

It is clear that $\Sigma \subseteq h^{-1}(0)$. In order to describe other zeros of h in Ω, we remark that $h(v, x) = 0$ if and only if $x \in \Sigma$ and $c_i(x, v) = 0$ for $i = 1, \dots, k, \infty$. This implies $v_\infty = 0$ and that for $i = 1, \dots, k$,

$$q_{N-i+1} \cdot (|v^i|)v_i + (1 - q_{N-i+1}(|v^i|))a_i(\pi(x))v_i = 0$$

Assume that for some $i \in \{1, \dots, k\}$, $v_i \neq 0$. Then it follows from the definition of $a_i(\pi(x))$ that $v_i = (z_i, 0, \dots, 0) \in \mathbb{C}^{d_i}$, $i \neq 0$, and $q_{N-i+1}(|v^i|) + (1 - q_i(|v^i|))\lambda^{\mu_i} = 0$, $\lambda = \eta^{-1}(\pi(x))$. Therefore, $q_{N-i+1}(|v^i|) = \frac{1}{2}$, $|v^i| =$

$\frac{N-i+1}{N-i+2}$, $\lambda^{\mu_i}=-1$. Let us denote by $\lambda_1, \ldots, \lambda_{|\mu_i|}$ the complex roots of the equation $\lambda^{\mu_i}=-1$. We claim that if (v,x) is a solution of the equation $h(v,x)=0$, then the vector v cannot have more than one nonzero component v_i. Indeed, suppose for contradiction, that v has two nonzero components v_i and v_j, $1\le i<j\le k$; then $|v^j|\ge|v^i|$ and $|v^i|=\frac{N-i+1}{N-i+2}>\frac{N-j+1}{N-j+2}=|v^j|$. This is clearly a contradiction. Consequently, we can classify all the solutions of the equation $h(v,x)=0$ as follows:

(i) The orbit Σ.

(ii) For every $i\in\{1,\ldots,k\}$, there are $|\mu_i|$ orbits of solutions $\Sigma_{i,\ell}:=\{x_\ell\}\times Gv_i$, where $x_\ell=\eta(\lambda_\ell)$, $v_i\ne 0$, $|v_i|=|v|=\frac{N-i+1}{N-i+2}$, and thus $|v|\ge\frac{1}{2}$ for all i.

Case I. We compute S^1-deg(h,U) in a neighborhood U of Σ. Let $U=B(\frac{1}{3})\oplus\Omega_0$, where $B(\frac{1}{3})=\{v\in V^*;\ |v|<\frac{1}{3}\}$. Since the mapping h transforms vectors $v\in V^*$, $|v|<\frac{1}{3}$, identically onto themselves, that is, $h(v,x)=h(x)+v$, $v\in V^*$, $x\in\Omega_0$, h is U-admissible, and $h|_U$ is G-normal. Therefore, it is normal. Consequently, S^1-deg$(h,U)=0$.

Case II. We fix an integer $i\in\{1,\ldots,k\}$ such that $\mu_i\ne 0$. Then there are μ_i-orbits of solutions $\Sigma_{i,\ell}$ to the equation $h(v,x)=0$ such that $v=v_i$, $|v_i|=\frac{N-i+1}{N-i+2}$, $v_i=(z_i,0,\ldots,0)\in\mathbb{C}^{d_i}$, and $x=\eta(\lambda)$, $\lambda^{\mu_i}=-1$. In order to compute the S^1-degree of h in a neighborhood of $\Sigma_{i,\ell}$, we need to apply the regular value formula.

Let D be a small neighborhood of $\Sigma_{i,\ell}$ such that $\Sigma_{i,\ell}=h^{-1}(0)\cap D$. The orbit $\Sigma_{i,\ell}=\{x_l\}\times Gv_i$ is diffeomorphic to S^1. We denote by $H_i:=\mathbb{Z}_i$ the isotropy group G_{v_i} and we consider the linear orthogonal slice S to the orbit $\{x_l\}\times Gv_i$ at $v_i^0=(t_i^0,0,\ldots,0)\in\mathbb{R}^{d_i}$, $t_i^0=\frac{N-i+1}{N-i+2}$. It is clear that

$$S=\hat{V}\oplus\mathbb{R}\,,\qquad \hat{V}=\hat{V}_i\oplus\bigoplus_{j\ne i}V_j\,,$$

where

$$\hat{V}_i=\{(t,z_2,\ldots,z_{d_i})\in\mathbb{C}^{d_i}; t\in\mathbb{R}, z_\ell\in\mathbb{C}, \ell=2,\ldots,d_i\}$$

Let $\tilde{A}:=Dh(v_i^0,x_l)|_S: S\to V\oplus\mathbb{R}$. We need to compute explicitly the H_i-equivariant operator $\tilde{A}$. The matrix representation of $Dh(v_i^0,x_l)$ is the following:

$$Dh(v_j^0, x) = \begin{array}{c} \begin{array}{cc} V^* & V^G \oplus \mathbb{R} \end{array} \\ \begin{bmatrix} D_v c(v_j^0, x) & D_x c(v_j^0, x) \\ 0 & Df_0(x) \end{bmatrix} \end{array} \begin{array}{c} V^* \\ V^G \end{array}$$

We consider the decomposition

$$V^G \oplus \mathbb{R} = T_{x_l}\Sigma \oplus (T_{x_l}\Sigma)^\perp$$

where $T_{x_l}\Sigma$ is oriented by η. Then

$$Df_0(x) = \begin{array}{c} \begin{array}{cc} \boldsymbol{T_x\Sigma} & \boldsymbol{(T_x\Sigma)^\perp} \end{array} \\ [0 \quad D_0] \end{array} \; V^G \qquad (6.3.3)$$

where $D_0 : (T_x\Sigma)^\perp \to \mathbb{R}$ is an isomorphism preserving the chosen orientations of $(T_x\Sigma)^\perp$ and V^G.

In the case $j \neq i$, the derivative $D_{v_j} c(v_i^0, x_l)$ is simply the identity operator. If $j = i$, the derivative $D_{v_i} c_i(v_i^0, x_l)$ has the following matrix representation under the fixed identification of V_i with $\mathbb{C}^{d_i}$:

$$\begin{array}{c} \begin{array}{cc} \mathbb{C} & \mathbb{C}^{d_i - 1} \end{array} \\ \begin{bmatrix} C & * \\ 0 & \mathrm{Id} \end{bmatrix} \end{array} \begin{array}{c} \mathbb{C} \\ \mathbb{C}^{d_i - 1} \end{array}$$

where the matrix C is the Jacobi matrix with respect to $z \in \mathbb{C}$ of the mapping $z \mapsto q_{N-i+1}(|z|)z + (1 - q_{N-i+1}(|z|))\lambda^{d_i} z$ at the point $z_0 = t_i^0 = \dfrac{N-i+1}{N-i+2}$, where $\lambda = \eta^{-1}(x_l)$. By direct computation, we obtain

$$C = \begin{bmatrix} \dfrac{-t_i^0}{\varepsilon_{N-i+1}} & 0 \\ 0 & 0 \end{bmatrix}$$

For the simplicity of computations, we choose the coordinate system in $V^G \oplus \mathbb{R} = T_x\Sigma \oplus (T_x\Sigma)^\perp$ such that the last coordinate vector vector is exactly the vector $\dot{\eta}(1)$. According to our choice of orientations, we can replace in (6.3.2), without loss of generality, the matrix D_0 by the identity matrix. We need to compute $D_{\dot{\eta}} c_j(v_j^0, \eta(\lambda_l))$. It follows directly from the chain rule that

$$D_{\dot{\eta}} c_i(v_i^0, \eta(\lambda_l)) = \begin{bmatrix} 0 \\ -\dfrac{\mu_i t_i^0}{2} \end{bmatrix}$$

Therefore, we have the following matrix representation of $\tilde{A} : S \to V \simeq S$:

$$\tilde{A} = \begin{array}{c} \begin{array}{cccc} \hat{V}_i & T_{x_l}\Sigma & (T_{x_l}\Sigma)^{\perp} & \bigoplus_{j\neq i} V_j \end{array} \\ \left[\begin{array}{cccc} K & 0 & 1 & * \\ 0 & -\dfrac{\mu_i t_i^0}{2} & * & * \\ 0 & 0 & D_0 & * \\ 0 & 0 & 0 & \tilde{a} \end{array}\right] \begin{array}{l} \hat{V}_i \\ \tilde{V}_i \\ V^G \\ \oplus_{j\neq i} V_j \end{array} \end{array}$$

where

$$K = \begin{array}{c} \begin{array}{cc} \mathbb{R} & \mathbb{C}^{d_i - 1} \end{array} \\ \left[\begin{array}{cc} -\dfrac{t_i^0}{\varepsilon_{N-i+1}} & 0 \\ 0 & \mathrm{Id} \end{array}\right] \begin{array}{l} \mathbb{R} \\ \mathbb{C}^{d_i-1} \end{array} \end{array}$$

$\tilde{a} = D_v c(x_l, v_i^0)$ and $V_i = \hat{V}_i \oplus \tilde{V}_i$. Since $\operatorname{sign}\det \tilde{A} = \operatorname{sign} \mu_i$, we have $\deg_i(h, D) = \operatorname{sign} \mu_i$. Finally, we may use the additivity property of the S^1-degree to obtain $\deg_i(h, \Omega) = \mu_i$, and the statement follows. □

EXERCISES

6.3.1 Prove Theorem 6.3.1.

6.3.2 Show that the projection P_j in (6.3.1) is exactly the projection described in Proposition 6.1.24.

6.3.3 Let V be an irreducible representation of a compact abelian Lie group G. Show that there is a closed subgroup H of G such that $G_x = H$ for all $x \in V \setminus \{0\}$, that is, the set $V \setminus \{0\}$ contains only one orbit type (H). Give an example to show that the above property does not hold for nonabelian groups.

6.3.4 Assume that $n \in \mathbb{N}$ and $V = \mathbb{C}$ is a representation of S^1, where the action of S^1 is given by $\gamma \cdot z = \gamma^n z$ for $\gamma \in S^1$ and $z \in V$. Let $f : V \oplus \mathbb{R}^2 \to V$ be defined by $f(z, u) = u^N z + g(z, u)$, where $N \in \mathbb{N}$, $(z, u) \in V \oplus \mathbb{R}^2$ with $\mathbb{R}^2$ being identified with $\mathbb{C}$, and g is an S^1-equivariant mapping such that $g(z, u) = o(|u^N|)$ uniformly with respect to z. We also define $\varphi(z, u) = (|z| - r) + |z|$, $r > 0$, and $F : V \oplus \mathbb{R}^2 \to V \oplus \mathbb{R}$ by $F(z, u) = (f(z, u), \varphi(z, u))$ for $(z, u) \in V \oplus \mathbb{R}^2$. Show that:

(i) For sufficiently small $r > 0$ and $\varepsilon > 0$, the mapping F is an

S^1-equivariant Ω-admissible mapping, where $\Omega = \{(z, u) \in V \oplus \mathbb{R}^2;\ |z| < r,\ |u| < \varepsilon\}$.

(ii) Use similar deformations as in the proof of Theorem 6.3.5 to show that $S^1\text{-deg}(F, \Omega) = \{\deg_k(F, \Omega)\}$, where

$$\deg_k(F, \Omega) = \begin{cases} N & \text{if } k = n \\ 0 & \text{otherwise} \end{cases}$$

6.4 S^1-EQUIVARIANT DEGREE IN INFINITE-DIMENSIONAL SPACES

Let W be a real isometric Banach representation of $G = S^1$. Let $\{\rho_n\}$, $\rho_n : S^1 \to S^1 \subseteq \mathbb{C}^*$, $n = 1, 2, \ldots$, be the sequence of all real irreducible representations of S^1. It is clear that all these representations are of the complex type. Therefore, the intrinsic dimension $n(\rho_n)$ of ρ_n is equal to 1. For every $n = 1, 2, \ldots$, we can define a linear mapping $P_n : W \to W$ by

$$P_n x := \int_G \chi_{\rho_n}(g) gx \, d\mu(g) = 2 \int_G \cos nt \ gx \, d\mu(g)$$

$g = \cos t + i \sin t \in G$, $x \in W$. The mapping $g \mapsto \chi_{\rho_n}(g)gx$ is continuous for each $x \in W$ and, thus, it is integrable. One can also verify, in the same way as in the final dimensional case, that P_n is a projection of W onto a subspace $W_n := P_n(W)$ and P_n satisfies:

(i) If $x \in W$ belongs to a subrepresentation of W that is irreducible and equivalent to ρ_n, then $P_n x = x$.

(ii) If $x \in W$ belongs to a subrepresentation of W that is irreducible and not equivalent to ρ_n, then $P_n x = 0$.

We will call $W_n := P_n(W)$ the *isotypical component of W* corresponding to ρ_n. So, every irreducible subrepresentation of W in W_n is equivalent to ρ_n and every irreducible subrepresentation of W in $(\text{Id} - P_n)(W)$ is not equivalent to ρ_n. It is important to notice that for every $x \in W_n$ with $x \neq 0$, $G_x = \mathbb{Z}_n$ and W_n can be equipped with a natural complex structure, similarly to the finite-dimensional case. We also put

$$W_\infty = P_\infty(W), \qquad P_\infty x := \int g(x) \, d\mu(g)$$

The subspace W_∞ is exactly W^G. Since the subspace $W_\infty \oplus \bigoplus_{n=1}^{\infty} W_n$ is a G-invariant subspace of W, $W = \overline{W_\infty \oplus \bigoplus_{n=1}^{\infty} W_n}$. That is, $W_\infty \oplus \bigoplus_{n=1}^{\infty} W_n$ is a dense subspace of W.

If $A : W \to W$ is a G-equivariant linear map, then $A(W_n) \subseteq W_n$ for all n. Moreover, $A_n := A|_{W_n} : W_n \to W_n$ is $\mathbb{C}$-linear with respect to the complex structure induced on W_n from the action of S^1 on W_n. Consequently, a linear mapping $A: W \to W$ is G-equivariant if and only if $A(W_n) \subset W_n$ for all $n = 1, 2, \ldots, \infty$, and $A_n := A|_{W_n}$ is $\mathbb{C}$-linear on W_n.

Theorem 6.4.1

Let $X \subseteq W \oplus \mathbb{R}^n$ be a G-invariant subset and $F: X \to W$ a G-equivariant compact mapping. Then for every $\varepsilon > 0$, there exists a G-equivariant finite-dimensional map $F_\varepsilon : X \to W$ such that

$$\|F_\varepsilon(x) - F(x)\| < \varepsilon \qquad \text{for all } x \in X$$

Proof. We denote by W_{fin} the set $W_\infty \oplus \bigoplus_{n=1}^{\infty} W_n$. It is clear that for every $x \in W_{\text{fin}}$, span$\{\rho x;\ g \in S^1\}$ is a finite-dimensional invariant subspace of W. Since $F(X)$ is relatively compact and W_{fin} is dense in W, there exists a finite set $N = \{w_1, \ldots, w_N\} \subseteq W_{\text{fin}}$ such that $F(X) \subseteq N + B_\varepsilon(0) =: N_\varepsilon$, where $B_\varepsilon(0) = \{w \in W;\ \|w\| < \varepsilon\}$. Let $\mu_i : N_\varepsilon \to \mathbb{R}$ denote the mapping

$$\mu_i(w) = \max\{0, \varepsilon - \|w - w_i\|\}\,, \qquad i = 1, \ldots, N, \quad w \in N_\varepsilon$$

and put

$$P_\varepsilon(w) = \frac{1}{\sum_{i=1}^{N} \mu_i(w)} \sum_{i=1}^{N} \mu_i(w) w_i\,, \qquad w \in N_\varepsilon$$

Define $\tilde{F}_\varepsilon : X \to W$ by $\tilde{F}_\varepsilon = P_\varepsilon(F(x))$, $x \in X$. Clearly, $\tilde{F}_\varepsilon$ is an ε-approximation of F such that $\tilde{F}_\varepsilon \subseteq \text{span}\{w_1, \ldots, w_n\}$. $\tilde{F}_\varepsilon$ is not necessarily equivariant. Since the values of the continuous function $f(g) = g\tilde{F}_\varepsilon(g^{-1}x)$ belong to the finite-dimensional G-invariant subspace $X_0 = \text{span}\{Gw_1, \ldots, Gw_n\}$, f is integrable. We then get the required G-equivariant ε-approximation of F by the integral

$$F_\varepsilon(x) = \int_G g\tilde{F}_\varepsilon(g^{-1}x)\, d\mu(g)\,, \qquad x \in X \quad \square \tag{6.4.1}$$

Let $\Omega \subseteq W \oplus \mathbb{R}$ be an open, bounded, G-invariant subset. A mapping $f : \overline{\Omega} \to W$ is called an *Ω-admissible G-map* if f is a G-equivariant compact field on $\overline{\Omega}$ such that $f(x) \neq 0$ for $x \in \partial\Omega$. We will also say that $h : [0, 1] \times \overline{\Omega} \to W$ is an Ω-admissible G-homotopy if h is a G-equivariant compact field on $[0, 1] \times \overline{\Omega}$ and $h(t, x) \neq 0$ for $(t, x) \in [0, 1] \times \partial\Omega$. For every given Ω-admissible G-map $f : \overline{\Omega} \to W$, we find an equivariant finite-dimensional map $F_\varepsilon : \overline{\Omega} \to W$ such that $f_\varepsilon = \text{Id} - F_\varepsilon$ is an approximation of f satisfying

$$\|f_\varepsilon(x) - f(x)\| < \inf\{\|f(y)\|;\ y \in \partial\Omega\}$$

for all $x \in \overline{\Omega}$. Let W_0 denote a G-invariant finite-dimensional subspace of W such that $F_\varepsilon(\overline{\Omega}) \subseteq W_0$. We define

$$S^1\text{-deg}(f, \Omega) = S^1\text{-deg}(f_\varepsilon|_{\overline{\Omega_0}}, \Omega_0) \tag{6.4.2}$$

where $\Omega_0 = \Omega \cap (W_0 \oplus \mathbb{R})$.

By applying standard arguments, we can verify that the above definition does not depend on the choice of the equivariant approximation f_ε as well as the invariant subspace W_0. Moreover, we have the following.

Theorem 6.4.2

Let W be an isometric, real Banach representation of S^1 and let $\Omega \subseteq W \oplus \mathbb{R}$ be an open, bounded, invariant subset. For every Ω-admissible G-map $f : \overline{\Omega} \to W$, we can assign an element $S^1\text{-deg}(f, \Omega) \in \bigoplus_{k=1}^{\infty} \mathbb{Z}$ satisfying the following properties:

(P1) *Existence: If $S^1\text{-deg}(f, \Omega) = \{\deg_k(f, \Omega)\} \neq 0$, that is, there is $k \in \{1, 2, \ldots,\}$ such that $\deg_k(f, \Omega) \neq 0$, then there is $x \in f^{-1}(0)$ such that $\mathbb{Z}_k \subseteq G_x$.*

(P2) *Homotopy Invariance: If h is an Ω-admissible G-equivariant homotopy, then $S^1\text{-deg}(h_t, \Omega)$ does not depend on $t \in [0, 1]$.*

(P3) *Excision: If $\Omega_0 \subseteq \Omega$ is an open, invariant subset such that $f^{-1}(0) \subseteq \Omega_0$, then*

$$S^1\text{-deg}(f, \Omega) = S^1\text{-deg}(f, \Omega_0)$$

(P4) *Additivity: If Ω_1 and Ω_2 are two open, invariant subsets of Ω such that $\Omega_1 \cap \Omega_2 = \varnothing$ and $f^{-1}(0) \subseteq \Omega_1 \cup \Omega_2$, then*

$$S^1\text{-deg}(f, \Omega) = S^1\text{-deg}(f, \Omega_1) + S^1\text{-deg}(f, \Omega_2)$$

We now develop an analog of the bijection theorem for equivariant cases and extend the S^1-equivariant degree to equivariant condensing fields.

Suppose that G is a compact Lie group and X is a real Banach isometric representation of G. Let $B \subseteq A$ be a pair of closed, bounded, invariant subsets of $X^{\infty+n} := X \times \mathbb{R}^n$. We use $\mathscr{A}^G(A, B)$, where $\mathscr{A}$ indicates one of the classes $\mathscr{K}$, $\mathscr{D}$, or $\mathscr{C}$ (defined in Chapter 4), to denote the class of all G-maps $F : A \to X$ such that:

(i) $F \in \mathscr{A}$.

(ii) $\pi(x) \neq F(x)$ for all $x \in B$, that is, F has no fixed point in B.

A G-map $H : A \times [0, 1] \to X$ is called a *homotopy* in $\mathscr{A}^G(A, B)$ if $H \in \mathscr{A}^G(A \times [0, 1], B \times [0, 1])$. We say that $F_0, F_1 \in \mathscr{A}^G(A, B)$ are *homotopic* in

$\mathscr{A}^G(X, A)$ if there is a homotopy H in $\mathscr{A}^G(A, B)$ such that $F_i = H(i, \cdot)$, $i = 0. 1$. We will denote two homotopic maps F_0 and $F_1 \in \mathscr{A}^G(A, B)$ by $F_0 \sim F_1$ in $\mathscr{A}^G(A, B)$. The homotopy relation $\sim$ is an equivalence relation on $\mathscr{A}^G(A, B)$, and in what follows we will denote by $\mathscr{A}^G[A, B]$ the set of all homotopy classes in $\mathscr{A}^G(A, B)$. Notice that the following inclusions:

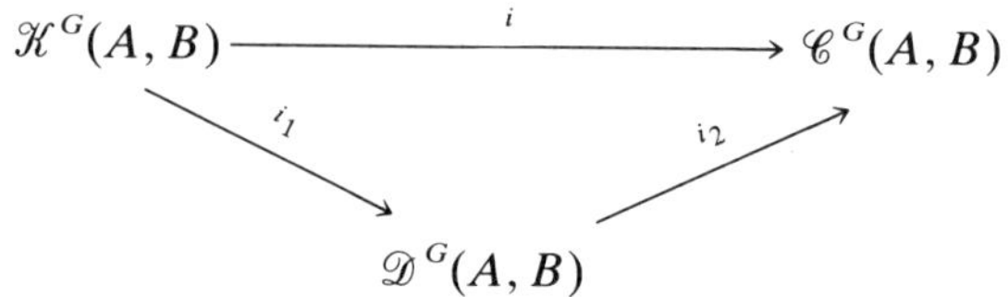

induce maps making the following diagram commutative:

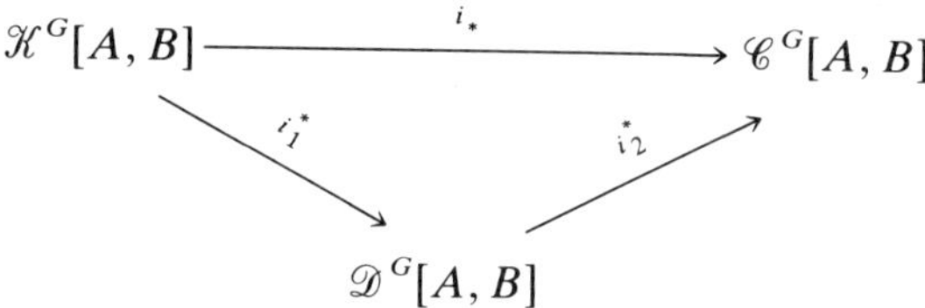

Theorem 6.4.3 (Equivariant Bijection Theorem)

For any closed, bounded, invariant pair (A, B) with $\varnothing \neq B \subseteq A \subseteq X^{\infty+n}$, the induced map

$$i_* : \mathscr{K}^G[A, B] \longrightarrow \mathscr{C}^G[A, B]$$

is bijective.

Proof. Clearly, it suffices to show that i_1^* and i_2^* are both bijective.

We begin with proving the surjectivity of i_1^*. Let $F \in \mathscr{D}^G(A, B)$ and let $0 \leq k < 1$ be a constant such that F is a μ-Lipschitzian G-map with constant k. We need to show that F is homotopic in $\mathscr{D}^G(A, B)$ to a compact G-map. For this purpose, we consider the following sequence of subsets of X:

$$Q_i := \overline{\mathrm{conv}(F(A), 0)}\ ;$$

$$Q_{i+1} := \overline{\mathrm{conv}(F(A \cap \pi^{-1}(Q_i)), 0)}\ , \qquad i = 1, 2, 3, \ldots$$

Obviously, $Q_1 \supseteq Q_2 \supseteq \cdots$ and every Q_i is a closed, bounded, invariant, and nonempty ($0 \in Q_i$) subset of X. Note that $\mu(Q_{i+1}) = \mu(F(A \cap \pi^{-1}(Q_i)) \leq k\mu(A \cap \pi^{-1}(Q_i)) \leq k\mu(Q_i)$ for all $i = 1, 2, \ldots$. It follows that $\mu(Q_i) \leq k^i\mu(A)$, $i = 1, 2, \ldots$. Thus, $\lim_{i\to\infty} \mu(Q_i) = 0$. Let $Q = \bigcap_{i=1}^{\infty} Q_i$. Then Q is an invariant, nonempty ($0 \in Q$), and closed subset such that $\mu(Q) = 0$ and $F(A \cap \pi^{-1}(Q)) \subseteq Q$.

Since Q is a nonempty, convex, compact, invariant set and $0 \in Q$, by the equivariant extension of Dugundji's theorem (Theorem 6.1.28), there exists a compact G-map $r: X \to Q$ that extends $\mathrm{Id}|_Q$. Therefore, $r: X \to Q$ is a G-equivariant retraction onto Q.

Define $H(x, t) = (1-t)F(x) + tr(F(x))$, $x \in A$, $t \in [0, 1]$. Clearly, H is a Darbó G-map. To prove $F \sim r \circ F$ in $\mathscr{D}^G(X, A)$, we need to show that $\pi(x) \neq H(x, t)$ for all $x \in B$ and $t \in [0, 1]$. Suppose the contrary, there exist $(x_0, t_0) \in B \times [0, 1]$ so that $\pi(x_0) = H(x_0, t_0)$. Since $F(x_0) \in Q_1$, $r(F(x_0)) \in Q \subseteq Q_1$, the convexity of Q_1 implies that $\pi(x_0) \in Q_1$. Thus, $x_0 \in A \cap \pi^{-1}(Q_1)$, which, in turn, gives $F(x_0) \in Q_2$ and $\pi(x_0) \in Q_2$. Inductively, we have $\pi(x_0) \in Q$ and $F(x_0) \in Q$. Since

$$r|_Q \equiv \mathrm{Id}|_Q$$

we have

$$\begin{aligned} \pi(x_0) &= (1-t_0)F(x_0) + t_0 r(F(x_0)) \\ &= (1-t_0)F(x_0) + t_0 F(x_0) = F(x_0) \end{aligned}$$

This leads to a contradiction of the assumption $F \in \mathscr{D}^G(A, B)$. So, F is homotopic to the compact map $r \circ F$ via the homotopy H in $\mathscr{D}^G(A, B)$. We then conclude that i_1^* is surjective.

We next prove that i_1^* is injective. Let F_0, $F_1 \in \mathscr{K}^G(A, B)$ and $H \in \mathscr{D}^G(A \times [0, 1], B \times [0, 1])$ be a G-homotopy from F_0 to F_1. We want to show that $F_0 \sim F_1$ in $\mathscr{K}^G(A, B)$. In fact, by the surjectivity of i_1^* for the pair $B \times [0, 1] \subseteq A \times [0, 1]$, we have a compact G-map $\overline{H}: A \times [0, 1] \to X$ and a G-homotopy $\tilde{H} \in \mathscr{D}^G(A \times [0, 1] \times [0, 1], B \times [0, 1] \times [0, 1])$ from H to $\overline{H}$ of the form

$$\tilde{H}(x, s, t) = (1-t)H(x, s) + t\overline{H}(x, s)$$

where $(x, s, t) \in A \times [0, 1] \times [0, 1]$. Since $H|_{A \times \{0,1\}}$ are compact G-maps, $\tilde{H}|_{A \times \{0,1\} \times [0,1]}$ are compact. Also, $\tilde{H}|_{A \times [0,1] \times \{1\}} = \overline{H}$ is an equivariant compact map. We define $H^*: A \times [0, 1] \to X$ by

$$H^*(x, t) = \begin{cases} \tilde{H}(x, 0, 3t) & \text{if } t \in [0, \frac{1}{3}],\ x \in A \\ \tilde{H}(x, 3t-1, 1) & \text{if } t \in [\frac{1}{3}, \frac{2}{3}],\ x \in A \\ \tilde{H}(x, 1, 3-3t) & \text{if } t \in [\frac{2}{3}, 1],\ x \in A \end{cases}$$

From the above construction, we can easily verify that H^* is a G-homotopy from F_0 to F_1 in $\mathscr{K}^G(A, B)$. This proves the injectivity of i_1^*.

We now prove that $i_1^*: \mathscr{D}^G[A, B] \to \mathscr{C}^G[A, B]$ is surjective. Assume that $F \in \mathscr{C}^G(A, B)$ and $\varepsilon = \inf\{\|\pi(x) - F(x)\|;\ x \in B\}$. Since F is bounded, $\|F(B)\| = \sup\{\|F(x)\|;\ x \in B\} < \infty$. Choose a constant $k > 0$ such that $1-$

$\varepsilon/\|F(B)\| < k < 1$. Then kF is a Darbó G-map and for all $x \in B$,

$$\|kF(x) - F(x)\| \le (1-k)\|F(B)\| < \varepsilon$$

Define $H : A \times [0,1] \to X$ by

$$H(x,t) = (1-t)F(x) + tkF(x)\,, \qquad (x,t) \in A \times [0,1]$$

Clearly, $H_t \in \mathscr{C}^G(A,B)$ for $t \in [0,1]$. Moreover, if $H(x_0,t_0) = \pi(x_0)$ for some $(x_0,t_0) \in B \times [0,1]$, then

$$\begin{aligned}\varepsilon &= \inf\{\|\pi(x) - F(x)\|; x \in B\} \le \|\pi(x_0) - F(x_0)\| \\ &= \|H(x_0,t_0) - F(x_0)\| = (1-k)t_0\|F(x_0)\| \le (1-k)\|F(B)\|\end{aligned}$$

contracting the choice of k. Therefore, F is G-homotopic to kF in $\mathscr{C}^G(A,B)$. This shows the surjectivity of i_2^*.

The injectivity of i_2^* can be proved by an argument similar to that for i_1^*, replacing "compact" and "Darbó" by "Darbó" and "condensing," respectively. This justifies the bijectivity of i_2^* and completes the proof. □

We now use the above equivariant bijection theorem to extend the notion of the S^1-equivariant degree to the class of condensing equivariant fields. From now on, we assume that G is the group S^1.

Let Ω be an invariant, open, bounded set in $X \times \mathbb{R}$ and let $F : \overline{\Omega} \to X$ be a condensing G-map such that $F \in \mathscr{C}^G(\overline{\Omega}, \partial\Omega)$. By Theorem 6.4.3, there exists $F_1 \in \mathscr{K}^G(\overline{\Omega}, \partial\Omega)$ such that $F \sim F_1$ in $\mathscr{C}^G(\overline{\Omega}, \partial\Omega)$. Moreover, if $F_2 \in \mathscr{K}^G(\overline{\Omega}, \partial\Omega)$ is another compact G-map such that $F \sim F_2$ in $\mathscr{C}^G(\overline{\Omega}, \partial\Omega)$, then $F_1 \sim F_2$ in $\mathscr{K}^G(\overline{\Omega}, \partial\Omega)$. Therefore, by the homotopy invariance of S^1-degree for equivariant compact fields, $S^1\text{-deg}(\pi - F_1, \Omega) = S^1\text{-deg}(\pi - F_2, \Omega)$. This justifies the following definition.

Definition 6.4.4 Let $F \in \mathscr{C}^G(\overline{\Omega}, \partial\Omega)$. Then the *$S^1$-equivariant degree* of the condensing field $f = \pi - F$ with respect to Ω is defined by the formula

$$S^1\text{-deg}(\pi - F, \Omega) := S^1\text{-deg}(\pi - F_1, \Omega)$$

where $F_1 \in \mathscr{K}^G(\overline{\Omega}, \partial\Omega)$ is a compact G-map such that $F \sim F_1$ in $\mathscr{C}^G(\overline{\Omega}, \partial\Omega)$.

We have observed that, by the homotopy invariance, the above definition does not depend on the choice of $F_1 \in \mathscr{K}^G(\overline{\Omega}, \partial\Omega)$. Moreover, it can be verified, by an argument similar to that in nonequivariant case, that the S^1-degree for condensing fields has the same standard properties (P1–P4) in Theorem 6.4.2.

Finally, for the convenience of applications to the global Hopf bifurcation theory of neutral equations, we introduce a variant of the S^1-degree for equivariant composite coincidence problems.

Let X and Y be two real, isometric Banach representations of $G := S^1$ and assume that $L : \mathrm{Dom}(L) \subseteq X \to Y$ is a closed, equivariant Fredholm operator of index zero, where $\mathrm{Dom}(L)$ is an invariant subspace of X. Let us recall that a compact linear operator $K : X \to Y$ such that $L + K : \mathrm{Dom}(L) \to Y$ is a bijection is called a *compact resolvent* of L. Let $R_K := (L + K)^{-1} : Y \to X$. R_K is a bounded linear operator. We denote by $CR(L)$ the set of all compact resolvents of L and we put

$$CR^G(L) := \{K \in CR(L);\ K \text{ is } G\text{-equivariant}\}$$

The set $CR^G(L)$ may be empty. However, in many applications the following assumption is always satisfied:

$$\text{(H1)} \qquad CR^G(L) \neq \varnothing$$

Let $\Omega \subseteq X^{\infty+n} = X \times \mathbb{R}^n$ be a bounded, invariant, open subset, and let $B : \overline{\Omega} \to X$ and $N : \overline{\Omega} \to Y$ be two G-equivariant maps. We consider the following nonlinear problem:

$$\text{(6.4.3)} \qquad \begin{cases} \text{Find } x \in \overline{\Omega} \text{ such that } \pi(x) - B(x) \in \mathrm{Dom}(L) \\ \text{and } L[\pi(x) - B(x)] = N(x) \end{cases}$$

Following the terminology introduced in Chapter 4, we call this nonlinear problem an *equivariant composite coincidence problem*. Let $K \in CR^G(L)$ be a fixed, compact, equivariant resolvent of L. We define

$$\text{(6.4.4)} \qquad \Theta_K(B, N) := B + R_K[N + K(\pi - B)] : \overline{\Omega} \to X$$

As compositions of equivariant maps, $\Theta_K(B, N)$ is equivariant. It can be verified that problem (6.4.3) is equivariant to the following fixed-point problem:

$$\text{(6.4.5)} \qquad \text{Find } x \in \overline{\Omega} \text{ such that } \pi(x) = \Theta_K(B, N)(x)$$

Assuming that the space X is equipped with a measure of noncompactness μ, we now introduce the following definition.

Definition 6.4.5 Let $B : \overline{\Omega} \to X$ and $N : \overline{\Omega} \to Y$ be two G-maps. The pair (B, N) is said to be an *L-condensing G-pair* if the map $\Theta_K(B, N) : \overline{\Omega} \to X$ defined by (6.4.4) is a condensing G-map.

It can be easily verified that the above definition does not depend on the choice of $K \in CR^G(L)$.

Let (A, A_0) be a pair of closed, bounded, and invariant subsets of $X \times \mathbb{R}$

with $A_0 \subseteq A$. We denote by $\mathscr{C}_L^G(A, A_0)$ the class of all L-condensing G-pairs (B, N), $B: A \to X$ and $N: A \to Y$, such that (6.4.5) has no solution in A_0, that is, $\pi(x) \neq \Theta_K(B, N)(x)$ for all $x \in A_0$.

Definition 6.4.6 Let $\Omega \subseteq X \times \mathbb{R}$ be an open, bounded, invariant subset. For every L-condensing G-pair $(B, N) \in \mathscr{C}_L^G(\overline{\Omega}, \partial\Omega)$, we define the S^1-*composite coincidence degree* of the pair (B, N) on Ω as the sequence of integers given by

$$S^1\text{-deg}_L([B, N], \Omega) := S^1\text{-deg}(\pi - \Theta_K(B, N), \Omega)$$

where $K \in CR^G(L)$ is fixed. We will also write that

$$S^1\text{-deg}_L([B, N], \Omega) := \{\deg_{k,L}([B, N], \Omega)\}_{k=1}^{\infty}$$

where $\deg_{k,L}([B, N], \Omega) := \deg_k(\pi - \Theta_K(B, N), \Omega), \quad k = 1, 2, \ldots.$

We should emphasize that this definition may depend on the choice of the equivariant resolvent K; each component $\deg_{k,L}([B, N], \Omega)$ is unique up to the sign.

It is now clear that the following standard properties are satisfied by the S^1-composite coincidence degree.

Theorem 6.4.7

The S^1-composite coincidence degree satisfies the following properties:

- **(P1)** *Existence: If S^1-$\deg_L([B, N], \Omega) = \{\deg_{k,L}([B, N], \Omega)\} \neq 0$, that is, $\deg_{k,L}([B, N], \Omega) \neq 0$ for some $k \geq 1$, then there exists $x \in \Omega^H$ such that $H = \mathbb{Z}_k$ and $L(\pi(x) - B(x)) = N(x)$.*
- **(P2)** *Homotopy Invariance: If $(B_t, N_t) \in \mathscr{C}_L^G(\overline{\Omega}, \partial\Omega)$, $t \in [0, 1]$, is a homotopy of G-pairs, that is, $\Theta_K(B_t, N_t)$ is a homotopy in $\mathscr{C}^G(\overline{\Omega}, \partial\Omega)$, then S^1-$\deg_L([B_t, N_t], \Omega)$ is independent of $t \in [0, 1]$.*
- **(P3)** *Excision: If $\Omega_0 \subseteq \Omega$ is an open, invariant subset and $(B, N) \in \mathscr{C}_L^G(\overline{\Omega}, \overline{\Omega \backslash \Omega_0})$, then*

$$S^1\text{-deg}_L([B, N], \Omega) = S^1\text{-deg}_L([B, N], \Omega_0)$$

- **(P4)** *Additivity: If Ω_1 and Ω_2 are two open, invariant subsets of Ω such* *that* $\Omega_1 \cap \Omega_2 = \emptyset$ *and* $F \in \mathscr{C}_L^G(\overline{\Omega}, \overline{\Omega \setminus (\Omega_1 \cup \Omega_2)})$, *then*

$$S^1\text{-deg}_L([B, N], \Omega) = S^1\text{-deg}_L([B, N], \Omega_1) + S^1\text{-deg}_L([B, N], \Omega_2)$$

EXERCISES

6.4.1 Show that the definition of the S^1-deg$(\text{Id} - F, \Omega)$ for the equivariant Ω-admissible compact fields $\text{Id} - F$ does not depend on the choice of a finite-dimensional equivariant ε-approximation of the map F.

6.4.2 Prove Theorems 6.4.2 and 6.4.7.

6.4.3 Give an example of a G-equivariant, closed Fredholm operator of index zero $L : \text{Dom}(L) \subseteq X \to Y$ such that $CR^G(L) = \emptyset$. Is it possible that $CR^G(L) = \emptyset$ when $\text{Dom}(L) = X = Y$?

6.5 BIBLIOGRAPHICAL NOTES

We introduce some elementary concepts and results from the Lie group theory, the representation theory, and equivariant topology. For complete discussions, we refer the reader to Adams (1969), Bredon (1972), Bröcker and tom Dieck (1985), Dinculeanu (1974), Hauschild (1974, 1977), Hochschild (1965), Kawakubo (1991), Kirillov (1976), Kuiper (1965), Lyubich (1988), Murayama (1983), Pontriagin (1946), Wawrzyńczyk (1984), Zhelobenko (1970), Zhelobenko and Shtern (1983), and tom Dieck (1987).

This chapter is related to an elemenary analytic presentation of the S^1-equivariant degree theory recently developed by Dylawerski et al. (1991). Our approach is different from the original one and uses the idea of regular normal approximations used in Gęba et al. (1994) and further developed in Krawcewicz and Xia (1996) and Xia (1994). In particular, the proof of the regular normal approximation theorem is based on the work of Krawcewicz and Xia (1996). The proof of Theorem 6.3.5 is based on the idea used in Krawcewicz et al. (1996). The infinite-dimensional extension of the S^1-degree is standard, and the generalization of the S^1-degree to Banach spaces uses the standard idea of Schauder approximations and averaging. The extension of the S^1-degree to composite coincidence problems is due to Krawcewicz et al. (1993).

It should be mentioned that the S^1-degree presented here is different from, although related to, the S^1-degree introduced by Ize et al. (1989, 1992). We also refer the reader to Gęba et al. (1985, 1994), and Krawcewicz and Vivi (1995). See, in addition, the work of Dancer (1985) and Dylawerski (1988).

Chapter Seven

Global Hopf Bifurcation Theory

This chapter provides an application of the S^1-equivariant degree theory to the global Hopf bifurcation problem of neutral functional differential equations.

We first establish an abstract global bifurcation theorem for an S^1-equivariant composite coincidence problem depending on two parameters, and then apply it to obtain a global Hopf bifurcation theorem for neutral equations. This theorem provides sharp information about the maximal continuation of periodic solutions. We will use an example from the lossless transmission line problem to demonstrate how to apply this global Hopf bifurcation theorem to establish the existence of large-amplitude periodic solutions.

7.1 HOPF BIFURCATION FOR COMPOSITE COINCIDENCE PROBLEMS

In this section we illustrate how to apply the S^1-composite coincidence degree to obtain some global bifurcation theorems of the Rabinowitz type for a class of S^1-equivariant nonlinear problems.

Assume that V and W are two real, isometric Banach representations of the group $G = S^1$, $X = V \times \mathbb{R}$, and $Y = W \times \mathbb{R}$. Suppose that

$$L_0 : \mathrm{Dom}(L_0) \subseteq V \rightarrow W$$

is a given equivariant, closed Fredholm operator of index zero such that $CR^G(L_0) \neq \emptyset$. It is clear that L_0 can be extended to a Fredholm operator of index zero $L : \mathrm{Dom}(L) \subseteq X \rightarrow Y$, where $\mathrm{Dom}(L) = \mathrm{Dom}(L_0) \times \mathbb{R}$ and $L(v, r) = (L_0 v, 0)$ for $(v, r) \in \mathrm{Dom}(L_0) \times \mathbb{R}$. It follows from $CR^G(L_0) \neq \emptyset$

that if we put $K = K_0 \oplus I_0$, where $K_0 \in CR^G(L_0)$ and I_0 is the identity mapping from $\mathbb{R}$ to $\mathbb{R}$, then $K \in CR^G(L)$, and consequently $CR^G(L) \neq \emptyset$.

Consider the following nonlinear problem:

$$(7.1.1) \qquad L_0[\pi_0(x) - B_0(x)] = N_0(x) , \qquad \pi_0(x) - B_0(x) \in \mathrm{Dom}(L_0)$$

where $\pi_0 : V \times \mathbb{R}^2 \to V$ is the natural projection, $B_0 : V \times \mathbb{R}^2 \to V$ and $N_0 : V \times \mathbb{R}^2 \to W$ are two equivariant mappings of class C^1 such that (B_0, N_0) is an L_0-condensing G-pair.

To describe our bifurcation problem, we assume that there exists a two-dimensional submanifold $M \subseteq V^G \times \mathbb{R}^2$ satisfying the following conditions:

(A) *For every* $x \in M$, $\pi_0(x) - B_0(x) \in \mathrm{Dom}(L_0)$ *and* $L_0[\pi_0(x) - B_0(x)] = N_0(x)$, *that is, x is a solution to Eq. (7.1.1).*

(B) *If* $(v_0, \lambda_0) \in M$ *with* $v_0 \in V^G$ *and* $\lambda_0 \in \mathbb{R}^2$, *then there exists an open neighborhood* U_{λ_0} *of* λ_0 *in* $\mathbb{R}^2$, *an open neighborhood* U_{v_0} *of* v_0 *in* V^G *and a* C^1*-map* $\eta : U_{\lambda_0} \to V^G$ *such that*

$$M \cap (U_{v_0} \times U_{\lambda_0}) = \{(\eta(\lambda), \lambda); \lambda \in U_{\lambda_0}\}$$

Since all points $(v, \lambda) \in M$ are solutions of (7.1.1), we call those points trivial solutions. All other solutions of (7.1.1) will be called *nontrivial*. A point $(v_0, \lambda_0) \in M$ is called a *bifurcation point* if in any neighborhood of (v_0, λ_0), there exists a nontrivial solution for (7.1.1).

The problem (7.1.1) is equivalent to the equation

$$v = \Theta_{K_0}(B_0, N_0)(v, \lambda) , \qquad (v, \lambda) \in V \times \mathbb{R}^2$$

where $\Theta_{K_0}(B_0, N_0) = B_0 + R_{K_0}[N_0 + K_0(\pi_0 - B_0)]$: $V \times \mathbb{R}^2 \to V$. We define a mapping $f : V \times \mathbb{R}^2 \to V$ by $f(v, \lambda) = v - \Theta_{K_0}(B_0, N_0)(v, \lambda)$. Clearly, f is an equivariant condensing field of class C^1.

Note that the derivative of a condensing map is not necessarily condensing [see Exercise (4.4.8)], but the derivative of a Darbó map is a Darbó map. We further assume:

(H1) $D_v f(v, \lambda) = \mathrm{Id} - \Theta_{K_0}(D_v B_0(v, \lambda), D_v N_0(v, \lambda)) : V \to V$ is a *condensing linear field* for all $(v, \lambda) \in M$.

Therefore, $D_v f(v, \lambda)$ is a Fredholm operator of index zero for $(v, \lambda) \in M$. Moreover, since $(v, \lambda) \in M$, $D_v f(v, \lambda)$ is also an equivariant operator. By the implicit function theorem, if $(v_0, \lambda_0) \in M$ is a bifurcation point, then the derivative $D_v f(v_0, \lambda_0)$ is not an isomorphism of V. Therefore, all bifurcation points of (7.1.1) are contained in the set

$$\Lambda := \{(v, \lambda) \in M;\ D_v f(v, \lambda) \notin GL(V)\}$$

In what follows, we call a point (v_0, λ_0) in Λ a *V-singular point*. Our goal is to find a bifurcation point for Eq. (7.1.1) that is equivalent to

$$f(v, \lambda) = 0\,, \qquad (v, \lambda) \in V \times \mathbb{R}^2 \tag{7.1.2}$$

As in the nonequivariant case, our approach to finding nontrivial solutions to (7.1.2) in a given, open, bounded, invariant neighborhood $U \subseteq V \times \mathbb{R}^2$ of a V-singular point $(v_0, \lambda_0) \in M$ is based on the notion of a complementing function introduced in Chapter 5.

A *complementing function* for Eq. (7.1.2) on the set U is an invariant function $\varphi : \overline{U} \to \mathbb{R}$ satisfying the condition that $\varphi(v, \lambda) < 0$ for all $(v, \lambda) \in \overline{U} \cap M$. Therefore, every solution to the system

$$\begin{cases} f(v, \lambda) = 0 \\ \varphi(v, \lambda) = 0\,, \qquad (v, \lambda) \in \overline{U} \end{cases} \tag{7.1.3}$$

is a nontrivial solution to (7.1.2). This leads to the S^1-map $F_\varphi : \overline{U} \to V \times \mathbb{R} =: X$, $F_\varphi(v, \lambda) = (f(v, \lambda), \varphi(v, \lambda))$. Clearly, we can replace the problem of finding a nontrivial solution to (7.1.1) in U by the problem of solving the equation $F_\varphi(v, \lambda) = 0$ for $(v, \lambda) \in U$. On the other hand, we can easily verify that (7.1.3) is equivalent to the following coincidence problem:

$$L[\pi(x) - B(x)] = N_\varphi(x)\,, \qquad x \in U \tag{7.1.4}$$

where π is the natural projection of $X \times \mathbb{R} = V \times \mathbb{R}^2$ onto $X := V \times \mathbb{R}$, $B(x) := (B_0(x), 0) \in X$ for $x \in X \times \mathbb{R}$, and $N_\varphi(x) := (N_0(x), \varphi(x)) \in Y := W \times \mathbb{R}$. Therefore, if we assume that (7.1.4) has no solution on ∂U, then the S^1-equivariant composite coincidence degree $S^1\text{-deg}_L([B, N_\varphi], U)$ is well defined and can be used to investigate the existence of nontrivial solutions of (7.1.1). In particular, the nontriviality of $S^1\text{-deg}_L([B, N_\varphi], U)$ will imply the existence of a nontrivial solution of (7.1.1) in U.

In the rest of this section, we will show that, for an isolated V-singular point $(v_0, \lambda_0) \in M$, it is possible to compute $S^1\text{-deg}_L([B, N_\varphi], U)$ in terms of derivatives $D_v B$ and $D_v N$ near (v_0, λ_0). As we will see, this computation plays a crucial role in developing a bifurcation theory for (7.1.1).

Let us denote by $V_\infty \oplus \bigoplus_{k=1}^{\infty} V_k$ the isotypical direct sum decomposition of V such that $V_\infty = V^G$ and for $k \geq 1$ and $x \in V_k \setminus \{0\}$, $G_x = \mathbb{Z}_k$.

Recall that all subspaces V_k, $k \geq 1$, admit a natural structure of a complex vector space, and with this structure of a complex vector space, an $\mathbb{R}$-linear operator $A : V_k \to V_k$ is equivariant if and only if it is a $\mathbb{C}$-linear operator.

We need some additional information about the sets $GL^G_{\mathscr{C}}(V_k)$ of all

G-equivariant automorphisms of V_k that are also condensing fields. In what follows, for a complex Banach space E we use $GL^{\mathbb{C}}_{\mathscr{C}}(E)$ [respectively, $GL^{\mathbb{C}}_{C}(E)$] to denote the space of all $\mathbb{C}$-linear automorphisms A of X such that $A = \mathrm{Id} - T$, where T is a linear condensing (respectively, compact) operator. Employing the same argument as that of Theorem 4.4.5, we get the following.

Theorem 7.1.1

Let C be a compact set and $a : C \to GL^{\mathbb{C}}_{\mathscr{C}}(E)$ a continuous mapping. Then there exists a direct sum decomposition $E = E^0 \oplus E_0$ and a homotopy $h : C \times [0, 1] \to GL^{\mathbb{C}}_{\mathscr{C}}(E)$ such that:

(i) $\dim_{\mathbb{C}} E_0 < \infty$.
(ii) $h(\lambda, 1)|_{E^0} = \mathrm{Id}|_{E^0}$ *for all* $\lambda \in C$.
(iii) $h(\lambda, 0) = a(\lambda)$ *for all* $\lambda \in C$.
(iv) $h(\lambda, 1)|_{E_0} : E_0 \to E_0$.

It is well known that there exists a sequence of subspaces $\{(E_n, E^n)\}$ of E such that for all $n = 1, 2, \ldots,$

(i) $\dim_{\mathbb{C}} E_n = n$.
(ii) $E_n \subseteq E_{n+1}$, $E^{n+1} \subseteq E^n$.
(iii) $E_n \oplus E^n = E$.

We now, for $m \geq n$, define embeddings $i_{n,m} : GL(n, \mathbb{C}) \to GL(m, \mathbb{C})$ by

$$i_{n,m}(A)(v_1, v_2) = (Av_1, v_2)$$

where $A \in GL(n, \mathbb{C})$ and $(v_1, v_2) \in \mathbb{C}^n \oplus \mathbb{C}^{m-n}$. Let $i_n : GL(n, \mathbb{C}) \cong GL^{\mathbb{C}}(E_n) \to GL^{\mathbb{C}}_{\mathscr{C}}(E)$ [respectively, $GL^{\mathbb{C}}_{c}(E)$] be defined by

$$i_n(A)(x_1, x_2) = (\tilde{A}x_1, x_2)$$

where $A \in GL(n, \mathbb{C})$, $\tilde{A}$ is the image of A under the isomorphism $GL(n, \mathbb{C}) \cong GL^{\mathbb{C}}(E_n)$, and $(x_1, x_2) \in E_n \oplus E^n$. Using the above embeddings, we obtain the following inclusions:

$$GL(1, \mathbb{C}) \subseteq \cdots \subseteq GL(n, \mathbb{C}) \subseteq GL(n+1, \mathbb{C}) \subseteq GL^{\mathbb{C}}_{c}(E) \subseteq GL^{\mathbb{C}}_{\mathscr{C}}(E)$$

Therefore, taking the direct limit, we have the following well-defined inclusion:

$$i : \lim_{n \to \infty} GL(n, \mathbb{C}) \to GL^{\mathbb{C}}_{\mathscr{C}}(E) \quad \text{[respectively, } GL^{\mathbb{C}}_{c}(E)]$$

where $\lim_{n\to\infty} GL(n, \mathbb{C}) = \bigcup_{n\geq 1} GL(n, \mathbb{C})$ is a topological space with the finest topology such that every inclusion $GL(n, \mathbb{C}) \to \lim_{n\to\infty} GL(n, \mathbb{C})$ is continuous.

Theorem 7.1.2

The inclusion i defined above is a weak homotopy equivalence, that is, the induced homomorphism i_* between their homotopy groups is an isomorphism. In particular, for all $k = 1, 2, \ldots,$

$$\lim_{n\to\infty} \pi_k(GL(n, \mathbb{C})) \cong \pi_k(GL^{\mathbb{C}}_{\mathscr{C}}(E)) \quad [\text{respectively, } \pi_k(GL^{\mathbb{C}}_c(E))]$$

Consequently, the isomorphisms $\deg : \pi_1(GL(n, \mathbb{C})) \to \mathbb{Z}$ *induce an isomorphism* $\deg : \pi_1(GL^{\mathbb{C}}_{\mathscr{C}}(E)) \to \mathbb{Z}$.

Proof. The proof of the first statement is basically the same as that of Theorem 4.4.6 and Lemma 4.4.7, replacing real Banach spaces by complex Banach spaces. The second statement follows from Theorem 6.1.38. □

Corollary 7.1.3

There exists a bijection

$$\deg : [S^1, GL^{\mathbb{C}}_{\mathscr{C}}(E)] \to \mathbb{Z}$$

Moreover, if $\dim_{\mathbb{C}} E < \infty$, *then*

$$\deg[\beta] = \deg(\det_{\mathbb{C}}(\beta(\cdot)), B_1(0))$$

where $\beta : S^1 \to GL^{\mathbb{C}}(E)$, $\det_{\mathbb{C}} : GL^{\mathbb{C}}(E) \to \mathbb{C}^* := \mathbb{C}\backslash\{0\}$ *is the determinant homomorphism*, $B_1(0) = \{z \in \mathbb{C}; |z| < 1\}$.

For additional information about the homotopic properties of $GL^{\mathbb{C}}_{\mathscr{C}}(E)$ and $GL^{\mathbb{C}}_c(E)$, we refer to Palais (1965) and Gęba (1968).

Let $(v_0, \lambda_0) \in M$ be an isolated V-singular point. We identify $\mathbb{R}^2$ with $\mathbb{C}$ and, for a sufficiently small number $\rho > 0$, we define $\alpha : D \to M$, $D := \{z \in \mathbb{C}; |z| \leq 1\}$ by putting

$$\alpha(z) = (\eta(\lambda_0 + \rho z), \lambda_0 + \rho z) \in V^G \times \mathbb{R}^2$$

Since $(v_0, \lambda_0) \in M$ is an isolated V-singular point, we can find a sufficiently small $\rho > 0$ such that $\alpha(D)$ contains only one V-singular point, namely, (v_0, λ_0). Now, the formula $\psi(z) := D_v f(\alpha(z))$, $z \in S^1 \subset D$, defines a continuous mapping $\psi : S^1 \to GL^G_{\mathscr{C}}(V)$. Therefore,

$$\psi = \psi_\infty \oplus \psi_1 \oplus \cdots \oplus \psi_k \oplus \cdots$$

where $\psi_k : S^1 \to GL^{\mathbb{C}}_{\mathscr{C}}(V_k)$ for $k = 1, \ldots,$ and $\psi_\infty : S^1 \to GL_{\mathscr{C}}(V_\infty)$. Notice that

$GL_{\mathscr{C}}(V_\infty)$ is a union of two connected components $GL^+_{\mathscr{C}}(V_\infty)$ and $GL^-_{\mathscr{C}}(V_\infty)$. We define

$$\varepsilon_0 = \varepsilon_0(v_0, \lambda_0) = \begin{cases} +1 & \text{if } \psi_\infty(S^1) \subseteq GL^-_{\mathscr{C}}(V_\infty) \\ -1 & \text{if } \psi_\infty(S^1) \subseteq GL^+_{\mathscr{C}}(V_\infty) \end{cases}$$

$$\mu_k(v_0, \lambda_0) = \deg([\psi_k]), \qquad k = 1, 2, \ldots$$

$$\mu(v_0, \lambda_0) = \{\mu_k(v_0, \lambda_0)\} \in \bigoplus_{k=1}^{\infty} \mathbb{Z}$$

The remaining part of this section is to describe the relationship between the numbers $\mu_k(v_0, \lambda_0) \in \mathbb{Z}$ and the S^1-composite coincidence degree S^1-$\deg_L([B, N_\varphi], U)$.

We consider a neighborhood U of (v_0, λ_0) defined by

$$B_M(v_0, \lambda_0; r, \rho) = \{(v, \lambda) \in V \times \mathbb{R}^2;\ |\lambda - \lambda_0| < \rho,\ \|v - \eta(\lambda)\| < r\}$$

where $r > 0$ is chosen so that:

(i) $f(v, \lambda) \neq 0$ for $(v, \lambda) \in \overline{B_M(v_0, \lambda_0; r, \rho)}$ with $|\lambda - \lambda_0| = \rho$ and $v \neq \eta(\lambda)$.

(ii) (v_0, λ_0) is the only V-singular point in $B_M(v_0, \lambda_0; r, \rho)$.

Such a neighborhood U is called a *special neighborhood* of (v_0, λ_0) determined by (r, ρ). The existence of a special neighborhood follows from the implicit function theorem. Moreover, we say that a continuous invariant function $\theta : \overline{U} \to R$ is an *almost complementing function* if:

(i) $\theta(\eta(\lambda), \lambda) = -|\lambda - \lambda_0|$ for all $(\eta(\lambda), \lambda) \in \overline{U} \cap M$.

(ii) $\theta(v, \lambda) = r$ if $\|v - \eta(\lambda)\| = r$.

The existence of such a function θ follows from Theorem 6.1.28. Note that if θ is an almost complementing function, then for each $\delta > 0$, $\varphi(v, \lambda) := \theta(v, \lambda) - \delta$ is negative on the set of trivial solutions $\overline{U} \cap M$. For a sufficiently small $\delta > 0$, F_φ and F_θ are homotopic in $\mathscr{C}^G(\overline{U}, \partial U)$. Therefore, by the homotopy invariance property, S^1-$\deg(F_\varphi, U) = S^1$-$\deg(F_\theta, U)$. Consequently, the nontriviality of the degree S^1-$\deg(F_\varphi, U)$ implies the existence of a nontrivial solution of (7.1.1) in U. It is also easy to verify that S^1-$\deg(F_\theta, U)$ does not depend on the choice of the complementing function θ.

Proposition 7.1.4

Suppose that $(v_0, \lambda_0) \in M$ is an isolated singular point and U is a special neighborhood of (v_0, λ_0). If S^1-$\deg(F_\theta, U) \neq 0$ for some almost complement-

ing function $\theta : \overline{U} \to \mathbb{R}$, *then* (v_0, λ_0) *is a bifurcation point for (7.1.1). More precisely, if* $\deg_k(F_\theta, U) \neq 0$ *for some* $k \in \mathbb{N}$, *then (7.1.1) has a sequence of nontrivial solutions* (v_n, λ_n) *such that* $\lim_{n\to\infty} (v_n, \lambda_n) = (v_0, \lambda_0)$ *and* $G_{v_n} \supseteq \mathbb{Z}_k$ *for all* $n = 1, 2, \ldots$.

Proof. As noted above, if $\delta > 0$ is sufficiently small, then the function $\varphi_\delta : \overline{U} \to \mathbb{R}$, defined by $\varphi_\delta(v, \lambda) = \theta(v, \lambda) - \delta$, $(v, \lambda) \in \overline{U}$, is a well-defined complementing function. Let $\theta_t(v, \lambda) = \theta(v, \lambda) - t\delta$, $t \in [0, 1]$. Then we can apply the homotopy invariance of the S^1-degree to obtain

$$S^1\text{-deg}(F_{\theta_\delta}, U) = S^1\text{-deg}(F_\theta, U)$$

Thus, S^1-deg$(F_\theta, U) \neq 0$ implies that the equation $F_{\theta_\delta}(v, \lambda) = 0$ has a solution in U and, hence, Eq. (7.1.1) has a nontrivial solution in U.

On the other hand, we can apply the linear homotopy between F_θ and $\overline{F}_\theta$, where

$$\overline{F}_\theta(v, \lambda) = ((\mathrm{Id} - D_v f(\eta(\lambda), \lambda)(v - \eta(\lambda)), \theta(v, \lambda)), \qquad (v, \lambda) \in \overline{U}$$

to obtain

$$S^1\text{-deg}(F_\theta, U) = S^1\text{-deg}(\overline{F}_\theta, U)$$

By the excision property of the S^1-degree, we know that S^1-deg$(\overline{F}_\theta, U)$ is independent of the choice of r and ρ. Consequently, the result follows. □

Let $(v_0, \lambda_0) \in M$ be a V-singular point. We make the following assumption:

(H2) $D_v f(v_0, \lambda_0)|_{V_\infty} : V_\infty \to V_\infty$ is an isomorphism.

The assumption (H2) ensures, by the implicit function theorem, that (v_0, λ_0) is not a bifurcation point of nontrivial stationary solutions of (7.1.1), that is, solutions in $V^G \times \mathbb{R}^2$.

Theorem 7.1.5

Let $(v_0, \lambda_0) \in M$ *be a V-singular point satisfying (H2),* $U = B_M(v_0, \lambda_0; r, \rho)$ *a special neighborhood, and* θ *an almost complementing function. Define* $N_\theta(x) = (N(x), \theta(x)) \in V \times \mathbb{R} =: X$, *where* $x \in \overline{U}$. *Then the* S^1*-composite coincidence degree* S^1-$\deg_L([B, N_\theta], U)$ *is well defined and*

$$S^1\text{-}\deg_L([B, N_\theta], U) = \varepsilon_0 \cdot \mu(v_0, \lambda_0)$$

That is,

$$\deg_{k,L}([B, N_\theta], U) = \varepsilon_0 \cdot \mu_k(v_0, \lambda_0), \qquad k = 1, 2, \ldots$$

Proof. We first use the properties of the S^1-degree to do some simplification of the mappings involved in the computations of $S^1\text{-deg}_L([B, N_\theta], U) = S^1\text{-deg}(F_\theta, U)$. Without loss of generality, we can replace F_θ by

$$F(v, \lambda) = \overline{F_{\theta_\delta}}(v, \lambda) = ((\mathrm{Id} - D_v f(\eta(\lambda), \lambda)(v - \eta(\lambda)), \theta(v, \lambda) - \delta)$$

where $\delta > 0$ is a sufficiently small number. Put $A(\lambda) := D_v f(\eta(\lambda), \lambda)$. By Theorem 7.1.1, we can assume that there is an invariant direct sum decomposition $V = V^0 \oplus V_0$ such that $\dim V_0 < \infty$, and for all $\lambda \in M \cap \overline{U}$, $A(\lambda)|_{V^0} = \mathrm{Id}|_{V^0}$. Consequently, by the construction of the degree, we obtain

$$S^1\text{-deg}_L([B, N_\theta], U) = S^1\text{-deg}(F|_{V_0 \times \mathbb{R}^2}, U \cap (V_0 \times \mathbb{R}^2))$$

This shows we can assume, without loss of generality, that the space V is finite-dimensional. In the next step of simplification, we use the diffeomorphism χ_0 from $\overline{U} \cap V^G \times \mathbb{R}^2$ into $V^G \times \mathbb{R}^2$ defined by $\chi_0(v, \lambda) = (v - \eta(\lambda), \lambda)$, which transforms $\overline{U} \cap M$ into $\{0\} \times \mathbb{R}^2$. It is clear χ_0 preserves the orientation of $V^G \times \mathbb{R}^2$. We extend χ_0 to a diffeomorphism χ from $\overline{U}$ into $V \times \mathbb{R}^2$ by $\chi(v_0 + v^0, \lambda) = \chi_0(v_0, \lambda) + v^0$, where $v_0 \in V^G$ and $v^0 \perp V^G$. Then, it follows from the construction that

$$S^1\text{-deg}(F \circ \chi^{-1}, \chi(U)) = S^1\text{-deg}(F, U)$$

This means we can assume, without loss of generality, that $M = \{0\} \times \mathbb{R}^2$ and $\lambda_0 = 0$.

Put $\Sigma = F^{-1}(0) \cap M := \{(0, \lambda); |\lambda| = \delta\}$. In order to simplify notations, we assume $\delta = 1$ and identify Σ with $S^1 \subseteq \mathbb{C} \cong \{0\} \times \mathbb{R}^2$. Let U_0 be a *tubular* neighborhood of Σ in $\{0\} \times \mathbb{R}^2$. That is, U_0 is a small neighborhood of Σ such that every element $\lambda \in U_0$ can be written uniquely as $\lambda = \sigma + \xi$, $\sigma \in \Sigma$, and ξ is normal to Σ at σ. It follows from th excision property that we can assume $U := B_V(1) \times U_0$, where $B_V(1) = \{v \in V; \|v\| < 1\}$. Next, the homotopy property ensures that we can assume, without loss of generality, that

$$F(v, \lambda) = (A(\sigma)v, \theta_0(\lambda)) \in V \times \mathbb{R}$$

where $\lambda \in U_0$, $v \in B_v(1)$, $\lambda = \sigma + \xi$, $\sigma \in \Sigma$, ξ is normal to Σ at σ, and

$$A(\sigma) = \mathrm{Id} - D_v f(0, \sigma)$$

$$\theta_0(\lambda) = 1 - |\lambda|$$

Indeed, the function $\theta(v, \lambda) = |v| - |\lambda|$ is an almost complementing function and the deformation

$$(\mathrm{Id} - D_v f(0, \sigma + t\xi)v, t|v| - |\lambda| + 1)$$

is a U-admissible homotopy.

It is now easy to check that all the assumptions of Theorem 6.3.5 are satisfied. Since the winding element of $\omega := A$ is exactly equal to $\mu(\lambda_0, v_0)$, the conclusion follows from Theorem 6.3.5. □

By Proposition 7.1.4 and Theorem 7.1.5, we obtain the following local bifurcation theorem of the Krasnosiel'skii type.

Theorem 7.1.6

Suppose that $L_0 : \mathrm{Dom}(L_0) \subset V \to W$ *is an equivariant, closed Fredholm operator of index zero such that* $CR^G(L_0) \neq \emptyset$, $B_0 : V \times \mathbb{R}^2 \to V$ *and* $N_0 : V \times \mathbb{R}^2 \to W$ *are two equivariant mappings of class* C^1 *such that* (B_0, N_0) *is an* L_0*-condensing G-pair satisfying (H1) and (H2). Assume, in addition, that* $M \subseteq V^G \times \mathbb{R}^2$ *is a two-dimensional submanifold satisfying conditions (A) and (B). If* $(v_0, \lambda_0) \in M$ *is an isolated V-singular point and there exists* $k \geq 1$ *such that* $\mu_k(v_0, \lambda_0) \neq 0$, *then* (v_0, λ_0) *is a bifurcation point for problem (7.1.1). More precisely, there exists a sequence* (v_n, λ_n) *of nontrivial solutions to (7.1.1) such that the isotropy group of* v_n *contains* $\mathbb{Z}_k$, *and* $(v_n, \lambda_n) \to (v_0, \lambda_0)$ *as* $n \to \infty$.

We now state and prove the following global bifurcation theorem of the Rabinowitz type.

Theorem 7.1.7

Suppose that L_0, B_0, N_0 *are as in Theorem 7.1.6, and suppose further that every V-singular point in M is isolated in M and M is complete. Let S denote the closure of the set of all nontrivial solutions of (7.1.1). Then, for each bounded, connected component C of S, the set* $C \cap M$ *is finite, i.e.* $C \cap M = \{(v_1, \lambda_1, \ldots, (v_q, \lambda_q)\}$, *and* $\sum_{i=1}^{q} \varepsilon_i \mu_k(v_i, \lambda_i) = 0$ *for every* $k = 1, 2, \ldots$, *where* $\varepsilon_i = \varepsilon(v_i, \lambda_i)$.

Proof. Clearly, every point of $C \cap M$ is a bifurcation point. Since V-singular points are isolated, the finiteness of $C \cap M$ follows from the compactness of $C \cap M$.

Choose r, $\rho > 0$ sufficiently small so that for each $i = 1, 2, \ldots, q$, $U_i = B_M(v_i, \lambda_i; r, \rho)$ is a special neighborhood of (v_i, λ_i) and $U_i \cap U_j = \emptyset$ if $i \neq j$. Let $U = U_1 \cup U_2 \cup \cdots \cup U_q$ and find a bounded, open set $\Omega_1 \subset V \times \mathbb{R}^2$ such that $C \backslash U \subseteq \Omega_1$ and $\Omega_1 \cap M = \emptyset$. Put $\Omega_2 = U \cup \Omega$. Then $C \subseteq \Omega_2$. Find an open, invariant subset $\Omega \subseteq V \times \mathbb{R}^2$ such that $C \subseteq \Omega \subseteq \Omega_2$ and $\partial\Omega \cap S = \emptyset$. Note that Ω is an open, bounded, invariant subset. We now choose $r_0 \in (0, r]$ and $\rho_0 \in (0, \rho)$ such that, for every $i = 1, 2, \ldots, q$,

(i) $B_M(v_i, \lambda_i; r_0, \rho_0) \subseteq \Omega$.
(ii) $U_i := B_M(v_i, \lambda_i; r_0, \rho_0)$ is a special neighborhood of (v_i, λ_i).

Set $U = U_1 \cup U_2 \cup \cdots \cup U_q$, and construct an invariant $\theta : \overline{\Omega} \to \mathbb{R}$ such that:

(iii) $\theta(v, \lambda) = -|\lambda - \lambda_i|$ if $(v, \lambda) \in U_i \cap M$.
(iv) $\theta(v, \lambda) = r_0$ if $(v, \lambda) \in \overline{\Omega} \backslash U$.

Let $F_\theta : \overline{\Omega} \to V \times \mathbb{R}$ be defined by

$$F_\theta(v, \lambda) := (f(v, \lambda), \theta(v, \lambda)) , \qquad (v, \lambda) \in V \times \mathbb{R}^2$$

where $f(v, \lambda) = v - \Theta_{K_0}(B_0, N_0)(v, \lambda)$. Then $F_\theta^{-1}(0) = f^{-1}(0) \cap \theta^{-1}(0)$. By (iii), $F_\theta^{-1}(0) \subseteq C$. Since $\partial\Omega \cap C = \varnothing$, $F_\theta^{-1}(0) \cap \partial\Omega = \varnothing$ and, hence, S^1-$\deg(F_\theta, \Omega)$ is well defined.

We now construct a homotopy $H : \overline{\Omega} \times [0, 1] \to V \times \mathbb{R}$ as follows:

$$H(v, \lambda, t) = (f(v, \lambda), (1-t)\theta(v, t) - t\rho_0) , \qquad (v, \lambda, t) \in \overline{\Omega} \times [0, 1]$$

By (iii), $H(v, \lambda, t) \neq 0$ for all $(v, \lambda, t) \in \partial\overline{\Omega} \times [0, 1]$. Notice that θ is invariant and f is equivariant. So, H is an S^1-homotopy. Since $H(v, \lambda, 0) = F_\theta(v, \lambda)$ and $H(v, \lambda, 1) = (f(v, \lambda), -\rho_0) \neq 0$ for all $x \in \overline{\Omega}$, from the existence and homotopy invariance of the S^1-degree, it follows that S^1-$\deg(F_\theta, \Omega) = 0$. But (i–iii) imply that $F_\theta^{-1}(0) = C \cap U$. Therefore, the excision property of the S^1-degree gives

$$S^1\text{-}\deg(F_\theta, \Omega) = S^1\text{-}\deg(F_\theta, U) = 0$$

On the other hand, by the additivity of the S^1-degree, we have

$$S^1\text{-}\deg(F_\theta, U) = S^1\text{-}\deg(F_\theta, U_1) + \cdots + S^1\text{-}\deg(F_\theta, U_q) = 0$$

Moreover, by Theorem 7.1.5, if r_0 and $\rho_0 > 0$ are sufficiently small, then

$$S^1\text{-}\deg(F_\theta, U_i) = S^1\text{-}\deg_L([B, N_\theta], U_i) = \varepsilon_i \cdot \mu(v_i, \lambda_i) , \qquad i = 1, 2, \ldots, q$$

Consequently, for every $k = 1, 2, \ldots,$

$$\varepsilon_1 \mu_k(v_1, \lambda_1) + \cdots + \varepsilon_q \mu_k(v_q, \lambda_q) = 0$$

This completes the proof. □

EXERCISES

Suppose that Σ is a compact subset of Λ such that $U_0 \cap \Lambda = \Sigma$ for an open, bounded $U_0 \subset M$. We will call such a subset Σ an *isolated singular set*. It is possible to find a smooth function $\gamma : M \to [0, 1]$ such that $\Sigma \subseteq \gamma^{-1}(0)$ and $M \backslash U_0 \subseteq \gamma^{-1}(1)$. By Theorem 1.3.9, there is a regular value $t_0 \in (0, 1)$ of the map γ. We put $U := \gamma^{-1}([0, t_0))$. It is clear that $\Sigma \subseteq U$ and the boundary $\partial U = C_1 \cup \ldots C_N$ of U is composed of a finite number of closed, regular curves C_i, where $C_i = \alpha_i(S^1)$ for a certain diffeomorphism $\alpha_i : S^1 \to M$, $i = 1, \ldots, N$. In addition, by the assumptions (A) and (B), the submanifold M is oriented. Consequently, we note that ∂U, which is an inverse image of a regular value of γ, is a one-dimensional oriented manifold. This means that each of the curves C_i has a natural orientation and we may assume that all the maps $\alpha_i : S^1 \to M$, $i = 1, \ldots, N$, are orientation-preserving diffeomorphisms. For each curve C_i, we can define an element $\mu(C_i) \in \bigoplus_{k=1}^{\infty} \mathbb{Z}$ by $\mu(C_i) = \{\mu_k(C_i)\}$, where $\mu_k(C_i) = \deg([\psi^i_k])$, $\psi^i_k(z) := D_v f(\alpha_i(z))|_{V_k}$. In what follows, we put $\mu(\Sigma) := \sum_{i=1}^{N} \varepsilon_i \mu(C_i)$, where $\varepsilon_i = 1$ if $\psi^i_\infty(z) \in GL^+_{\mathscr{C}}(V_\infty)$ or $\varepsilon_i = -1$ if $\psi^i_\infty(z) \in GL^-_{\mathscr{C}}(V_\infty)$ for $z \in C_i$.

7.1.1 Define a *special neighborhood* $B_M(\Sigma, U, r)$ of the isolated singular set Σ by

$$B_M(\Sigma, U, r) = \{(v, \lambda) \in V \times \mathbb{R}^2; \quad (\eta(\lambda), \lambda) \in U, \|v - \eta(\lambda)\| < r\}$$

and let $\theta : \overline{B_M(\Sigma, U, r)} \to \mathbb{R}$ be an equivariant function such that $\theta(v, \lambda) < 0$ for $(v, \lambda) \in \overline{U}$ and $\theta(v, \lambda) > 0$ for $\|v - \eta(\lambda)\| = r$. Show that for a sufficiently small $r > 0$, the map $F_\theta(v, \lambda) = (f(v, \lambda), \theta(v, \lambda))$ is $B_M(\Sigma, U, r)$-admissible and

$$S^1\text{-deg}(F_\theta, B_M(\Sigma, U, r)) = \mu(\Sigma)$$

7.1.2 Show that if $S^1\text{-deg}(F_\theta, B_M(\Sigma, U, r) \neq 0$, then there exists a bifurcation point $(v_0, \lambda_0) \in \Sigma$ for the problem (7.1.1). More precisely, there exists a sequence (v_n, λ_n) of nontrivial solutions to (7.1.1) such that $(v_n, \lambda_n) \to (v_0, \lambda_0)$ as $n \to \infty$.

7.1.3 Assume that there is a sequence of isolated compact singular sets Σ_l, $l = 1, 2, \ldots$, such that $\bigcup_l K_l = \Lambda$. Let S denote the closure of the set of all nontrivial solutions of (7.1.1) and C be a bounded component of S (i.e., C is open and closed in S). We say that C is (Σ_l)-*compatible* if for every set Σ_l such that $C \cap \Sigma_l \neq \emptyset$, we have $S \cap \Sigma_l \subseteq C \cap \Sigma_l$. Show that the following generalized global bifurcation relations hold for the component C:

$$\sum_{l} \mu(\Sigma_l) = 0$$

7.2 GLOBAL HOPF BIFURCATION THEORY OF NEUTRAL EQUATIONS

In this section we apply Theorems 7.1.6 and 7.1.7 to Hopf bifurcation problems for functional differential equations of the neutral type.

We start this section with the notion of a *formal crossing number* and a corresponding computation formula. Suppose $\Omega_1 := (0, b) \times (\beta - c, \beta + c) \subseteq \mathbb{R}^2$, where $b, c > 0$ and $\beta \neq 0$. Let $h : [-a, a] \times \overline{\Omega} \to \mathbb{R}^2$, $a > 0$, be a continuous function satisfying the following conditions:

(C1) $h(\alpha, x, y) \neq 0$ for all $\alpha \in [-a, a]$ and $(x, y) \in \partial\Omega \backslash \{(0, y);\ y \in (\beta - c, \beta + c)\}$.

(C2) Let $(x, y) \in \overline{\Omega}$, if $h(\pm a, x, y) = 0$ then $x \neq 0$.

For every $\alpha \in [-a, a]$, we put $h_\alpha(x, y) = h(\alpha, x, y) \in \overline{\Omega}$. The conditions (C1) and (C2) imply that $h_\pm := h_{\pm a}|_{\overline{\Omega}} : \overline{\Omega} \to \mathbb{R}^2$ has no zero on the boundary $\partial\Omega$. Consequently, we can define the following integers:

$$\gamma_\pm := \deg(h_\pm, \Omega)$$

where $\deg(\cdot, \Omega)$ denotes the Brouwer degree on $\Omega \subseteq \mathbb{R}^2$. We define the *formal crossing number* for the family $\{h_\alpha\}$, $h_\alpha = h(\alpha, \cdot, \cdot)$, by

$$\gamma := \gamma_- - \gamma_+$$

Then we have the following conclusion.

Lemma 7.2.1

Suppose that $h : [-a, a] \times \overline{\Omega} \to \mathbb{R}^2$ is continuous and satisfies conditions (C1) and (C2). We put $\Omega_1 := (-a, a) \times (\beta - c, \beta + c)$ and define the function $\psi_h : \overline{\Omega}_1 \to \mathbb{R}^2$ by $\psi_h(\alpha, y) = h(\alpha, 0, y)$, $\alpha \in [-a, a]$, $y \in [\beta - c, \beta + c]$. Then $\psi_h(\alpha, y) \neq 0$ for $(\alpha, y) \in \partial\Omega_1$ and $\deg(\psi_h, \Omega_1) = \gamma$.

Proof. The conclusion that $\psi_h(\alpha, y) \neq 0$ for $(\alpha, y) \in \partial\Omega_1$ follows immediately from (C1) and (C2). To prove that $\deg(\psi_h, \Omega_1) = \gamma$, we consider the parallelepiped $P := [-a, a] \times [0, b) \times [\beta - c, \beta + c]$. It is easy to see that there exists an orientation-preserving homeomorphism $\mathscr{P} : P \to P$ such that it transforms the faces $P_- := \{-a\} \times [0, b] \times [\beta - c, \beta + c]$, $P_+ := \{a\} \times [0, b] \times [\beta - c, \beta + c]$, and $P_1 := [-a, a] \times \{0\} \times [\beta - c, \beta + c]$ onto the face

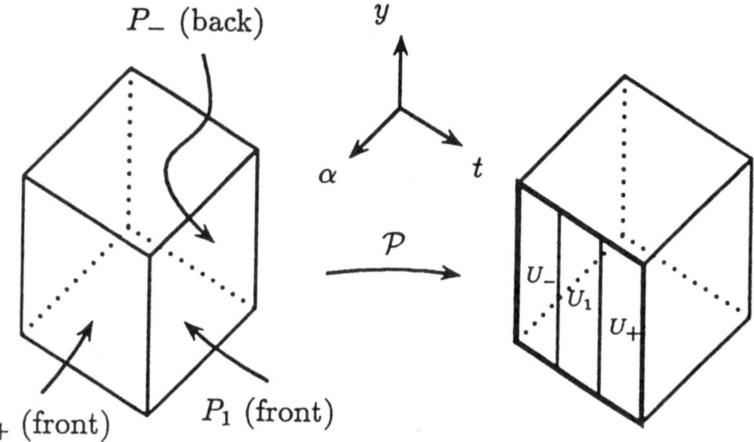

FIGURE 7.2.1. An orientation-preserving homeomorphism of the parallelepiped.

P_1 (see Figure 7.2.1). We denote $\mathscr{P}(P_+)$ by U_+, $\mathscr{P}(P_-)$ by U_-, and $\mathscr{P}(P_1)$ by U_1.

Next, we consider the composition

$$\tilde{h} := h \circ \mathscr{P}^{-1} : P \to \mathbb{R}^2$$

By the construction, for each $t \in [0, b]$, the mapping $\tilde{h}_t : P_1 \to \mathbb{R}^2$ defined by $\tilde{h}_t(\alpha, 0, y) = \tilde{h}(\alpha, t, y)$ for $(\alpha, t, y) \in P$ has no zero in ∂P_1. Therefore, $\tilde{h} : [0, b] \times P_1 \to \mathbb{R}^2$ may be viewed as a homotopy $\tilde{h}_t$ from $\tilde{h}_0 : P_1 \to \mathbb{R}^2$ to $\tilde{h}_b : P_1 \to \mathbb{R}^2$. Moreover, by condition (C1), $\tilde{h}_b$ has no zero in P_1. Consequently, $\deg(\tilde{h}_b, P_1) = 0$. This implies $\deg(\tilde{h}_0, P_1) = 0$.

On the other hand, since $\mathscr{P}^{-1}$ is a homeomorphism, in view of orientation-preserving and reversing properties of $\mathscr{P}^{-1}$ on the boundary ∂P, we have

$$\deg(\tilde{h}|_{\bar{U}_+}, U_+) = \deg(h_a, \Omega)$$

$$\deg(\tilde{h}|_{\bar{U}_-}, U_-) = -\deg(h_{-a}, \Omega)$$

$$\deg(\tilde{h}|_{\bar{U}_1}, U_1) = \deg(\psi_h, \Omega_1)$$

By applying the additivity and excision properties of the Brouwer degree, we obtain

$$\begin{aligned} 0 &= \deg(\tilde{h}_0, P_1) \\ &= \deg(\tilde{h}|_{\bar{U}_+}, U_+) + \deg(\tilde{h}|_{\bar{U}_1}, U_1) + \deg(\tilde{h}|_{\bar{U}_-}, U_-) \end{aligned}$$

$$= \deg(h_a, \Omega) + \deg(\psi_h, \Omega_1) - \deg(h_{-a}, \Omega)$$
$$= \deg(\psi_h, \Omega_1) + \gamma_+ - \gamma_-$$

This completes the proof. □

We now turn to the Hopf bifurcation problem for neutral equations. Denote by $C(\mathbb{R}; \mathbb{R}^n)$ the Banach space of continuous, bounded functions from $\mathbb{R}$ to $\mathbb{R}^n$ equipped with the usual supremum norm $\|\varphi\| = \sup_{v \in \mathbb{R}} |\varphi(v)|$ for $\varphi \in C(\mathbb{R}; \mathbb{R}^n)$, where $|\cdot|$ denotes the Euclidean norm on $\mathbb{R}^n$. If $\varphi \in C(\mathbb{R}; \mathbb{R}^n)$ and $t \in \mathbb{R}$, then $\varphi_t \in C(\mathbb{R}; \mathbb{R}^n)$ is defined by $\varphi_t(v) = \varphi(t + v)$ for $v \in \mathbb{R}$.

Consider the following one-parameter family of neutral equations:

$$\frac{d}{dt}[x(t) - b(x_t, \alpha)] = F(x_t, \alpha) \tag{7.2.1}$$

where $F, b : C(\mathbb{R}; \mathbb{R}^n) \times \mathbb{R} \to \mathbb{R}^n$ are two continuous mappings. We assume that the mapping b satisfies the following Lipschitz condition:

$$|b(\varphi, \alpha) - b(\psi, \alpha)| \leq k\|\varphi - \psi\|, \qquad \varphi, \psi \in C(\mathbb{R}; \mathbb{R}^n) \tag{7.2.2}$$

where $0 \leq k < 1$ is a given constant. We further assume that there exists a number $a \geq 0$ such that the maps F and b have continuous factorizations $\hat{F}$, $\hat{b} : C((-\infty, a]; \mathbb{R}^n) \times \mathbb{R} \to \mathbb{R}^n$, that is, we assume that there is the following commutative diagram:

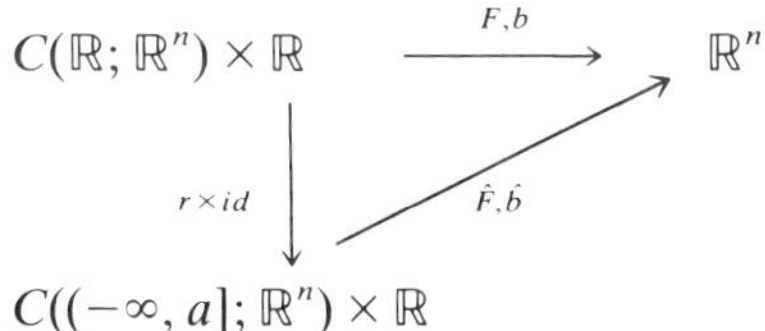

where $r : (\mathbb{R}; \mathbb{R}^n) \to C((-\infty, a]; \mathbb{R}^n)$ is simply the restriction operator. This allows us to consider neutral equations with both infinite delays and possibly also with finite advances and, therefore, permits us to include functional differential equations of anticipatory type. We refer the interested reader to Chukwu (1992), Hale (1977), and Salamon (1984) for various examples.

In what follows, we will assume that $\hat{F}$ and $\hat{b}$ are both C^2-maps. However, we should emphasize that, for the local bifurcation result, this assumption can be replaced by the weaker assumption requiring continuous differentiability only at the set of constant solutions.

For a point $x_0 \in \mathbb{R}^n$, we will denote by the same symbol the constant function $x_0(t) \equiv x_0$. Since $(x_0)_t \equiv x_0$ for all $t \in \mathbb{R}$, if $F(x_0, \alpha_0) = 0$, then x_0 is a solution of (7.2.1). We call such points $(x_0, \alpha_0) \in \mathbb{R}^n \times \mathbb{R}$ *stationary points* of (7.2.1). We say that a stationary point (x_0, α_0) is *nonsingular* if the restriction of F to the subspace $\mathbb{R}^n \times \mathbb{R} \subseteq C(\mathbb{R}; \mathbb{R}^n) \times \mathbb{R}$, still denoted by F, has the derivative with respect to $x \in \mathbb{R}^n$, denoted by $D_x F(x_0, \alpha_0)$, which is an isomorphism.

For a stationary point of (7.2.1), the linearization of Eq. (7.2.1) at (x_0, α_0) leads to the following *characteristic equation*:

$$\det_{\mathbb{C}} \Delta_{(x_0,\alpha_0)}(\lambda) = 0 \tag{7.2.3}$$

where $\Delta_{(x_0,\alpha_0)}(\lambda)$ is an $n \times n$ complex matrix defined by

$$\Delta_{(x_0,\alpha_0)}(\lambda) := \lambda[\mathrm{Id} - S(x_0, \alpha_0)(e^{\lambda \cdot} \cdot)] - T(x_0, \alpha_0)(e^{\lambda \cdot}) : \mathbb{C}^n \to \mathbb{C}^n$$

$$S(x_0, \alpha_0) := D_\varphi \hat{b}(x_0, \alpha_0) : C((-\infty, \alpha]; \mathbb{C}^n) \to \mathbb{C}^n$$

$$T(x_0, \alpha_0) := D_\varphi \hat{F}(x_0, \alpha_0) : C((-\infty, \alpha]; \mathbb{C}^n) \to \mathbb{C}^n$$

and

$$(e^{\lambda \cdot} \cdot)(v, x) = e^{\lambda v} x \qquad \text{for } (v, x) \in \mathbb{R} \times \mathbb{C}^n$$

Since we allow only finite advances, formula (7.2.3) is well defined for all complex numbers λ such that $\mathrm{Re}\,\lambda \geq 0$. Moreover, the function $\lambda \mapsto \det_{\mathbb{C}} \Delta(x_0, \alpha_0)(\lambda)$ is analytic in the semiplane $\mathrm{Re}\,\lambda > 0$.

A solution λ_0 to Eq. (7.2.3) is called a *characteristic value* of the stationary point (x_0, α_0). So, (x_0, α_0) is a nonsingular stationary point if and only if zero is not a characteristic value of (x_0, α_0).

We say that a nonsingular stationary point (x_0, α_0) is a *center* if its set of purely imaginary characteristic values is nonempty and discrete, and we say that (x_0, α_0) is an *isolated center* if it is the only center in some neighborhood of (x_0, α_0) in $\mathbb{R}^n \times \mathbb{R}$.

For a local Hopf bifurcation, we make the following assumption:

(H1) (7.2.1) has an isolated center $(x_0, \alpha_0) \in \mathbb{R}^n \times \mathbb{R}$

Under assumption (H1), (x_0, α_0) is a nonsingular stationary point, that is, $D_x F(x_0, \alpha_0) : \mathbb{R}^n \to \mathbb{R}^n$ is an isomorphism. Hence, by the implicit function theorem, there is a differentiable curve $x(\alpha) \in \mathbb{R}^n$ for α in a δ-neighborhood of α_0 such that $(x(\alpha), \alpha)$ is a stationary point for each α and $x(\alpha_0) = x_0$. Therefore, we can define the following mappings:

$$S(\alpha) := S(x(\alpha), \alpha) = D_\varphi \hat{b}(x(\alpha), \alpha) : C((-\infty, a]; \mathbb{C}^n) \to \mathbb{C}^n$$

$$T(\alpha) := T(x(\alpha), \alpha) = D_\varphi \hat{F}(x(\alpha), \alpha) : C((-\infty, a]; \mathbb{C}^n) \to \mathbb{C}^n$$

In addition, for all complex numbers λ such that $\mathrm{Re}\,\lambda \geq 0$ and $\alpha \in [\alpha_0 -$

$\delta, \alpha_0 + \delta]$ with a sufficiently small number $\delta > 0$, we define

$$\Delta_\alpha(\lambda) := \Delta_{(x(\alpha),\alpha)}(\lambda)$$

By the hypothesis (H1), there is a $\beta_0 > 0$ such that $\det_{\mathbb{C}} \Delta_{(x_0,\alpha_0)}(i\beta_0) = 0$ and if $0 < |\alpha - \alpha_0| \le \delta$, then $i\mathbb{R} \cap \{\lambda \in \mathbb{C};\ \det_{\mathbb{C}} \Delta_\alpha(\lambda) = 0,\ \mathrm{Re}\,\lambda \ge 0\} = \varnothing$.

We can choose constants $b = b(\alpha_0, \beta_0) > 0$ and $c = c(\alpha_0, \beta_0) > 0$ such that the closure of $\Omega := (0, b) \times (\beta_0 - c, \beta_0 + c) \subseteq \mathbb{R}^2 \cong \mathbb{C}$ contains no other zero of $\det_{\mathbb{C}} \Delta_{\alpha_0}(\lambda)$. Since $\det_{\mathbb{C}} \Delta_\alpha(\lambda)$ is analytic in $\lambda \in \Omega$ and continuous in α, if $\delta > 0$ is sufficiently small, then there is no zero of $\det_{\mathbb{C}} \Delta_{\alpha_0 \pm \delta}(\lambda)$ in $\partial\Omega$. So we can define the numbers

$$\gamma_\pm(x_0, \alpha_0, \beta_0) = \deg(\det_{\mathbb{C}} \Delta_{\alpha_0 \pm \delta}(\cdot), \Omega)$$

Definition 7.2.2 The *crossing number* of (x_0, α_0, β_0) is defined as

$$\gamma(x_0, \alpha_0, \beta_0) = \gamma_-(x_0, \alpha_0, \beta_0) - \gamma_+(x_0, \alpha_0, \beta_0)$$

By the definition, the crossing number counts the number of characteristic values (with multiplicities) *escaping* from the region Ω as α crosses α_0 from left to right.

Now we can state our first main result, the local Hopf bifurcation theorem. We should remark that the bifurcations points we are considering here are in the Fuller space $C(\mathbb{R}; \mathbb{R}^n) \times \mathbb{R} \times (0, \infty)$, not the phase space $C(\mathbb{R}; \mathbb{R}^n) \times \mathbb{R}$.

Theorem 7.2.3

Suppose that the hypothesis (H1) is satisfied. If $\gamma(x_0, \alpha_0, \beta_0) \ne 0$, then there exists a bifurcation of nonconstant, periodic solutions from (x_0, α_0). More precisely, there exists a sequence $\{(x_n(t), \alpha_n, \beta_n)\}$ such that $\alpha_n \to \alpha_0$, $\beta_n \to \beta_0$, $x_n \to x_0$ as $n \to \infty$ and $x_n(t)$ is a nonconstant periodic solution of Eq. (7.2.1.) with period $2\pi/\beta_n$.

Proof. First of all, we introduce the period as an additional parameter so that we can work in the space of continuous functions of fixed period 2π. Making a change of variable $x(t) = z(\beta t)$, we obtain

$$\frac{d}{dt}[z(t) - b(z_{t,\beta}, \alpha)] = \frac{1}{\beta} F(z_{t,\beta}, \alpha) \tag{7.2.4}$$

where $z_{t,\beta}(\theta) = z(t + \beta\theta)$ for $\theta \in \mathbb{R}$. Obviously, $z(t)$ is a 2π-periodic solution of (7.2.4) if and only if $x(t)$ is a $(2\pi/\beta)$-periodic solution of (7.2.1).

Let $S^1 = \mathbb{R}^1/2\pi\mathbb{Z}$, $V = C(S^1; \mathbb{R}^n)$, $W = L^2(S^1; \mathbb{R}^n)$, and $\mathscr{D}(\alpha_0, \beta_0) =$

$(\alpha_0 - \delta, \alpha_0 + \delta) \times (\beta_0 - c, \beta_0 + c)$. Define the maps

$$\begin{cases} L_0 z(t) := \dot{z}(t)\,, & z \in \mathrm{Dom}(L_0) = H^1(S^1; \mathbb{R}^n) \\ B_0(z, \alpha, \beta)(t) := b(z_{t,\beta}, \alpha)\,, & z \in V \\ N_0(z, \alpha, \beta)(t) := \dfrac{1}{\beta} F(z_{t,\beta}, \alpha)\,, & z \in V \end{cases}$$

for $(\alpha, \beta) \in \mathscr{D}(\alpha_0, \beta_0)$ and $t \in \mathbb{R}$. The spaces V and W are isometric Banach representations of the group $G = S^1$, where S^1 acts by shifting the argument.

The operator $L_0 : \mathrm{Dom}(L_0) \subseteq V \to W$ is an equivariant, closed Fredholm operator of index zero such that $\mathrm{Ker}\, L_0 = \mathbb{R}^n \subseteq V$ is the subspace of constant functions. Let $K_0 x = 1/(2\pi) \int_0^{2\pi} x(t)\, dt$, that is, $K_0 : V \to W$ is the composition of the inclusion $V \subseteq W$ with the orthogonal projection of W onto the subspace $\mathbb{R}^n \subseteq W$. Clearly, K_0 is an equivariant, compact resolvent of the operator L_0, that is, $K_0 \in CR^G(L_0)$. Moreover, by Sobolev inequalities, the operator $R_{K_0} : L^2(S^1; \mathbb{R}^n) \to C(S^1; \mathbb{R}^n)$ is compact. This implies that N_0 is L_0-compact. Therefore, the assumption (7.2.2) implies that (B_0, N_0) is an L_0-condensing pair. Since both B_0 and N_0 are G-equivariant, it follows that (B_0, N_0) is an L_0-condensing G-pair.

Therefore, finding a $(2\pi/\beta)$-periodic solution for Eq. (7.2.1) is equivalent to finding a solution of the following composite coincidence problem:

$$\text{(7.2.5)} \quad \begin{cases} \text{Find } (z, \alpha, \beta) \in V \times \mathbb{R}^2 \text{ such that } z - B_0(z, \alpha, \beta) \in \mathrm{Dom}(L_0) \\ \text{and } L_0[z - B_0(z, \alpha, \beta)] = N_0(z, \alpha, \beta) \end{cases}$$

Now we can use the ideas described in Section 7.1. First of all, we transform problem (7.2.5) to the equivalent fixed-point problem

$$\text{(7.2.6)} \qquad z = \Theta_{K_0}(B_0, N_0)(z, \alpha, \beta)\,, \qquad (z, \alpha, \beta) \in V \times \mathbb{R}^2$$

where $\Theta_{K_0}(B_0, N_0) : V \times \mathbb{R}^2 \to V$ is given by $\Theta_{K_0}(B_0, N_0) = B_0 + R_{K_0}[N_0 + K_0(\pi_0 - B_0)]$ and $\pi_0 : V \times \mathbb{R}^2 \to V$ is the natural projection. We then define the mapping $f : V \times \mathbb{R}^2 \to V$ by

$$f(z, \alpha, \beta) = z - \Theta_{K_0}(B_0, N_0)(z, \alpha, \beta)$$

which is an equivariant condensing field of class C^1. Consequently, finding a 2π-periodic solution of Eq. (7.21) is equivalent to finding a solution of the problem.

$$\text{(7.2.7)} \qquad f(z, \alpha, \beta) = 0\,, \qquad (z, \alpha, \beta) \in V \times \mathbb{R}^2$$

Moreover, it is easy to see that, for every constant function $x \in \mathbb{R}^n$, the derivative $D_z f(x, \alpha, \beta)$ is an equivariant linear operator on V that is also a condensing field given by

$$D_x f(x, \alpha, \beta) = \mathrm{Id} - \Theta_{K_0}(D_z B_0, D_z N_0)(z, \alpha, \beta)$$
$$= \mathrm{Id} - D_z B_0(x, \alpha, \beta) - R_{K_0}[D_z N_0(x, \alpha, \beta)$$
$$+ K_0(\pi_0 - D_z B_0(x, \alpha, \beta))]$$

By assumption (H1), (x_0, α_0, β_0) is an isolated center. Thus, we can define the two-dimensional submanifold $M \subseteq V^G \times \mathbb{R}^2$ by taking

$$M := \{(x_0(\alpha), \alpha, \beta) : \alpha \in (\alpha_0 - \delta, \alpha_0 + \delta),\ \beta \in (\beta_0 - c, \beta_0 + c)\}$$

M satisfies assumptions (A) and (B) of the previous section. Moreover, the point (x_0, α_0, β_0) is the only V-singular point in M and

$$D_z f(x(\alpha), \alpha, \beta) = z(t) - S(\alpha) z_{t,\beta} - R_{K_0}[(1/\beta) T(\alpha) z_{t,\beta}$$
$$+ K_0(z(t) - S(\alpha) z_{t,\beta})]$$

It is well known that V has the following isotypical direct decomposition $V = \bigoplus_{k=0}^{\infty} V_k$, where $V_0 = V^G$ and for $k \geq 1$, V_k is the subspace spanned by $\cos kt \cdot \varepsilon_j$ and $\sin kt \cdot \varepsilon_j$, $j = 1, 2, \ldots, n$. Here and in what follows, $\{\varepsilon_1, \ldots, \varepsilon_n\}$ denotes the standard basis of $\mathbb{R}^n$. From now on, we will identify V_k, $k = 1, 2, \ldots,$ with the linear space over the complex numbers spanned by $\exp(ikt)\varepsilon_j$, $j = 1, \ldots, n$.

Put $\Psi(\alpha, \beta) = D_z f(x(\alpha), \alpha, \beta) : V \to V$, $(\alpha, \beta) \in \mathcal{D}(\alpha_0, \beta_0)$. Since $\Psi(\alpha, \beta)$ is S^1-equivariant, $\Psi(\alpha, \beta)(V_k) \subset V_k$ for all $k = 0, 1, 2, \ldots,$ and therefore we can define $\Psi_k : \mathcal{D}(\alpha_0, \beta_0) \to L(V_k, V_k)$ by $\Psi_k(\alpha, \beta) := \Psi(\alpha, \beta)|_{V_k}$.

By direct verification, we have for $k = 1, 2, \ldots,$

$$\Psi(\alpha, \beta)(e^{ikt}\varepsilon_j) = \frac{e^{ikt}}{i\beta k}\{(i\beta k)[\varepsilon_j - S(\alpha)(e^{ik\beta} e_j)] - T(\alpha)(e^{ik\beta}\varepsilon_j)\}$$

Therefore, the matrix representation of $\Psi_k(\alpha, \beta)$ with respect to the ordered basis $(e^{ikt}\varepsilon_1, \ldots, e^{ikt}\varepsilon_n)$ is exactly $1/(\beta ik)\Delta_\alpha(ik\beta)$. Now we can compute $\mu_1(x_0, \alpha_0, \beta_0)$ defined in Theorem 7.1.5. First, we identify $\mathcal{D}(\alpha_0, \beta_0)$ with S^1 and define $H : \mathcal{D}(\alpha_0, \beta_0) \to \mathbb{R}^2 \cong$ by

$$H(\alpha, \beta) = \det_{\mathbb{C}} \Delta_\alpha(i\beta)$$

The number $\mu_1(x_0, \alpha_0, \beta_0)$, defined in Theorem 7.2.5, can be expressed by $\deg(H, \Omega)$ as follows:

$$\mu_1(x_0, \alpha_0, \beta_0) = \varepsilon \cdot \deg(H, \mathcal{D}(\alpha_0, \beta_0))$$

where $\varepsilon = \operatorname{sign} \det \Psi_0(\alpha, \beta)$ for $(\alpha, \beta) \in \partial\mathcal{D}(\alpha_0, \beta_0)$. We define the function

$k : [\alpha_0 - \delta, \alpha_0 + \delta] \times \overline{\Omega} \to \mathbb{R}^2 \cong \mathbb{C}$ by

$$k(\alpha, u, v) = \det_{\mathbb{C}} \Delta_\alpha(u + iv)$$

where $\Omega = (0, b) \times (\beta_0 - c, \beta_0 + c)$, $b = b(\alpha_0, \beta_0) > 0$. The choice of b, c, and δ guarantees that k satisfies all conditions of Lemma 7.2.1. So, we obtain

$$\deg(H, \mathscr{D}(\alpha_0, \beta_0)) = \gamma(x_0, \alpha_0, \beta_0)$$

By the assumption, $\gamma(x_0, \alpha_0, \beta_0) \neq 0$. Thus, $\mu_1(x_0, \alpha_0, \beta_0) = \varepsilon \cdot \gamma(x_0, \alpha_0, \beta_0) \neq 0$. This implies that, by Theorem 7.1.6, (x_0, α_0, β_0) is a bifurcation point. Consequently, there exists a sequence $\{x_n(t), \alpha_n, \beta_n\}$ such that $\alpha_n \to \alpha_0$ and $\beta_n \to \beta_0$ as $n \to \infty$, and $x_n(t)$ is a periodic solution of (7.2.1) for $\alpha = \alpha_n$ with period $2\pi/\beta_n$ and $x_n \to x_0$ in sup-norm as $n \to \infty$. This completes the proof. □

Note that in Theorem 7.2.3, we did not claim that $2\pi/\beta_n$ is the minimal period of $x_n(t)$. To establish the relationship between $2\pi/\beta_n$ and the minimal period of $x_n(t)$, we need the following result on positive lower bounds for the periods of periodic solutions of neutral equations.

Lemma 7.2.4

Consider Eq. (7.2.1) in $I \times S$, where I is an open interval of $\mathbb{R}$, S is a subset of $C((-\infty, a]; \mathbb{R}^n)$. Assume b and F satisfy the following Lipschitz conditions:

$$|F(\varphi, \alpha) - F(\psi, \alpha)| \leq L\|\varphi - \psi\|$$

$$|b(\varphi, \alpha) - b(\psi, \alpha)| \leq k\|\varphi - \psi\|$$

for $\varphi, \psi \in S$, and $\alpha \in I$, where L and k are positive constants and $k < 1$. If $x(t)$ is a nonconstant periodic solution of (7.2.1) with $x_t \in S$ for all $t \in \mathbb{R}$, then the minimal period p of $x(t)$ satisfies $p \geq 4(1-k)/L$.

The similar results for ordinary and retarded differential equations have been established in Lasota and Yorke (1971), Li (1975), Vidossich (1976), and Yorke (1969). The analog for neutral equations was obtained by Krawcewicz et al. (1993). The proof was based on the following result due to Vidossich (1976).

Lemma 7.2.5

Let X be a Banach space, $V : \mathbb{R} \to X$ a p-periodic function with the following properties:

(i) *V is integrable and* $\int_0^p V(t)\,dt = 0$.

(ii) *There exists* $U \in L^1([0, p/2]; \mathbb{R}_+)$ *such that*

$$|V(t) - V(s)| \leq U(t-s)$$

for almost all s, t *with* $0 \leq s \leq t \leq p$ *and* $t - s \leq p/2$. *Then*

$$p \sup_{t\in\mathbb{R}} |V(t)| \leq 2 \int_0^{p/2} U(t)\,dt$$

We now state the proof of Lemma 7.2.4.

Proof. Let $s \leq t$ and $D(x_t, \alpha) = x(t) - b(x_t, \alpha)$. We have

$$\begin{aligned} |D(x_t, \alpha) - D(x_s, \alpha)| &= |x(t) - x(s) + b(x_s, \alpha) - b(x_t, \alpha)| \\ &\geq |x(t) - x(s)| - k\|x_t - x_s\| \end{aligned}$$

which gives, for all τ,

$$|D(x_{t+\tau}, \alpha) - D(x_{s+\tau}, \alpha)| \geq |x(t+\tau) - x(s+\tau)| - k\|x_{t+\tau} - x_{s+\tau}\|$$

By the mean value theorem, the above inequality yields

$$|x(t+\tau) - x(s+\tau)| \leq k\|x_{t+\tau} - x_{s+\tau}\| + |D'(x_\xi, \alpha)|(t-s)$$

for some $\xi \in [s+\tau, t+\tau]$. By taking the supremum with respect to τ on two sides of the above inequality and using the periodicity of $x(t)$, one obtains

$$\|x_t - x_s\| \leq k\|x_t - x_s\| + \sup_{\xi\in\mathbb{R}} |D'(x_\xi, \alpha)|(t-s)$$

It follows that

$$\|x_t - x_s\| \leq \frac{1}{1-k} \sup_{\xi\in\mathbb{R}} |D'(x_\xi, \alpha)|(t-s)$$

On the other hand,

$$\begin{aligned} |D'(x_t, \alpha) - D'(x_s, \alpha)| &= |F(x_t, \alpha) - F(x_s, \alpha)| \\ &\leq L\|x_t - x_s\| \end{aligned}$$

Therefore,

$$|D'(x_t, \alpha) - D'(x_s, \alpha)| \leq \frac{L}{1-k} \sup_{\xi \in \mathbb{R}} |D'(x_\xi, \alpha)|(t-s)$$

Let $V(t) = D'(x_t, \alpha)$ and $U(t) = L/(1-k) \sup_{\xi \in \mathbb{R}} |D'(x_\xi, \alpha)|t$. Then, applying Lemma 7.25, we get

$$p \sup_{t \in \mathbb{R}} |D'(x_t, \alpha)| \leq 2 \int_0^{p/2} \frac{L}{1-k} \sup_{\xi \in \mathbb{R}} |D'(\xi, \alpha)| s \, ds$$

Hence, $p \geq 4(1-k)/L$ and the proof is complete. □

The following result provides a result regarding the relationship between the minimal period of $x_n(t)$ in Theorem 7.2.3 and the characteristic values of (7.2.1). The result was first established by Mallet-Paret and Yorke (1982) for ordinary differential equations and was extended to neutral equations by Wu (1993a).

Lemma 7.2.6

Suppose that there exists a sequence of real numbers $\{\alpha_n\}_{n=1}^{\infty}$ such that:

(i) *For each n, (7.2.1) with $\alpha = \alpha_n$ has a nonconstant, periodic solution $x_n(t)$ with the minimal period $T_n > 0$.*

(ii) $\lim_{n\to\infty} \alpha_n = \alpha_0 \in \mathbb{R}$, $\lim_{n\to\infty} T_n = T_0 < \infty$, *and* $\lim_{n\to\infty} x_n(t) = x_0 \in \mathbb{R}^n$ *uniformly for $t \in \mathbb{R}$.*

Then (x_0, α_0) is a stationary solution of (7.2.1) and $\pm i2\pi/T_0$ are characteristic values of (7.2.1) with $\alpha = \alpha_0$.

Proof. Throughout the proof, we use $Db(\varphi, \alpha)$ and $DF(\psi, \alpha)$ to denote $D_\varphi b(\varphi, \alpha)$ and $D_\varphi F(\varphi, \alpha)$, respectively. We first show that $T_0 > 0$. Since F is continuously differentiable, there exists $\varepsilon_0 > 0$ such that

$$\|DF(\varphi, \alpha) - DF(x_0, \alpha)\| < 1$$

if $\|\varphi - x_0\| < \varepsilon_0$ and $|\alpha - \alpha_0| < \varepsilon_0$. Consequently, $|F(\varphi, \alpha) - F(\psi, \alpha) \leq L_0\|\varphi - \psi\|$ if $\|\varphi - x_0\| < \varepsilon_0$, $\|\psi - x_0\| < \varepsilon_0$, and $|\alpha - \alpha_0| < \varepsilon_0$, where $L_0 = |DF(\alpha_0, x_0)| + 1$. Therefore, under assumption (ii), we can apply Lemma 7.2.4 to conclude $T_n \geq 4(1-k)/L_0$ for sufficiently large n, from which it follows that $T_0 \geq 4(1-k)/L_0$.

Next, we show that the following linear equation:

$$\frac{d}{dt}[y(t) - Db(x_0, \alpha_0)y_t] = DF(x_0, \alpha_0)y_t \tag{7.2.8}$$

has a periodic solution. For $\tau \in (0, 1)$, define

$$\varepsilon_{n,\tau} = \max_{t \in \mathbb{R}} |x_n(t + \tau T_n) - x_n(t)|$$

$$y_n(t) - \varepsilon_{n,\tau}^{-1}[x_n(t + \tau T_n) - x_n(t)], \qquad t \in \mathbb{R}$$

$y_n(t)$ satisfies the following equation:

$$\frac{d}{dt}[y_n(t) - Db(x_0, \alpha_0)(y_n)_t - \delta_{1n}(t)] = DF(x_0, \alpha_0)(y_n)_t + \delta_{2n}(t) \tag{7.2.9}$$

where

$$\delta_{1n}(t) = \varepsilon_{n,\tau}^{-1}[b((x_n)_{t+\tau}, \alpha_n) - b((x_n)_t, \alpha_n) - Db(x_0, \alpha_0)((x_n)_t - (x_n)_t)]$$

$$\delta_{2n}(t) = \varepsilon_{n,\tau}^{-1}[F((x_n)_{t+\tau}, \alpha_n) - F((x_n)_t, \alpha_n) - DF(x_0, \alpha_0)((x_n)_{t+\tau} - (x_n)_t)]$$

Noting that $|y_n(t)| \leq 1$ for $t \in \mathbb{R}$, we have

$$|\delta_{1n}(t)| \leq \int_0^1 |Db((z_n)_t(\theta), \alpha_n) - Db(x_0, \alpha_0)| \, d\theta \to 0$$

$$|\delta_{2n}(t)| \leq \int_0^1 |Db((z_n)_t(\theta), \alpha_n) - Db(x_0, \alpha_0)| \, d\theta \to 0$$

as $n \to \infty$ uniformly for $t \in \mathbb{R}$, where

$$(z_n)_t(\theta) = \theta(x_n)_t + (1 - \theta)(x_n)_{t+\tau}, \qquad \theta \in [0, 1], \quad t \in \mathbb{R}$$

For any given $t \geq t'$, integrating (7.2.1) from t' to t, we get

$$x_n(t) - x_n(t') = b((x_n)_t, \alpha_n) + \int_{t'}^{t} F((x_n)_s, \alpha_n) \, ds$$

from which and the Lipschitz condition (7.22), it follows that

$$|x_n(t) - x_n(t')| \leq \frac{L_0}{1 - k}(t - t')$$

for sufficiently large n. Consequently,

$$\|(z_n)_t(\theta) - (z_n)_{t'}(\theta)\| \le \frac{L_0}{1-k}(t-t'), \qquad \theta \in [0, 1]$$

for sufficiently large n. On the other hand,

$$|Db((z_z)_{t'}(\theta), \alpha_n) - Db(x_0, \alpha_0)| \le \frac{1-k}{2}$$

provided that n is sufficiently large. So,

$$|\delta_{1n}(t) - \delta_{1n}(t')| \le \left| \int_0^1 [Db((z_n)_t(\theta), \alpha_n) - Db((z_n)_{t'}(\theta), \alpha_n)]\, d\theta (y_n)_t \right|$$

$$+ \left| \int_0^1 [Db((z_n)_{t'}(\theta), \alpha_n) - Db(x_0, \alpha_0)]\, d\theta((y_n)_t - (y_n)_{t'}) \right|$$

$$\le [|D^2 b(x_0, \alpha_0)| + 1] \frac{L_0}{1-k}(t-t') + \frac{1-k}{2}\|(y_n)_t - (y_n)_{t'}\|$$

for sufficiently large n.

We now integrate (7.2.9) from t' to t to obtain

$$|y_n(t) - y_n(t')| \le k\|(y_n)_t - (y_n)_{t'}\| + [|D^2 b(x_0, \alpha_0)| + 1] \frac{L_0}{1-k}(t-t')$$

$$+ \frac{1-k}{2}\|(y_n)_t - (y_n)_{t'}\| + (|DF(x_0, \alpha_0)| + 1)(t-t')$$

from which it follows that

(7.2.10) $$\max_{t>t'} |y_n(t) - y_n(t')| \le M(t-t')$$

where

$$M = \frac{2}{1-k}\left[\frac{|D^2 b(x_0, \alpha_0)| + 1}{1-k} L_0 + |DF(x_0, \alpha_0)| + 1\right]$$

Therefore, $\{y_n\}_{n=1}^{\infty}$ has a convergent subsequence, denoted again by y_n for simplicity. Let $y_\tau(t) = \lim_{n\to\infty} y_n(t)$. Then $y_\tau(t)$ is a periodic solution of (7.2.8) with a period T_0. Since the maximal value of $|y_n(t)|$ is 1 and the average value of each y_n is zero, the same is true for y_τ. So, y_τ is a nonconstant periodic solution.

Denote by T_τ the minimal period of $y_\tau(t)$. Then $T_0 = kT_\tau$ for some positive integer k. If $k = 1$, then we are done. If not, we wish to find a collection of solutions of (7.2.8) having T_0 as the least common multiple of their minimal periods. To do so, we first show that τT_0 is not an integer multiple of T_τ. Indeed, if τT_0 is an integer multiple of T_τ, then $y_n(t+k\tau T_0)=y_n(t)$ implies that

$$\begin{aligned}0&=[y_n(t+\tau T_n)-y_n(t)]\varepsilon_{n,\tau}^{-1}+[y_n(t+2\tau T_n)-y_n(t+\tau T_n)]\varepsilon_{n,\tau}^{-1}\\&\quad+\cdots+[y_n(t+k\tau T_n)-y_n(t+(k-1)\tau T_n)]\varepsilon_{n,\tau}^{-1}\\&\quad\to y_\tau(t)+y_\tau(t+\tau T_0)+\cdots+y_\tau(t+(k-1)\tau T_0)=ky_\tau(t)\end{aligned}$$

a contradiction of the fact that $y_\tau(t)$ is nonconstant.

Since for every rational $\tau\in(0,1)$, there is a periodic solution y_τ of (7.2.8) whose minimal period divides T_0, but does not divide τT_0, we may choose some collection $\{z_j\}_{j=0}^{k-1}$ of solutions of (7.2.8) such that T_0 is the smallest number that is a multiple of their minimal periods. It then follows that for almost any choice of real numbers $\{c_j\}_{j=0}^{k-1}$, $\sum_{j=0}^{k-1} c_jz_j(t)$ is a periodic solution of (7.2.8) with the minimal period T_0. This completes the proof. □

Now, we can describe the relation between $2\pi/\beta_n$ and the minimal period of $x_n(t)$ in Theorem 7.2.3.

Theorem 7.2.7

In Theorem 7.2.3, the minimal period of $x_n(t)$ is convergent to the set

$$\left\{\frac{2\pi}{m\beta_0};\ \pm im\beta_0 \text{ are characteristic values of } (x_0,\alpha_0)\right\}$$

In particular, if $\pm im\beta_0$ are not characteristic values of (x_0,α_0) for any integer $m>1$, then $2\pi/\beta_n$ is the minimal period of $x_n(t)$ and $2\pi/\beta_n\to 2\pi/\beta_0$.

Proof. Let T_n denote the minimal period of $x_n(t)$. Then there exists a positive integer m_n such that $2\pi/\beta_n=m_nT_n$. Since $T_n\le 2\pi/\beta_n\to 2\pi/\beta_0$ as $n\to\infty$, $\{T_n\}_{n=1}^{\infty}$ has a convergent subsequence $\{T_{n_k}\}_{k=1}^{\infty}$. Let $T_0=\lim_{k\to\infty}T_{n_k}$. By Lemma 7.2.6, $\pm i2\pi/T_0$ are characteristic values of (x_0,α_0). On the other hand, since $2\pi/\beta_{n_k}\to 2\pi/\beta_0$, $T_{n_k}\to T_0$ as $k\to\infty$, m_{n_k} is identical to a constant m for sufficiently large k. Consequently, $2\pi/\beta_0=mT_0$. Thus, $T_{n_k}\to 2\pi/m\beta_0$ and $\pm im\beta_0$ are characteristic values of (x_0,α_0). This completes the proof. □

We now consider the global bifurcation problem. First, we assume that F is bounded on bounded subsets. Next, we make the following hypothesis:

(H2) All stationary points of (7.2.1) are not singular and all centers of (7.2.1) are isolated.

Proceeding as in the proof of Theorem 7.2.3, we reformulate the problem (7.2.1) as a problem of finding 2π-periodic solutions to the following equation:

$$\frac{d}{dt}[y(t) - b(y_{t,1/p}, \alpha)] = pF(y_{t,1/p}, \alpha) \tag{7.2.11}$$

where $y_{t,1/p}(v) = y(t + v/p)$, $v \in \mathbb{R}$. The only difference is that now we regard the *period* p as an additional parameter. Using the same notations as in the proof of Theorem 7.2.3, we define

$$\tilde{B}_0(y, \alpha, p) = B_0(y, \alpha, 1/p)$$

$$\tilde{N}_0(y, \alpha, p) = N_0(y, \alpha, 1/p)\,, \qquad y \in V$$

Then we obtain the following composite coincidence problem that is equivalent to (7.2.11):

$$L_0[y - \tilde{B}_0(y, \alpha, p)] = \tilde{N}_0(y, \alpha, p)\,, \qquad p > 0 \tag{7.2.12}$$

We can also put $\tilde{f}(y, \alpha, p) = f(y, \alpha, 1/p)$. Under assumption (H2), zero is a regular value of the restriction $\tilde{f}_0 := \tilde{f}|_{V^G \times \mathbb{R} \times \mathbb{R}_+} : V^G \times \mathbb{R} \times \mathbb{R}_+ \to V^G$. Consequently, $\tilde{f}_0^{-1}(0) = M$ is a two-dimensional submanifold of $V^G \times \mathbb{R}^2$ such that $M \subseteq \tilde{f}^{-1}(0)$. It is clear that M satisfies conditions (A) and (B) in the last section. We will call M the set of *trivial* (*stationary*) periodic solutions of (7.2.1).

Let $\mathcal{S}$ denote the closure of the set of all nontrivial periodic solutions of $\tilde{f}(y, \alpha, p) = 0$ in the space $V \times \mathbb{R} \times \mathbb{R}_+$. Then it follows from Lemma 7.2.4 that it is impossible that $(y_0, \alpha_0, 0)$ belongs to this set. Therefore, without loss of generality, we can assume that problem (7.2.12) is well posed on the whole space $V \times \mathbb{R}^2$. Then, as an immediate consequence of Theorem 7.1.7, we obtain the following.

Theorem 7.2.8

Under assumption (H2), if $(y_0, \alpha_0, p_0) \in M$ is a bifurcation point of (7.2.12), then either the connected component $\mathscr{C}(y_0, \alpha_0, p_0)$ of (y_0, α_0, p_0) in $\mathcal{S}$ is unbounded, or the number of bifurcation points belonging to $\mathscr{C}(y_0, \alpha_0, p_0)$ is finite, that is,

$$\mathscr{C}(y_0, \alpha_0, p_0) \cap M = \{(y_0, \alpha_0, p_0), (y_1, \alpha_1, p_1), \ldots, (y_q, \alpha_q, p_q)\}$$

where $p_i \in \mathbb{R}_+$, $(y_i, \alpha_i, p_i) \in M$, $i = 0, 1, \ldots, q$. *Moreover, in the latter case, we have the following equality:*

$$\sum_{i=1}^{q} \gamma(y_i, \alpha_i, 1/p_i) = 0$$

EXERCISES

The purpose of this set of exercises is to draw some corollaries from Theorem 7.2.8. Suppose that (x_0, α_0) is an isolated center of (7.2.1) such that $\det \Delta_{\alpha_0}(i\beta_0) = 0$. Furthermore, we assume the following factorization:

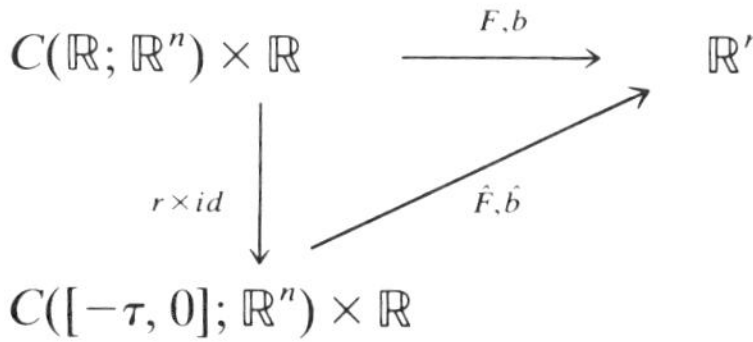

where $\tau \geq 0$ is a constant and $r : C(\mathbb{R}; \mathbb{R}^n) \to C([-\tau, 0]; \mathbb{R}^n)$ is the restriction operator. Then we can cover $i\beta_0$ by a closed disk $D(i\beta_0)$ such that $D(i\beta_0)$ contains only one characteristic value (i.e., $i\beta_0$) of the stationary point (x_0, α_0). Define

$$\text{Mult}_\alpha(i\beta_0) = \{z \in D(i\beta_0); \det \Delta_\alpha(z) = 0\}$$

If there is an open interval Λ_{α_0} containing α_0 such that the real part of $z \in \text{Mult}_\alpha(i\beta_0)$ is nonzero for $\alpha \in \Lambda_{\alpha_0} - \{\alpha_0\}$, then for α sufficiently close to α_0, the number of elements in $\text{Mult}_\alpha(i\beta_0)$ (counted with algebraic multiplicity) is the same as the algebraic multiplicity of $i\beta_0$ as a zero of $\det \Delta_{\alpha_0}(\lambda) = 0$. It can be verified that $\gamma_\pm(x_0, \alpha_0, \beta_0)$ is the number of elements of $\text{Mult}_\alpha(i\beta_0)$ (counted with algebraic multiplicity) whose real parts are positive for $\alpha > \alpha_0$ or $\alpha < \alpha_0$, respectively. In terms of the crossing number $\gamma(x_0, \alpha_0, \beta_0)$, we can define the following three invariants:

$$r(x_0, \alpha_0, \beta_0) = \sum_{\substack{m \in \mathbb{N} \\ \det \Delta_{\alpha_0}(\text{im}\, \beta_0) = 0}} \gamma(x_0, \alpha_0, m\beta_0)$$

$$\Phi(x_0, \alpha_0) = -\frac{\varepsilon}{2} \sum_{\substack{\beta > 0 \\ \det \Delta_{\alpha_0}(i\beta) = 0}} \gamma(x_0, \alpha_0, \beta_0)$$

$$H(b, F) = - \sum_{\substack{(x_0, \alpha_0, \beta_0) \in \mathbb{R}^n \times \mathbb{R} \times (0, \infty) \\ \det \Delta_{\alpha_0}(i\beta_0) = 0}} \frac{\varepsilon}{2} \gamma(x_0, \alpha_0, \beta_0)$$

where $\varepsilon = (-1)^{E(\alpha_0)}$ and $E(\alpha_0)$ is the sum of the numbers (counted with algebraic multiplicity) of the characteristic values of (7.2.1) whose real parts are positive. $\gamma(x_0, \alpha_0, \beta_0)$, $\Phi(x_0, \alpha_0)$, and $H(b, F)$ are called the *index*, *center index*, and *Hopf index*, respectively. They are introduced by Nussbaum (1978) for retarded equations, Mallet-Paret and Yorke (1982) for ordinary differential equations, and Fiedler (1985) for parabolic equations.

7.2.1 Prove that if $\gamma(x_0, \alpha_0, \beta_0) \neq 0$ and $\mathscr{C}(x_0, \alpha_0, 2\pi/\beta_0)$ is bounded, then $\mathscr{C}(x_0, \alpha_0, 2\pi/\beta_0)$ must contain another bifurcation point. This is an analog of the well-known Alexander–Yorke theorem [see Alexander and Yorke (1978)] on the global bifurcation of periodic orbits. Their results were refined by Chow and Mallet-Paret (1978), using Fuller's index for periodic solutions of autonomous systems.

7.2.2 Prove that if the set of centers is bounded and $H(b, F) \neq 0$, then there exists $(x_0, \alpha_0, \beta_0) \in \mathbb{R}^n \times \mathbb{R} \times (0, \infty)$ such that $\mathscr{C}(x_0, \alpha_0, 2\pi/\beta_0)$ is unbounded. This represents an analog of the global bifurcation theorem obtained by Mallet-Paret (1977) and Fiedler (1985) using generic approximation techniques.

7.2.3 Prove that if $r(x_0, \alpha_0, \beta_0) \neq 0$, then there must be an integer k such that $\gamma(x_0, \alpha_0, k\beta_0) \neq 0$. Prove also that $\gamma_k(x_0, \alpha_0, \beta_0) = \varepsilon\gamma(x_0, \alpha_0, k\beta_0)$ with $\varepsilon = \pm 1$ independent of the choice (x_0, α_0, β_0). Use the above to show that if $r(x_0, \alpha_0, \beta_0) \neq 0$ and $\mathscr{C}(x_0, \alpha_0, 2\pi/\beta_0)$ is bounded, then $\mathscr{C}(x_0, \alpha_0, 2\pi/\beta_0)$ must contain another bifurcation point. This result was essentially discovered by Nussbaum (1978), using a strong approximation of the operator of translation along trajectories of retarded equations with respect to a natural set of projections.

7.3 GLOBAL CONTINUATION OF PERIODIC SOLUTIONS: AN EXAMPLE

In this section we apply the Hopf bifurcation theorem to investigate the maximal continuation of nonconstant periodic solutions of the following difference-differential equation of the neutral type:

$$\text{(7.3.1)} \qquad \frac{d}{dt}[x(t) - qx(t-r)] = -ax(t) - bqx(t-r) - g(x(t)) + qg(x(t-r))$$

where $q \in [0, 1)$, a, b, $r > 0$ are given constants and $g : \mathbb{R} \to \mathbb{R}$ is continuously differentiable with $g(0) = g'(0) = 0$. Such an equation arises in the network shown in Figure 7.3.1, where the section between zero and 1 is a lossless transmission line with specific inductance L_s and capacitance C_s.

The voltage v across this line and the current i flowing through it are

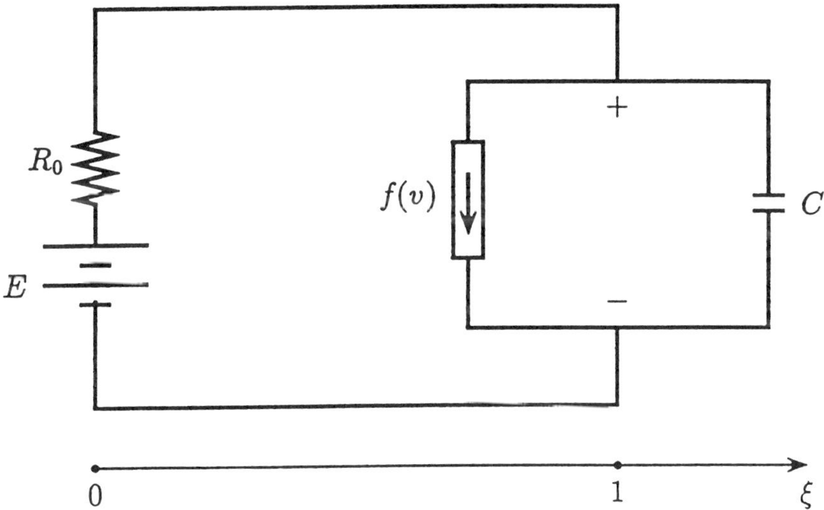

FIGURE 7.3.1. A lossless transmission line.

functions of ξ and t and obey the following partial differential equations:

$$\begin{cases} \dfrac{\partial v}{\partial \xi} = -L_s \dfrac{\partial i}{\partial t} \\ \dfrac{\partial i}{\partial \xi} = -C_s \dfrac{\partial v}{\partial t}, \qquad t \geq 0, \quad 0 < \xi < 1 \end{cases}$$

subject to the boundary condition

$$\begin{cases} E - v(0,t) - R_0 i(0,t) = 0 \\ -C \dfrac{dv(1,t)}{dt} = -i(1,t) + f(v(1,t)) \end{cases}$$

Under certain conditions, one can follow the procedure in Exercise 4.5.1 to show that $x(t) = v(1,t) - v^*$ satisfies (7.3.1) for some constants a, b, q, and r, and for a function $g: \mathbb{R} \to \mathbb{R}$ defined by

$$g(x) = \frac{1}{C}[f(x + v^*) - f(x)]$$

where v^* is the voltage at the equilibrium state. In Brayton (1966, 1967), Lopes (1975), and Slemrod (1971), the existence and nonexistence of oscillatory phenomena in this lossless transmission line were studied. In particular, the existence of nonconstant periodic solutions was discussed in Brayton (1967) by converting the neutral equation into an equivalent nonlinear integral equation and demonstrating the convergence of an appropriate iteration scheme. Under certain conditions, Brayton proved the existence of a pair of characteristic values crossing the pure imaginary axis as q crosses a certain bifurcation value and, thus, obtained a periodic solution with small amplitude when q is sufficiently close to the bifurcation value.

Our major focus here is on the existence and multiplicity of periodic solutions with *large amplitudes* when q is *far away* from the bifurcation value. Our technical steps consist of: (i) locating characteristic values in order to find Hopf bifurcation points via linearization; (ii) computing the crossing numbers associated with each bifurcation point in order to show the sum of these crossing numbers over any continuum is nonzero; (iii) verifying the existence of a priori bounds for the periods and amplitudes of nonconstant periodic solutions so that any continuum is trapped in a certain strip.

First of all, we follow the idea of Chow and Mallet-Paret (1978) to exclude periodic solutions of period $2r$ for the nonlinear neutral equation

$$\frac{d}{dt}[x(t) - qx(t-r)] = F(q, x(t), x(t-r)) \tag{7.3.2}$$

where $r > 0$ and q are given constants, and $F : \mathbb{R}^3 \to \mathbb{R}$ is locally Lipschitzian.

Lemma 7.3.1

If $|q| < 1$, *then (7.3.2) has no nonconstant periodic solution of period* $2r/m$ *for any positive integer* m.

Proof. Clearly, it suffices to prove that (7.3.2) has no nonconstant periodic solution of period $2r$. By way of contradiction, we assume that there is a nonconstant periodic solution $x(t)$ of period $2r$. Set $y(t) = x(t-r)$. Then $(x(t), y(t))$ satisfies the following system of ordinary differential equations:

$$\begin{cases} \dfrac{d}{dt}[x(t) - qy(t)] = F(q, x(t), y(t)) \\ \dfrac{d}{dt}[y(t) - qx(t)] = F(q, y(t), x(t)) \end{cases}$$

Put

$$\begin{cases} u(t) = x(t) - qy(t) \\ v(t) = y(t) - qx(t) \end{cases}$$

Then we get

$$\begin{cases} x(t) = \dfrac{u(t) + qv(t)}{1 - q_2} \\ y(t) = \dfrac{qu(t) + v(t)}{1 - q^2} \end{cases}$$

Therefore, $(u(t), v(t))$ is a solution to the following system of ordinary differential equations:

$$\begin{cases} \dot{u}(t) = F\left(q, \dfrac{u(t) + qv(t)}{1 - q^2}, \dfrac{qu(t) + v(t)}{1 - q^2}\right) \\ \dot{v}(t) = F\left(q, \dfrac{qu(t) + v(t)}{1 - q^2}, \dfrac{u(t) + qv(t)}{1 - q^2}\right) \end{cases}$$

Clearly, the diagonal $\Delta \cong \{(u, v) \in \mathbb{R}^2;\ u = v\}$ is invariant for the above system of ordinary differential equations. Since any vector field on $\Delta \cong \mathbb{R}$ cannot have nonconstant periodic solutions, $(u(t), v(t)) \not\in \Delta$ for all t. So, without loss of generality, we may assume that

$$u(t) < v(t) \qquad \text{for all } t \in \mathbb{R} \tag{7.3.3}$$

Replacing t by $t - r$ in (7.3.3), we get

$$u(t - r) < v(t - r) \qquad \text{for all } t \in \mathbb{R} \tag{7.3.4}$$

However, we have

$$v(t - r) = y(t - r) - qx(t - r) = x(t - 2r) - qy(t) = x(t) - qy(t) = u(t)$$

and

$$u(t - r) = x(t - r) - qy(t - r) = y(t) - qx(t - 2r) = y(t) - qx(t) = v(t)$$

Therefore, from (7.3.4) it follows that

$$v(t) < u(t) \qquad \text{for all } t \in \mathbb{R}$$

This contradicts (7.3.3) and thus completes the proof. □

We now concentrate on the special form (7.3.1). Clearly, zero is a solution of (7.3.1) for any given q. The characteristic equation of the stationary point $(q, 0)$ is

$$(\lambda + a)e^{\lambda r} - q(\lambda - b) = 0 \tag{7.3.5}$$

The following lemma summarizes useful information about the characteristic equation.

Lemma 7.3.2

If $0 < a < b$, then we have the following:

(i) *The equation*

$$\tan(\beta r) = \frac{(a+b)\beta}{\beta^2 - ab} \tag{7.3.6}$$

has infinitely many positive solutions $0 < \beta_1 < \beta_2 < \cdots < \beta_n < \cdots \to \infty$ *as* $n \to \infty$, *such that:*

(a) $\dfrac{2r}{n+1} < \dfrac{2\pi}{\beta_n} < \dfrac{2r}{n} \leq 2r$ *for* $n \geq 1$, *if* $\sqrt{ab} = \dfrac{\pi}{2r}$.

(b) $\dfrac{2\pi}{\beta_1} > 2r, \dfrac{2r}{n} < \dfrac{2\pi}{\beta_n} < \dfrac{2r}{n-1} \leq 2r$ *for* $2 \leq n \leq m$ (*if* $m \geq 2$), *and*

$\dfrac{2r}{n+1} < \dfrac{2\pi}{\beta_n} < \dfrac{2r}{n} \leq 2r$ *for* $n \geq m+1$, *if* $\sqrt{ab} = \dfrac{\pi}{2r} + \dfrac{m\pi}{r}$ *for some positive integer* m.

(c) $\dfrac{2\pi}{\beta_1} > 2$ *and* $\dfrac{2r}{n} < \dfrac{2\pi}{\beta_n} < \dfrac{2r}{n-1} \leq 2r$ *for* $n \geq 2$, *if* $\dfrac{r\sqrt{ab}}{\pi} - \frac{1}{2}$ *is not an integer.*

(ii) $\pm i\beta_n$ *are characteristic values of the stationary points* $(q_{\pm n}, 0)$, *where*

$$q_{\pm n} = \pm \left[\frac{\beta_n^2 + a^2}{\beta_n^2 + b^2} \right]^{1/2}$$

Moreover, if $q > 0$ *and* $q \neq q_n$, $n = 1, 2, \ldots$, *then there exists no pure imaginary characteristic value of the stationary point* $(q, 0)$.

(iii) *Let* $\lambda_n(q) = u_n(q) + iv_n(q)$ *be the root of (7.3.5) for* q *close to* q_n *such that* $u_n(q_n) = 0$ *and* $v_n(q_n) = \beta_n$. *Then* $\dfrac{d}{dq} u_n(q)|_{q=q_n} > 0$.

Proof. Denote by Γ_1 and Γ_2 the graphs in the region $\{(\beta, z); \beta > 0\}$ of the $\beta - z$ plane for the functions $z = \dfrac{(a+b)\beta}{\beta^2 - ab}$ and $z = \tan(\beta r)$, respectively. If $\sqrt{ab} < \dfrac{\pi}{2r}$, we can easily observe that these graphs have infinitely many intersection points (β_n, z_n) such that

$$\frac{(n-1)\pi}{r} < \beta_n < \frac{\pi}{2r} + \frac{(n-1)\pi}{r}$$

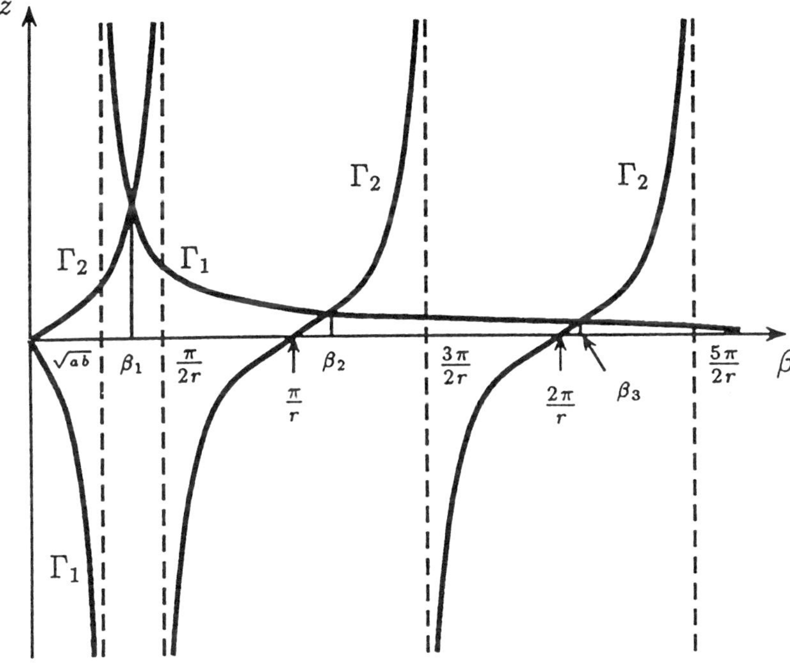

FIGURE 7.3.2. The graphs Γ_1 and Γ_2 in the case where $\sqrt{ab} < \frac{\pi}{2r}$.

for $n = 1, 2, \ldots$. Therefore, $\frac{2\pi}{\beta_1} > 4r$ and

$$\frac{2r}{n} < \frac{2\pi}{\beta_n} < \frac{2r}{(n-1)} \leq 2r$$

for $n \leq 2$. See Figure 7.3.2 for details.

If $\frac{\pi}{2r} + \frac{m\pi}{r} < \sqrt{ab} < \frac{\pi}{2r} + \frac{(m+1)\pi}{r}$ for some nonnegative integer m, then as Figure 7.3.3 shows, Γ_1 and Γ_2 have infinitely many intersection points (β_n, z_n) such that

$$(n-1)\frac{\pi}{r} + \frac{\pi}{2r} < \beta_n < \frac{\pi}{2r} + (n-1)\frac{\pi}{r} + \frac{\pi}{2r} = \frac{n\pi}{r}$$

for $n = 1, \ldots, m$, and

$$\frac{m+k-1}{r}\pi = \frac{k-1}{r}\pi + \frac{m\pi}{r} < \beta_{m+k} < \frac{m\pi}{r} + \frac{\pi}{2r} + \frac{(k-1)\pi}{r}$$
$$= \frac{m+k-1/2}{r}\pi$$

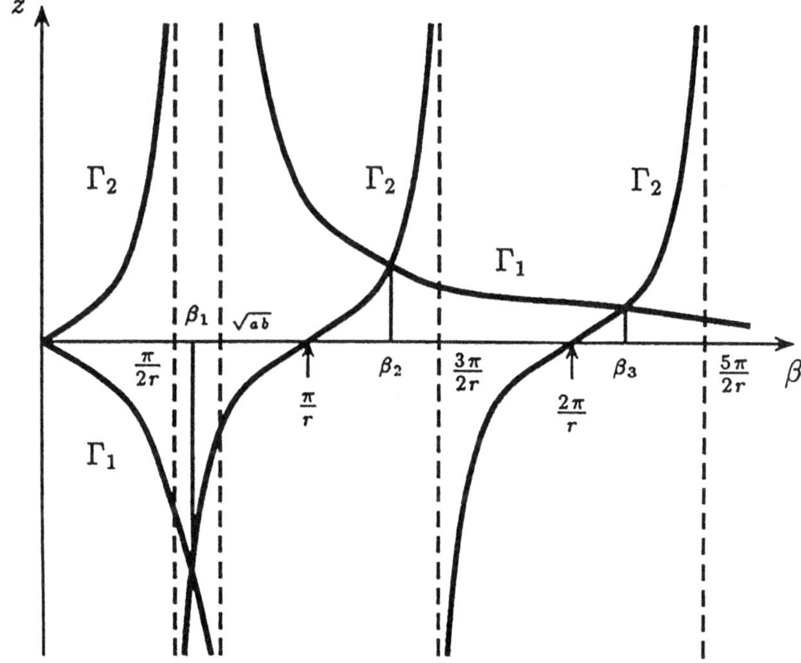

FIGURE 7.3.3. The graphs Γ_1 and Γ_2 in the case where $\frac{\pi}{2r} + \frac{m\pi}{r} < \sqrt{ab} < \frac{\pi}{2r} + \frac{(m+1)\pi}{r}$.

for $k = 1, 2, \ldots$. In particular, we have $2r < \frac{2\pi}{\beta_1} < 4r$,

$$\frac{2r}{n} < \frac{2\pi}{\beta_n} < \frac{2r}{n - 1/2} < \frac{2r}{n-1} \le 2r \qquad \text{for } n = 2, \ldots, m \quad (\text{if } m \ge 2)$$

and

$$\frac{2r}{m+k} < \frac{2r}{m+k-1/2} < \frac{2r}{\beta_{m+k}} < \frac{2r}{m+k-1} \qquad \text{for } k = 1, 2, \ldots$$

If $\sqrt{ab} = \frac{\pi}{2r}$, then as Figure 7.3.4 shows, Γ_1 and Γ_2 have infinitely many intersection points (β_n, z_n) such that

$$\frac{(n-1)\pi}{r} + \frac{\pi}{r} < \beta_n < \frac{3\pi}{2r} + \frac{(n-1)\pi}{r}$$

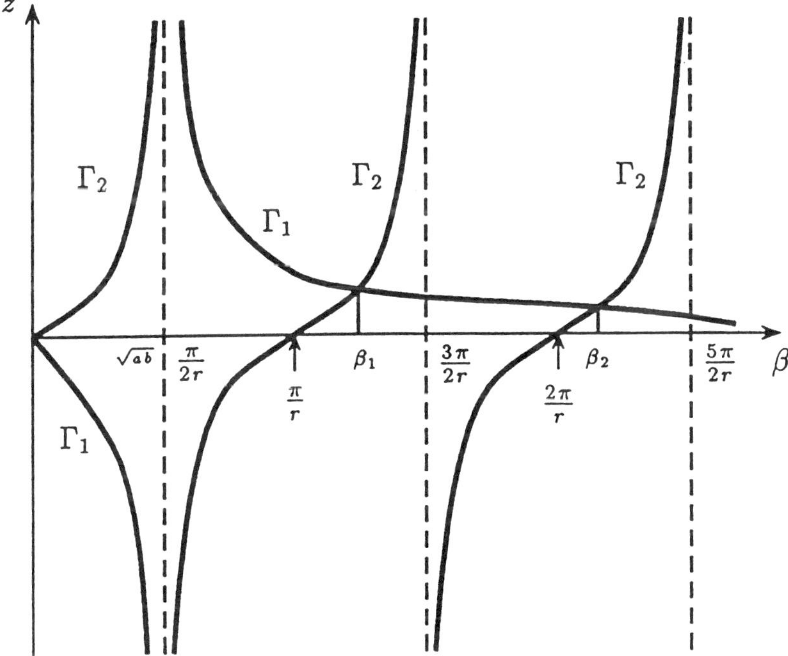

FIGURE 7.3.4. The graphs Γ_1 and Γ_2 in the case where $\sqrt{ab} = \dfrac{\pi}{2r}$.

for $n = 1, 2, \ldots$. In particular, $\frac{4}{3}r < \dfrac{2\pi}{\beta_1} < 2r$, and

$$\frac{2r}{n+1} < \frac{2\pi}{\beta_n} < \frac{2r}{n} \leq 2r \qquad \text{for } n = 2, 3, \ldots$$

If $\sqrt{ab} = \dfrac{\pi}{2r} + \dfrac{m\pi}{r}$ for a positive integer m, than as Figure 7.3.5 shows, Γ_1 and Γ_2 have infinitely many intersection points (β_n, z_n) such that

$$(n-1)\frac{\pi}{r} + \frac{\pi}{2r} < \beta_n < \frac{n\pi}{r}$$

for $n = 1, 2, \ldots, m$, and

$$\begin{aligned}\frac{m+k}{r}\pi &< \frac{m\pi}{r} + \frac{\pi}{2r} + \frac{\pi}{2r} + \frac{(k-1)\pi}{r} < \beta_{m+k} \\ &< \frac{m\pi}{r} + \frac{\pi}{2r} + \frac{(k-1)\pi}{r} + \frac{\pi}{r} = \frac{(m+k+1/2)}{r}\pi\end{aligned}$$

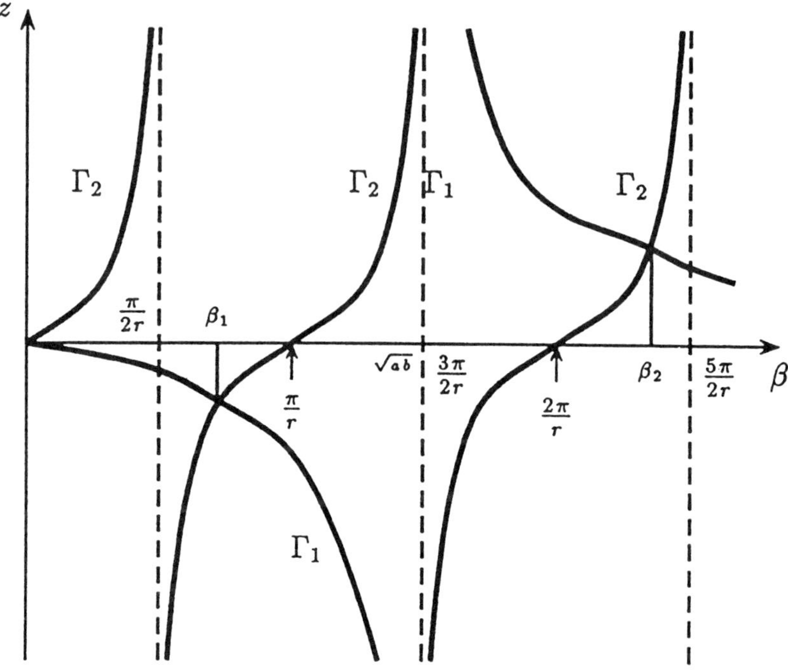

FIGURE 7.3.5. The graphs Γ_1 and Γ_2 in the case where $\sqrt{ab} = \frac{\pi}{2r} + \frac{m\pi}{r}$.

for $k = 1, 2, \ldots$. In particular, $2r < \frac{2\pi}{\beta_1} < 4r$,

$$\frac{2r}{n} < \frac{2\pi}{\beta_n} < \frac{2r}{n - 1/2} < \frac{2r}{n-1}$$

for $n = 2, \ldots, m$ (if $m \geq 2$), and

$$\frac{2r}{m+k+1} < \frac{2\pi}{\beta_{m+k}} < \frac{2r}{m+k}$$

for $k = 1, 2, \ldots$. This completes the proof for (i).

To prove (ii), we let $\lambda = i\beta$ in (7.3.5) and obtain

$$(i\beta + a)e^{i\beta r} = q(i\beta - b) \tag{7.3.7}$$

which is equivalent to

$$\begin{cases} -a\cos\beta r + \beta \sin\beta r = qb \\ \beta\cos\beta r + a\sin\beta r = q\beta \end{cases} \tag{7.3.8}$$

That is,

$$\begin{cases} \tan \beta r = \beta \dfrac{a+b}{\beta^2 - ab} \\ \beta^2 = \dfrac{q^2 b^2 - a^2}{1-q^2} \end{cases} \tag{7.3.9}$$

from which (ii) follows immediately.

Finally, we prove (iii). Rewriting (7.3.5) according to its real part and imaginary part, we obtain

$$\begin{cases} [u_n(q)+a]e^{u_n(q)r}\cos(v_n(q)r) - v_n(q)e^{u_n(q)r}\sin(v_n(q)r) = q[u_n(q)-b] \\ [u_n(q)+a]e^{u_n(q)r}\sin(v_n(q)r) + v_n(q)e^{u_n(q)r}\cos(v_n(q)r) = qv_n(q) \end{cases}$$

Differentiating both sides of the above equations and using (7.3.9), we get

$$\begin{cases} A_n u'(q_n) - B_n v'(q_n) = -b \\ B_n u'(q_n) + A_n v'(q_n) = \beta_n \end{cases}$$

where

$$\begin{cases} A_n = -\dfrac{q_n}{q_n^2+\beta_n^2}[a^2 br + \beta_n^2 br + ab + a^2] \\ B_n = \dfrac{q_n \beta_n}{q_n^2+\beta_n^2}[r(q_n^2+\beta_n^2) + a + b] \end{cases}$$

from which it follows that

$$u'(q_n) = \frac{1}{A_n^2 + B_n^2}(-bA_n + \beta_n B_n) > 0$$

This completes the proof. □

According to the above result, $(q_n, 0)$ is an isolated center of (7.3.1) and the crossing number $\gamma(q_n, 0, \beta_n) = -1$. By Theorem 7.2.3, $(q_n, 0)$ is a bifurcation point and, hence, a nonconstant periodic solution with a period close to $\dfrac{2\pi}{\beta_n}$ bifurcates from the stationary point $(q_n, 0)$. To investigate the maximal continuum of such a periodic solution, we need the following a priori bound on the amplitude of arbitrary nonconstant periodic solutions.

Lemma 7.3.3

Suppose that $0<a<b$ *and:*

(i) $\dfrac{g(x)}{x}>0$ *if* $x\neq 0$.

(ii) $g(x)$ *is nondecreasing.*

(iii) $\dfrac{g(x)}{x}\to\infty$ *as* $x\to\pm\infty$.

(iv) For any $q_0\in[a/b,1)$, *we have*

$$\sup_{a/b\leq q\leq q_0}\ \limsup_{x\to\pm\infty}\frac{g(qx)-qg(x)}{qx}<-(a+b)$$

Then for any $k\in[a/b,1)$, *there exists* $M=M(k)>0$ *such that if* $q\in[a/b,k]$ *and* $x(t)$ *is a periodic solution of (7.3.1), then* $|x(t)|\leq M$ *for all* $t\in\mathbb{R}$.

Proof. We prove the existence of M such that $x(t)\leq M$. The existence of M such that $x(t)\geq -M$ can be proved analogously.

First of all, we notice that if there exists $t\in\mathbb{R}$ such that

$$x(t)-qx(t-r)=\max_{s\in\mathbb{R}}[x(s)-qx(s-r)]$$

then for any fixed $s\in\mathbb{R}$, we have

$$x(s)\leq qx(s-r)+[x(t)-qx(t-r)]$$

Replacing s by $s-r$ in the above inequality, we obtain

$$x(s)\leq q^2x(s-2r)+(q+1)[x(t)-qx(t-r)]$$

Continuing the above process m-times, we get

$$x(s)\leq q^mx(s-mr)+\frac{1-q^m}{1-q}[x(t)-qx(t-r)]$$

Therefore, letting $m\to\infty$, we get

$$x(s)\leq\frac{x(t)-qx(t-r)}{1-q}\qquad\text{for all } s\in\mathbb{R}$$

In particular,

$$x(t)\leq\frac{x(t)-qx(t-r)}{1-q}$$

from which it follows that $x(t)\geq x(t-r)$.

Furthermore, we know that if x is a periodic solution of (7.3.1), then x is differentiable. In particular, $\frac{d}{dt}[x(t) - qx(t-r)] = 0$. This implies that

$$ax(t) + bqx(t-r) = -g(x(t)) + qg(x(t-r)) \tag{7.3.10}$$

We now distinguish two cases.

CASE 1. $x(t) > 0$. In this case, $x(t-r)$ must be negative. For otherwise, if $x(t-r) \geq 0$, then the left-hand side of (7.3.10) is positive, but the right-hand side

$$-g(x(t)) + qg(x(t-r)) = g(x(t))\left[-1 + q\frac{g(x(t-r))}{g(x(t))}\right] < 0$$

due to the nondecreasing property of g.

We now rewrite (7.3.10) as

$$ax(t) + g(x(t)) = qx(t-r)\left[\frac{g(x(t-r))}{x(t-r)} - b\right] \tag{7.3.11}$$

Since $x(t) > 0$ and $x(t-r) < 0$, we have $\frac{g(x(t-r))}{x(t-r)} < b$. By assumption $\lim_{z\to-\infty}\frac{g(z)}{z} = \infty$, we can find a constant $M_1 > 0$ such that $x(t-r) \geq -M_1$. Substituting this result into (7.3.11), we get

$$0 \leq ax(t) + g(x(t)) \leq \max_{-M_1 \leq z \leq 0} k[g(z) - bz]$$

from which we can find a constant $M_2 > 0$ such that $x(t) \leq M_2$. Therefore,

$$x(t) - qx(t-r) \leq M_2 + kM_1$$

This, in turn, implies $x(s) \leq \frac{M_2 + kM_1}{1-k}$ for all $s \in \mathbb{R}$.

CASE 2. $x(t) \leq 0$. In this case, $x(t-r) \leq x(t) \leq 0$. If $x(t) - qx(t-r) \leq 0$, then we are done. If $x(t) - qx(t-r) > 0$, then $x(t) \geq qx(t-r)$. Therefore, from (7.3.10), we get

$$qg(x(t-r)) - bqx(t-r) \geq aqx(t-r) + g(qx(t-r))$$

That is,

$$\frac{g(qx(t-r)) - qg(x(t-r))}{qx(t-r)} \geq -(a+b)$$

Therefore, by assumption (iv), there exists $M_3 > 0$ such that $x(t-r) \geq -M_3$.

Repeating the argument in the last part of case 1, we can find a constant M such that $x(s) \leq M$ for all $s \in \mathbb{R}$. This completes the proof. □

Remark 7.3.4 One can easily verify that all conditions (i–iv) are satisfied by the function $g(x) = x^3$ that was widely used in the problem of transmission lines.

Now we are in a position to state our main result in this section.

Theorem 7.3.5

Suppose $0 < a < b$ and g satisfies conditions (i–iv) of Lemma 7.3.3:

(i) *If $\sqrt{ab} = \frac{\pi}{2r}$, then for any $n \geq 1$ and $q \in (q_n, 1)$, system (7.3.1) has n periodic solutions $x_{k,\alpha}$ whose periods $p_{k,q}$ satisfy $\frac{2r}{k+1} < p_{k,q} < \frac{2r}{k}$ for $k = 1, \ldots, n$.*

(ii) *If $\sqrt{ab} = \frac{\pi}{2r} + \frac{m\pi}{r}$ for some positive integer m, then for any $n \geq 2$ and $q \in (q_n, 1)$, system (7.3.1) has $n-1$ periodic solutions whose periods $p_{k,q}$ satisfy $\frac{2r}{k} < p_{k,q} < \frac{2r}{k-1}$ for $2 \leq k \leq m$ (if $m \geq 2$) and $\frac{2r}{k+1} < p_{k,q} < \frac{2r}{k}$ for $m+1 \leq k \leq n$.*

(iii) *If $\frac{r\sqrt{ab}}{\pi} - \frac{1}{2}$ is not a nonnegative integer, then for any $n \geq 2$ and $q \in (q_n, 1)$, system (7.3.1) has $n-1$ periodic solutions $x_{k,\alpha}$ whose periods $p_{k,\alpha}$ satisfy $\frac{2\pi}{k} < p_{k,\alpha} < \frac{2r}{k-1}$ for $k = 2, \ldots, n$.*

Proof. We prove (iii) only. The other cases can be proved analogously. For any given positive integer n, we consider the following neutral equation:

$$\text{(7.3.12)} \qquad \frac{d}{dt}[x(t) - Q_n(\alpha)x(t-r)] = -ax(t) - bQ_n(\alpha)x(t-r) - g(x(t)) + Q_n(\alpha)g(x(t-r))$$

where

$$Q_n(\alpha) = \frac{q_{n+1} + a/b}{\pi}\left(\arctan \alpha + \frac{\pi}{2}\right) - \frac{a}{b}$$

Clearly, $Q_n(\alpha)$ is an increasing function such that $\lim_{\alpha \to \infty} Q_n(\alpha) = q_{n+1}$ and $\lim_{\alpha \to -\infty} Q_n(\alpha) = -a/b$. Therefore, if we let

$$B(\alpha\ \varphi) = Q_n(\alpha)\varphi(-r)$$

for $\varphi \in C([-r, 0]; \mathbb{R})$, then B satisfies a Lipschitz condition with the Lipschitz constant $k = q_{n+1} < 1$.

Under assumption (i) of Lemma 7.3.3, we can easily show that for any $\alpha \in (-\infty, \infty)$, $(\alpha, 0)$ is the only stationary point of (7.3.12). Moreover, using (7.3.5), we can prove that zero is never a characteristic value since $Q_n(\alpha) > -a/b$. Therefore, all stationary points of (7.3.12) are not singular. Let $\alpha_k = Q_n^{-1}(q_k)$ for $k = 1, \ldots, n$. Then $(\alpha_k, 0)$ are isolated centers of (7.3.12), $k = 1, \ldots, n$. Except at these isolated centers, there are no pure imaginary characteristic values of (7.3.12). Moreover, by (iii) of Lemma 7.3.2, the crossing number of $(\alpha_k, 0, \beta_k)$ satisfies $\gamma(\alpha_k, 0, \beta_k) = -1$ for $k = 1, 2, \ldots, n$. We now fix k and consider the connected component $\mathscr{C}\left(\alpha_k, 0, \frac{2\pi}{\beta_k}\right)$. Applying Theorem 7.2.8, we know that $\mathscr{C}\left(\alpha_k, 0, \frac{2\pi}{\beta_k}\right)$ is unbounded.

We also notice that $\frac{2r}{k} < \frac{2\pi}{\beta_k} < \frac{2r}{k-1}$. Therefore, by Lemma 7.3.1 and Lemma 7.3.3, there exists a constant $M_n = M_n(q_{n+1}) > 0$ such that

$$\mathscr{C}\left(\alpha_k, 0, \frac{2\pi}{\beta_k}\right) \subset BC(M_n) \times \mathbb{R} \times \left[\frac{2r}{k}, \frac{2r}{k-1}\right]$$

where

$$BC(M_n) = \{y \in BC(\mathbb{R}; \mathbb{R}); \sup_{t \in \mathbb{R}} |y(t)| < M_n\}$$

On the other hand, since $Q_m(\alpha)$ is increasing from $-a/b$ to q_{n+1}, there exists z_n such that $Q_n(z_n) = 0$. At $\alpha = z_n$, (7.3.12) becomes the following ordinary differential equation:

$$\frac{d}{dt} x(t) = -ax(t) - g(x(t))$$

which clearly has no nonconstant periodic solutions. So, we further conclude that

$$\mathscr{C}\left(\alpha_k, 0, \frac{2\pi}{\beta_k}\right) \subset BC(M_n) \times (z_n, \infty) \times \left[\frac{2r}{k}, \frac{2r}{k-1}\right]$$

Therefore, since $\mathscr{C}\left(\alpha_k, 0, \frac{2\pi}{\beta_k}\right)$ is unbounded, the projection of $\mathscr{C}\left(\alpha_k, 0, \frac{2\pi}{\beta_k}\right)$ onto the parameter (α) space must be unbounded above. This implies that for every $\alpha > \alpha_k$, system (7.3.12) has a nonconstant periodic solution with a period in $\left(\frac{2r}{k}, \frac{2r}{k-1}\right)$. This, in turn, implies that for every $q \in (q_k, q_{n+1})$, system (7.3.1) has a nonconstant periodic solution with a period in $\left(\frac{2r}{k}, \frac{2r}{k-1}\right)$. This completes the proof. □

Theorem 7.3.5 guarantees the existence of periodic solutions of periods less than $2r$. These periodic solutions are called *rapidly oscillatory* solutions. The existence of *slowly oscillatory* periodic solutions (i.e., solutions whose periods are larger than $2r$) has been discussed in Wu (1993a).

EXERCISES

7.3.1 Discuss the local bifurcation and global continuation of periodic solutions for the nonlinear difference-differential equation of the neutral type:

$$\frac{d}{dt}[x(t) - qx(t-r)] = -ax(t) - bqx(t-r) - g[x(t) - qx(t-r)]$$

where a, b, $r > 0$, and $q \in (0, 1)$ are given constants, $g : \mathbb{R} \to \mathbb{R}$ is continuously differentiable, and $\inf_{x \neq 0} \frac{g(x)}{x} > -a$.

7.3.2 Consider a system of neutral equations

$$\begin{aligned}\frac{d}{dt}[x_i(t) - qx_i(t-r)] = {} & -ax_i(t) - bqx_i(t-r) - g[x_i(t) - qx_i(t-r)] \\ & + \beta[x_{i-1}(t) - qx_{i-1}(t-r) + x_{i+1}(t) \\ & - qx_{i+1}(t-r) - 2x_i(t) + 2qx_i(t-r)]\end{aligned}$$

where, $1 \leq i \leq n \pmod n$, $\beta > 0$, and all other constants and functions are given in Exercise 7.3.1. A solution $x(t) = (x_1(t), \ldots, x_n(t))$ is called a *phase-locked oscillation* if there exists $p > 0$ such that $x(t) = x(t+p)$ and $x_{i-1} = x_i\left(t - \frac{p}{n}\right)$, $i \pmod n$. Show that $x(t)$ is a phase-locked oscillation if and only if $y(t) = x_1(t)$ satisfies

$$\begin{aligned}\frac{d}{dt}[y(t) - qy(t-r)] = {} & -ay(t) - bqy(t-r) - g[y(t) - qy(t-r)] \\ & + \beta\left[y\left(t - \frac{p}{n}\right) - qy\left(t - \frac{p}{n} - r\right) + y\left(t + \frac{p}{n}\right)\right. \\ & \left. - qy\left(t + \frac{p}{n} - r\right) - 2y(t) + 2qy(t-r)\right]\end{aligned}$$

Use this result to obtain a bifurcation theory for phase-locked oscillations.

7.4 BIBLIOGRAPHICAL NOTES

In this chapter we presented an illustration of the application of the S^1-degree to global bifurcation problems for a class of two-parameter coincidence equations, and demonstrate the strength of the developed bifurcation theory in the study of Hopf bifurcations of neutral equations. The approach follows the ideas of Gęba and Marzantowicz (1991) and all results are due to Erbe et al. (1992), and Krawcewicz et al. (1993). See also Wu (1993a, b).

For other interesting discussions of the Hopf bifurcation problem using various approaches, we refer to Alexander and Yorke (1978), Alligood et al. (1981, 1983), Alligood and Yorke (1983), Chow and Mallet-Paret (1977, 1978), Chow et al. (1978), Claeyssen (1980), Crandall and Rabinowitz (1977), Fiedler (1985, 1988), Fuller (1967), Hale and De Oliviera (1980), Hassard et al. (1981), Ize (1985, 1993), Kazarinoff et al. (1978), Mallet-Paret (1977), Mallet-Paret and Yorke (1982), Marsden and McCracken (1976), Nussbaum (1975, 1976, 1978), De Oliviera (1980), Rustichini (1989), Staffans (1987), and Stech (1985a, b). In addition, see Brayton (1966, 1967), Lopes (1975), and Slemrod (1971) for earlier work on the sustained oscillation of neutral equations, and we refer the interested reader to Dancer (1980, 1982) for bifurcation in the presence of symmetries.

Chapter Eight

Equivariant Degree of Dold–Ulrich

In this chapter we develop the equivariant degree based on the *equivariant fixed-point index* introduced by Dold (1982, 1983, 1984) and Ulrich (1988).

Our definition of this equivariant degree is based on the idea of regular normal approximation of equivariant mappings that was initially introduced in Gęba et al. (1994) and further developed in Krawcewicz et al. (1996), Krawcewicz and Xia (1996), and Xia (1994). We will present several examples to illustrate the computation of the G-degree for several classical Lie groups. We will also provide some applications to differential equations and time-reversible systems.

8.1 PRELIMINARIES ON G-ACTIONS

Throughout this section we assume that G is a compact Lie group. We need to extend the notations introduced in Sections 6.1 and 6.2.

Let us start with some basic facts from transformation group theory. We will state most results without proofs and refer the reader to Bredon (1972) and Kawakubo (1991), as well as Murayama (1983), for more details.

Definition 8.1.1 Let X be a Hausdorff topological space. Recall that a *topological transformation group* is a triple (G, X, φ), where $\varphi : G\times X\to X$ is a continuous map such that:

(i) $\varphi(g, \varphi(h, x)) = \varphi(gh, x)$ for all g, $h\in G$ and $x\in X$.

(ii) $\varphi(e, x) = x$ for all $x\in X$, where e is the identity of G.

The map φ is called an *action* of G on X and the space X, under the action of G, a *G-space*. We shall use the notation $g(x)$ or gx for $\varphi(g, x)$. Similarly, for $K\subset G$ and $A\subset X$, we put $K(A) = \{g(x); g\in K, x\in A\}$. A set $A\subset X$ is said to be *G-invariant*, or simply *invariant*, if $G(A) = A$. It is easy to see that $G(A)$ is compact if A is compact.

For any $x \in X$, the subgroup $G_x = \{g \in G;\ gx = x\}$ of G is called the *isotropy group* of x and the invariant subspace $Gx := \{gx;\ g \in G\}$ of X is called the *orbit* of x. Clearly, G_x is closed in G. We denote by X/G the set of all orbits in X. There is a canonical projection $\pi : X \to X/G$, $x \to G(x)$. We provide X/G with the quotient topology with respect to the projection π and call it the *orbit space*. Under this topology, X/G is also Hausdorff and the projection π is a closed map.

Let X and Y be two G-spaces. A continuous map $f : X \to Y$ is called a *G-equivariant map*, or simply a *G-map*, if $f(gx) = gf(x)$ for all $g \in G$ and $x \in X$. An equivariant homeomorphism is called a *G-homeomorphism*. We say two G-spaces X and Y are *G-homeomorphic* if there exists a G-homeomorphism $f : X \to Y$. A continuous map $f : X \to Y$ satisfying $f(gx) = f(x)$ for all $g \in G$ and $x \in X$ is said to be an *invariant map*.

The above concepts can be carried over to the case where $X = M$ is a smooth manifold. It is well known that if G acts on M by diffeomorphisms, in this case the action $\varphi : G \times M \to M$ is smooth, and M is called a *smooth G-manifold* or simply *G-manifold*.

An orbit Gx can be described as the space G/H of left cosets gH of H in G, where $H = G_x$ is the isotropy group of x. The *homogeneous space* G/H is equipped with the quotient topology induced by the canonical map $\pi : G \to G/H$.

The following result describes the space G/H.

Proposition 8.1.2

Let H be a closed subgroup of G and $\pi : G \to G/H$ be the canonical projection map. Then G/H has a structure of a smooth G-manifold with $\dim G/H = \dim G - \dim H$, *where the G-action of G on G/H is given by left translations, that is, $\varphi(g, g'H) := L_g(g'H) := gg'H$, and π is smooth.*

Now suppose that X is a G-space and $x \in X$. We have a natural map $l_x : G/G_x \to Gx$ defined by $l_x(gG_x) = gx$. The following proposition describes the orbit Gx.

Proposition 8.1.3

Under the natural action of G on G/G_x, we have the following:

(i) *$l_x : G/G_x \to Gx$ is a G-homeomorphism.*

(ii) *$l_x : G/G_x \to Gx$ is a G-diffeomorphism if X is a G-manifold. Therefore, the orbit Gx is a G-invariant submanifold.*

Proof. See Kawakubo (1991). □

Recall that two closed subgroups H and K are *conjugate* in G, denoted by $H \sim K$, if $H = gKg^{-1}$ for some $g \in G$. Clearly, $\sim$ is an equivalence relation. The equivalence class of H is called a *conjugacy class* of H in G and will be denoted by (H).

From Proposition 8.1.2, we see that the quotient space G/G_x describes the orbit Gx. Notice that $G_{gx} = gG_xg^{-1}$, that is, G_{gx} and G_x are conjugate. Therefore, the conjugacy class (G_x) contains all the information about the orbit Gx. This leads us to a consideration of the set $O(G)$ of all conjugacy classes of closed subgroups of G, which is called the set of all *orbit types* of G. Moreover, given a G-space X and $x \in X$, the conjugacy class (G_x) will be called the *orbit type* of x.

The set $O(G)$ can be partially ordered by the following relation $\leq$: let α, $\beta \in O(G)$ with $\alpha = (H)$ and $\beta = (K)$; then

$$\alpha \leq \beta \Leftrightarrow \exists g \in G \; gKg^{-1} \subset H$$

Let X be a G-space and H a closed subgroup of G. We consider the following subsets of X:

$$\begin{aligned}
X^H &:= \{x \in X; G_x \supseteq H\} \\
X_H &:= \{x \in X; G_x = H\} \\
X^{[H]} &:= \bigcup \{X^K; H \subsetneq K\} \\
X^{(H)} &:= \{x \in X; (G_x) \leq (H)\} \\
X_{(H)} &:= \{x \in X; (G_x) = (H)\} \\
X^{\{H\}} &= GX^{[H]}
\end{aligned}$$

We will need the following fact.

Proposition 8.1.4

Let X be an ANR-space and a G-space. Then for every closed subgroup $H \subset G$, the sets X^H, $X^{(H)}$, $X^{\{H\}}$, $X^{[H]}$, and $X_{(H)}$ are ANRs.

Proof. See Murayama (1983). □

Denote by $N(H)$ the *normalizer* of the closed subgroup H of G. Then $N(H)$ is a closed subgroup of G and, hence, is a Lie group. Therefore, by Proposition 8.1.1, the *Weyl group* $W(H) := N(H)/H$ is a Lie group.

The following proposition summarizes the elementary properties of the above subsets of X.

Proposition 8.1.5

Let X and Y be G-spaces, V an orthogonal representation of G, and H a closed subgroup of G. Then, we have the following:

(i) X^H *is a closed $N(H)$-space, and consequently a $W(H)$-space, where the $W(H)$-action is given by $\varphi(gH, x) = gx$ for $g \in N(H)$ and $x \in X^H$.*
(ii) V^H *is a linear subspace of V.*
(iii) X_H *is an open, dense subset of X^H, and $W(H)$ acts freely on X_H.*
(iv) $X^{(H)} = GX^H$ *and* $X_{(H)} = GX_H$. *In particular, $X^{(H)}$ is closed.*
(v) *If (H) is a minimal orbit type that occurs in X, then $X_{(H)}$ is closed in X.*
(vi) *Any G-map $f : X \to Y$ induces $W(H)$-maps $f^H : X^H \to Y^H$, where $f^H := f|_{X^H}$.*

Proof. See Kawakubo (1991). □

We will also need the following result.

Proposition 8.1.6

Let $K \subseteq H$ be two closed subgroups of G; then the orbit space $(G/H)^K/N(K)$ of the left translation action of $N(K)$ on $(G/H)^K$ is finite.

Proof. See Bredon (1972). □

Definition 8.1.7 A vector bundle (E, X, p) is called a *G-vector bundle* if E, X are G-spaces, p is a G-map, and for each $g \in G$ and $x \in X$, the map $g : p^{-1}(x) \to p^{-1}(gx)$ is a linear isomorphism. A G-vector bundle (E, X, p) is said to be *smooth* if X and E are G-manifolds and p is smooth. Two G-vector bundles (E, X, p) and (E', X, p') are called *isomorphic* if there exists a G-map $f : E \to E'$ such that $p' \circ f = p$ and $f|_{E_x} : E_x \to E'_x$ is a linear isomorphism for each $x \in X$. Such a G-map f is called a *G-bundle isomorphism*.

Example 8.1.8 Suppose that M is a G-manifold. Let $x \in M$ and denote by $T_x(M)$ the tangent space of M at x. Then the *tangent (vector) bundle*

$$T(M) := \bigcup_{x \in M} T_x(M)$$

with $p : T(M) \to M$, $p(T_x(M)) = x$, $x \in M$, can be endowed with a G-vector bundle structure if we define a G-action $\varphi : G \times T(M) \to T(M)$ by $\varphi(g, v) = g_* v$, $(g, v) \in G \times T(M)$, where $g_* : T(M) \to T(M)$ is the tangent map of the diffeomorphism $g : M \to M$. Under the action φ, $T(M)$ is a G-manifold and

$g_* : T_x(M) \to T_{gx}(M)$ is a linear isomorphism. We call $(T(M), M, p)$ the *tangent G-vector bundle* of M.

Recall that a *Riemannian metric* $\langle \cdot,\cdot \rangle$ on M is a smooth metric on the tangent bundle $T(M)$. By using the Haar integral, one can see that on the tangent G-vector bundle $T(M)$ there exists a *smooth G-invariant Riemannian metric* $\langle \cdot,\cdot \rangle$.

Let $A \subset M$ be a G-invariant submanifold of M. Then we have the G-vector subbundle $T(A)$ of the restriction $T(M)|_A$ of the tangent G-vector bundle $T(M)$ of M. With respect to the above G-invariant Riemannian metric, there exists an orthogonal complement of $T(A)$ in $T(M)|_A$

$$\nu(A) := T(A)^{\perp}$$

which is called the *normal G-vector bundle* of A in M. Notice that $\nu(A)$ is canonically endowed with a G-invariant Riemannian metric.

Recall that two G-maps $f_0, f_1 : Y \to X$ are *G-homotopic* if there exists a G-map $H : Y \times [0, 1] \to X$ such that $f_i = H(\cdot, i)$, $i = 0, 1$, where G acts on $[0, 1]$ trivially.

Definition 8.1.9 Let H be a closed subgroup of G and A an H-space. Define an H-action on $G \times A$ by $\varphi(h, (g, a)) = (gh^{-1}, ha)$ for $h \in H$, $a \in A$. The orbit space $G \times_H A := (G \times A)/H$ of this H-action is called the *twisted product* of G and A. We denote by $[g, a]$ the *H-orbit* of (g, a).

Remark 8.1.10 The twisted product $G \times_H A$ is a G-space with the G-action

$$\tilde{\varphi} : G \times (G \times_H A) \to G \times_H A$$

defined by $\tilde{\varphi}(g', [g, a]) = [g'g, a]$. The following properties of $G \times_H A$ are straightforward:

(i) If A is a G-space, then $G \times_G A$ is G-homeomorphic to A.

(ii) If H is a subgroup of G and A is an H-space, then $(G \times_H A)/G$ is homeomorphic to A/H.

We refer the reader to Kawakubo (1991) for the proof of the following result that is fundamental in the study of the structure of a G-manifold.

Theorem 8.1.11 (The Slice Theorem)

Let M be a G-manifold. For any $x \in M$ with $H = G_x$, there exists a unique H-representation V and a G-diffeomorphism $f : G \times_H V \to M$ onto an open neighborhood of the orbit $G(x)$ such that $f([g, 0]) = gx$. Moreover, the normal G-vector bundle ν of the G-invariant submanifold $G(x)$ in M is *isomorphic to*

$$p : G\times_H V \to G/H$$

as smooth G-vector bundles, where $p([g, v]) = [g]$, $[g]$ *denotes the H-orbit of* $g \in G$ *and H acts on G by* $h \cdot g = gh^{-1}$ *for* $h \in H$.

In view of the above theorem, the H-invariant image $S := f([e, v])$ of V under the G-diffeomorphism f above will be called a *slice* of Gx at x and $f(G\times_H V)$ in M will be called a *tube about the orbit Gx*.

As a consequence of the Slice theorem, we have the following result concerning the orbit types around the orbit Gx.

Proposition 8.1.12

Let M be a G-manifold and $x \in M$. *Then there exists a neighborhood U of the orbit Gx in M such that* $(G_x) \le (G_y)$ *for all* $y \in U$.

Proof. Let S be a slice of Gx at x. Then $G(S)$ is an open neighborhood of Gx. For any $y \in G(S)$, $y = gs$ for some $g \in G$ and $s \in S$. Now S is a G_x-representation. One has $G_s \subset G_x$. This implies that $G_y = gG_s g^{-1}$, that is, $(G_x) \le (G_s) = (G_y)$. The corollary follows by letting $U = G(S)$. □

For a G-manifold M, there exists generally more than one orbit type. Recall that $M_{(H)}$ denotes the union of all orbits of the type (H). By definition, $M_{(H)}$ is G-invariant and M decomposes as $M = \bigcup_{(H)\in O(G)} M_{(H)}$ with $M_{(H)} \cap M_{(K)} = \varnothing$ if $(H) \ne (K)$.

Theorem 8.1.13

Let M be a G-manifold and H a closed subgroup of G. Then $M_{(H)}$ *is a G-invariant submanifold of M. Moreover,* $M_{(H)}$ *is closed if* (H) *is a minimal orbit type. Furthermore,* $M_{(H)}$ *is open and dense in M if* M/G *is connected and* (H) *is a maximal orbit type. Consequently,* $M_{(H)}/G$ *is a manifold and* M_H *is a free* $W(H)$*-submanifold of M.*

Proof. The set $M_{(H)}$ is a submanifold if and only if each point of $x \in M_{(H)}$ has an open neighborhood U such that $M_{(H)} \cap U$ is a submanifold of U. By Theorem 8.1.11, we may assume that $U = G\times_H V$, where V is a linear representation of H. The isotropy group of $[g, v] \in G\times_H V$ is $G_{[g,v]} = gH_v g^{-1}$, $H_v \subset H$; thus, $(G_{[g,v]}) = (H)$ if and only if $H_v = H$, that is, $v \in V^H$. But V^H is a subspace of V and hence $G\times_H V^H$, which is equal to $M_{(H)} \cap U$, is a submanifold of $G\times_H V$. We should point out that the manifold $M_{(H)}$ may have components of different dimension. If (H) is a maximal orbit type, $V^H = V$ and, hence, $M_{(H)} \cap U = U$. Thus, $M_{(H)}$ is open in M. It is clear that $U/G \simeq V$, thus, $M_{(H)}/G$ is a manifold.

For more details, we refer the reader to Kawakubo (1991) or tom Dieck (1987). □

When M is compact, as the following theorem shows, there are only a finite number of disjoint, nonempty, invariant subsets $M_{(H)}$ in the decomposition $M = \bigcup_{(H)\in O(G)} M_{(H)}$. For a subset A (not necessarily invariant) of a G-manifold M, we denote by $\mathscr{J}(A)$ the number of orbit types occurring in A, that is, $\mathscr{J}(A) := |\{(G_x); x \in A\}|$.

Theorem 8.1.14

The following statements hold true:

- **(i)** *If A is a compact set of a G-manifold M, then $\mathscr{J}(A) < \infty$. In particular, if M is a compact G-manifold, then $\mathscr{J}(M) < \infty$.*
- **(iii)** *If V is a G-representation vector space, then $\mathscr{J}(A) < \infty$.*

Proof. See Kawakubo (1919). □

EXERCISES

8.1.1 Let G be a compact Lie group and G_0 a connected component of G containing the identity. Show that G_0 is a closed normal subgroup of G.

8.1.2 Let X be a G-space and $H \subseteq G$ a closed subgroup. Show that X^H is a closed subspace of X.

8.1.3 Let X be a G-space and $H \subseteq G$ a closed subgroup. Show that X^H is $N(H)$-invariant, that is, X^H is a $N(H)$-space.

8.1.4 Determine the isotropy group at the point $(1, 0, \ldots, 0) \in \mathbb{R}^n$ of the natural $GL(n, \mathbb{R})$-action on $\mathbb{R}^n$.

8.1.5 Let H be a closed, normal subgroup of the group G and X a G-space with one orbit type (H). Show that $G_x = H$ for every $x \in X^H$.

8.1.6 Let H be a discrete, normal subgroup of a compact, connected Lie group G. Show that H is contained in the center $Z(G)$ of G, that is, $Z(G) := \{h \in G; gh = hg \text{ for all } g \in G\}$.

8.2 BURNSIDE RING A(G)

In this section we introduce the definition of the *Burnside ring* $A(G)$ and provide several examples.

Let $\Phi(G)$ denote the set of conjugacy classes (H) such that $N(H)/H$ is finite. We denote by $A(G)$ the free abelian group generated by $(H)\in\Phi(G)$. There is a *multiplication* operation on $A(G)$ that induces a structure of a ring with identity on $A(G)$. In order to define the multiplication operation, we remark that for (H), (K), $(L)\in\Phi(G)$,

$$\begin{aligned}(G/H\times G/K)_{(L)}/G &\cong (G/H\times G/K)_L/N(L)\\ &\subset (G/H\times G/K)^L/N(L)\\ &= (G/H^L\times G/K^L)/(N(L)/L)\end{aligned}$$

Since the spaces G/H^L and G/K^L consist of finitely many $N(L)/L$orbits (by Proposition 81.4), $N(L)/L$ is finite, and both G/H^L and G/K^L are finite. Consequently, the set $(G/H\times G/K)_{(L)}/G$ is finite.

The multiplication table of the generators (H) is given by the relation

$$(H)\cdot(K)=\sum_{(L)\in\Phi(G)} n_L(L) \tag{8.2.1}$$

where n_L denotes the number of elements in the set $(G/H\times G/K)_{(L)}/G$, that is,

$$n_L := |(G/H\times G/K)_{(L)}/G|$$

The ring $A(G)$ is called the *Burnside ring of* G.

Assume that (L) is an orbit type in $(G/H\times G/K)$; then there is a point $p\in(G/H\times G/K)$ such that the isotropy group G_p of p is exactly L. We can assume, without loss of generality, that $p=(H,kK)$; thus,

$$\begin{aligned}L &= \{g\in G;\ gH=H \text{ and } gkK=kK\}\\ &= H\cap kKk^{-1}\end{aligned}$$

This implies that all the orbit types (L) in $(G/H\times G/K)$ can be characterized as the conjugacy classes of the intersections $H\cap kKk^{-1}$, where $k\in G$.

We are going to show that some information on the number n_L can be obtained from a purely group-theoretic argument. Given closed subgroups L and H of the group G, we put

$$N(L,H)=\{g\in G;\ gLg^{-1}\subset H\}$$

Lemma 8.2.1

$N(L,H)$ is a closed subset of G and, hence, a compact set.

Proof. Since G is a manifold, it is metrizable. Given an element $g\in\overline{N(L,H)}$, then there is a sequence $\{g_n\}\in N(L,H)$ converging to g. It follows from the continuity of the multiplication that for every $h \in L$, we have $\lim_{n\to\infty} g_n h g_n^{-1} = ghg^{-1}$. Since $g_n h g_n^{-1} \in H$ and H is closed, we obtain $ghg^{-1} \in H$. Since this is true for arbitrary $h \in L$, we have $gLg^{-1} \subseteq H$. This implies that $g \in N(L, H)$ and, consequently, $N(L, H) = \overline{N(L, H)}$. Since G is compact, $N(L, H)$ is compact as well. □

Lemma 8.2.2

If we define a G-action on G by $(h, g) \mapsto hg$, then $N(L, H)$ is an $N(H)$-invariant subset of G.

Proof. For any $h \in N(H)$ and $g \in N(L, H)$, we have $hgL(hg)^{-1} = h(gLg^{-1})h^{-1} \subset hHh^{-1}$, that is, $hg \in N(L, H)$. □

Since the space $N(L, H)$ is a compact H-space and the natural projection $\pi : N(L, H) \to N(L, H)/H$ is continuous, the orbit spaces $N(L, H)/H$ and $N(L, H)/N(H)$ are compact. It should be pointed out that the set $N(L, H)$ is not a group in general [see, e.g., Ihring and Golubitsky (1984)].

Proposition 8.2.3

The correspondence $H(g) \mapsto g^{-1}H$ gives an homeomorphism $\Phi : N(L, H)/H \to (G/H)^L$.

Proof. First, we show that if $g \in N(L, H)$, then $g^{-1}H \in (G/H)^L$. Since $ghg^{-1} \in H$ for $g \in N(L, H)$ and $h \in L$, we have $ghg^{-1}H = H$ and, thus, $g^{-1}H \in (G/H)^L$. On the other hand, if $g_1, g_2 \in N(L, H)$ and $H(g_1) = H(g_2)$, then there is $h \in H$ such that $g_2 = hg_1$ and, consequently, $g_1^{-1}H = g_2^{-1}H$. This means that Φ is well defined and injective.

Let $g \in G$. Observe that $gH \in (G/H)^L$ if and only if $g^{-1} \in N(L, H)$. Thus, given $g \in G$ with $gH \in (G/H)^L$, we have $g^{-1} \in N(L, H)$ and $\Phi(H(g^{-1})) = gH$, that is, Φ is surjective. For the verification that Φ is a continuous, open mapping, we refer the reader to Kawakubo (1991). □

Corollary 8.2.4

For a closed subgroup H of a compact group G, we have $N(H, H) = N(H)$.

Let us introduce, following Ihring and Golubitsky (1984), the integer $n(L, H)$ that denotes the number of conjugate copies of L contained in the subgroup H:

$$n(L, H) = \left| \frac{N(L, H)}{N(H)} \right|$$

We remark that if $\alpha = (L)$ is a minimal orbit type in $(G/H \times G/K)$, then we have the equality

$$(G/H \times G/K)_L/N(L) = (G/H \times G/K)^L/N(L)$$

and in order to compute the number of elements in $(G/H \times G/K)^L/N(L) = (G/H^L \times G/K^L)/N(L)$, we need to describe the set $(G/H)^L \times (G/K)^L$ that, by Proposition 8.2.3, is homeomorphic to $N(L, H)/H \times N(L, K)/K$.

The action of $W(L)$ is free on the space $(G/H)^L \times (G/K)^L$ that can be identified with $N(L, H)/H \times N(L, K)/K$. On the other hand, $N(H)$ and $N(K)$ act freely on $N(L, H)$ and $N(L, K)$, respectively. Therefore, the number of elements in $(G/H^L \times G/K^L)/W(L)$ is equal to

$$n_L = n(L, H) \cdot n(L, K) \frac{|W(H)| \cdot |W(K)|}{|W(L)|},$$

where $n(L, H) = |N(L, H)/N(H)|$ and $n(L, K) = |N(L, K)/N(K)|$. We can write the above result as the following.

Proposition 8.2.5

Let $\alpha = (L)$ be a minimal orbit type in $(G/H \times G/K)$. Then

$$n_L = n(L, H) \cdot n(L, K) \frac{|W(H)| \cdot |W(K)|}{|W(L)|} \tag{8.2.2}$$

where $n(L, H) = |N(L, H)/N(H)|$ and $n(L, K) = |N(L, K)/N(K)|$.

In the case where G is an abelian group, formula (8.2.1) simplifies to

$$(H) \cdot (K) = n_{H \cap K}(H \cap K)$$

where $n_{H \cap K}$ is equal to the number of all $(H \cap K)$-orbits in $G/H \times G/K$. From this we may say that the integer n_L in formula (8.2.1) represents the number of elements in $(G/H \times G/K)_{(L)}/G$, that is, it is the *number of G-orbits in $G/H \times G/K$ of the orbit type* (L). In other words, $n_L = (G/H \times G/K)/G = (G/H \times G/K)/W(H \cap K)$.

The computation of the Burnside ring $A(G)$, in many cases, may be quite complicated and require extensive information of the subgroup structure of G. However, there is another description of the Burnside ring, due to tom Dieck (1987), which uses the fact that $A(G)$ may be isomorphically mapped onto a subring of the ring $C(G) := C(\Phi(G); \mathbb{Z})$ of continuous functions from $\Phi(G)$ into the discrete space $\mathbb{Z}$. In order to present this description, we need some facts about the topology of the space $\Phi(G)$.

Let $S(G)$ denote the set of closed subgroups of the compact Lie group G. As the group G is a metric space, we can equip $S(G)$ with the usual Hausdorff metric. So, $S(G)$ becomes a compact metric space such that the action $G \times S(G) \to S(G)$ defined by $(g, H) \mapsto gHg^{-1}$ is continuous. Moreover, the orbit space $S(G)/G$, which is exactly $O(G)$, is a countable (and thus a totally disconnected) Hausdorff space such that $\Phi(G)$ is a closed (and thus compact) subspace of $O(G)$ [see tom Dieck (1987) for more details]. It can be shown that for every $(H) \in \Phi(G)$, the function $z_H : \Phi(G) \to \mathbb{Z}$ defined by

$$z_H((L)) = |(G/H)^L| = |N(L, H)/H| = n(L, H) \cdot |W(H)| \tag{8.2.3}$$

for $(L) \in \Phi(G)$, is continuous and thus belongs to $C(G)$. Therefore, we can define a $\mathbb{Z}$-homomorphism $\varphi : A(G) \to C(G)$, which is defined on the generators $(H) \in \Phi(G)$ by $\varphi((H)) = z_H$. The following theorem due to tom Dieck (1987) expresses the fact that φ is, in fact, an injective ring homomorphism.

Theorem 8.2.6

The map $\varphi : A(G) \to C(G)$ is a well defined and injective ring homomorphism. Moreover, $C(G)$ is a free abelian group with basis $\{|W(H)|^{-1} z_H; (H) \in \Phi(G)\}$.

Proof. See tom Dieck (1987). □

Since the image $\varphi(A(G))$ may be identified with $A(G)$, it is possible to describe the ring structure of $A(G)$ just by computing the generators z_L, for $(L) \in \Phi(G)$, and then by expressing the products $z_H \cdot z_K$ as a linear combination of the generators z_L. We will illustrate this process by an example regarding the ring structure in $A(SO(3))$. It is possible to describe explicitly the image $\varphi(A(G))$ in the ring $C(G)$. Assume that (H), $(K) \in \Phi(G)$ are such that H is a normal subgroup of K and K/H is cyclic. We denote by K/H^* the set of generators of K/H, and by $|K/H^*|$ the number of generators of K/H. We put

$$l(H, K) = \begin{cases} |N(H)/N(H) \cap N(K)||K/H^*|, & \text{if } K/H \text{ is cyclic} \\ 0 & \text{otherwise} \end{cases}$$

If H is a normal subgroup of K, then $K \subset N(H)$ and we can denote by $(K)_H$ the $N(H)$-conjugacy class of K. Now we can formulate the following result describing the image $\varphi(A(G))$ in $C(G)$.

Theorem 8.2.7

A function $z \in C(G)$ belongs to $\varphi(A(G))$ if and only if for all $(H) \in \Phi(G)$, the congruence relation

$$\sum_{(K)_H} l(H, K) z(K) \equiv 0 \mod |W(H)|$$

is satisfied, where the sum is taken over all $N(H)$-conjugacy classes $(K)_H$ such that H is normal in K.

Proof. See tom Dieck (1987). □

In order to illustrate the multiplication operation in the Burnside ring $A(G)$, we discuss the Burnside rings of the groups $\mathbb{Z}_2 \times \mathbb{Z}_2$, $O(2)$, and $SO(3)$.

Example 8.2.8 Let us consider the group $G = \mathbb{Z}_2 \oplus \mathbb{Z}_2$. The orbit types in $\Phi_0(G)$ are denoted by (G), $(\mathbb{Z}_2 \oplus 1) = (\{(1,1),(-1,1)\})$, $(1 \oplus \mathbb{Z}_2) = (\{1,1),(1,-1)\})$, $(\Delta) = (\{(1,1),(-1,-1)\}$ and $(\mathbb{1}) = (\{1,1\})$. According to formula (8.2.1), we have the following multiplication table in $A_0(G)$:

	(G)	**$(\mathbb{Z}_2 \oplus 1)$**	**$(1 \oplus {}_2\mathbb{Z})$**	**(Δ)**	**$(\mathbb{1})$**
(G)	(G)	$(\mathbb{Z}_1 \oplus 1)$	$(1 \oplus \mathbb{Z}_1)$	(Δ)	$(\mathbb{1})$
$(\mathbb{Z}_2 \oplus 1)$	$(\mathbb{Z}_1 \oplus 1)$	$2(\mathbb{Z}_1 \oplus 1)$	$(\mathbb{1})$	$(\mathbb{1})$	$2(\mathbb{1})$
$(1 \oplus \mathbb{Z}_2)$	$(1 \oplus \mathbb{Z}_1)$	$(\mathbb{1})$	$2(1 \oplus \mathbb{Z}_1)$	$(\mathbb{1})$	$2(\mathbb{1})$
(Δ)	(Δ)	$(\mathbb{1})$	$(\mathbb{1})$	$2(\Delta)$	$2(\mathbb{1})$
$(\mathbb{1})$	$(\mathbb{1})$	$2(\mathbb{1})$	$2(\mathbb{1})$	$2(\mathbb{1})$	$4(\mathbb{1})$

Example 8.2.9 In this example, we consider the *orthogonal group* $O(2)$ of order 2 that consists of all 2×2 matrices A satisfying $AA^t = \mathrm{Id}$. We distinguish the following subgroups of $O(2)$:

(i) The *special orthogonal group* $SO(2)$ consisting of all planar rotations:

$$r_\theta = \begin{bmatrix} \cos\theta & -\sin\theta \\ \sin\theta & \cos\theta \end{bmatrix}$$

(ii) The *cyclic group* $\mathbb{Z}_n$ of order n, which is the subgroup of 2×2 matrices generated by the rotation

$$r_{\frac{2\pi}{n}}.$$

(iii) The *dihedral group* D_n of order $2n$, which is generated by $\mathbb{Z}_n$ together with the reflection

$$\kappa = \begin{bmatrix} 1 & 0 \\ 0 & -1 \end{bmatrix}.$$

It can be verified that every closed proper subgroup of $O(2)$ is conjugated to one of the above subgroups, and consequently we have the following

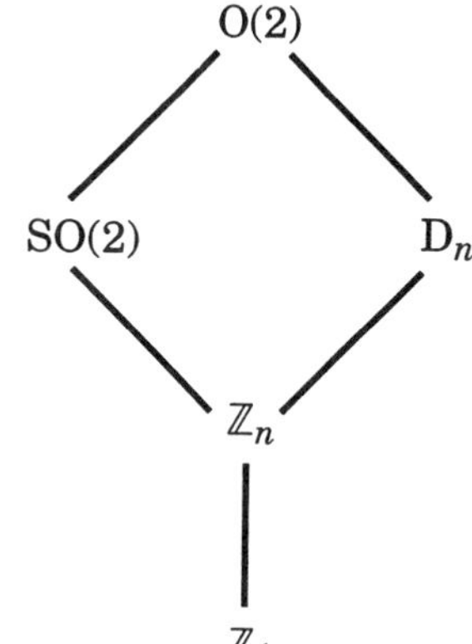

FIGURE 8.2.1. Isotropy lattice of $O(2)$.

orbit types in $O(2)$:

$$\mathcal{O}(O(2)) = \{(O(2)), (SO(2)), (\mathbb{Z}_n), (D_n); n = 1, 2, \ldots\}$$

We represent the lattice of conjugacy subgroups of $O(2)$ by Figure 8.2.1, where a line down from the subgroup H to the subgroup K denotes the relation $(H) \leq (K)$.

In order to describe the generators of $A(O(2))$, we notice that $\kappa r_\theta = r_{-\theta}\kappa$. This implies that a rotation r_θ belongs to $N(D_n)$ if and only if $r_{-\theta}\kappa\mathbb{Z}_n r_\theta = \kappa\mathbb{Z}_n$; hence, $r_{2\theta} \in \mathbb{Z}_n$ and, consequently, $N(D_n) = D_{2n}$. For the other subgroups of $O(2)$, we have $N(SO(2)) = O(2)$ and $N(\mathbb{Z}_n) = O(2)$. Thus, $\Phi(O(2)) = \{(O(2)), (SO(2)), (D_n), n = 1, 2, 3, \ldots\}$ is the set of the generators of $A(O(2))$.

Let us illustrate how to compute the multiplication table for the Burnside ring $A(O(2))$. Consider the subgroups $H = D_n$ and $K = D_k$. It is clear that $\mathbb{Z}_l \subset H \cap g^{-1}Kg := L$, where $l = gcd(n, k)$. Therefore, one can easily conclude from the lattice of orbit types of $O(2)$ that the only possible orbit types (L) in $G/H \times G/K$ are (D_l) and $(\mathbb{Z}_n)$. By direct computation, we obtain for $L = D_l$, $G/H^L = N(H)/H = \mathbb{Z}_2$ and $G/K^L = N(K)/K = \mathbb{Z}_2$. Thus, the free $N(L)/L$-space, where $N(L)/L = \mathbb{Z}_2$, is equivalent to $\mathbb{Z}_2 \times \mathbb{Z}_2$. Therefore, $n_L = 4/2 = 2$. The multiplication table for the elements of $\Phi(O(2))$ is given in Table 8.2.1.

Table 8.2.1. The multiplication table for $A(O(2))$

	$(O(2))$	$(SO(2))$	D_k
$(O(2))$	$(O(2))$	$(SO(2))$	(D_k)
$(SO(2))$	$(SO(2))$	$2(SO(2))$	0
(D_n)	(D_n)	0	$2(D_l)$

Where $l = gcd(n, k)$.

Example 8.2.10 We consider the group $SO(3)$ of all 3×3 orthogonal matrices of determinant 1. We will state some facts and their consequences and leave the details to Chossat et al. (1990).

It can be shown that every proper, closed subgroup of $SO(3)$ is conjugate to one of the following subgroups:

(i) *The subgroup of* $SO(3)$, consisting of all 3×3 matrices

$$\begin{bmatrix} A & 0 \\ 0 & \text{sign det } A \end{bmatrix},$$

$A \in O(2)$, which can be identified with the orthogonal group $O(2)$.

(ii) The following subgroups of $O(2)$: $SO(2)$, D_n, $n = 2, 3, \dots, \mathbb{Z}_n$, $n = 1, 2, 3, \dots$, which were discussed in Example 8.2.9. The subgroup D_1 is conjugated to $\mathbb{Z}_2$ and, therefore, is not included in the list. On the other hand, the group D_2 is traditionally called the *Klein group* of order 4 and is denoted by V_4.

(iii) The *exceptional* groups: *tetrahedral* $\boldsymbol{T}$, *octahedral* $\boldsymbol{O}$, *and icosahedral* $\boldsymbol{I}$, which are "rotational" symmetry groups of the regular tetrahedron, octahedron (or cube), and icosahedron (or dodecahedron), respectively. It is well known that the group $\boldsymbol{T}$ is isomorphic to the alternating group A_4 of all even permutations of $\{1, 2, 3, 4\}$, the group $\boldsymbol{O}$ is isomorphic to the symmetric group S_4 of all permutations of $\{1, 2, 3, 4\}$, and the group $\boldsymbol{I}$ is isomorphic to the alternating group A_5.

The subgroups of $SO(3)$ can be represented by a lattice of conjugacy classes (see Figure 8.2.2).

We will also need to identify the subgroups of $SO(3)$ with finite Weyl's groups (see Table 8.2.2). We have the following generators of $A(SO(3))$:

$$\Phi(SO(3)) = \{(SO(3), O(2), A_4, A_5, S_4, (D_n); n = 1, 2, 3, \dots\}$$

We shall illustrate how to compute the complete multiplication table for the ring $A(SO(3))$. For the purpose of the computation, we will need Table 8.2.3, borrowed from Chossat et al. (1990), listing the integers $n(L, H)$ for the subgroups of $SO(3)$.

As all the computations of the multiplication table for $A(SO(3))$ follow the same pattern, we will only give the details of the computation for the product $(H) \cdot (K)$, where $H = S_4$ and $K = A_5$. It is clear from the subgroup lattice that the only possible orbit types in $G/H \times G/K$, which belong to $A(SO(3))$, are $(L_1) = (A_4)$, $(L_2) = (D_3)$, and $(L_3) = (V_4)$. Consequently, $(S_4) \cdot (A_5) = n_1(A_4) + n_2(D_3) + n_3(V_4)$. Since (A_4) and (D_3) are evidently the minimal orbit types in $G/H \times G/K$, we can use formula (8.2.2) to

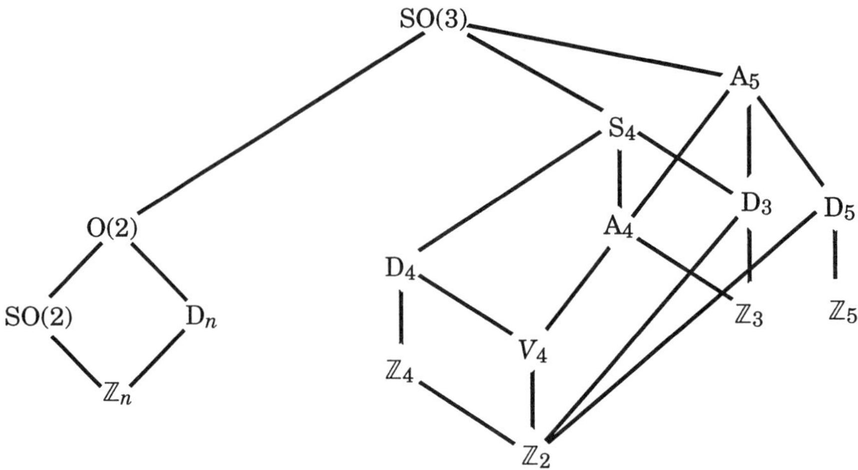

FIGURE 8.2.2. Isotropy lattice of SO(3).

compute the numbers n_1 and n_2. We have

$$n_1 = n(A_4, S_4) \cdot n(A_4, A_5) \cdot \frac{|W(S_4)| \cdot |W(A_5)|}{|W(A_4)|} = 2 \cdot 1 \cdot \frac{1 \cdot 1}{2} = 1$$

$$n_2 = n(D_3, S_4) \cdot n(D_3, A_5) \cdot \frac{|W(S_4)| \cdot |W(A_5)|}{|W(D_3)|} = 2 \cdot 2 \cdot \frac{1 \cdot 1}{2} = 2$$

where $n(A_4, S_4)$, $n(A_4, A_5)$, $n(D_3, S_4)$ and $n(D_3, A_5)$ are given in Table 8.2.3.

Table 8.2.2. Subgroups and related Weyl's groups in $SO(3)$

H	$N(H)$	$W(H)$
$O(2)$	$O(2)$	$\mathbb{Z}_1$
$SO(2)$	$O(2)$	$\mathbb{Z}_2$
$D_n, n \geq 3$	D_{2n}	$\mathbb{Z}_2$
V_4	S_4	S_3
$\mathbb{Z}_n, n \geq 2$	$O(2)$	$O(2)$
$\mathbb{Z}_1$	$SO(3)$	$SO(3)$
S_4	S_4	$\mathbb{Z}_1$
A_5	A_5	$\mathbb{Z}_1$
A_4	S_4	$\mathbb{Z}_2$

Table 8.2.3. The number of conjugate copies of L contained in a subgroup B of $SO(3)$

L	H	$n(L, H)$	Condition
$SO(2)$	$O(2)$	1	
D_m	D_n	1	$n\|m \in \mathbb{Z}, m \geq 3$
V_4	D_n	3	n even
D_m	$O(2)$	1	$m \geq 3$
V_4	$O(2)$	3	
D_3	S_4	2	
D_4	S_4	2	
D_3	A_5	2	
D_5	A_5	2	
V_4	A_4	1	
V_4	S_4	4	
V_4	A_5	2	
A_4	S_4	1	
A_4	A_5	2	

In order to compute the coefficient n_3, we will use the ring inclusion φ: $A(SO(3)) \to C(SO(3))$, which is defined on generators $(H) \in A(SO(3))$ by $\varphi((H)) = z_H$, where z_H is given by formula (8.2.3). Since we want to compute the product $(S_4) \cdot (A_5)$, it follows from Theorem 8.2.6 that

$$\begin{aligned}\varphi((S_4) \cdot (A_5)) &= z_{S_4} \cdot z_{A_5} = n_1 z_{A_4} + n_2 z_{D_3} + n_3 z_{V_4} \\ &= z_{A_4} + 2z_{D_3} + n_3 z_{V_4}\end{aligned}$$

Consequently, we need to compute the values of the functions z_H for $(H) \in \Phi(SO(3))$, which we will express in a tabular form, and to find the number n_3 by inspection.

For two positive integers n and k, we define $\left|\begin{matrix} n \\ k \end{matrix}\right|$ by

$$\left|\begin{matrix} n \\ k \end{matrix}\right| = \begin{cases} 1 & \text{if } k \text{ divides } n \\ 0 & \text{otherwise} \end{cases}$$

and $[n \| k_1, \ldots, k_r] = 1$ if $n \in \{k_1, \ldots, k_r\}$, and zero otherwise.

Table 8.2.4 can be computed directly by using the values of $n(L, H)$ given in Table 8.2.3.

By using the information from Table 8.2.4, we obtain Table 8.2.5. Consequently,

$$(S_4) \cdot (A_5) = (A_4) + 2(D_3) + (V_4)$$

Table 8.2.4. Computation of z_H for $(H) \in A(SO(3))$

z_H (L)	$SO(3)$	$O(2)$	$SO(2)$	A_4	A_5	S_4	V_4	$D_n, n \geq 3$
$z_{SO(3)}$	1	1	1	1	1	1	1	1
$z_{O(2)}$	0	1	1	0	0	0	3	1
$z_{SO(2)}$	0	0	2	0	0	0	0	0
z_{S_4}	0	0	0	1	0	1	4	$2[n\|3,4]$
z_{A_5}	0	0	0	2	1	0	2	$2[n\|3,5]$
z_{A_4}	0	0	0	2	0	0	2	0
z_{D_5}	0	0	0	0	0	0	0	$2[n\|5]$
z_{D_4}	0	0	0	0	0	0	6	$2[n\|4]$
z_{D_3}	0	0	0	0	0	0	0	$2[n\|3]$
z_{V_4}	0	0	0	0	0	0	6	0
z_{D_m}	0	0	0	0	0	0	$6\left\|{m \atop 2}\right\|$	$2\left\|{m \atop n}\right\|$

Where $m \geq 3$.

Table 8.2.5. Computation of $z_{S_4} \cdot z_{A_5}$

	$SO(3)$	$O(2)$	$SO(2)$	A_4	A_5	S_4	V_4	D_3	$D_n, n \geq 3$
z_{A_4}	0	0	0	2	0	0	2	0	0
$2z_{D_3}$	0	0	0	0	0	0	0	4	0
z_{A_4}	0	0	0	2	0	0	6	0	0
$z_{S_4} \cdot z_{A_5}$	0	0	0	2	0	0	8	4	0

In order to compute the other products $(H) \cdot (K)$ in $A(SO(3))$, one needs to follow exactly the steps in which the product $z_H \cdot z_K$ has to be expressed as a linear combination of the functions z_L. We summarize the obtained multiplication formulae for $A(SO(3))$ in Tables 8.2.6 and 8.2.7 that follow.

Table 8.2.6. The first multiplication table for $A(SO(3))$

	$(O(2))$	(S_4)	(A_5)	(A_4)
$(O(2))$	$(O(2)) + (V_4)$	$(D_4) + (D_3) + (V_4)$	$(D_5) + (D_3) + (V_4)$	(V_4)
(S_4)	$(D_4) + (D_3) + (V_4)$	$(S_4) + (D_4) + (D_3) + (V_4)$	$(A_4) + 2(D_3) + (V_4)$	$(A_4) + (V_4)$
(A_5)	$(D_5) + (D_3) + (V_4)$	$(A_4) + 2(D_3) + (V_4)$	$(A_5) + (A_4) + (D_3) + (D_5)$	$2(A_4)$
(A_4)	(V_4)	$(A_4) + (V_4)$	$2(A_4)$	$2(A_4)$
(D_5)	(D_5)	0	$2(D_5)$	0
(D_4)	$(D_4) + 2(V_4)$	$2(D_4) + 2(V_4)$	$2(V_4)$	$2(V_4)$
(D_3)	(D_3)	$2(D_3)$	$2(D_3)$	0
(V_4)	$3(V_4)$	$4(V_4)$	$2(V_4)$	$2(V_4)$
(D_n), $n \geq 3$	$(D_n) + 2\left\|{n \atop 2}\right\|(V_4)$	$2\left\|{n \atop 4}\right\|(D_4) + 2\left\|{n \atop 3}\right\|(D_3) + 2(2 - \left\|{n \atop 4}\right\|)\left\|{n \atop 2}\right\|(V_4)$	$2\left\|{n \atop 5}\right\|(D_5) + 2\left\|{n \atop 3}\right\|(D_3) + 2\left\|{n \atop 2}\right\|(V_4)$	$2\left\|{n \atop 2}\right\|(V_4)$

Table 8.2.7. The second multiplication table for $A(SO(3))$

	$SO(2)$	(D_5)	(D_4)	(D_3)	(V_4)
$(O(2))$	$(SO(2))$	(D_5)	$(D_4)+2(V_4)$	(D_3)	$3(V_4)$
$SO(2)$	$2(SO(2))$	0	0	0	0
(S_4)	0	0	$2(D_4)+2(V_4)$	$2(D_3)$	$4(V_4)$
(A_5)	0	$2(D_5)$	$2(V_4)$	$2(D_3)$	$2(V_4)$
(A_4)	0	0	$2(V_4)$	0	$2(V_4)$
(D_5)	0	$2(D_5)$	0	0	0
(D_4)	0	0	$2(D_4)+4(V_4)$	0	$6(V_4)$
(D_3)	0	0	0	$2(D_3)$	0
(V_4)	0	0	$6(V_4)$	0	$6(V_4)$
(D_n)	0	$2\left\lvert\begin{smallmatrix} n\\ 5\end{smallmatrix}\right\rvert(D_5)$	$2\left\lvert\begin{smallmatrix} n\\ 4\end{smallmatrix}\right\rvert(D_4)+2\left(3-\left\lvert\begin{smallmatrix} n\\ 4\end{smallmatrix}\right\rvert\right)\left\lvert\begin{smallmatrix} n\\ 2\end{smallmatrix}\right\rvert(V_4)$	$2\left\lvert\begin{smallmatrix} n\\ 3\end{smallmatrix}\right\rvert(D_3)$	$6\left\lvert\begin{smallmatrix} n\\ 2\end{smallmatrix}\right\rvert(D_4)$

We also have

$$(D_n)\cdot(D_k)=2[l](D_l)+2(3-[l])[l]|2](V_4)$$

where $l=gcd(n,k)$ is the greatest common divisor of n and k, and

$$[l]=\begin{cases}0 & \text{if } l<3\\ 1 & \text{if } l\geq 3\end{cases}$$

EXERCISES

8.2.1 Let G be a finite group. Two finite G-sets X and Y are said to be *equivalent*, if there exists a G-equivariant bijection $\varphi: X\to Y$. Denote by $\tilde{A}(G)$ the set of all equivalence classes $[X]$ of finite G-spaces X. The set $\tilde{A}(G)$ can be equipped with two operations: the *addition:* $[X]+[Y]:=[X\cup Y]$, where $\cup$ denotes the disjoint union, and the *multiplication:* $[X]\cdot[Y]:=[X\times Y]$. Show that the set $\tilde{A}(G)$ equipped with the above two operations is a semiring with unity. Let $\bar{A}(G)$ denote the completion of $\tilde{A}(G)$ to a ring. Show that the Burnside ring $A(G)$ is isomorphic to $\bar{A}(G)$.

8.2.2 Compute the multiplication table for the Burnside ring $A(\mathbb{Z}_n)$.

8.2.3 Consider the dihedral group $\Gamma=D_n$ for $n>2$. Show that if n is an odd number, then $\Phi(D_n)=\{(D_k),\ (\mathbb{Z}_k);\ k|n\}$, and if n is even, then $\Phi(D_n)=\{(D_k),\ (\tilde{D}_k),\ (\mathbb{Z}_k);\ k|n\}$ where $\tilde{D}_k=\mathbb{Z}_k\cup\kappa\xi_n\mathbb{Z}_k,\ \xi_n=e^{\frac{2i\pi}{n}}$, and $\kappa=\begin{bmatrix}1 & 0\\ 0 & -1\end{bmatrix}$. Compute $W(K)$ for all subgroups $K\in\Phi(D_n)$.

8.2.4 Show that we have the following multiplication table for the Burnside ring $A(D_n)$:

	(D_m) $2m \nmid n$	(D_m) $2m \mid n$	$(\tilde{D}_m)$ $2m \nmid n$	$(\tilde{D}_m)$ $2m \mid n$	(Z_m)
(D_k) $2k \nmid n$	$(D_l)+ \frac{nl-mk}{2mk}(\mathbb{Z}_l)$	$(D_l)+ \frac{nl-mk}{2mk}(\mathbb{Z}_l)$	$\frac{ln}{2mk}(\mathbb{Z}_l)$	$\frac{ln}{2mk}(\mathbb{Z}_l)$	$\frac{nl}{km}(\mathbb{Z}_l)$
(D_k) $2k \mid n$	$(D_l)+ \frac{nl-mk}{2mk}(\mathbb{Z}_l)$	$2(D_l)+ \frac{nl-2mk}{2mk}(\mathbb{Z}_l)$	$\frac{ln}{2mk}(\mathbb{Z}_l)$	$\frac{ln}{2mk}(\mathbb{Z}_l)$	$\frac{nl}{km}(\mathbb{Z}_l)$
$(\tilde{D}_k)$ $2k \nmid n$	$\frac{ln}{2mk}(\mathbb{Z}_l)$	$\frac{ln}{2mk}(\mathbb{Z}_l)$	$(\tilde{D}_l)+ \frac{nl-mk}{2mk}(\mathbb{Z}_l)$	$(\tilde{D}_l)+ \frac{nl-mk}{2mk}(\mathbb{Z}_l)$	$\frac{nl}{km}(\mathbb{Z}_l)$
$(\tilde{D}_k)$ $2k \mid n$	$\frac{ln}{2mk}(\mathbb{Z}_l)$	$\frac{ln}{2mk}(\mathbb{Z}_l)$	$(\tilde{D}_l)+ \frac{nl-mk}{2mk}(\mathbb{Z}_l)$	$2(\tilde{D}_l)+ \frac{nl-2mk}{2mk}(\mathbb{Z}_l)$	$\frac{nl}{km}(\mathbb{Z}_l)$
$(\mathbb{Z}_k)$	$\frac{nl}{km}(\mathbb{Z}_l)$	$\frac{nl}{km}(\mathbb{Z}_l)$	$\frac{nl}{km}(\mathbb{Z}_l)$	$\frac{nl}{km}(\mathbb{Z}_l)$	$\frac{2nl}{km}(\mathbb{Z}_l)$

Where $l = gcd(m, k)$, $m \mid n$ and $k \mid n$.

8.2.5 Compute the multiplication table for the Burnside ring $A(S^1 \times D_4)$.

8.3 *NORMAL APPROXIMATIONS AND G-DEGREE OF DOLD–ULRICH*

In this section we use the idea of normal approximations to introduce an analytic definition of an equivariant degree that is based on the equivariant fixed-point index of Dold–Ulrich.

Let V be a real, orthogonal, finite-dimensional representation of G.

Definition 8.3.1 For an open, bounded, invariant set Ω of V, an equivariant continuous map $f: V \to V$ is said to be *Ω-admissible* if $f(x) \neq 0$ for all $x \in \partial\Omega$. An equivariant continuous map $h: [0, 1] \times V \to V$, where G acts on $[0, 1]$ trivially, is called an *equivariant homotopy*. An equivariant homotopy $h: [0, 1] \times V \to V$ is said to be *Ω-admissible* if $h(t, x) \neq 0$ for $x \in \partial\Omega$. For an admissible homotopy $h: [0, 1] \times V \to V$, we say that h_0 and h_1 are *Ω-homotopic*, where $h_t: V \to V$ for $t \in [0, 1]$ is defined by $h_t(x) = h(t, x)$, $x \in V$.

Remark 8.3.2 We should point out that for technical reasons, we consider in this section G-maps that are defined on the whole space V or $[0, 1] \times V$. For an open, bounded, invariant set Ω of V, we may say that a G-map $f: \overline{\Omega} \to V$ is *Ω-admissible* if $f(x) \neq 0$ for all $x \in \partial\Omega$. In this case, by the equivariant version of Dugundji's extension theorem (Theorem 6.1.28), there exists a G-equivariant extension $\bar{f}: V \to V$ of the G-map f that is Ω-admissible according to Definition 8.3.1.

We consider the space $W := \mathbb{R} \times V$, where G acts on $\mathbb{R}$ trivially. Given $\alpha \in O(G)$, it is known from Theorem 8.1.13 that $V_\alpha := \{x \in V; (G_x) = \alpha\}$ and $W_\alpha := \{x \in W; (G_x) := \alpha\}$ are G-invariant submanifolds of V and W, respectively. We consider the normal bundles $\nu(V_\alpha)$ and $\nu(W_\alpha)$ to V_α in V and to W_α in W, respectively, that will be denoted by $\gamma_\alpha : N^\alpha \to V_\alpha$ and $\gamma_\alpha : N^\alpha \to W_\alpha$. Let D be a compact invariant subset of V_α (respectively, of W_α). We put

$$\mathcal{N}(D, \varepsilon) := \{(v, w) \in N^\alpha; v \in D, w \in N_v^\alpha, |w| \le \varepsilon\}$$

where $|\cdot|$ is an invariant norm induced by the standard, invariant, Euclidean metric in V. The set $\mathcal{N}(D, \varepsilon)$ is the total space of the ε-disk subbundle of the restricted vector bundle $\nu(D) := \nu(V_\alpha)|_D$. Consider the G-map $\mu : \mathcal{N}(D, \varepsilon) \to V$ given by $\mu(v, w) = v + w$. Then, by the implicit function theorem, for a sufficiently small $\varepsilon > 0$, the G-map μ is a G-diffeomorphism onto its image $\mu(\mathcal{N}(D, \varepsilon))$ that will be called an *α-normal neighborhood of* D. In particular, each element x of $\mu(\mathcal{N}(D, \varepsilon))$ can be written uniquely as $x = v + w$, where $v \in D$ and $w \in N_v^\alpha$ with $|w| < \varepsilon$.

Recall that for an invariant subset X of V, we denote by $J(X)$ the set of all orbit types in X.

Definition 8.3.3 Let $\Omega \subset V$ be an open, bounded, invariant set and $f : V \to V$ be an Ω-admissible G-map. We say that f is normal in Ω, if for every $x \in f^{-1}(0) \cap \Omega$, the following *$\alpha$-normality* condition at x with $\alpha = (H)$, $H = G_x$, is satisfied:

(i) There exists $\delta_x > 0$ such that for all $w \in N_x^\alpha$ with $|w| < \delta_x$,

$$f(x + w) = f(x) + w = w$$

Similarly, an Ω-admissible G-homotopy $h : [0, 1] \times V \to V$ is called a *normal homotopy* in Ω, if for every $(t, x) \in h^{-1}(0) \cap ([0, 1] \times \Omega)$, the following *$\alpha$-normality* condition at (t, x) with $\alpha = (H)$, with $H = G_{(t,x)}$, is satisfied.

(ii) There exists $\delta_{(t,x)} > 0$ such that for all $w \in N_{(t,x)}^\alpha$ with $|w| < \delta_{(t,x)}$,

$$h(t, x + w) = h(t, x) + w = w$$

It should be clear that in the case of an equivariant homotopy $h : [0, 1] \times$

$V \to V$, we have $[0, 1] \times \{0\} \subset W^H$ for all $\alpha = (H)$. Therefore, $N_v^\alpha \subset \{0\} \times V = V$. Consequently, condition (ii) of Definition 8.3.3 is "well posed."

Remark 8.3.4 Condition (i) [respectively, condition (ii)] of Definition 8.3.3 can be replaced with a more "practical" condition that requires only that $f(x + w) = f(x) + n(x)w$ [respectively, $h(t, x + w) = h(t, x) + n(t, x)w$], where $n(x) > 0$. This latter condition will be more convenient in computing the degree. In what follows, we will often say that such a map f is *almost normal*. In this case, there is an Ω-admissible homotopy h_t between f and a normal map $\hat{f}$ such that $f^{-1}(0) \cap \Omega = h_t^{-1}(0) \cap \Omega$ for all $t \in [0, 1]$.

For technical reasons, it is appropriate to modify slightly the definition of a normal G-map f in Ω. We have the following.

Definition 8.3.5 Let Ω be an open, bounded, invariant subset of V and $f : V \to V$ an Ω-admissible G-map. We say that f is *strongly normal* in Ω if for all $\alpha \in J(f^{-1}(0) \cap \Omega)$, there is an α-normal neighborhood $C(\alpha) = \mu(\mathcal{N}(D_\alpha, \varepsilon_\alpha)) \subset \Omega$ of some $D_\alpha \subset V_\alpha$ such that:

(i) $C(\alpha) \cap C(\beta) = \emptyset$ for $\alpha \neq \beta$.

(ii) $\bigcup_{\alpha \in J(f^{-1}(0) \cap \Omega)} C(\alpha)$ is a compact neighborhood of $f^{-1}(0) \cap \Omega$.

(iii) For every $\alpha \in J(f^{-1}(0) \cap \Omega)$, the G-map f is *α-normal* in $C(\alpha)$, that is,

$$f(v + w) = f(v) + w \quad \text{for all} \quad v + w \in C(\alpha)\,, \quad w \in N_v^\alpha \qquad (*)$$

Clearly, every strongly normal G-map is normal.

The following result says that normal in Ω G-maps are dense in the space of Ω-admissible G-maps.

Theorem 8.3.6 (Normal Approximation Theorem)

Let $\Omega \subset V$ be an open, bounded, invariant set and $f : V \to V$ be an Ω-admissible equivariant map. Then for every $\eta > 0$, there exists an Ω-normal G-map $f_0 : V \to V$ such that $\sup_{x \in \Omega} |f_0(x) - f(x)| < \eta$.

Proof. We will construct a strongly normal in Ω G-map f_0 that satisfies $\sup_{x \in \Omega} |f_0(x) - f(x)| < \eta$.

Let D_1 be a compact invariant neighborhood of $f^{-1}(0)$ in Ω and $\delta > 0$ be given so $K := \overline{B(D_1, \delta)} \subset \Omega$, where

$$B(D_1, \delta) = \{x \in V; \operatorname{dist}(D_1, x) < \delta\}$$

Set $m = |J(K)|$. We extend the partial order in $J(K)$ to a total order

$\alpha_1 < \alpha_2 < \cdots < \alpha_k < a_{k+1} < \cdots < \alpha_m$. Assume that $0 < \varepsilon < \delta$. We consider the minimal orbit type $\alpha := \alpha_1 \in J(K)$. Let $D := D_1 \cap V_\alpha \subset \operatorname{Int} K$ and $\tilde{D} := \{x \in D_j \operatorname{dist}(x, f^{-1}(0)) \leq \operatorname{dist}(x, V_\alpha \setminus D)\}$. Without loss of generality, we assume that $D \neq \varnothing$. It follows from Theorem 8.1.13 that D is an invariant compact set. This gives an α-normal neighborhood $N_0 := \mu(\mathcal{N}(D, \varepsilon)) \subset \operatorname{Int} K$ of D, where $\mu : \mathcal{N}(D, \varepsilon) \to \Omega$ is a diffeomorphism onto its image N_0 for a suitable small $\varepsilon > 0$. There exists an invariant Urysohn function $\gamma : V \to [0, 1]$ such that $\gamma(x) = 1$ for $x \in \mu(\mathcal{N}(\tilde{D}, \varepsilon/2))$ and $\gamma(x) = 0$ for $x \in V \setminus N_0$. Define now a G-map $f_1 : V \to V$ as follows:

$$f_1(x) := \begin{cases} f(x), & x \in V \setminus N_0 \\ \gamma(x)[f(v) + w] + [1 - \gamma(x)]f(x), & x = v + w \in N_0 \end{cases}$$

Since we can choose $\varepsilon > 0$ to be as small as we wish, we can assume that

$$\sup_{v + w \in N_0} |f(v) - f(v + w)| < \frac{\eta}{m} - \varepsilon$$

Consequently,

$$\begin{aligned} \sup_{x \in \Omega} |f_1(x) - f(x)| &= \sup_{x \in N_0} |f_1(x) - f(x)| \\ &= \sup_{x = v + w \in N_0} |\gamma(x)[f(v) - f(x)] + \gamma(x) w| \\ &\leq \sup_{x = v + w \in N_0} |f(v) - f(x)| + \varepsilon < \frac{\eta}{m} \end{aligned}$$

This implies that $f_1 : V \to V$ is Ω-admissible and is α-normal in $C(\alpha) := \mu(\mathcal{N}(\tilde{D}, \varepsilon/2))$ and $C(\alpha) \supset f_1^{-1}(0) \cap W_\alpha$.

Assume now that for $k \geq 1$, we have already constructed a G-map $f_k : V \to V$ satisfying the following conditions:

(i) For every $\alpha \leq \alpha_k$, there is an α-normal neighborhood $C(\alpha) \subset \operatorname{Int} K$ such that the G-map f_k is α-normal in $C(\alpha)$.

(ii) For every $\alpha, \beta \leq \alpha_k$, $C(\alpha) \cap C(\beta) = \varnothing$ if $\alpha \neq \beta$.

(iii) Every $\alpha \in J(f_k^{-1}(0) \setminus \bigcup_{i=1}^{k} C(\alpha_i))$ is greater than α_k.

(iv) $\sup_{x \in \Omega} |f_k(x) - f(x)| < k \dfrac{\eta}{m}$.

From the above, $A := f_k^{-1}(0) \setminus \bigcup_{i=1}^{k} C(\alpha_i)$ is an invariant compact subset of $\operatorname{Int} K$. There is a compact G-neighborhood D_{k+1} of A in $\Omega_{k+1} := \operatorname{Int} K \setminus \bigcup_{i=1}^{k} C(\alpha_i)$ such that any $\alpha \in \mathcal{J}(D_{k+1})$ is greater than α_k. Assume that $\delta_{k+1} > 0$ is such that $\overline{B(D_{k+1}, \delta_{k+1})} \subset \Omega_{k+1}$, and consider $D := V_{\alpha_{k+1}} \cap D_{k+1}$. The set D is compact by Theorem 8.1.13 since α_{k+1} is the minimal orbit type in D_{k+1}. Then again, for a sufficiently small $0<\varepsilon<\delta_{k+1}$, the G-map $\mu : \mathcal{N}(D, \varepsilon) \to \Omega$ is a diffeomorphism onto its image $N_0 := \mu(\mathcal{N}(D, \varepsilon)) \subset \Omega_{k+1} \subset \Omega$, that is, $C(\alpha_{k+1}) := N_0$ gives an α_{k+1}-normal neighborhood. We choose again an invariant Uryshon function $\gamma : V \to [0, 1]$ such that $\gamma(x) = 1$ for $x \in \mu(\mathcal{N}(\tilde{D}, \varepsilon/2))$ and $\gamma(x) = 0$ for $x \in V \backslash N_0$, where $\tilde{D} := \{x \in D; \operatorname{dist}(x, f_k^{-1}(0)) \le \operatorname{dist}(x, V_\alpha \backslash D)\}$. Define the following G-map $f_{k+1} : V \to V$ by

$$f_{k+1}(x) := \begin{cases} f_k(x), & x \in V \backslash N_0 \\ \gamma(x)[f_k(v) + w] + [1 - \gamma(x)] f_k(x), & x = v + w \in N_0 \end{cases}$$

Since we can choose $\varepsilon > 0$ as small as we wish, we can also assume that

$$\sup_{v+w \in N_0} |f_k(v) - f_k(v + w)| < \frac{\eta}{m} - \varepsilon$$

$$\sup_{x \in \Omega} |f_{k+1}(x) - f_k(x)| < \frac{\eta}{m}$$

Consequently,

$$\begin{aligned} \sup_{x \in \Omega} |f_{k+1}(x) - f(x)| &\le \sup_{x \in \Omega} |f_{k+1}(x) - f_k(x)| + \sup_{x \in \Omega} |f_k(x) - f(x)| \\ &< \frac{\eta}{m} + k \frac{\eta}{m} = (k + 1) \frac{\eta}{m} \end{aligned}$$

By definition, $f_{k+1} : V \to V$ is Ω-admissible and is α-normal in $C(\alpha) := \mu(\mathcal{N}(\tilde{D}, \varepsilon/2))$, $\alpha \le \alpha_{k+1}$. Therefore, by the induction principle we obtain a G-map $f_0 := f_m$ and the conclusion follows. □

Remark 8.3.7 It follows from the construction in the proof of Theorem 8.3.6 that if the mapping f is normal in a subset $A \subset f^{-1}(0) \cap \Omega$, then it is possible to construct a normal G-map $f_0 : V \to V$ such that:

(i) $\sup_{x \in \Omega} |f_0(x) - f(x)| < \eta$.

(ii) $f_0(x) = f(x) = 0$ for all $x \in A$.

Theorem 8.3.6 can be easily extended to Ω-admissible homotopies.

Theorem 8.3.8 (Normal Approximation Theorem for Homotopies) *Let $\Omega \subset V$ be an open, bounded, invariant set and $h : [0, 1] \times V \to V$ be an Ω-admissible homotopy. Then for every $\eta > 0$, there exists a normal in Ω G-homotopy $\tilde{h} : [0, 1] \times V \to V$ such that $\sup_{(t,x) \in [0,1] \times \Omega} |\tilde{h}(t, x) - h(t, x)| < \eta$. In addition, if $h_0 := h(0, \cdot)$ and $h_1 := h(1, \cdot)$ are normal in Ω, then we can construct such a homotopy so that $\tilde{h}_0 = h_0$ and $\tilde{h}_1 = h_1$.*

Proof. The proof follow the same idea as that of Theorem 8.3.6. □

Definition 8.3.9 Let $\Omega \subset V$ be an open, bounded, invariant set and $f: V \to V$ be an Ω-admissible G-map. We say that f is a *regular normal* in Ω G-map if:

(i) f is of class C^1.
(ii) f is a normal in Ω G-map.
(iii) For every $\alpha \in J(f^{-1}(0) \cap \Omega)$, $\alpha = (H)$, zero is a regular value of

$$f_H := f_{|\Omega_H} : \Omega_H \to V^H$$

Similarly, an Ω-admissible homotopy $h : [0, 1] \times V \to V$ is called a *regular normal* in Ω *homotopy* if:

(i) h is of class C^1.
(ii) h is a normal in Ω homotopy.
(iii) For every $\alpha \in J(h^{-1}(0) \cap [0, 1] \times \Omega)$, $\alpha = (H)$, zero is a regular value of the maps h_H, $(h_0)_H$, and $(h_1)_H$, where

$$h_H := h_{|[0,1] \times \Omega_H} : [0, 1] \times \Omega_H \to V^H$$

$$(h_0)_H := h_{0|\Omega_H} : \Omega_H \to V^H$$

$$(h_1)_H := h_{1|\Omega_H} : \Omega_H \to V^H$$

Lemma 8.3.10

Let $\Omega \subset V$ be an open, bounded, invariant set and $f: V \to V$ be an Ω-admissible G-map. Then for every $\varepsilon > 0$, there exists an Ω-admissible G-map $f_0 : V \to V$ of class C^1 such that $\sup_{x \in \Omega} |f(x) - f_0(x)| < \varepsilon$.

Proof. Let K denote $\overline{\Omega}$. Since K is an invariant, compact set such that $K \supset f^{-1}(0) \cap \Omega$, by using smooth Uryshon functions, we can find a C^1-map $\bar{f}_0 : V \to V$ (not necessarily equivariant) such that

$$\sup_{x \in K} |f(x) - \bar{f}_0(x)| < \varepsilon$$

Put

$$f_0(x) := \int_G g\bar{f}_0(g^{-1}x)\, dg\,, \qquad x \in \Omega$$

Then f_0 is of class C^1 and

$$|f(x) - f_0(x)| \leq \int_G |f(g^{-1}x) - f_0(g^{-1})|\, dg < \varepsilon$$

as desired. □

Lemma 8.3.11

Let $\Omega \subset V$ be an open, bounded, invariant set and $f: V \to V$ be an Ω-admissible G-map. Then for every $\eta > 0$, there exists a normal in Ω G-map $f_0 : V \to V$ of class C^1 such that $\sup_{x \in \Omega} |f(x) - f_0(x)| < \eta$.

Proof. It is sufficient to notice that in the proof of Theorem 8.3.6, one can choose smooth Uryshon functions, and therefore if f is of class C^1, the resulting normal in Ω approximation f_0 is also of class C^1. □

In what follows, we will construct regular normal in Ω approximations for a given Ω-admissible map f. From Theorem 8.3.6, and Lemmas 8.3.10 and 8.3.11, the problem is reduced to finding a C^1 normal in Ω approximation that also satisfies (iii) of Definition 8.3.9.

Theorem 8.3.12 (REGULAR NORMAL APPROXIMATION THEOREM)

Let $\Omega \subset V$ be an open, bounded, invariant set and $f: V \to V$ be an Ω-admissible G-map. Then for every $\eta > 0$, there exists a regular normal in Ω G-map $f_0 : V \to V$ such that $\sup_{x \in \Omega} |f_0(x) - f(x)| < \eta$. In addition, if $h : [0,1] \times V \to V$ is an Ω-admissible G-homotopy, then for every $\eta > 0$, there exists a regular normal in Ω G-homotopy $\tilde{h} : [0,1] \times V \to V$ such that $\sup_{(t,x) \in [0,1] \times \Omega} |h(t,x) - \tilde{h}(t,x)| < \eta$.

Proof. By Theorem 8.3.6, there exists a strongly normal in Ω G-map $\tilde{f} : \Omega \to V$ such that $\tilde{f}$ is Ω-admissible and $\sup_{x \in \Omega} |\tilde{f}(x) - f(x)| < \eta/2$. By Lemmas 8.3.10 and 8.3.11, we can assume that $\tilde{f}$ is of class C^1. Since a map $u : V^H \to V^H$ has only regular zeros in Ω if and only if the associated map $(\mathrm{Id}, u) : V^H \to V^H \times V^H$ is transverse to the zero section $V^H \times \{0\}$ of the bundle $\pi_1 : V^H \times V^H \to V^H$ with $\pi_1(x, y) = x$, we consider the following diagram of vector bundle maps:

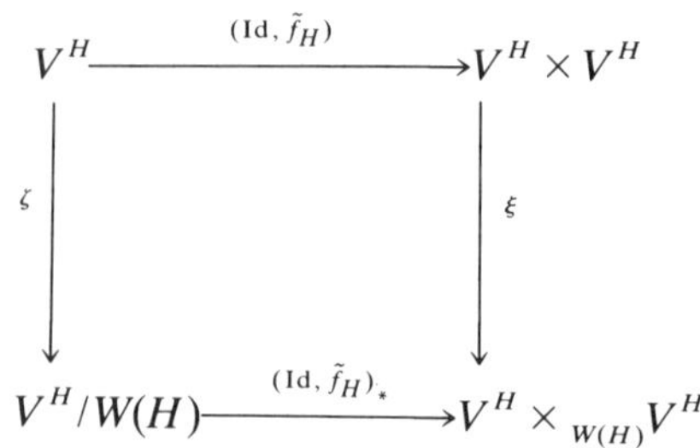

According to the transversality theorem, there exists a smooth homotopy F^* (relative to $\partial\Omega^H/W(H)$) between $(\mathrm{Id}, \tilde{f}_H)_*$ and a map that is transverse to the zero section in $V^H \times_{W(H)} V^H$. The composition of the bundle isomorphism $\zeta \times \mathrm{Id}_I \simeq (F_0^*)^*(\xi) \times \mathrm{Id}_I \simeq (F^*)^*_{\sim}(\xi)$ with the projection $(F^*)^*(\xi) \to \xi$ is a $W(H)$-homotopy F of $(\mathrm{Id}, \tilde{f}_H)$ over F^*, where $I = [0, 1]$. See Ulrich (1988) (proof of Theorem 3.2, step 1) for more details. Thus, we may approximate $\tilde{f}_H$, $\alpha = (H) \in J(\tilde{f}^{-1}(0) \cap \Omega)$, by a $W(H)$-map $\hat{f}_H$ that comes transversally to zero. We now extend the $W(H)$-map $\hat{f}_H$ to a G-map $\hat{f}_{(H)}$ defined on the set $C(\alpha)_{(H)}$. Next, by using condition (iii) of Definition 8.3.5, we extend $\hat{f}_{(H)}$ (as an identity in normal directions) to a G-map $\hat{f}_\alpha$ on the set $C(\alpha)$. Finally, with the help of smooth Uryshon functions, we can "glue" the G-map $\hat{f}_\alpha$ to a G-map $\tilde{f}$ on the set $C(\alpha)$, and the G-map thus obtained, which is an approximation of $\tilde{f}$, satisfies condition (iii) of Definition 8.3.9 on the set $C(\alpha)$. Since the same "adjustment" of $\tilde{f}$ can be made on every $C(\alpha')$, $\alpha' \in J(\tilde{f}^{-1}(0) \cap \Omega)$, the above construction shows that $\tilde{f}$ can be approximated by a regular Ω-normal G-map with an arbitrarily high accuracy.

The proof for the case of the homotopy h is similar. □

We finally remark that, according to the definition, if f is a regular normal G-map, then for every orbit type (H), the map f_H is transverse to $0 \in V^H$ on Ω. However, it does not imply that zero is a regular value of the mapping $f_{|\Omega}$. In fact, it may happen that the mapping $f_{(H)} := f_{|V_{(H)}}$ has zero as a critical value. Consequently, the above regular normal approximation theorem cannot be considered as a version of the *equivariant transversality theorem* in the usual sense [see Ulrich (1980)].

Let $(H) \in \Phi(G)$, that is, $W(H)$ is finite, and let V be a finite-dimensional orthogonal representation of G, $\Omega \subset V$ be an open, bounded, invariant set. Our goal is to associate with every G-equivariant Ω-admissible map $f: V \to V$ an element $G\text{-deg}(f, \Omega) \in A(G)$, which will be called the *G-degree* of the map f with respect to Ω. The properties of the G-degree, expressed in the next theorem, are completely parallel to those of the classical Brouwer degree. However, its value belongs to the Burnside ring $A(G)$ and thus is not an integer.

Theorem 8.3.13

For every Ω-admissible map $f: V \to V$, where Ω is an open, bounded, invariant subset of V, we can assign an element $G\text{-deg}(f, \Omega) \in A(G)$ such that the following properties are satisfied:

(P1) *Existence: If $G\text{-deg}(f, \Omega) = \sum n_\alpha \neq 0$, that is, there is an $\alpha \in \Phi(G)$ such that $n_\alpha \neq 0$, then there exists $x \in f^{-1}(0) \cap \Omega$ such that $(G_x) \leq \alpha$.*

(P2) *Homotopy Invariance: If* $h:[0,1]\times V\to V$ *is an* Ω*-admissible homotopy, then* G-deg(h_t,Ω) *does not depend on* $t\in[0,1]$.

(P3) *Excision: If* $\Omega_0\subseteq\Omega$ *is an open and invariant subset and* $f^{-1}(0)\cap\Omega\subseteq\Omega_0$, *then* G-deg$(f,\Omega)=G$-deg(f,Ω_0).

(P4) *Additivity: If* Ω_1 *and* Ω_2 *are two open invariant subsets of* Ω *such that* $\Omega_1\cap\Omega_2=\varnothing$ *and* $f^{-1}(0)\cap\Omega\subset\Omega_1\cup\Omega_2$, *then*

$$G\text{-deg}(f,\Omega)=G\text{-deg}(f,\Omega_1)+G\text{-deg}(f,\Omega_2)$$

(P5) *Multiplicativity: Let* V *and* W *be two orthogonal representations of* G, $\Omega\subset V$ *and* $U\subset W$ *two invariant, open, bounded subsets,* $f:V\to V$ *(respectively,* $g:W\to W$*) an* Ω*-admissible (respectively,* U*-admissible) map. Then the map* $F:V\times W\to V\times W$, *defined by* $F(x,y)=(f(x),g(y))$, $(x,y)\in V\times W$, *is* $\Omega\times U$*-admissible and we have*

$$G\text{-deg}(F,\Omega\times\mathcal{U})=G\text{-deg}(f,\Omega)\cdot G\text{-deg}(g,U)$$

Proof. In order to construct the G-degree, we consider an Ω-admissible G-map $f:V\to V$. Suppose that $0<\varepsilon<\frac{1}{4}\inf_{x\in\partial\Omega}|f(x)|$ and let $\hat{f}$ be a regular normal ε-approximation of f, that is, $\sup_{x\in\Omega}|f(x)-\hat{f}(x)|<\varepsilon$. Then $h(t,x)=tf(x)+(1-t)\hat{f}(x)$ is an Ω-admissible G-homotopy between f and $\hat{f}$. We define the G-degree G-deg$(f,\Omega):=G$-deg$(\hat{f},\Omega)=\sum_{\alpha\in\Phi(G)}n_\alpha\cdot\alpha$, where the numbers n_α can be computed as follows: For $\alpha=(H)$ with $\alpha\in J(\hat{f}^{-1}(0))\cap\Phi(G)$, the restriction of $\hat{f}$ to the subspace V_H induces a $W(H)$-equivariant map $\hat{f}_H:V_H\to V^H$, $\hat{f}_H(x)=f(x)$, $x\in V_H$, such that $\hat{f}_H^{-1}(0)\subset\Omega_H$ is composed of a finite number of (finite) $W(H)$-orbits, say, $W(H)x_1$, $\dots$, $W(H)x_k$. We denote by S_i a slice to the orbit $W(H)x_i$ at the point x_i, $i=1,\dots,k$, that is, S_i is an open neighborhood of x_i such that $S_i\cap W(H)x_i=\{x_i\}$. The slice S_i has the same orientation as the space V^H. Now we can define

$$n_\alpha=\sum_{i=1}^{k}\text{sign det }D\hat{f}^H(x_i)$$

In order to show that the above definition does not depend on the choice of a regular normal in Ω ε-approximation $\hat{f}$ of the map f, we assume that $\tilde{f}$ and $\hat{f}$ are two regular Ω-normal ε-approximations of f. Define a G-homotopy by $h(t,x):=t\hat{f}(x)+(1-t)\tilde{f}(x)$. The homotopy h is Ω-admissible.

Indeed, for all $x \in \partial\Omega$ and $t \in [0, 1]$, we have

$$\begin{aligned}|h(t,x)| &= |tf(x) + (1-t)f(x) + t(\hat{f}(x) - f(x)) + (1-t)(\tilde{f}(x) - f(x))| \\ &\geq |f(x)| - t|\hat{f}(x) - f(x)| - (1-t)|\tilde{f}(x) - f(x)| \\ &\geq \inf_{x \in \partial\Omega} |f(x)| - t\varepsilon - (1-t)\varepsilon \geq 3\varepsilon > 0\end{aligned}$$

Without loss of generality, we may assume $h : \mathbb{R} \times V \to V$. Let $a : \mathbb{R} \to [0, 1]$ be a function of class C^∞ such that

$$a(t) = \begin{cases} 0, & t \leq \frac{1}{3} \\ 1, & t > \frac{2}{3} \end{cases}$$

Define a new Ω-admissible G-homotopy h^* between $\hat{f}$ and $\tilde{f}$ by $h^*(t, x) = h(a(t), x)$, where $t \in \mathbb{R}$ and $x \in V$. By the regular normal approximation theorem, there is a regular normal in Ω G-homotopy $\tilde{h}$ such that $\sup_{(t,x)\in[0,1]\times\Omega} |h^*(t, x) - \tilde{h}(t, x)| < \varepsilon$, $\tilde{h}_t = \hat{f}$ for $t \leq \frac{1}{3}$, and $\tilde{h}_t = \tilde{f}$ for $t \geq \frac{2}{3}$. Let U be an invariant neighborhood of $[0, 1] \times \overline{\Omega}^H$ in $[0, 1] \times V^H$. Then $M := \tilde{h}_H^{-1}(0) \cap U_H$ is a one-dimensional submanifold of U_H. The submanifold M is naturally oriented (see the proof of Lemma 2.3.1). Since $\tilde{h}$ is an Ω-admissible and regular normal in Ω homotopy, the only points belonging to $M \cap \partial([0, 1] \times \Omega_H) \subset \mathbb{R} \times V^H$ are the orbits $(0, W(H)x_i)$ with $x_i \in \hat{f}^{-1}(0) \cap \Omega_H$, and $(1, W(H)x_j)$ with $x_j \in \tilde{f}^{-1}(0) \cap \Omega_H$. An oriented curve, as a connected component of M, that starts at $(0, x_i)$ or $(1, x_j)$ is either entering or exiting the region $[0, 1] \times \Omega_H$ through that point. The assumption that $\tilde{h}$ is Ω-normal prevents the components of M from leaving the region $[0, 1] \times \Omega_H$ by passing through a point in $(0, 1) \times \Omega^H$ having a smaller orbit type. Consequently, any oriented curve contained in $[0, 1] \times \Omega_H$ must be one of the following types:

(i) It contains points $(0, x_i)$ and $(1, x_j)$, and $\operatorname{sign}\det D\hat{f}_H(x_i) = \operatorname{sign}\det D\tilde{f}_H(x_j)$.

(ii) It contains points $(0, x_i)$ and $(0, x_i^*)$ with x_i, $x_i^* \in \hat{f}_H^{-1}(0) \cap \Omega_H$, where $x_i^* \not\in W(H)x_i$, and $\operatorname{sign}\det D\hat{f}_H(x_i) = -\operatorname{sign}\det D\hat{f}_H(x_i^*)$.

(iii) It contains points $(1, x_j)$ and $(1, x_j^*)$ with x_j, $x_j^* \in \tilde{f}_H^{-1}(0) \cap \Omega_H$, where $x_j^* \not\in W(H)x_j$, and $\operatorname{sign}\det D\tilde{f}_H(x_j) = -\operatorname{sign}\det D\tilde{f}_H(x_j^*)$.

By using a similar argument as in the proof of Lemma 2.3.1, we can establish $\deg(\hat{f}, \Omega_H) = \deg(\tilde{f}, \Omega_H)$.

The verification of the existence, homotopy, excision, and additivity properties is similar to that for Brouwer's degree and we leave it as an exercise. For the verification of the multiplicativity property of G-degree, we refer the interested reader to Ulrich (1988). □

The following result shows that our definition of the G-degree coincides with that of Dold–Ulrich given in Ulrich (1980).

Theorem 8.3.14

For every Ω-admissible map $f:\Omega\to V$, where Ω is an open, bounded, invariant subset of V, we have

$$G\text{-deg}(f,\Omega)=\sum_{(H)\in\Phi(G)} m_{(H)}(F)\cdot(H) \tag{8.3.1}$$

where $F(x)=x-f(x)$ for $x\in\Omega$ and

$$m_{(H)}(F)=[I(F^H)-I(F^{[H]})]/|W(H)| \tag{8.3.2}$$

where $I(F^H)$ and $I(F^{[H]})$ denote the fixed-point indices of $F^H:\Omega^H\to V^H$ and $F^{[H]}:\Omega^{[H]}\to V^{[H]}$, respectively.

Proof. By Theorem 3.6.4, we observe that the right-hand side of (8.3.1) does not depend on admissible homotopies and satisfies the additivity and excision properties. Consequently, we can assume, without loss of generality, that f is a regular normal mapping such that $f^{-1}(0)\cap\Omega\subset\Omega_{(H)}$ consists of a single orbit, where $(H)\in\Phi(G)$ is the minimal orbit type in Ω. Since $f^H:\Omega^H\to V^H$ is $W(H)$-equivariant and $W(H)$ is finite, $(f^H)^{-1}(0)=W(H)x_0$ is also a finite set. Moreover, since f^H is $W(H)$-equivariant, we have $f^H(gx_0)=gf^H(x_0)$ for every $g\in W(H)$ and, therefore, $Df^H(gx_0)\circ g=g\circ Df^H(x_0)$. Consequently, $\det Df^H(gx_0)=\det Df^H(x_0)$ for all $g\in W(H)$. This implies that $I(F^H)=n_{(H)}|W(H)|$, where $n_{(H)}$ is the (H)-component of G-deg(f,Ω). Consequently, since $\Omega^{[H]}=\varnothing$, we have $m_{(H)}=n_{(H)}$. Assume now that $H\subsetneq K$. Since $f=\text{Id}-F$ is (H)-normal, it follows from the product property of the fixed-point index that $I(F^K)=I(F^H)$. In order to compute $I(F^{[K]})$, we remark that $\Omega_H\subset\Omega^{[K]}$. If $\Omega^{[K]}\backslash\Omega^H=\varnothing$, then clearly $m_{(K)}=0$ and, thus, $n_{(H)}=m_{(H)}=0$. Therefore, assume that $\overline{\Omega^{[K]}\backslash\Omega^H}\neq\varnothing$. Since f is (H)-normal, $F^K(u,v)=(\varphi(u),0)$, where $(u,v)\in\overline{\Omega^K}$, $u\in\Omega^H$, v is orthogonal to Ω^H, and $\varphi:\overline{\Omega^H}\to V^H$. Let $r:\overline{\Omega^K}\to\overline{\Omega^{[K]}}$ denote a retraction onto $\overline{\Omega^{[K]}}$. Then by the definition of the fixed-point index for ANR-spaces, $I(F^{[K]}):=I(\tilde F)$, where $\tilde F=F\circ r:\overline{\Omega^K}\to V^K$. We claim that $H_t:=tF^K+(1-t)\tilde F$ is an admissible homotopy (without fixed points on the boundary of Ω^K). Indeed, if $H_t(u,v)=(u,v)$, then it implies that $v=0$ and, thus, $(u,v)\in\overline{\Omega^H}$. Since r restricted to $\overline{\Omega^H}$ acts as an identity, $\varphi(u)=u$, the homotopy H_t has the same fixed points as F^H. Consequently, by the homotopy property of the fixed-point index, we obtain $I(F^{[K]})=I(F^K)=I(F^H)$. Thus, $m_{[K]}=0$ and the statement follows. □

In order to extend the definition of the equivariant degree of Dold–

Ulrich to infinite-dimensional spaces, we consider an isometric Banach representation W of the Lie group G.

Lemma 8.3.15

Let ρ be an irreducible real representation of G. Define a linear mapping $P: W \to W$ by the following formula:

$$Px = n(\rho) \int_G \chi_\rho(g)\varphi(g)x \, dg \,, \qquad x \in W$$

where $n(\rho)$ denotes the intrinsic dimension of the representation ρ. Then P is an equivariant bounded projection of W and:

(i) *If $x \in W$ belongs to the representation space of an irreducible subrepresentation of W that is equivalent to ρ, then $Px = x$.*
(ii) *If $x \in W$ belongs to the representation space of an irreducible subrepresentation of W that is not equivalent to ρ, then $Px = 0$.*

Proof. The proof is identical to that of its finite-dimensional version, Proposition 6.1.24. □

The above result indicates that the image $P(W)$ is the *isotypical component* of W corresponding to ρ. That is, every irreducible subrepresentation of W with the representation space in $P(W)$ is equivalent to ρ and every irreducible subrepresentation of W with the representation space in $(\mathrm{Id} - P)(W)$ is not equivalent to ρ. Moreover, it follows directly from the definition of the isotypical component $P(W)$ that for every $x \in P(W)$, the orbit Gx is contained in a finite-dimensional subspace of $P(W)$.

By the completeness theorem of Peter–Weyl (Theorem 6.1.29), the characters of irreducible, complex, nonequivalent representations of G form an orthogonal set in $L^2(G; \mathbb{C})$. Therefore, there is only a denumerable number of real irreducible representations of G. We will denote these irreducible representations by $\rho_1, \rho_2, \ldots$ and their characters by $\chi_1, \chi_2, \ldots$. For a positive integer n, we define the projection $P_n : W \to W$ by

$$P_n x = n(\rho_n) \int_G \chi_n(g)\varphi(g)x \, dg \,, \qquad x \in W$$

and put

$$W_n = P_n(W) \,, \qquad W_0 = W^G$$

$$W_\infty = \bigoplus_{n=0}^{\infty} W_n$$

The following result is an immediate consequence of Lemma 8.3.15.

Lemma 8.3.16

The subspaces W_n are the isotypical components of W such that:

(i) *The subspace $W_\propto = \bigoplus_{n=0}^{\infty} W_n$ is dense in W.*

(iii) *If $A : W \to W$ is an equivariant linear mapping, then $A(W_n) \subseteq W_n$ for $n = 0, 1, 2, \ldots$.*

Let W_{fin} denote the set of all points $x \in W$ such that the orbit Gx is contained in a finite-dimensional invariant subspace of W. Since $W_\propto \subset W_{\text{fin}}$, it follows from Lemma 8.3.16 that W_{fin} is dense in W.

We are now in a position to define the G-degree for compact fields by applying the standard method of finite-dimensional approximations. The following result plays a crucial role in such a construction.

Theorem 8.3.17

Let X be a G-space and $F : X \to W$ a G-equivariant compact mapping. Then for any $\varepsilon > 0$, there exists an equivariant finite-dimensional map $F_\varepsilon : X \to W$ such that

$$\|F_\varepsilon(x) - F(x)\| < \varepsilon \qquad \textit{for all } x \in X$$

Proof. The proof is the same as that for the case $G = S^1$ (Theorem 6.4.1) and, therefore, is omitted. □

Definition 8.3.18 Assume that Ω is an open, bounded, invariant subset of W. An equivariant continuous map $f : W \to W$ is *Ω-admissible* if:

(i) $f(x) \neq 0$ for all $x \in \partial\Omega$.

(ii) $F := \text{Id} - f : W \to W$ is a compact mapping, where $\text{Id} : W \to W$ denotes the identity operator on W.

For every given Ω-admissible $f : W \to W$, we can find an equivariant finite-dimensional map $F_\varepsilon : W \to W$ such that

$$\|F_\varepsilon(x) - F(x)\| < \inf\{\|f(y)\|;\ y \in \partial\Omega\}$$

and

$$F_\varepsilon(W) \subseteq \hat{W} \subseteq W\,, \qquad \text{where } \hat{W} \text{ is a } G\text{-invariant subspace of } W \text{ with } \dim \hat{W} < \infty$$

We now define

$$G\text{-deg}(f, \Omega) := G\text{-deg}((\mathrm{Id} - F_\varepsilon)|_{\hat{W}}, \Omega \cap \hat{W})$$

By applying standard arguments, one can verify that the above definition of G-deg(f, Ω) does not depend on the choice of the equivariant approximation F_ε as well as the invariant subspace $\hat{W}$, and that the defined G-deg(f, Ω) possesses all standard properties of the equivariant degree listed in Theorem 8.3.13 for finite-dimensional mappings.

We can apply the equivariant bijection theorem (Theorem 6.4.3) to extend the notion of the G-degree to the class of condensing equivariant fields, exactly in the same way as the S^1-degree was extended to the class of condensing fields in Section 6.4.

Let Ω be an invariant, open, bounded set in W and let $F : \overline{\Omega} \to W$ be a condensing G-map such that $F \in \mathscr{C}^G(\overline{\Omega}, \partial\Omega)$. By Theorem 6.4.4, there exists $F_1 \in \mathscr{K}^G(\overline{\Omega}, \partial\Omega)$ such that $F \sim F_1$ in $\mathscr{C}^G(\overline{\Omega}, \partial\Omega)$. Moreover, if $F_2 \in \mathscr{K}^G(\overline{\Omega}, \partial\Omega)$ is another compact G-map such that $F \sim F_2$ in $\mathscr{C}^G(\overline{\Omega}, \partial\Omega)$, then $F_1 \sim F_2$ in $\mathscr{K}^G(\overline{\Omega}, \partial\Omega)$. Therefore, by the homotopy invariance of G-degree for equivariant compact fields, G-deg$(\mathrm{Id} - F_1, \Omega) = G$-deg$(\mathrm{Id} - F_2, \Omega)$. This justifies the following definition.

Definition 8.3.19 Let $F \in \mathscr{C}^G(\overline{\Omega}, \partial\Omega)$. Then the *G-degree* of the condensing field $f = \mathrm{Id} - F$ with respect to Ω is defined by the formula

$$G\text{-deg}(\mathrm{Id} - F, \Omega) := G\text{-deg}(\mathrm{Id} - F_1, \Omega)$$

where $F_1 \in \mathscr{K}^G(\overline{\Omega}, \partial\Omega)$ is a compact G-map such that $F \sim F_1$ in $\mathscr{C}^G(\overline{\Omega}, \partial\Omega)$.

It can be verified that the G-degree for condensing fields has the standard properties (P1–P5) in Theorem 8.3.13.

EXERCISES

8.3.1 Let $V = \mathbb{R}^3$ be the representation of the group $G = \mathbb{Z}_2 \oplus \mathbb{Z}_2$, where the action of G on V is given by the formula $(\gamma_1, \gamma_2) \cdot (x, y, z) := (\gamma_1 x, \gamma_2 y, \gamma_1\gamma_2 z)$ for $(\gamma_1, \gamma_2) \in \mathbb{Z}_2 \oplus \mathbb{Z}_2$ and $(x, y, z) \in \mathbb{R}^3 = V$. Let $\Omega = \{v \in \mathbb{R}^3;\ |v| < 1\}$ and $f : \overline{\Omega} \to V$ be given by $f(v) = -v$, where $v \in \overline{\Omega}$. Find a regular normal in Ω Ω-admissible approximation of f.

8.3.2 Let $V = \mathbb{C}$ be the (real) representation of $G = D_n$, where the action of the group G is defined by $r_\varphi \cdot z = e^{ki\varphi} z$, $\kappa \cdot z = \bar{z}$, where $z \in \mathbb{C} = V$, and the elements r_φ and κ were as defined in Example 8.2.2. Let Ω denote the unit ball in V. Find a regular normal in Ω Ω-admissible approximation of $f = -\mathrm{Id} : V \to V$.

8.4 EXAMPLES OF COMPUTATIONS OF G-DEGREE

In this section we illustrate how to compute G-degree for some equivariant maps.

Example 8.4.1 In this example, we compute the degree of the operator $-\mathrm{Id}$ on the unit ball relative to the following representation V of the group $G = \mathbb{Z}_2 \oplus \mathbb{Z}_2$, where $\mathbb{Z}_2 = \{1, -1\} \subset \mathbb{C}$. We assume $V = \mathbb{R}^3$ and define the action of G on V as follows:

$$(\gamma_1, \gamma_2) \cdot (x, y, z) := (\gamma_1 x, \gamma_2 y, \gamma_1\gamma_2 z), \qquad (\gamma_1, \gamma_2) \in \mathbb{Z}_2 \oplus \mathbb{Z}_2, \quad (x, y, z) \in \mathbb{R}^3$$

We put $\Omega = \{v \in V; |v| < 1\}$ and let $f : \overline{\Omega} \to V$ be given by $f(v) = -v$, $v \in \overline{\Omega}$. It is clear that $f^{-1}(0) = \{0\}$.

As discussed in Example 8.2.1, the generators of $A(G)$ are (G), $(\mathbb{Z}_2 \oplus 1) = (\{(1,1), (-1,1)\})$, $(1 \oplus \mathbb{Z}_2) = (\{1,1), (1,-1)\})$, $(\Delta) = (\{(1,1), (-1,-1)\})$ and $(\mathbb{1}) = (\{1,1\})$, that is, $\Phi_0(G) = \{(\mathbb{Z}_2 \oplus 1), (1 \oplus \mathbb{Z}_2, (\Delta), (\mathbb{1})\}$. As an "almost" regular normal approximation, we take the map $g : \Omega \to V$ given by

$$g(x, y, z) = (-x(x - \tfrac{1}{2})(x + \tfrac{1}{2}), -y(y - \tfrac{1}{2})(y + \tfrac{1}{2}), -z(z - \tfrac{1}{2})(z + \tfrac{1}{2}))$$

where $(x, y, z) \in \Omega$. The G-map g is not exactly a regular normal G-map in the sense of Definition 8.3.9. However, the only violation of the normality property is that g, instead of acting in normal directions as identity, acts as $c \cdot \mathrm{Id}$ with $c > 0$.

We have the following orbits of zeros in the set $g^{-1}(0)$:

Orbit of Zeros Gv	**Orbit Type (G_v)**	**sign $Dg(v)\vert_{S_v}$**
$\{(0,0,0)\}$	(G)	$+1$
$\{(0,-\frac{1}{2},0), (0,\frac{1}{2},0)\}$	$(\mathbb{Z}_2 \oplus 1)$	-1
$\{(-\frac{1}{2},0,0), (\frac{1}{2},0,0)\}$	$(1 \oplus \mathbb{Z}_2)$	-1
$\{(0,0,-\frac{1}{2}), (0,0,\frac{1}{2})\}$	(Δ)	-1
$\{(-\frac{1}{2},-\frac{1}{2},0), (\frac{1}{2},-\frac{1}{2},0), (-\frac{1}{2},\frac{1}{2},0), (\frac{1}{2},\frac{1}{2},0)\}$	$(\mathbb{1})$	$+1$
$\{(-\frac{1}{2},0,-\frac{1}{2}), (\frac{1}{2},0,-\frac{1}{2}), (-\frac{1}{2},0,\frac{1}{2}), (\frac{1}{2},0,\frac{1}{2})\}$	$(\mathbb{1})$	$+1$
$\{(0,-\frac{1}{2},-\frac{1}{2}), (0,\frac{1}{2},-\frac{1}{2}), (0,-\frac{1}{2},\frac{1}{2}), (0,\frac{1}{2},\frac{1}{2})\}$	$(\mathbb{1})$	$+1$
$\{(\frac{1}{2},\frac{1}{2},\frac{1}{2}), (-\frac{1}{2},\frac{1}{2},-\frac{1}{2}), (\frac{1}{2},-\frac{1}{2},-\frac{1}{2}), (-\frac{1}{2},-\frac{1}{2},\frac{1}{2})\}$	$(\mathbb{1})$	-1
$\{(-\frac{1}{2},-\frac{1}{2},-\frac{1}{2}), (\frac{1}{2},-\frac{1}{2},\frac{1}{2}), (-\frac{1}{2},\frac{1}{2},\frac{1}{2}), (\frac{1}{2},\frac{1}{2},-\frac{1}{2})\}$	$(\mathbb{1})$	-1

Consequently, we obtain

$$G\text{-deg}(f, \Omega) = (G) - (\mathbb{Z}_2 \oplus 1) - (1 \oplus \mathbb{Z}_2) - (\Delta) + (\mathbb{1})$$

This method of direct computation of the G-degree can be simplified if one uses the multiplicativity property of the G-degree. We remark that the representation space V can be decomposed into a direct sum of three invariant one-dimensional subspaces:

$$V^{(\mathbb{Z}_2\oplus 1)} := \{v \in V; (G_v) \leq (\mathbb{Z}_2 \oplus 1)\}$$
$$V^{(1\oplus\mathbb{Z}_2)} := \{v \in V; (G_v) \leq (1 \oplus \mathbb{Z}_2)\}$$
$$V^{(\Delta)} := \{v \in V; (G_v) < (\Delta)\}$$

The subspaces V^{α_i}, $i = 1, 2, 3$, are the isotypical components of the space V and we have $V = V^{(\mathbb{Z}_2\oplus 1)} \oplus V^{(1\oplus\mathbb{Z}_2)} \oplus V^{(\Delta)}$ and $f(V^{\alpha_i}) \subset V^{\alpha_i}$ for $i = 1, 2, 3$. Since the mapping f can be represented as the product $f = f_1 \times f_2 \times f_3$, where $f_i = f_{|V^{\alpha_i}}$, $i = 1, 2, 3$, it follows from the multiplicativity property that

$$G\text{-deg}(f, \Omega) = G\text{-deg}(f_1, \Omega_1) \cdot G\text{-deg}(f_2, \Omega_2) \cdot G\text{-deg}(f_3, \Omega_3)$$

where $\Omega_i = \Omega \cap V^{\alpha_i}$, $i = 1, 2, 3$. Note that for every $i = 1, 2, 3$, $f_i(t) = -t$. By using an almost regular normal approximation $g_i(t) = -t(t - \frac{1}{2})(t + \frac{1}{2})$, we obtain $G\text{-deg}(f_i, \Omega_i) = G\text{-deg}(g_i, \Omega_i) = (G) - \alpha_i$, $i = 1, 2, 3$. Consequently, by using the multiplication table for the generators of $A(\mathbb{Z}_2 \oplus \mathbb{Z}_2)$ presented in Example 8.2.1, we obtain

$$\begin{aligned} G\text{-deg}(f, \Omega) &= ((G) - (\mathbb{Z}_2 \oplus 1))((G) - (1 \oplus \mathbb{Z}_2))((G) - (\Delta)) \\ &= ((G) - (\mathbb{Z}_2 \oplus 1) - (1 \oplus \mathbb{Z}_2) + (\mathbb{1}))((G) - (\Delta)) \\ &= (G) - (\mathbb{Z}_2 \oplus 1) - (1 \oplus \mathbb{Z}_2) - (\Delta) + (\mathbb{1}) . \end{aligned}$$

In the second example, we show that the G-degree is indeed a more sensitive topological invariant than the usual (nonequivariant) Brouwer degree. More precisely, we present a homotopically trivial mapping on a unit ball in $\mathbb{R}^3$ that satisfies a symmetry property with respect to a $\mathbb{Z}_2$ action, but its $\mathbb{Z}_2$-degree is nonzero.

Example 8.4.2 Let $V = \mathbb{C} \times \mathbb{R}$ be the representation of $\mathbb{Z}_2$, on which the group $G = \mathbb{Z}_2$ acts by $\gamma \cdot (z, t) := (z, \gamma t)$, where $(z, t) \in \mathbb{C} \times \mathbb{R}$, $\gamma \in \mathbb{Z}_2$. We define a mapping $f: V \to V$ by $f(z, t) = (z^2 - \frac{1}{4}, \operatorname{Re} zt)$, where $(z, t) \in \mathbb{C} \oplus \mathbb{R} = V$. It is clear that the mapping f is G-equivariant.

First of all, we compute the Brouwer degree $\deg(f, \Omega)$, where Ω denotes the unit ball in V. Since zero is a regular value of f, and $f^{-1}(0) = \{(-\frac{1}{2}, 0), (\frac{1}{2}, 0)\} \subset \mathbb{C} \oplus \mathbb{R}$, we obtain by direct computation, $\det Df(\pm\frac{1}{2}, 0) = \pm\frac{1}{2}$. Therefore, $\deg(f, \Omega) = 0$.

Now, we compute the G-degree of the mapping f in Ω. For this purpose, we consider two open sets: $\Omega_- := \{(z, t) \in V; |z + \frac{1}{2}| < \frac{1}{2}, |t| < 1\}$ and $\Omega_+ := \{(z, t) \in V; |z - \frac{1}{2}| < \frac{1}{2}, |t| < 1\}$. It follows from the excision property that

$$\mathbb{Z}_2\text{-deg}(f,\Omega)=\mathbb{Z}_2\text{-deg}(f,\Omega_-)+\mathbb{Z}_2\text{-deg}(f,\Omega_+)$$

Since both zeros of f belong to V^G, in order to deform the mapping f into a normal mapping, we need to modify the mapping f so that the modified version transforms a vector (z,t), where t is assumed to be sufficiently small, into a vector whose component orthogonal to V^G (i.e., component in $\{0\}\times\mathbb{R}\subseteq V$) is exactly $(0,t)$. As remarked before, this requirement on the modified version of f can be weakened. That is, we can consider a mapping that is "almost" normal in the sense of Remark 8.3.4. More precisely, as the mapping f restricted to Ω_+ transforms the vectors (z,t) onto vectors $(z^2-\frac{1}{4}, \operatorname{Re} zt)$, where $\operatorname{Re} z>0$, it is an almost regular normal mapping. This implies $\mathbb{Z}_2\text{-deg}(f,\Omega_+)=n\cdot(\mathbb{Z}_2)$, where n is simply $\operatorname{sign}\det Df_+(\frac{1}{2})=1$, $f_+:=f_{|\mathbb{C}}$ is the restriction of the mapping f to the subspace V^G, that is, $\mathbb{Z}_2\text{-deg}(f,\Omega_+)=(\mathbb{Z}_2)$. On the other hand, the mapping f restricted to the set Ω_- is not almost normal. We define a new mapping $\hat{f}:\overline{\Omega_-}\to V$ by the formula $\hat{f}(z,t)=(z^2-\frac{1}{4}, \operatorname{Re} zt(t^2-\frac{1}{16}))$. This mapping has two orbits of zeros in $\Omega_-:(-\frac{1}{2},0)$ $(-\frac{1}{2},\pm\frac{1}{4})$. It is easy to see that $\hat{f}$ satisfies the requirement for an almost regular normal mapping. Consequently, $\mathbb{Z}_2\text{-deg}(f,\Omega_-)=\mathbb{Z}_2\text{-deg}(\hat{f},\Omega_-)=n_1\cdot(\mathbb{Z}_2)+n_2\cdot(\mathbb{Z}_1)$, where n_1 is sign det $D\hat{f}_-(-\frac{1}{2})=1$, $\hat{f}_-$ denotes the restriction of $\hat{f}$ to the subspace V^G, and $n_2=\operatorname{sign}\det D\hat{f}(-\frac{1}{2},\frac{1}{4})=-1$. Therefore, we obtain $\mathbb{Z}_2\text{-deg}(f,\Omega)=2(\mathbb{Z}_2)-(\mathbb{Z}_1)$.

Example 8.4.3 Let us consider the following two-dimensional real representation V of the group $O(2)$ defined by

$$r_\varphi\cdot z=e^{ki\varphi}z\,,\qquad \kappa\cdot z=\bar{z}$$

where $z\in\mathbb{C}=V$, and the elements r_φ and κ were defined in Example 8.2.9. Let $f:V\to V$ denote the operator $-\mathrm{Id}$, which is evidently $O(2)$-equivariant. We wish to compute $O(2)\text{-deg}(f,\Omega)$, where Ω denotes the unit ball in the space V. It is clear that f is not Ω-normal. Indeed, an element z of the representation V may have only one of two orbit types, namely, $(O(2))$ if $z=0$, or (D_k) if $z\neq 0$. Since $(O(2))$ is the minimal orbit type in V, and the representation V is absolutely irreducible, $-\mathrm{Id}$ cannot be connected by a path in $GL^G(V)$ to the identity operator Id. Consequently, in order to compute $O(2)\text{-deg}(f,\Omega)$, we need to correct it to a normal in Ω map. Let $\gamma:\mathbb{R}\to\mathbb{R}$ be the function

$$\gamma(t)=\begin{cases}1\,, & t\leq\frac{1}{3}\\ -3t+2\,, & \frac{1}{3}<t\leq\frac{2}{3}\\ 0\,, & \frac{2}{3}\leq t\end{cases}$$

Then the mapping $g(z) := (2\gamma(|z|) - 1)z$, where $z \in V$, is a normal in Ω $O(2)$-map, homotopic to f. The set $g^{-1}(0)$ consists of two $O(2)$-orbits: $\{0\}$ and $\{z; |z| = \frac{1}{2}\}$, of the orbit types $(O(2))$ and (D_k), respectively. Clearly, $O(2)\text{-deg}(f, \Omega) = O(2)\text{-deg}(g, \Omega) = n_1 \cdot (O(2)) + n_2 \cdot (D_k)$, where n_1 is evidently equal to 1. In order to compute the integer n_2, we consider the subspace V^H, where $H = D_k$, which is the real axis in $V = \mathbb{C}$, and the orbit $W(H) = \{\pm \frac{1}{2}\}$, where $W(H) = \mathbb{Z}_2$. We consider a slice S to the orbit $W(H)\frac{1}{2}$ at the point $z = \frac{1}{2}$, for example, $S := \{(t, 0) \in \mathbb{C};\ \frac{1}{3} < t < \frac{2}{3}\}$. Consequently, $n_2 = \text{sign} \frac{d}{dt} g^H(\frac{1}{2}) = \text{sign}\, 2\gamma'(\frac{1}{2}) = \text{sing}(-6) = -1$. Therefore, $O(2)\text{-deg}(f, \Omega) = (O(2)) - (D_k)$.

Example 8.4.4 It is well known that any nontrivial, real, irreducible representation of the group $O(2)$ is a representation of the types described in Example 8.4.3, or it is a one-dimensional representation $V = \mathbb{R}$ on which the action of the group $O(2)$ is defined by $r_\varphi \cdot t = t$, $\kappa \cdot t = -t$. The representation V contains only two orbit types: $(O(2))$ that is the orbit type of $t = 0$, and $SO(2)$ that is the orbit type of elements $t \neq 0$. Suppose that $f: V \to V$ is given by the formula $f(t) = -t$. We shall compute $O(2)$-deg(f, Ω), where Ω denotes the interval $(-1, 1)$ in V. By using the same function γ as in Example 8.4.3, we correct the map f to a normal map $g(t) = (2\gamma(|t|) - 1)t$, and obtain $O(2)\text{-deg}(f, \Omega) = O(2)\text{-deg}(g, \Omega) = (O(2)) - (SO(2))$.

Example 8.4.5 Let V denote an arbitrary, finite-dimensional, real representation of the group $O(2)$, and let $A: V \to V$ be an arbitrary $O(2)$-equivariant isomorphism of the space V. We will compute $O(2)$-deg(A, Ω), where Ω denotes the unit ball in V.

The space V can be decomposed into a direct sum of its isotypical components $V = V^G \oplus V_0 \oplus V_1 \oplus \cdots \oplus V_n$, where V_0 is a direct sum of all irreducible one-dimensional subrepresentations of V that are isomorphic to the representation of $O(2)$ described in Example 8.4.4, and V_k, with $k \geq 1$, is the direct sum of all irreducible subrepresentations of V that are isomorphic to the representation of $O(2)$ described in Example 8.4.3. The isomorphism A preserves the isotypical components, that is, $A(V_i) = V_i$. Since all nontrivial, real, irreducible representations of $O(2)$ are absolutely irreducible, the operator $A_i := A_{|V_i}$ can be represented by a real $n_i \times n_i$ matrix, where n_i denotes the number of irreducible components of V_i, in the following way: We assume that all the irreducible components of V_i are identified with one irreducible representation $\tilde{V}_i$ of $O(2)$ and, therefore, the operator A_i, which maps the sth irreducible component of V_i onto the rth irreducible component of V_i, is simply $a^i_{rs} \cdot \text{Id}$ and, consequently, we can represent the operator A_i by the real matrix $[a^i_{rs}]$. As a consequence of this representation, we can identify $GL^G(V_i)$, where $G = O(2)$, with $GL(n_i, \mathbb{R})$. As $GL(n_i, \mathbb{R})$ has two open, connected components $GL^-(n_i, \mathbb{R})$ and

$GL^+(n_i, \mathbb{R})$, we can connect the operator A_i by a homotopy with the identity operator $\mathrm{Id}_{|V_i}$ or with the operator J_i having the following matrix representation:

$$J_i = \begin{bmatrix} -1 & 0 & \cdots & 0 \\ 0 & 1 & \cdots & 0 \\ \vdots & \vdots & \ddots & \vdots \\ 0 & 0 & \cdots & 1 \end{bmatrix}$$

Assume that indeed A_i can be connected by a path with the operator J_i; then $O(2)$-deg(A_i, Ω_i), where Ω_i denotes the unit ball in V_i, is equal to the degree $O(2)$-deg(J_i, Ω_i). On the other hand, by the multiplicativity property, $O(2)$-deg$(J_i, \Omega_i) = O(2)$-deg$(-\mathrm{Id}, \tilde{\Omega}_i)$, where Ω_i denotes the unit ball in the irreducible representation $\tilde{V}_i$. Finally, by the results obtained in Examples 8.4.3 and 8.4.4, $O(2)$-deg$(-\mathrm{Id}, \tilde{\Omega}_i) = (O(2)) - (D_i)$ if $i > 0$, and $O(2)$-deg$(-\mathrm{Id}, \tilde{\Omega}_i) = (O(2)) - (SO(2))$ if $i = 0$.

In order to give a complete formula for the degree $O(2)$-deg(A, Ω), we need to introduce some notations. We put

$$\eta_i(A) = \begin{cases} -1 & \text{if } \det[a^i_{rs}] < 0 \\ 0 & \text{if } \det[a^i_{rs}] > 0 \end{cases}$$

where $[a^i_{rs}]$ is the matrix representation of A_i. We also set $\varepsilon(A) = \mathrm{sign} \det A^G$, where $A^G = A|_{V^G}$. Then, it follows from the multiplicativity property that

$$O(2)\text{-deg}(A, \Omega) = \varepsilon(A)((O(2)) + \eta_0(A)(SO(2))) \prod_{i=1}^{n} ((O(2)) + \eta_i(A)(D_i))$$

where the product is taken in the Burnside ring $A(O(2))$.

As an immediate application of this formula, we have the following regular value formula for $O(2)$-deg(f, Ω), where $f: V \to V$ is a C^1-differentiable Ω-admissible G-map such that zero is a regular value of f.

Theorem 8.4.6

Let V be a real finite-dimensional representation of the group $G = O(2)$, Ω be a bounded, invariant, open subset of V, and $f: V \to V$ be a C^1-differentiable Ω-admissible G-map such that zero is a regular value of $f_{|\Omega}$, and $f^{-1}(0) \cap \Omega := \{x_1, x_2, \ldots, x_N\} \subset \Omega^G$. Then

$$O(2)\text{-deg}(f, \Omega) =$$

$$\sum_{j=1}^{N} \varepsilon(Df(x_j))((O(2)) + \eta_0(Df(x_j)(SO(2))) \prod_{i=1}^{n} ((O(2)) + \eta_i(Df(x_j)(D_i))$$

To present the next example, we recall a classical result that $SO(3)$ has precisely one real irreducible representation, up to isomorphism, in each odd dimension $2k+1$. These irreducible representations can be described with the help of vector spaces W_k of homogeneous polynomials $p : \mathbb{R}^3 \to \mathbb{R}$ of degree k. The group $SO(3)$ acts on W_k by $(Ap)(x) := p(A^{-1}x)$, where $A \in SO(3)$. For $k = 0$, the representation W_0 is a one-dimensional trivial representation of $SO(3)$. For $k = 1$, the representation W_1 is the natural representation of $SO(3)$ in $\mathbb{R}^3$. However, for $k \geq 2$, the representation W_k is not irreducible. In order to describe the irreducible representations of $SO(3)$, we denote by ρ the polynomial $\rho(x) = x_1^2 + x_2^2 + x_3^2$, where $x = (x_1, x_2, x_3)$, and we define a linear (injective operator) $j_k : W_{k-2} \to W_k$, $k > 2$, by $j_k(p) = p\rho$, where $p \in W_{k-2}$. Put $J_k = j_k(W_{k-2})$ and let V_k be the space spanned by the spherical harmonics of degree k, that is, the elements $p \in W_k$ for which $\Delta p = 0$, where Δ denotes the Laplacian operator. The subspace V_k is also invariant under $SO(3)$. We have the following classical result.

Theorem 8.4.7

The representations V_k are precisely the irreducible representations of $SO(3)$ such that:

(i) $W_k = J_k \oplus V_k$, *that is,* V_k *is an* $SO(3)$*-invariant complement of* J_k.
(ii) $\dim V_k = 2k+1$.
(iii) *The representations* V_k *are absolutely irreducible.*

Proof. See Bröcker and tom Dieck (1984) and Golubitsky et al. (1988). □

In addition, we will need the following result whose proof can also be found in Golubitsky et al. (1988).

Theorem 8.4.8

The dimensions $d(H) := \dim V_k^H$ *of the fixed-point subspaces* V_k^H *in the representation space* V_k *are:*

(i) $d(\mathbb{Z}_m) = 2[k/m] + 1$ *for* $m \geq 1$.
(ii) $d(D_m) = [k/m]$ *if* k *is odd, and* $d(D_m) = [k/m] + 1$ *for even* k.
(iii) $d(SO(2)) = 1$.
(iv) $d(O(2)) = 0$ *for odd* k *and* $d(O(2)) = 1$ *for even* k.
(v) $d(A_4) = 2[k/3] + [k/2] - k + 1$.
(vi) $d(S_4) = [k/4] + [k/3] + [k/2] - k + 1$.
(vii) $d(A_5) = [k/5] + [k/3] + [k/2] - k + 1$.

Example 8.4.9 In this example, we consider the operator $-\mathrm{Id}$ on the representations V_1, V_2, V_3, V_4, and V_5 described above. We want to compute $SO(3)$-deg$(-\mathrm{Id}, \Omega_i)$, where Ω_i denotes the unit ball in V_i, $i = 1, 2, 3, 4, 5$. We will need the following lattice of the isotropy groups in V_i, where each isotropy group H is written in the form $H^{(\dim V_i^H)}$:

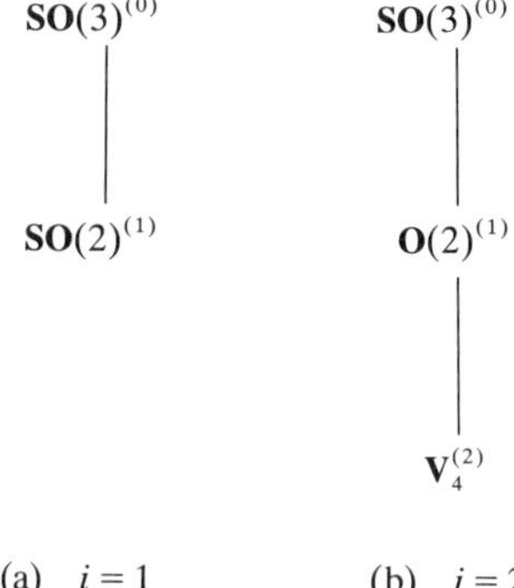

(a) $i = 1$ (b) $i = 2$

In order to compute the isotropy lattices for the representations V_i, we apply Theorem 8.4.8 to compute the dimensions of the fixed-point spaces V_i^H and we recognize the isotropy groups by comparing these dimensions.

We now compute $SO(3)$-deg$(-\mathrm{Id}, \Omega_1)$. Using the function $\gamma(t)$ given by

$$\gamma(t) = \begin{cases} 1 & \text{if } t \leq \frac{1}{3} \\ -3t + 2 & \text{if } \frac{1}{3} \leq t \leq \frac{2}{3} \\ 0 & \text{if } \frac{2}{3} \leq t \end{cases}$$

we correct the map $f = -\mathrm{Id}$ into a normal map $g(x) = (2\gamma(|x|) - 1)x$. The restriction of g to the one-dimensional subspace V^H, where $H = SO(2)$, has exactly one $W(H) = \mathbb{Z}_2$-orbit of critical points. Therefore, by direct computation we obtain $SO(3)$-deg$(-\mathrm{Id}, \Omega_1) = SO(3)$-deg$(g, \Omega_1) = (SO(3)) - (SO(2))$.

In a similar way, we can compute degree $SO(3)$-deg$(-\mathrm{Id}, \Omega_2)$. Again, we correct the map $-\mathrm{Id}$ to a normal map $g(x) = (2\gamma(|x|) - 1)x$. The space V^H, where $H = O(2)$, is one-dimensional, and since $W(O(2)) = \mathbb{Z}_1$, the mapping g^H has exactly two critical points (more precisely, two orbits of critical points). In the case $H = \mathbf{V_4}$, we have $\dim V^H = 2$, $F(x) = 2x$. Therefore, in this case, $I(F^H) = 1$. In order to determine the set $\Omega^{[H]}$, we need to find all the subgroups $K \in (O(2))$ such that $\mathbf{V_4} \subset K$. It is easy to verify that for the group

$$\mathbf{V_4} = \left\{ \begin{bmatrix} 1 & 0 & 0 \\ 0 & 1 & 0 \\ 0 & 0 & 1 \end{bmatrix}, \begin{bmatrix} -1 & 0 & 0 \\ 0 & 1 & 0 \\ 0 & 0 & -1 \end{bmatrix}, \begin{bmatrix} -1 & 0 & 0 \\ 0 & -1 & 0 \\ 0 & 0 & 1 \end{bmatrix}, \begin{bmatrix} 1 & 0 & 0 \\ 0 & -1 & 0 \\ 0 & 0 & -1 \end{bmatrix} \right\}$$

there are exactly three subgroups K_1, K_2, $K_3 \in (O(2))$ such that $\mathbf{V_4} \subset K_i$,

$i = 1, 2, 3$. Thus, $I(F^{[H]}) = -5$ and $m_{(H)} = (1+5)/6 = 1$. Consequently, we obtain $SO(3)\text{-deg}(-\text{Id}, \Omega_2) = SO(3)\text{-deg}(g, \Omega_2) = (SO(3)) - 2(O(2)) + (V_4)$.

We have obtained

$$SO(3)\text{-deg}(-\text{Id}, \Omega_1) = (SO(3)) - (SO(2))$$

$$SO(3)\text{-deg}(-\text{Id}, \Omega_2) = (\text{SO}(3)) - 2(\text{O}(2)) + (V_4)$$

where the subgroups $SO(S)$ are $O(2)$ were maximal isotropy groups for V_1 and V_2, respectively.

In what follows, we will compute $SO(3)\text{-deg}(-\text{Id}, \Omega_i)$ also for $i = 3$, 4, and 5. We start with the lattice of isotropy subgroups for V_3:

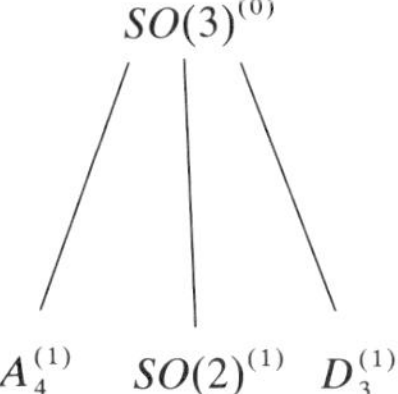

We may compute $SO(3)\text{-deg}(-\text{Id}, \Omega_3)$ in a similar way to that above, that is, we correct the map $-\text{Id}$ to a normal map $g(x) = (2\gamma(|x|) - 1)x$ and compute the degree as follows. The space V^H, for $H = A_4$, $SO(2)$, and D_3 is, in this case, one-dimensional, and since $W(H) = \mathbb{Z}_2$, we know that g^H has exactly two critical points (i.e., one orbit of critical points). Consequently, $SO(3)\text{-Deg}(-\text{Id}, \Omega_3) = (SO(3)) - (A_4) - (SO(2)) - (D_3)$. We can also apply formula (8.3.2) to compute this degree $SO(3)\text{-deg}(-\text{Id}, \Omega_3)$. Since the only orbit types (H) in Ω_3 that belong to $\Phi(SO(3))$ are $(SO(3))$, (A_4), $(SO(2))$, and (D_3), we have

$$SO(3)\text{-deg}(-\text{Id}, \Omega_3) = m_0(SO(3)) + m_1(A_4) + m_2(SO(2)) + m_3(D_3)$$

where for $F(x) = 2x$, we have

$$m_0 = I(F^H), \qquad H = SO(3)$$

$$m_i = [I(F^H) - I(F^{[H]})]/|W(H)|, \qquad i = 1, 2, 3$$

where $H = A_4$, $SO(2)$, and D_3, respectively. For $H = SO(3)$, we have $F^H : \{0\} \to \{0\}$; thus, $I(F^H) = 1$ and so $m_0 = 1$. For $H = A_4$, $SO(2)$, or D_3, we have $F^H : \Omega^H \to V^H$, and since V^H is one dimensional, $I(F^H) = -1$. On the other hand, $\Omega^{[H]} = \{0\}$; thus, $F^{[H]} : \{0\} \to \{0\}$ and $I(F^{[H]}) = 1$. Consequently, $m_i = (-1-1)/|W(H)| = -2/2 = -1$. So, again we obtain

$$SO(3)\text{-deg}(-\text{Id}, \Omega_3) = (SO(3)) - (A_4) - (SO(2)) - (D_3)$$

The reduced lattice of isotropy groups for the representation V_4 is next illustrated:

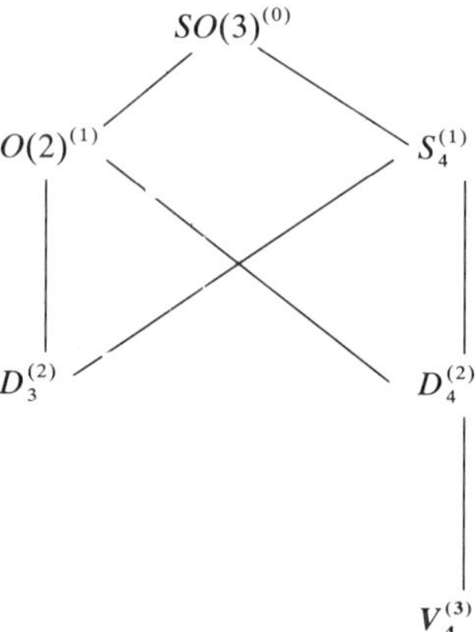

In order to compute $SO(3)$-deg$(-\mathrm{Id}, \Omega_4)$, we apply formulas (8.3.1) and (8.3.2). In this case,

$$SO(3)\text{-deg}(-\mathrm{Id}, \Omega_4) = \sum_{(H)} m_{(H)} \cdot (H)$$

where $(H) \in \{(SO(3)), (O(2)), (S_4), (D_3), (D_4), (V_4)\}$. For $H = SO(3)$, we obtain trivially $m_{(H)} = 1$. For $H = O(2)$ or S_4, since $\dim V^H = 1$, we obtain $I(F^H) - I(F^{[H]}) = -1 - 1 = -2$. In this case, $W(H) = \mathbb{Z}_1$. Thus, $m_{(H)} = -2$. Now we have to compute $m_{(H)}$ for $H = D_3$ or D_4. In this case, $\dim V^H = 2$, and since $F(x) = 2x$, we obtain $I(F^H) = 1$. Now we need to determine the set $\Omega^{[H]}$. First, we need to find all the subgroups K such that $H \subset K$ and K is conjugated to $O(2)$ (respectively, to S_4). We will use here some well-known facts about the maximal torus of a compact Lie group. Let us recall that a subgroup Γ of a Lie group G is called a *torus* if it is isomorphic to a product group $S^1 \times \cdots \times S^1$. A maximal subgroup of G isomorphic to some torus is called a *maximal torus*. Any two maximal tori are conjugate, and if G is a connected compact Lie group, then any element is contained in at least one maximal torus. In $SO(3)$, each nontrivial element is contained in precisely one maximal torus isomorphic to $SO(2)$. This property may be seen from the fact that such an element has a unique eigenvector with eigenvalue $+1$. The rotations about this eigenvector form the maximal torus. Clearly, the maximal torus also contains all powers of this element. In the case of the group D_3 or D_4, it is clear that there is only one maximal torus, namely, $SO(2)$ that contains $\mathbb{Z}_3$ and $\mathbb{Z}_4$. Consequently, if $K \in (O(2))$ and D_3 or D_4 is contained in K, then $K = O(2)$. In order to determine all the subgroups $K \in (S_4)$, which in fact are symmetry groups of an octahedron (or cube), such that K contains D_3 and D_4, we notice that D_3 or D_5 completely determines the position of the octahedron, and thus there is only one such group K, namely, S_4. This implies that

$$\Omega^{[H]} = \Omega^H \cap [V^{O(2)} \cup V^{S_4}]$$

We can identify $\Omega^{[H]}$ with the subset of a plane composed of two lines transversally intersecting at the origin. We have $F(x) = 2x$, and since $\Omega^{]H]}$ is an ENR, we can compute that $I(F^{[H]}) = -3$. Consequently, for $H = D_3$ or D_4, $m_{(H)} = (1+3)/|W(H)| = 4/2 = 2$.

In the case where $H = \mathbf{V_4}$, we have $\dim V^H = 3$, $F(x) = 2x$. Therefore, in this case, $I(F^H) = -1$. In order to determine the set $\Omega^{[H]}$, we need to find all the subgroups $K \in (D_4)$ such that $\mathbf{V_4} \subset K$. Since the rotations of K belong to a unique maximal torus, it is easy to verify that for the group $\mathbf{V_4}$, there are exactly three different planes of rotations, which implies that there are exactly three subgroups K_1, K_2, $K_3 \in (D_4)$ such that $V_4 \subset K_i$, $i = 1, 2, 3$. It may be verified that $V^{K_1} \cap V^{K_2} \cap V^{K_3} = V^{S_4}$. Indeed, the group S_4 of the orientation-preserving symmetries of the octahedron (the cube) contains three subgroups of symmetries of the parallel faces in the octahedron, which are exactly K_1, K_2, and K_3. On the other hand, it is clear from the lattice of subgroups of $SO(3)$ that any two of the subgroups K_1, K_2, and K_3 generate S_4. Therefore, $V^{K_i} \cap V^{K_j} = V^{S_4}$ for $i \neq j, i, j \in \{1, 2, 3\}$. Consequently, $\Omega^{[H]}$ may be identified with the subset of $\mathbb{R}^3$ consisting of three planes intersecting along a single line (containing the origin). We need to compute $I(F^{[H]})$, where $F^{[H]} : \Omega^{[H]} \to \Omega^{[H]}$ is given by $F^{[H]}(x) = 2x$. Let X denote the subspace of $\mathbb{R}^2$ consisting of three lines intersecting at the origin. Then $\Omega^{[H]}$ may be identified with the product $X \times \mathbb{R}$ and $F^{[H]} = F_1 \times F_2$, where $F_i(x) = 2x$, $i = 1, 2$, $F_1 : X \to X$, and $F_2 : \mathbb{R} \to \mathbb{R}$. Then by the product property, $I(F^{[H]}) = I(F_1) \cdot I(F_2)$. Since $I(F_1) = -5$ and $I(F_2) = -1$, we obtain $I(F^{[H]}) = 5$. Consequently, $m_{(H)} = (-1-5)/6 = -1$. Finally,

$$SO((3)\text{-deg}(-\text{Id}, \Omega_4) = (SO(3)) - 2(O(2)) - 2(S_4) + 2(D_4) + 2(D_3) - (\mathbf{V_4})$$

Let us now compute $SO(3)$-deg$(-\text{Id}, \Omega_5)$. We have the following reduced lattice of isotropy subgroups for V_5:

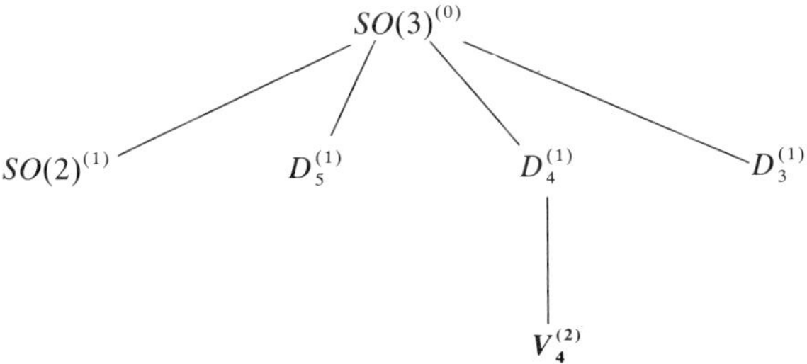

It is easy to see that for $H = SO(2)$, D_5, D_4, or D_3, we have $m_{(H)} = 1$.

For $H = V_4$, we have already computed that there are exactly three subgroups $K_1, K_2, K_3 \in (D_4)$. We have $I(F^H) = 1$. Since $\Omega^{[H]}$ may be identified with the set of three lines in a plane passing through the origin and since $F(x) = 2x$, we have $I(F^{[H]}) = -5$. Consequently, $m_{[H]} = (1 - (-5))/6 = 1$. This implies

$$SO(3)\deg(-\mathrm{Id}, \Omega_5) = (SO(3)) - (SO(2)) - (D_5) - (D4) - (D_3) + (\mathbf{V_4})$$

In what follows, we will be interested in computing the G-degree in a special case where G is the group $O(2)$. One can easily verify that if V is a finite-dimensional complex representation of $G = O(2)$, then V considered as a real representation has the following isotypical decomposition:

$$V = V^G \oplus \bigoplus_{k=0}^{N} V_k$$

where $O(2)/SO(2) \simeq \mathbb{Z}_2 = \{1, -1\}$ acts on V_0 by multiplication and V_k can be represented as $\mathbb{C}^n \oplus \mathbb{C}^n$ on which $O(2)$ acts by

$$T_\alpha \begin{bmatrix} z_1 \\ z_2 \end{bmatrix} := \begin{bmatrix} e^{ik\alpha} z_1 \\ e^{ik\alpha} z_2 \end{bmatrix}, \qquad \alpha \in SO(2), \quad z_1, z_2 \in \mathbb{C} \tag{8.4.1}$$

and

$$\kappa \begin{bmatrix} z_1 \\ z_2 \end{bmatrix} := \begin{bmatrix} \overline{z_1} \\ -\overline{z_2} \end{bmatrix}, \qquad z_1, z_2 \in \mathbb{C}^n \tag{8.4.2}$$

In what follows, we will call such a representation $\mathbb{C}^n \oplus \mathbb{C}^n$ a *special k-representation of* $O(2)$ and we will denote it by V_k^n.

Let A be a given real $2n \times 2n$ matrix. Define $\psi : V_k^n \to V_k^n$ by

$$\psi(z) = z - iAz, \qquad z \in V_k^n$$

We assume the following:

($\boldsymbol{\alpha}$**1**) $AR = -RA$, where $R = \begin{bmatrix} \mathrm{Id} & 0 \\ 0 & -\mathrm{Id} \end{bmatrix} \in \mathbb{R}^{2n \times 2n}$.
($\boldsymbol{\alpha}$**2**) Ψ is nonsingular.

It can be easily shown that (α1) is equivalent to $\kappa(iAz) = iA(\kappa z)$ for $z \in V_k^n$. So, ψ is $O(2)$-equivariant and $O(2)$-$\deg(\psi, \Omega)$ is well defined, where

$$\Omega = B_0(1) := \{z \in V_k^n;\ \|z\| < 1\}$$

The following result gives a computation of $O(2)$-$\deg(\psi, \Omega)$ in terms of eigenvalues of A.

Proposition 8.4.10

Under assumptions (α1) *and* (α2), *we have* $O(2)\text{-deg}(\psi, \Omega) = ((O(2)) - \nu(A)(D_k)$, *where* $\nu(A) = \frac{1}{2}(1-(-1)^p)$ *and* p *is the number of real numbers* $\lambda > 1$ *such that* λi *is an eigenvalue of* A, *that is,*

$$p = |\{\lambda > 1;\ \lambda i \in \sigma(A)\}| \quad \textit{(counted with multiplicity)}$$

Proof. We first remark that $\mathcal{J}(\Omega) \cap \Phi(O(2))$ contains only two orbit types, $(O(2))$ and (D_k). Clearly, the $(O(2)$-component $n_{(O(2))}$ of $O(2)$-$\deg(\psi, \Omega) = n_{(O(2))} \cdot (O(2)) + n_{(D_k)} \cdot (D_k)$ is equal to 1.

In order to compute $n_{(D_k)}$, we will use Ulrich's formula given in Theorem 8.3.14. We have

$$\begin{aligned}(V_k^n)^H &= \left\{ \begin{bmatrix} z^+ \\ z^- \end{bmatrix} \in V_k^n;\ \begin{bmatrix} z^+ \\ z^- \end{bmatrix} = \begin{bmatrix} \overline{z}^+ \\ -\overline{z}^- \end{bmatrix} \right\} \\ &= \left\{ \begin{bmatrix} z^+ \\ z^- \end{bmatrix} \in V_k^n;\ \operatorname{Im} z^+ = \operatorname{Re} z^- = 0 \right\}\end{aligned}$$

and we can identify $(V_k^n)^H$ with $\mathbb{C}^n = \{x + iy; x, y \in \mathbb{R}^n\}$, where $z^+ = x$ and $z^- = iy$. Since $N(D_k) = D_{2k}$, we have $|W(D_k)| = 2$. It is also clear that $\Omega^{[H]} = \{0\}$; thus, $I(\Psi^{[H]}) = 1$, where $\psi = \operatorname{Id} - \Psi : \overline{\Omega} \to V_k^n$. Since $\psi_H = \psi_{|\Omega^H}$ is given by

$$\psi_H \begin{bmatrix} x \\ iy \end{bmatrix} = [\operatorname{Id} - iA] \begin{bmatrix} x \\ iy \end{bmatrix}, \qquad \begin{bmatrix} x \\ iy \end{bmatrix} \in \Omega^H$$

we have $I(\Psi^H) = \deg(\psi^H, \Omega^H) = (-1)^p$. Consequently, $n_{(D_k)} = -\nu(A)$. This completes the proof. □

In order to present the computational formula for $O(2)$-representations, we need the following.

Definition 8.4.11 A representation V is called a *special representation* if

$$V = V_0^{n_0} \oplus V_1^{n_1} \oplus \cdots \oplus V_k^{n_k}$$

is a direct sum of the special i-representation of $O(2)$ for $i = 1, 2, \ldots, k$, and $V_0^{n_0} = \mathbb{R}^{n_0} \oplus \mathbb{R}^{n_0}$ is a *special* 0*-representation* of $O(2)$, that is, for $x = \begin{bmatrix} x_1 \\ x_2 \end{bmatrix} \in V_0^{n_0}$, $T_\alpha x = x$, $\mu \begin{bmatrix} x_1 \\ x_2 \end{bmatrix} = \begin{bmatrix} x_1 \\ -x_2 \end{bmatrix}$, where T_α, $\mu \in O(2)$.

Denote by $S_0 = \begin{bmatrix} 0 & \operatorname{Id} \\ \operatorname{Id} & 0 \end{bmatrix} : \mathbb{R}^{n_0} \times \mathbb{R}^{n_0} \to \mathbb{R}^{n_0} \times \mathbb{R}^{n_0}$ and $R_j = \begin{bmatrix} \operatorname{Id} & 0 \\ 0 & -\operatorname{Id} \end{bmatrix}$: $\mathbb{R}^{n_j} \times \mathbb{R}^{n_j} \to \mathbb{R}^{n_j \times n_j}$. Assume that $\{A_j\}_{j=0}^k$ is a family of real matrices such that:

($\boldsymbol{\beta}$1) A_j is a $2n_j \times 2n_j$ matrix and $A_j R_j = -R_j A_j$ for $j = 0, 1, \ldots, k$.

(β2) A_0 is nonsingular.
(β3) $\mathrm{Id} - iA_j$ is nonsingular for $j = 1, 2, \ldots, k$.

Define the linear operator $\psi : V \to V$ by

$$\psi = \psi_0 \oplus \psi_1 \oplus \cdots \oplus \psi_k$$

where:

(i) $\psi_j : V_j^{n_j} \to V_j^{n_j}, \quad j = 0, 1, \ldots, k.$
(ii) $\psi_0 = -A_0 S_0$.
(iii) $\psi_j = \mathrm{Id} - iA_j$ for $j = 1, \ldots, k$.

Under assumptions (β1–β3), Ψ is an $O(2)$-equivariant isomorphism and we have the following theorem.

Theorem 8.4.12

Let

$$\Omega = \{z \in V;\ \|z\| < 1\}$$

$$\varepsilon = \operatorname{sign}\det(-A_0^{12}), \qquad A_0 = \begin{bmatrix} 0 & A_0^{12} \\ A_0^{21} & 0 \end{bmatrix}$$

$$p_j = |\{\lambda > 1;\ i\lambda \in \sigma(A_j)\}| \ \textit{(counted with multiplicity)}, \qquad j = 1, \ldots, k$$

Then

$$O(2)\text{-deg}(\Psi, \Omega) = \varepsilon \prod_{j=1}^{k} ((O(2)) - \nu(A_j)(D_j))$$

where $\nu(A_j) := \dfrac{1 - (-1)^{p_j}}{2}$.

Proof. It follows from (P5) that

$$O(2)\text{-deg}(\psi, \Omega) = \varepsilon \prod_{j=1}^{k} O(2)\text{-deg}(\psi_j, \Omega_j)$$

where $\Omega_j = \Omega \cap V_j^{n_j}$, and the product is taken in $A(O(2))$. Consequently, the result follows from Proposition 8.4.12. □

EXERCISES

8.4.1 Let X be a Banach space, $\Omega \subset X$ an open and bounded set, and

$F:\overline{\Omega}\to X$ a compact mapping such that $F(x)\neq x$ for all $x\in\partial\Omega$. Assume that $G=\mathbb{Z}_p$, where p is a prime, acts on X through linear isometries, $\Omega\subset X$ is invariant, and F is equivariant. Let $F^G:\overline{\Omega^G}\to X^G$ denote the restriction of F to the fixed-point set $\Omega^G=\overline{\Omega}\cap X^G$. Show that

$$\deg(\mathrm{Id}-F,\Omega)\equiv\deg(\mathrm{Id}-F^G,\Omega^G)\pmod p$$

where deg denotes the Leray–Schauder degree [see Bartsch (1993)].

8.4.2 Let V be an orthogonal representation of the group $G=SO(3)$ and $A:V\to V$ an G-equivariant linear isomorphism of V. Use the isotypical decomposition of V to find a general formula expressing G-deg(A,Ω), where Ω denotes the unit ball in V, as a function of G-deg$(-\mathrm{Id},\Omega_i)$, where Ω_i denotes the unit ball in V_i, $i=1,2,3,4,\ldots$

8.5 G-EQUIVARIANT BIFURCATION PROBLEMS

In this section we will discuss the local and global equivariant bifurcation problem in an infinite-dimensional, isometric, Banach representation of a compact Lie group G.

Let W be a real, Banach, isometric representation of G. Consider the following parametrized family of fixed-point problem:

$$x=F(x,\alpha)\,,\qquad (x,\alpha)\in W\times\mathbb{R} \tag{8.5.1}$$

where $F:W\times\mathbb{R}\to W$ is a given equivariant, completely continuous C^1-mapping satisfying the following conditions:

(H1) There exists a one-dimensional submanifold $\mathscr{C}\subseteq W^G\times\mathbb{R}$ such that $x=F(x,\alpha)$ for every $(x,\alpha)\in W^G\times\mathbb{R}$, and that for every $(x_0,\alpha_0)\in W^G\times\mathbb{R}$, there exist an open neighborhood U_{x_0} of x_0 in W^G, an open interval $U_{\alpha_0}=(\alpha_0-\varepsilon,\ \alpha_0+\varepsilon)\subseteq\mathbb{R}$ for some $\varepsilon>0$, and a C^1-mapping $\eta:U_{\alpha_0}\to W^G$ such that

$$\mathscr{C}\cap(U_{x_0}\times U_{\alpha_0})=\{(\eta(\alpha),\alpha);\,\alpha\in U_{\alpha_0}\}$$

A point $(x_0,\alpha_0)\in\mathscr{C}$ is called a *bifurcation point* if, in any neighborhood of (x_0,α_0), there exists a *nontrivial solution* of (8.5.1), that is, a solution of (8.5.1) that is not in $\mathscr{C}$. Let

$$\Lambda=\{(x,\alpha)\in\mathscr{C};\,\mathrm{Id}-DF(x,\alpha)\not\in GL(W)\}$$

where $DF(x,\alpha)$ denotes the derivative of F with respect to $x\in W$, evaluated at (x, α). It follows from the implicit function theorem that if $(x_0, \alpha_0) \in \mathscr{C}$ is a bifurcation point, then (x_0, α_0) is a *singular point*, that is, $(x_0, \alpha_0) \in \Lambda$.

Assume now that $(x_0, \alpha_0) \in \Lambda$ is an isolated singular point. Then we can find $r, \rho > 0$ such that

$$U(r, \rho) := \{(x, \alpha) \in W \times \mathbb{R};\ \|x - \eta(\alpha)\| < r,\ |\alpha - \alpha_0| < \rho\}$$

satisfies:

(i) $(\alpha_0 - \rho, \alpha_0 + \rho) \subseteq U_{\alpha_0}$.
(ii) $U(r, \rho) \cap \Lambda = \{(x_0, \alpha_0)\}$.
(iii) $x \neq F(\alpha, x)$ for $(x, \alpha) \in \overline{U(r, \rho)}$ with $|\alpha - \alpha_0| = \rho$ and $x \neq \eta(\alpha)$.

We call $U(r, \rho)$ a *special neighborhood* of (x_0, α_0). For a given special neighborhood $U(r, \rho)$ of (x_0, α_0), we can construct a G-invariant function $\varphi : W \times \mathbb{R} \to \mathbb{R}$ such that:

(i) $\varphi(\eta(\alpha), \alpha) = -|\alpha - \alpha_0|$ if $\alpha \in (\alpha_0 - \rho, \alpha_0 + \rho)$.
(ii) $\varphi(x, a) = r$ if $\|x - \eta(\alpha)\| = r$ and $\alpha \in (\alpha_0 - \rho, \alpha_0 + \rho)$.
(iii) $\varphi(x, \alpha_0) = \|x - x_0\|$ if $\|x - \eta(\alpha_0)\| \leq r$.

Such a function is called a *complementing function*. Clearly, if $\varphi : \overline{U(r, \rho)} \to \mathbb{R}$ is a complementing function, then the system

$$\begin{cases} x = F(x, \alpha) \\ \varphi(x, \alpha) = 0 \end{cases} \tag{8.5.2}$$

has no solution on $\partial U(r, \rho)$. Consequently, $G\text{-deg}(f_\varphi, U(r, \rho))$ is well defined, where

$$f_\varphi(x, \alpha) = (x - F(x, \alpha), \varphi(x, \alpha)), \qquad (x, \alpha) \in W \times \mathbb{R}$$

is a mapping from $V := W \times \mathbb{R}$ into itself. Following the same arguments as those in Section 5.1, we can show that $G\text{-deg}(f_\varphi, U(r, \rho))$ does not depend on the choice of φ. Moreover, we have the following.

Proposition 8.5.1

Assume that $(x_0, \alpha_0) \in \Lambda$ is an isolated singular point, $U(r, \rho)$ a special neighborhood of (x_0, α_0), and $\varphi : W \times \mathbb{R} \times \mathbb{R}$ a complementing function. If $G\text{-deg}(f_\varphi, U(r, \rho)) \neq 0$, then (x_0, α_0) is a bifurcation point of (8.5.1). More precisely, if $G\text{-deg}(f_\varphi, U(r, \rho)) = \sum_\gamma n_\gamma\, \gamma$ and $n_\gamma \neq 0$ for some $\gamma = (H)$ with $H \leq G$ and $\dim W(H) = 0$, then (8.5.1) has a sequence of nontrivial solution (x_n, α_n) such that $\lim_{n \to \infty} (x_n, \alpha_n) = (x_0, \alpha_0)$ and $(G_{x_n}) \leq (H)$ for $n = 1, 2, \ldots$.

We now develop a computational formula for $G\text{-deg}(f_\varphi, U(r, \rho))$. Let

$$\Omega^{\pm}(\gamma, \rho) = \{x \in W;\ \|x - \eta(\alpha_0 \pm \rho)\| < r\}$$

Since $U(r, \rho)$ is a special neighborhood of (x_0, α_0), the mapping $f_\pm : W \to W$ defined by

$$f_\pm(x) := x - F(x, \alpha_0 \pm \rho)\,, \qquad x \in W$$

is $\Omega^+(r, \rho)$-admissible. So, $G\text{-deg}(f_\pm, \Omega^{\pm}(r, \rho))$ is well defined.

Theorem 8.5.2

Under the assumptions of Proposition 8.5.1, we have

$$G\text{-deg}(f_\varphi, U(r, \rho)) = G\text{-deg}(f_-, \Omega^-(r, \rho)) - G\text{-deg}(f_+, \Omega^+(r, \rho))$$

Proof. We can assume, without loss of generality, that W is a finite-dimensional orthogonal representation of G, $V := W \times \mathbb{R}$, $\mathscr{C} = \{(0, \alpha) \in W \times \mathbb{R}; \alpha \in \mathbb{R}\}$, $(x_0, \alpha_0) = (0, 0)$, $U(r, \rho) = U(1, 1)$, and the complementing function φ satisfies

$$\varphi(x, \alpha) \begin{cases} > 0 & \text{if } |\alpha| < 1,\ \|x\| = 1 \\ < 0 & \text{if } |\alpha| = 1,\ \|x\| < 1 \end{cases}$$

It follows from the regular normal approximation theorem that we can assume the mappings $f_\pm(x) := f(x, \pm 1)$ are regular normal in $B_1(0)$ mappings, the mapping f is a regular normal in $U(1, 1)$ homotopy, and f_φ is also a normal in $U(1, 1)$ mapping. In order to simplify the notations, we put $\mathscr{U} = U(1, 1)^H$ and $B := B_1(0)^H$, where $(H) \in \Phi(G)$. Then the sets $(f_\pm^H)^{-1}(0) \cap B$ are finite and composed of regular zeros. Put $(f_-^H)^{-1}(0) \cap B = W(H)\{x_1, \ldots, x_k\}$, and $(f_+^H)^{-1} \cap B = W(H)\{y_1, \ldots, y_m\}$. Then, $M := (f^H)^{-1}(0) \cap \mathscr{U}$ is a one-dimensional submanifold with $\partial M = D_0 \cup D_1 \cup D_2$, where

$$D_0 = W(H)\{(x_1, -1), \ldots, (x_k, -1)\}$$
$$D_1 = W(H)\{(y_1, 1), \ldots, (y_m, 1)\}$$
$$D_2 = W(H)\{(z_1, t_1), \ldots, (z_l, t_l)\}$$
$$\text{where } t_j \in (-1, 1), \qquad z_j \in \partial B$$

Let

$$\mathscr{U}_+ := \{(x, t) \in \mathscr{U};\ \varphi(x, t) > 0\}$$
$$\mathscr{U}_- := \{(x, t) \in \mathscr{U};\ \varphi(x, t) < 0\}$$
$$\mathscr{U}_0 := \{(x, t) \in \mathscr{U};\ \varphi(x, t) = 0\}$$

The set $D=(f_\varphi^H)^{-1}(0)\cap\mathcal{U}$ is finite, so we can put $D=\{(v_1,\tau_1),\ldots,(v_s,\tau_s)\}$. Every connected component of M passing through some $(v,\tau)\in D$ either leaves $\mathcal{U}_-$ and enters $\mathcal{U}_+$, or leaves $\mathcal{U}_+$ and enters $\mathcal{U}_-$. That is, such a component cannot bounce against $\mathcal{U}_0$ inside $\mathcal{U}_-$ or $\mathcal{U}_+$. Indeed, if (v,τ) is such a bounce-off point, then the derivative of f^H in the normal to $\mathcal{U}_0$ direction at (v,τ) is zero; consequently, $Df_\varphi^H(v,\tau)$ could not be an isomorphism, contradicting the assumption that f_φ is regular and normal. Let us fix a point $(v,\tau)\in D$ and consider a connected component M_0 of M passing through (v,τ). If M_0 has a nonempty boundary, then ∂M_0 contains exactly two points of the set ∂M and we have only the following four possibilities:

(i) $M_0\subset\mathcal{U}$, that is, $\partial M_0=\varnothing$.
(ii) M_0 joins two points from the same set D_i, $i=1,2,3$.
(iii) M_0 joins two points, one from D_0 and another from D_1.
(iv) M_0 joins two points, one from $D_0\cup D_1$ and another from D_2.

Assume for simplicity, that M_0 has only one intersection point (v,τ) with $\mathcal{U}_0$. In case (iv), we note that:

(a) If M_0 joins a point $(x_i,-1)\in D_0$ with a point from D_2, then $\operatorname{sign}\det Df_-^H(x_i,-1)=\operatorname{sign}\det Df_\varphi^H(v,\tau)$.
(b) If M_0 joins a point $(y_j,1)\in D_1$ with a point from D_2, then $\operatorname{sign}\det Df_+^H(y_j,1)=-\operatorname{sign}\det Df_\varphi^H(v,\tau)$.

Consequently, we have

$$\sum_{(v,\tau)\in D\cap M_0}\operatorname{sign}\det Df_\varphi^H(v,\tau)=\begin{cases}\operatorname{sign}\det Df_-^H(x_i) & \text{in case (a)}\\ -\operatorname{sign}\det Df_+^H(y_j) & \text{in case (b)}\end{cases}$$

Similarly, we can show that in cases (i), (ii), and (iii),

$$\sum_{(v,\tau)\in D\cap M_0}\operatorname{sign}\det Df_\varphi^H(v,r)=0$$

Consequently

$$\sum_{(v,\tau)\in D}\operatorname{sign}\det Df_\varphi^H(v,\tau)=\sum_{(v,_i,-1)\in D_0}\operatorname{sign}\det Df_-^H(x_i)-\sum_{(y_j,1)\in D_1}\operatorname{sign}\det Df_+^H(y_j)$$

This completes the proof. □

Definition 8.5.3 Suppose that $(x_0, \alpha_0) \in \Lambda$ is an isolated singular point and $U(r, \rho)$ is a special neighborhood. We define the *crossing index* $i(x_0, \alpha_0) \in A(G)$ by

$$i(x_0, \alpha_0) := G\text{-deg}(F_-, \mathcal{U}^-(r, \rho)) - G\text{-deg}(F_+, \mathcal{U}^+(r, \rho))$$

Theorem 8.5.4

Assume that $\mathcal{C}$ is complete and all singular points in $\mathcal{C}$ are isolated. Let $\mathcal{J}$ denote the closure of the set of all nontrivial solutions of (8.5.1). If $\mathcal{E} \subseteq \mathcal{J}$ is a bounded connected component of $\mathcal{J}$, then the number of bifurcation points belonging to $\mathcal{E}$ is finite and

$$\sum_{(x,\alpha)\in\mathcal{E}\cap\Lambda} i(x, \alpha) = 0 \qquad \text{in } A(G)$$

Proof. Clearly, $\mathcal{E} \cap \Lambda$ is finite. Let

$$\mathcal{E} \cap \Lambda = \{(x_1, \alpha_1), \ldots, (x_n, \alpha_m)\}$$

We choose r and $\rho > 0$ such that

(**i**) $U_i(r, \rho)$ is a special neighborhood of (x_i, α_i), $i = 1, \ldots, m$.
(**ii**) $U_i(r, \rho) \cap U_j(r, \rho) = \varnothing$ for $i \neq j$.

Let $U = U_1(r, \rho) \cup U_2(r, p) \cdots \cup U_m(r, \rho)$. Since $\mathcal{E}$ is bounded, we can choose a bounded open subset $\mathcal{U}_1 \subseteq W \times \mathbb{R}$ such that $\mathcal{E} \backslash U \subseteq \mathcal{U}_1$ and $\mathcal{U}_1 \cap \mathcal{C} = \varnothing$ (see Lemma 5.2.1). Let $\mathcal{U}_2 = U \cup \mathcal{U}_1$. Then there exists an open invariant subset $\mathcal{U}$ of $W \times \mathbb{R}$ such that $\mathcal{E} \subseteq \mathcal{U} \subseteq \mathcal{U}_2$ and $\partial\mathcal{U} \cap \mathcal{J} = \varnothing$.

Applying the implicit function theorem, we can find r_0 and ρ_0 so small that $0 < r_0 < r$, $0 < \rho_0 < \rho$, $U_i(r_0, \rho_0) \subseteq \mathcal{U}$, and $\mathcal{J} \cap U_i(r_0, \rho) \subseteq U_i(r_0, \rho_0)$ for $i = 1, \ldots, m$. Set $U_i = U_i(r_0, \rho)$, $i = 1, \ldots, m$, and $V = U_1 \cup U_2 \cdots \cup U_m$. Let $\varphi : \overline{\mathcal{U}} \cup \overline{V} \to \mathbb{R}$ be a continuous invariant function such that

$$\varphi(x, \alpha) = -|\alpha - \alpha_i| \qquad \text{and} \qquad (x, \alpha) \in \mathcal{C} \cap U_i \tag{8.5.3}$$

and

$$\varphi(x, \alpha) = r \qquad \text{if } (x, \alpha) \in \overline{\Omega} \backslash V$$

Define $F_\theta : \overline{\Omega} \cup \overline{V} \to W \times \mathbb{R}$ by

$$F_\theta(x, \alpha) = (x - F(x, \alpha), \theta(x, \alpha))$$

(8.5.3) implies that $F_\theta^{-1}(0) \subseteq \mathcal{J}$. Since $\partial\Omega \cap \mathcal{J} = \varnothing$, F_θ is Ω-admissible and, consequently, G-deg(F_θ, Ω) is well defined. Now, we consider the following equivariant homotopy:

$$H(x, \alpha, t) = (x - F(x, \alpha), \varphi(x, \alpha, t))$$

where

$$\varphi(x, \alpha, t) = (1 - t)\theta(x, \alpha)\,, \qquad (x, \alpha, t) \in \overline{\Omega} \times [0, 1]$$

Clearly, $H(x, \alpha, 0) = F_\theta$, $H(x, \alpha, 1) \neq 0$ for all $(x, \alpha) \in \overline{\Omega}$ and H is Ω-admissible. Therefore, $G\text{-deg}(F_\theta, \Omega) = 0$. On the other hand, $F_\theta^{-1}(0) \subseteq \mathcal{J} \cap V \subseteq \mathcal{J} \cap \Omega$. So,

$$G\text{-deg}(F_\theta, \Omega) = G\text{-deg}(F_\theta, \Omega \cap V) = G\text{-deg}(F_\theta, V) = 0$$

Therefore, by the additivity property,

$$0 = G\text{-deg}(F_\theta, V) = \sum_{i=1}^{m} G\text{-deg}(F_\theta, U_i)$$

from which and Theorem 8.5.2 the conclusion follows. This completes the proof. □

EXERCISES

8.5.1 Since the statement of Theorem 8.5.2 is true in the case of the trivial action of the group G (see Theorem 5.1.7), use the Ulrich formula (8.3.1) to prove Theorem 8.5.2.

8.5.2 Let V be a finite-dimensional representation of a compact Lie group G, $\Omega \subseteq V$ a bounded invariant set, and $f: V \to V$ a continuous equivariant Ω-admissible map. Use the Ulrich formula (8.3.1) to show that $G\text{-deg}(f, \Omega) \neq 0$ if and only if there is a closed subgroup $H \subset G$ such that $\deg(f^H, \Omega^H) \neq 0$. Use this fact to formulate a global bifurcation theorem for equivariant maps using only the Brouwer degree.

8.6 PERIODIC SOLUTIONS OF TIME-REVERSIBLE FUNCTIONAL SYSTEMS

Let n be a given positive integer. Denote by C_{2n} the Banach space of bounded continuous functions from $\mathbb{R}$ into $\mathbb{R}^{2n}$ equipped with the usual sup-norm. In what follows, we will employ the following notations:

(i) For $\varphi \in C_{2n}$ and $t \in \mathbb{R}$, $\varphi_t \in C_{2n}$ is defined by $\varphi_t(\theta) = \varphi(t + \theta)$ for $\theta \in \mathbb{R}$.

(ii) $T : C_{2n} \to C_{2n}$ is a bounded linear operator defined by $T\varphi(\theta) = \varphi(-\theta)$ for $\theta \in \mathbb{R}$ and $\varphi \in C_{2n}$.

(iii) $R : \mathbb{R}^{2n} \to \mathbb{R}^{2n}$ denotes the $2n \times 2n$ matrix $\begin{bmatrix} \mathrm{Id} & 0 \\ 0 & -\mathrm{Id} \end{bmatrix}$, and for every $\varphi \in C_{2n}$, $R\varphi \in C_{2n}$ is defined by

$$(R\varphi)(\theta) = R\varphi(\theta) \qquad \text{for } \theta \in \mathbb{R}$$

Consider the following one-parameter family of functional differential equations:

$$\dot{x}(t) = f(x_t, \alpha) \tag{8.6.1}$$

where $x \in \mathbb{R}^{2n}$, $f : C_{2n} \times \mathbb{R} \to \mathbb{R}^{2n}$ is continuously differentiable, $f(0, \alpha) = 0$ for $\alpha \in \mathbb{R}$, and

$$-Rf(\varphi, \alpha) = f(RT\varphi, \alpha) \qquad \text{for } \varphi \in C_{2n}, \quad \alpha \in \mathbb{R} \tag{8.6.2}$$

A functional differential equation satisfying (8.6.2) is said to be *time-reversible*.

Example 8.6.1 If $f(\varphi, \alpha) = g(\varphi(0), \alpha)$ for $\varphi \in C_{2n}$ and $\alpha \in \mathbb{R}$, where $g : \mathbb{R}^{2n} \times \mathbb{R} \to \mathbb{R}^{2n}$ is continuously differentiable, then (8.6.1) is a family of ordinary differential equations. In this case, (8.6.1) is time-reversible if and only if

$$g(Rz, \alpha) = -Rg(z, \alpha) \qquad \text{for } (z, \alpha) \in \mathbb{R}^{2n} \times \mathbb{R}$$

Example 8.6.2 If $f(\varphi, \alpha) = h(\varphi(0), \varphi(1), \alpha)$ for $\varphi \in C_{2n}$ and $\alpha \in \mathbb{R}$, where $h : \mathbb{R}^{2n} \times \mathbb{R}^{2n} \times \mathbb{R}^{2n} \times \mathbb{R} \to \mathbb{R}^{2n}$ is continuously differentiable, then (8.6.1) is a family of difference-differential equations. It can be verified that (8.6.1) is time-reversible if and only if

$$-Rh(u, v, w, \alpha) = h(Ru, Rw, Rv, \alpha)$$

for $u, v, w \in \mathbb{R}^{2n}$, and $\alpha \in \mathbb{R}$.

Example 8.6.3 If $f(\varphi, \alpha) = h\left(\varphi(0), \int_{-\infty}^{0} K_1(s)\varphi(s)\, ds, \int_{0}^{\infty} K_2(s)\, ds, \alpha\right)$ for $(\varphi, \alpha) \in C_{2n} \times \mathbb{R}$, where $h : \mathbb{R}^{2n} \times \mathbb{R}^{2n} \times \mathbb{R}^{2n} \times \mathbb{R} \to \mathbb{R}^{2n}$ is continuously differentiable, $K_1 : (-\infty, 0] \to \mathbb{R}^{2n \times 2n}$ and $K_2 : [0, \pi) \to \mathbb{R}^{2n \times 2n}$ are continuous, then (8.6.1) is a family of integro-differential equations. Clearly, (8.6.1) is time-reversible if and only if

$$-Rh\left(\varphi(0), \int_{-\infty}^{0} K_1(s)\varphi(s)\,ds, \int_{0}^{\infty} K_2(s)\varphi(s)\,ds, \alpha\right)$$
$$= h\left(R\varphi(0), \int_{0}^{\infty} K_1(-s)R\varphi(s)\,ds, \int_{-\infty}^{0} K_2(-s)R\varphi(s)\,ds, \alpha\right)$$

for $\varphi \in C_{2n}$ and $\alpha \in \mathbb{R}$.

We are interested in the existence of a bifurcation of 2π-periodic solutions from the trivial solution $(0, \alpha)$. To provide a fixed-point setting for this problem, we identify S^1 with $R/2\pi\mathbb{Z}$ and put

$$W := H^1(S^1; \mathbb{R}^{2n}), \qquad E := L^2(S^1; \mathbb{R}^{2n})$$

where $H^1(S^1; \mathbb{R}^{2n})$ denotes the first Sobolev space of 2π-periodic $2n$-vector functions on $\mathbb{R}$. Define $L : W \to E$, $F : W \times \mathbb{R} \to E$, and $K : W \to E$ by

$$\begin{cases} Lx(t) &= \dot{x}(t), \\ F(x, \alpha)(t) = f(x_t, \alpha), \\ Kx(t) &= \dfrac{1}{2\pi}\displaystyle\int_0^{2\pi} x(s)\,dx, \quad x \in W,\ t \in S^1,\ \alpha \in \mathbb{R}. \end{cases}$$

Clearly, we have:

(i) $(x, \alpha) \in W \times \mathbb{R}$ is a 2π-periodic solution of (8.6.1) if and only if

(8.6.3) $$Lx = F(x, \alpha)$$

(ii) F is completely continuous.

(iii) K is a compact operator and $L + SK : W \to E$ is invertible, where $S = \begin{bmatrix} 0 & \mathrm{Id} \\ \mathrm{Id} & 0 \end{bmatrix} \in R^{2n \times 2n}$.

Let $R_{SK} = (L + SK)^{-1} : E \to W$. Then (8.6.2) is equivalent to

(8.6.4) $$x = N(x, \alpha), \qquad (x, \alpha) \in W \times \mathbb{R}$$

where

(8.6.5) $$N(x, \alpha) = R_{SK}[F(x, \alpha) + SKx]$$

We now introduce an action of $O(2)$ on W and E. Let $r_\theta = \begin{bmatrix} \cos\theta & -\sin\theta \\ \sin\theta & \cos\theta \end{bmatrix} \in O(2)$. Then r_θ acts on $x \in W$ or $x \in E$ by shifting the argument by θ, that is,

$$(r_\theta x)(t) = x(t+\theta)\,, \qquad t, \theta \in S^1$$

For $\kappa = \begin{bmatrix} 1 & 0 \\ 0 & -1 \end{bmatrix} \in O(2)$, κ acts on $x = \begin{pmatrix} x_1 \\ x_2 \end{pmatrix} \in W$ by

$$\kappa \begin{pmatrix} x_1 \\ x_2 \end{pmatrix}(t) = \begin{pmatrix} x_1(-t) \\ -x_2(-t) \end{pmatrix}, \qquad t \in S^1$$

and κ acts on $x = \begin{pmatrix} x_1 \\ x_2 \end{pmatrix} \in E$ by

$$\kappa \begin{pmatrix} x_1 \\ x_2 \end{pmatrix}(t) = \begin{pmatrix} -x_1(-t) \\ x_2(-t) \end{pmatrix}, \qquad t \in S^1$$

Clearly, W and E are both orthogonal representations of $G = O(2)$. Moreover, F, L, and SK are $O(2)$-equivariant. In particular,

$$\begin{cases} \kappa F(x, \alpha)(t) = -RF(x_{-t}, \alpha) \\ F(\kappa x, \alpha)(t) = F(RTx_{-t}, \alpha) \qquad \text{for } t \in S^1, \alpha \in \mathbb{R} \end{cases}$$

Identify the subspace of C_{2n} consisting of all constant functions with $\mathbb{R}^{2n}$. Let $\hat{f} = f|_{\mathbb{R}^{2n} \times \mathbb{R}}$. We assume:

(A1) $D\hat{f}(0, \alpha) : \mathbb{R}^{2n} \to \mathbb{R}^{2n}$ is an isomorphism for every $\alpha \in \mathbb{R}$, where $D\hat{f}(0, \alpha)$ denotes the derivative of $\hat{f}$ with respect to the argument in $\mathbb{R}^{2n}$, evaluated at $(0, \alpha)$.

Under the above assumption, the set

$$\mathscr{C} = \{(0, \alpha); \alpha \in \mathbb{R}\} \subseteq \mathbb{R}^{2n} \times \mathbb{R}$$

satisfies (H1) of the previous section with $\eta(\alpha) = 0$ for $\alpha \in \mathbb{R}$.

Let

$$\Lambda : \{(0, \alpha) \in \mathscr{C};\ \mathrm{Id} - DN(0, \alpha) \notin GL(W)\}$$

It is not difficult to show that $(0, \alpha) \in \Lambda$ if and only if there exists $k \in \mathbb{Z}$ such that $\lambda = i2k\pi$ is a solution to the following *characteristic equation* at α:

$$\det{}_{\mathbb{C}}\, \Delta_\alpha(\lambda) = 0 \tag{8.6.6}$$

where $\Delta_\alpha(\lambda) : \mathbb{C}^{2n} \to \mathbb{C}^{2n} := \mathbb{R}^{2n} + i\mathbb{R}^{2n}$ is defined by

$$\Delta_\alpha(\lambda) = \lambda\, \mathrm{Id} - D_\varphi f(0, \alpha)(e^{\lambda \cdot}\, \mathrm{Id})$$

$$D_\varphi f(0, \alpha)(e^{\lambda \cdot}\, \mathrm{Id}) = (D_\varphi f(0, \alpha)(e^{\lambda \cdot} \varepsilon_1), \ldots, D_\varphi f(0, \alpha)(e^{\lambda \cdot} \varepsilon_{2n}))$$

$D_\varphi f(0, \alpha)$ denotes the derivative of $f: C_{2n} \times \mathbb{R} \to \mathbb{R}^{2n}$ with respect to $\varphi \in C_{2n}$, evaluated at $(0, \alpha)$, $\{\varepsilon_1, \ldots, \varepsilon_{2n}\}$ is the standard basis of $\mathbb{R}^{2n}$.

Assume that $(0, \alpha_0)$ is an isolated point in Λ, $\mathcal{U}(r, \rho)$ a special neighborhood of $(0, \alpha_0)$, and $\theta: \overline{\mathcal{U}(r, \rho)} \to \mathbb{R}$ a complementing function. Define $N_\theta: W \times \mathbb{R} \to W \times \mathbb{R}$ and $N^\pm: \overline{B_r(0)} \subseteq W \to W$ by

$$N_\theta(x, \alpha) = (x - N(x, \alpha), \theta(x, \alpha)), \qquad (x, \alpha) \in W \times \mathbb{R}$$

$$N^\pm(x) = x - N(x, \alpha_0 \pm \rho), \qquad x \in \overline{B_r(0)} := \{y \in W; \|y\| \leq r\}$$

Then, by Theorem 8.5.2, we have

$$O(2)\text{-deg}(N_\theta, \mathcal{U}(r, \rho)) = i(0, \alpha_0)$$

where

$$i(0, \alpha_0) = O(2)\text{-deg}(N^-, B_r(0)) - O(2)\text{-deg}(N^+, B_r(0)) \tag{8.6.7}$$

By Proposition 8.5.1, $i(0, \alpha_0) \neq 0$ implies that $(0, \alpha_0)$ is a bifurcation point.

To compute $i(0, \alpha_0)$, we first consider the irreducible representations of W and E. Since $SO(2) \subseteq O(2)$ acts on W and E by shifting arguments, the $SO(2)$-isotypical decompositions of W and E are as follows:

$$W = W_0 \oplus W_1 \oplus \cdots \oplus W_k \oplus \ldots$$

$$E = E_0 \oplus E_1 \oplus \cdots \oplus E_k \oplus \ldots$$

where $W_0 = E_0 \simeq \mathbb{R}^{2n}$ is the subspace of C_{2n} consisting of constant mappings, and for $k > 0$, both W_k and E_k are spanned by functions $\cos(kt)x$ and $\sin(kt)y$ for $x, y \in \mathbb{R}^{2n}$. It is convenient to identify the function $\cos(kt)x - \sin(kt)y$ with the vector $x + iy \in \mathbb{R}^{2n} + i\mathbb{R}^{2n} = \mathbb{C}^{2n}$. Note that subspaces W_k and E_k have a natural complex structure, where the multiplication by i is defined by

$$i \cdot \varphi_k = r_{\frac{\pi}{2k}} \varphi_k, \qquad \varphi_k \in W_k, \text{ or } E_k$$

Under the above identification, we have

$$\begin{aligned} i(x + iy) &= i[\cos(kt)x - \sin(kt)y] \\ &= \cos\left(kt + \frac{\pi}{2}\right)x - \sin\left(kt + \frac{\pi}{2}\right)y \\ &= \left[\cos(kt)\cos\frac{\pi}{2} - \sin(kt)\sin\frac{\pi}{2}\right]x - \left[\sin(kt)\cos\frac{\pi}{2} + \cos(kt)\sin\frac{\pi}{2}\right]y \\ &= -\cos(kt)y - \sin(kt)y \\ &= -y + ix \end{aligned}$$

That is, the usual multiplication by i in $\mathbb{C}^{2n}$ coincides with the multiplication by i in W_k and E_k.

The subspaces W_k and E_k are invariant with respect to the action of $O(2)$. For $k \in \mathbb{Z}$ and $z = x + iy \in W_k$, where $x = \begin{bmatrix} x_1 \\ x_2 \end{bmatrix}$, $y = \begin{bmatrix} y_1 \\ y_2 \end{bmatrix}$, and $z = \begin{bmatrix} z_1 \\ z_2 \end{bmatrix}$, we have

$$\begin{aligned} \kappa \begin{bmatrix} z_1 \\ z_2 \end{bmatrix} &= \kappa \begin{bmatrix} \cos(kt)x_1 - \sin(kt)y_1 \\ \cos(kt)x_2 - \sin(kt)y_2 \end{bmatrix} \\ &= \begin{bmatrix} \cos(-kt)x_1 - \sin(-kt)y_1 \\ -\cos(-kt)x_2 + \sin(-kt)y_2 \end{bmatrix} \\ &= \begin{bmatrix} \cos(kt)x_1 + \sin(kt)y_1 \\ -[\cos(kt)x_2 + \sin(kt)y_2] \end{bmatrix} = \begin{bmatrix} \bar{z}_1 \\ -\bar{z}_2 \end{bmatrix} \end{aligned}$$

Similarly, for $z = \begin{bmatrix} z_1 \\ z_2 \end{bmatrix} \in E_k$, we have

$$\kappa \begin{bmatrix} z_1 \\ z_2 \end{bmatrix} = \begin{bmatrix} -\bar{z}_1 \\ z_2 \end{bmatrix}$$

In particular, W_k is a special k-representation of $O(2)$ after the identification $W_k \simeq \mathbb{C}^n \times \mathbb{C}^n$.

For $k = 0$, W_0 and E_0 can be identified with $\mathbb{R}^{2n}$, where the group $SO(2)$ acts trivially and $\kappa \in O(2)$ acts by

$$\kappa \begin{bmatrix} x_1 \\ x_2 \end{bmatrix} = \begin{bmatrix} x_1 \\ -x_2 \end{bmatrix} \quad \text{if } \begin{bmatrix} x_1 \\ x_2 \end{bmatrix} \in W_0$$

$$\kappa \begin{bmatrix} x_1 \\ x_2 \end{bmatrix} = \begin{bmatrix} -x_1 \\ x_2 \end{bmatrix} \quad \text{if } \begin{bmatrix} x_1 \\ x_2 \end{bmatrix} \in E_0$$

In particular, W_0 is a 0-special representation of $O(2)$.

Let us now look at how $\mathrm{Id} - DN(0, \alpha) : W \to W$ acts on subspaces W_0 and W_k for $k \in \mathbb{Z}$. For $W_0 = \mathbb{R}^{2n}$, we have

$$\begin{aligned} DN(0, \alpha)|_{W_0} &= SDF(0, \alpha)|_{W_0} + \mathrm{Id}|_{W_0} \\ &= SD\hat{f}(0, \alpha)|_{W_0} + \mathrm{Id}|_{W_0} \end{aligned}$$

Consequently,

$$\mathrm{Id} - DN(0, \alpha)|_{W_0} = -SD\hat{f}(0, \alpha) : \mathbb{R}^{2n} \to \mathbb{R}^{2n}$$

For $k > 0$, we can easily show that

$$DN(0, \alpha)|_{W_k} = L^{-1} DF(0, \alpha)|_{W_k}$$

Denote by $e^{ikt}z = e^{ikt}(x + iy)$ the element $\cos(kt)x - \sin(kt)y$ in W_k. Then

$$e^{ikt}(e^{ik\theta}z) = e^{ik(t+\theta)}z$$

So,

$$\begin{aligned} L^{-1}DF(0,\alpha)(e^{ik(t+\cdot)}z) &= \frac{1}{ik}Df(0,\alpha)(e^{ikt}\cdot e^{ik\cdot}z) \\ &= \frac{1}{ik}e^{ikt}Df(0,\alpha)(e^{ik\cdot}z) \end{aligned}$$

Therefore,

$$\begin{aligned} \mathrm{Id} - DN(0,\alpha)|_{W_k} &= \mathrm{Id} - \frac{1}{ik}Df(0,\alpha)(e^{ik\cdot}\mathrm{Id}) \\ &= \mathrm{Id} - i\left[\frac{-1}{k}Df(0,\alpha)(e^{ik\cdot}\mathrm{Id})\right] \end{aligned}$$

Moreover,

$$\begin{aligned} i\cdot Df(0,\alpha)(\kappa e^{ik(t+\cdot)}z) &= i\cdot Df(0,\alpha)RSe^{ik(t+\cdot)}z \\ &= Df(0,\alpha)e^{-ik(t+\pi/2k)}RSe^{ik\cdot}z \\ &= -ie^{-ikt}Df(0,\alpha)RSe^{ik\cdot}z \end{aligned}$$

and

$$\begin{aligned} \kappa(i\cdot Df(0,\alpha)e^{ik(t+\cdot)}z) &= \kappa(e^{ik(t+\pi/2)}Df(0,\alpha)e^{ik\cdot}z) \\ &= ie^{-ikt}RDf(0,\alpha)e^{ik\cdot}z \end{aligned}$$

Since $-RDf(0,\alpha) = Df(0,\alpha)RS$, we have

$$i\cdot Df(0,\alpha)(\kappa e^{ik(t+\cdot)}z) = \kappa(i\cdot Df(0,\alpha)e^{ik(t+\cdot)}z)$$

We are now in a position to state our local bifurcation theorem for time-reversible functional differential equations.

Theorem 8.6.4

Assume that:

- **(i)** *(A1) is satisfied.*
- **(ii)** $(0,\alpha_0)\in\Lambda := \{(0,\alpha);\ \det_{\mathbb{C}}\Lambda_\alpha(ik) = 0\}\subseteq\mathbb{R}^{2n}\times\mathbb{R}$ *is an isolated point in* Λ.
- **(iii)** *There exists* $k\in\mathbb{N}$ *so that* $m_k(\alpha_0) := m_k^+(\alpha_0) - m_k^-(\alpha_0) = 1$ *(mod 2),*

where

$$m_k^{\pm}(\alpha_0) := |\{\lambda > 1; \det_{\mathbb{C}}[ik\lambda \,\mathrm{Id} - Df(0, \alpha_0 \pm \varepsilon)(e^{ik\cdot}\mathrm{Id})] = 0\}|,$$

where $\varepsilon > 0$ is sufficiently small so that

$$\{(0, \alpha);\ \alpha_0 - \varepsilon \le \alpha \le \alpha_0 + \varepsilon\} \cap \Lambda - \{(0, \alpha_0)\}$$

Then $(0, \alpha_0)$ is a bifurcation point of (8.6.1). More precisely, there exists a sequence of 2π-periodic functions $x_p : \mathbb{R} \to \mathbb{R}^{2n}$ and a sequence of real numbers α_p such that:

(a) *$\alpha_p \to \alpha_0$, $x_p(t) \to 0$ uniformly for $t \in \mathbb{R}$ as $p \to \infty$.*

(b) *$x_p(t)$ is a 2π-periodic solution of (8.6.1) with $\alpha = \alpha_p$.*

(c) *$(G_{x_p}) \le (D_k)$ for all $p = 1, 2, \ldots$.*

Proof. This is an immediate consequence of Theorems 8.4.13 and 8.5.2 with $i(0, \alpha_0) = O(2)\text{-deg}(N^-, B_r(0) - O(2)\text{-deg}(N^+, B_r(0))$ and

$$O(2)\text{-deg}(N^{\pm}, B_r(0)) = \varepsilon_0((O(2)) - \nu_0^{\pm}(SO(2)) \prod_{k>0} ((O(2)) + \varepsilon_k^{\pm}(D_k))$$

where

$$\varepsilon(\alpha) = \text{sign}\det(-A_0^{12}(\alpha_0))$$

$$\nu_k^{\pm} = \frac{1 - (-1)^{m^{\pm}k(\alpha_0)}}{2}$$

$$D\hat{f}(0, \alpha) = \begin{bmatrix} 0 & A^{12}(\alpha_0) \\ A_0^{21}(\alpha) & 0 \end{bmatrix}$$

$$m_0^{\pm}(\alpha_0) = |\{\lambda < 0;\ \lambda \in \sigma(-A_0^{21})\}|$$

This completes the proof. □

Remark 8.6.5 If (8.6.1) is a family of time-reversible systems of ordinary differential equations (see Example 8.6.1), then under certain generic conditions, the characteristic equation (8.6.1) has two pairs of eigenvalues $i \pm \sqrt{\gamma(\alpha)}$ and $-i \pm \sqrt{\gamma(\alpha)}$ with $\gamma(\alpha_0) = 0$ and $\gamma'(\alpha_0) \ne 0$ [see, e.g., Vanderbauwhede (1982)]. That is, as α crosses α_0, two pairs of purely imaginary eigenvalues collide and then split off the imaginary axis. In this case, one can easily show that $m_1(\alpha_0) = 1 \pmod 2$. Consequently, $(0, \alpha_0)$ is a bifurcation point of (8.6.1). However, it should be mentioned that Theorem 8.6.4 does not require any genericity and nonresonance conditions.

Theorem 8.6.6

Assume that:

(i) *(A1) is satisfied.*
(ii) Λ *has no accumulation points.*

Let $\mathcal{J} \subseteq C_{2n} \times \mathbb{R}$ *denote the closure of the set of all nontrivial solutions of (8.6.1). Then for every bounded, connected component* $\mathscr{E}$ *of* $\mathcal{J}$, $\mathscr{E} \cap \Lambda$ *is a finite set and*

$$\sum_{(0,\alpha)\in \mathscr{E}\cap\Lambda} i(0, \alpha) = 0$$

Proof. This is an immediate consequence of Theorem 8.5.4. □

8.7 BRANCHING LEMMA FOR EQUIVARIANT BIFURCATIONS

Assume that $W = \mathbb{R}^n$ is an orthogonal representation of a compact Lie group G. We denote by V the representation $W \oplus \mathbb{R}$.

In what follows, we will assume that M is a one-dimensional, smooth, complete submanifold of $W^G \oplus \mathbb{R}$. For $x \in M$, we denote by T_xM the tangent space to M at x, and by N_xM the normal space to M at x. So, we have $W \oplus \mathbb{R} = T_xM \oplus N_xM$.

We denote by $\mathscr{C}_M$ the class of all continuous equivariant maps $f : W \oplus \mathbb{R} \to W$ such that:

(i) f is differentiable at every point $x \in M$ and the derivative $Df(x)$ depends continuously on $x \in M$.
(ii) $M \subset f^{-1}(0)$, that is, for all $x \in M$, we have $f(x) = 0$.

Let $f \in \mathscr{C}_M$; we define the set of *M-singular points of f* by

$$\Lambda(f) := \{x \in M;\ D_v f(x) : N_xM \to W \text{ is not an isomorphism}\}$$

where $D_v f(x) := Df(x)|_{N_xM} : N_xM \to W$.

We also define

$$\mathscr{C}^1_M := \{f \in \mathscr{C}_M;\ f \text{ is of class } C^1 \text{ on the set } W \oplus \mathbb{R}\backslash\Lambda(f)\}$$

Suppose that $f \in \mathscr{C}^1_M$. The condition (ii) above implies that the manifold M is contained in the solution set of the equation

$$f(x) = 0\,, \qquad x \in W \oplus \mathbb{R} \tag{8.7.1}$$

As usual, we call points from M *trivial solutions* of Eq. (8.7.1), and all other solutions of (8.7.1) *nontrivial solutions*. A point $x_0 \in M$ is a *bifurcation point* of (8.7.1) if in every neighborhood of x_0, there exists a nontrivial solution of (8.7.1). We will denote by $\mathcal{B}(f)$ the subset of M of all bifurcation points of (8.7.1). It follows from the implicit function theorem that if $x_0 \in M$ is a bifurcation point of (8.7.1), then $x_0 \in \Lambda(f)$, that is, $\mathcal{B}(f) \subset \Lambda(f)$.

Definition 8.7.1 Assume that $A \subset W \oplus \mathbb{R}$ is a compact invariant subset and $f \in \mathcal{C}^1_M$. We say that f is a *(equivariant) normal b-map* relative to A if for every $x \in [f^{-1}(0) \backslash M] \cap A$, there exists $\varepsilon_x > 0$ such that the following α-normality condition for $\alpha = (G_x)$ is satisfied:

$$f(x+h) = f(x) + h \qquad \text{for all } h \in N_x(V_\alpha) \text{ with } |h| < \varepsilon_x$$

We will denote by $\mathcal{G}_M(A)$ the set of all normal b-maps relative to A.

Since the space $\mathcal{C}^1_M$ is a subset of the complete, locally convex space $C(W \oplus \mathbb{R}; W)$ of continuous maps φ from $W \oplus \mathbb{R}$ to W, equipped with the topology induced by seminorms $p_K(\varphi) = \sup_{x \in K} |\varphi(x)|$, where K is a compact subset of $W \oplus \mathbb{R}$, $\mathcal{C}^1_M$ has the induced topology.

We have the following.

Theorem 8.7.2 (Normal Approximation Theorem)

Assume that $A \subset W \oplus \mathbb{R}$ is a compact invariant subset and $f \in \mathcal{C}^1_M$. Then for every $\eta > 0$, there exists $\tilde{f} \in \mathcal{G}_M(A)$ such that:

(i) $\sup_{x \in W \oplus \mathbb{R}^n} |\tilde{f}(x) - f(x)| < \eta$.

(ii) $\Lambda(f) = \Lambda \tilde{f})$.

(iii) $\mathcal{B}(f) = \mathcal{B}(\tilde{f})$.

Consequently, the set $\mathcal{G}_M(A)$ is dense in $\mathcal{C}^1_M$.

Proof. We define a continuous function $\varepsilon : V \to \mathbb{R}$ by $\varepsilon(x) := [\text{dist}(x, M)]^2$. Since M is an invariant manifold, the function ε is also invariant. We put

$$\varepsilon_1(x) = \frac{\varepsilon(x)}{n_{11}}$$

where n_{11} is a positive integer to be specified later. We put $Z_1 :=$

$[f^{-1}(0)\backslash \mathcal{M}] \cap A$ and define

$$D_1 = \{x \in V\backslash M;\ \mathrm{dist}(x, Z_1) < \varepsilon_1(x)\}$$

$$L_1 = \{x \in V\backslash M;\ \mathrm{dist}(x, Z_1) < 2\varepsilon_1(x)\}$$

$$A_1 = \left\{x \in V\backslash M;\ \mathrm{dist}(x, Z_1) < \frac{\varepsilon_1(x)}{2}\right\}$$

It is clear that D_1, L_1, and A_1 are open subsets of $V\backslash M$. Moreover, if $x \in \overline{L_1} \cap M$, then x is a bifurcation point of (8.7.1). Indeed, let $x_n \in L_1$ be a sequence such that $x_n \to x$. By definition,

$$\mathrm{dist}(x_n, Z_1) < \frac{2[\mathrm{dist}(x_n, M)]^2}{n_{11}} \to 0\,.$$

Therefore, there is a sequence $y_n \in Z_1$ such that $|x_n - y_n| \to 0$. Hence, $|x - y_n| \le |x - x_n| + |x_n - y_n| \to 0$, which implies that x is a bifurcation point.

We extend the partial order in $J(V\backslash M)$ to a total order $\alpha_1 < \alpha_2 < \cdots < \alpha_k < \alpha_{k+1} < \cdots < \alpha_m$, and we consider the minimal orbit type $\alpha = \alpha_1$.

Put $\tilde{D}_1 := D_1 \cap V_\alpha \subset L_1$ and $\tilde{A}_1 = A_1 \cap V_\alpha \subset L_1$. If $\tilde{A}_1 = \varnothing$, then we put $f_1(x) = f(x)$. We now assume $\tilde{A}_1 \neq \varnothing$. Since the function $x \mapsto \mathrm{dist}(x, Z_1)$ is an invariant function, the open sets D_1, L_1, and A_1 are also invariant and, consequently, $\tilde{D}_1$ is an open invariant subset of V_α. Thus, there exists a continuous invariant function $\nu_1 : \tilde{D}_1 \to \mathbb{R}_+$ such that

$$\nu_1(x) \le \frac{\varepsilon(x)}{2}$$

and the mapping $\mu : N(V_\alpha) \to V$, $\mu(v, w) = v + w$, restricted to the set $N(\tilde{D}_1, \nu_1)$, is a G-imbedding into L_1. We put $N_1 := \mu(N(\tilde{D}_1, \nu_1))$. Let $\gamma_1 : V\backslash M \to [0, 1]$ be an invariant C^∞-function such that $\gamma_1(x) = 1$ for $x \in \mu(N(\tilde{A}_1, \nu_1/2))$, and $\gamma_1(x) = 0$ for $x \in V\backslash(M \cup N_1)$. We define $f_1 : V \to W$ by

$$f_1(x) := \begin{cases} f(x)\,, & \text{for } x \in V\backslash N_1 \\ \gamma_1(x)(f(v) + w) + (1 - \gamma_1(x))f(x)\,, & \text{for } x = v + w \in N_1 \end{cases}$$

where $x = v + w$ denotes the decomposition with $v \in \tilde{D}_1$ and $w \in N_v(\tilde{D}_1)$.

We put $\delta_1 := \max_{x \in \tilde{D}_1} \nu_1(x)$. We now assume that the number n_{11} is chosen sufficiently large so that the corresponding function $\nu_1(x)$ can be chosen in such a way that

$$\sup_{v + w \in N_1} |f(v) - f(v + w)| \le \frac{\eta}{m} - \delta_1$$

This is evidently possible: The above estimation is made on a relatively compact (in V) neighborhood of the set Z_1, on which the function f is zero;

therefore, the existence of such a number n_{11} follows from the continuity of f. Consequently,

$$\begin{aligned}\sup_{x\in V} |f_1(x)-f(x)| &= \sup_{x\in N_1} |f_1(x)-f(x)| \\ &= \sup_{x=v+w\in N_1} |\gamma_1(x)(f(v)-f(x))+\gamma_1(x)w| \\ &\le \sup_{x=v+w\in N_1} |f(v)-f(x)|+\delta_1 \\ &\le \frac{\eta}{m}-\delta_1+\delta_1=\frac{\eta}{m}\end{aligned}$$

This implies that $f_1 : V \to W$ is a well-defined η/m-approximation of f. It is also clear that f_1 is a C^1-function on $V\backslash M$. Moreover, f_1 is invariant and satisfies the following conditions: *For every* $x \in [f^{-1}(0)\backslash M] \cap A$ *such that* $(G_x) = \alpha$, *and that for every* $w \in T_x V_\alpha$ *with* $|w| \le \frac{\nu_1(x)}{2}$, *we have* $f_1(x+w) = f_1(x) + w$. On the other hand, the set $[f_1^{-1}(0)\backslash M] \cap V_\alpha \cap A$ is contained in $\tilde{A}_1$. That means that in the first step of our construction, we have "corrected" the map f to a new map f_1 such that it satisfies α_1-normality condition.

We want to check that $f_1 \in \mathscr{C}^1_M$. We will show that f_1 if differentiable at every point $x \in M$ and that the derivative $Df_1(x)$ depends continuously on $x \in M$. It is clear that $f_1(x) = f(x)$ for all $x \in \overline{Z_1} \cap M$ and, therefore, we need only to check the differentiability at points $x \in \overline{Z_1} \cap M$. We will show that for $x \in \overline{Z_1} \cap M$, $Df_1(x) = Df(x)$. For this purpose, assume that $x + h \in N_1$ and let $x + h = v + w$, where $w \in T_v \tilde{D}_1$. Then we have

$$\begin{aligned}&|f_1(x+h)-Df(x)h| \\ &\quad= |\gamma_1(x+h)(f(v)+w)+(1-\gamma_1(x+h))f(x+h)-Df(x)h| \\ &\quad\le |f(x+h)-Df(x)h|+|\gamma_1(x+h)(f(v)+w-f(x+h))| \\ &\quad\le |f(x+h)-Df(x)h|+|f(v)+w-f(x+h)| \\ &\quad\le |f(x+h)-Df(x)h| \\ &\quad+|f(v)-Df(x)(v-x)-(f(x+h)-Df(x)h)+Df(x)(v-x-h)| \\ &\quad\le 2|f(x+h)-Df(x)h|+|f(v)-Df(x)(v-x)|+|Df(x)||w| \\ &\quad\le 2o(|h|)+o(|v-x|)+|Df(x)||w|\end{aligned}$$

Let $H = G_v$. Since $x \in V^G$, $x - v \in V_H \subset V_\alpha$; thus, $x - v$ is orthogonal to w.

This implies that $|x-v|^2+|w|^2=|h|^2$. We may assume that $|h|<1$. Then $\text{dist}(v, M) \leq |x-v|<1$ and $|w|<1$. Let $y \in \overline{M}$ be such that $\text{dist}(x+h, M)=|x+h-y|$. Then

$$|v-y| \leq |x+h-y|+|w|=\text{dist}(x+h, M)+|w|$$

Thus,

$$\text{dist}(v, M) \leq \inf_{y \in M} |v-y| \leq \text{dist}(x+h, M)+|w|$$

Also since $|w| \leq \dfrac{[\text{dist}(v, M)]^2}{2} \leq \dfrac{\text{dist}(v, M)}{2}$, we have $\text{dist}(v, M) \leq \text{dist}(x+h, M)+\dfrac{\text{dist}(v, M)}{2}$ and, hence, $\text{dist}(v, M) \leq 2\,\text{dist}(x+h, M)$. Consequently,

$$|w| \leq \frac{[\text{dist}(v, M)]^2}{2} \leq 2[\text{dist}(x+h, M)]^2 \leq |h|^2$$

from which we obtain

$$|f_1(x+h)-Df(x)h|=o(|h|)$$

This means that f_1 is differentiable at every point $x \in M$. Since $Df_1(x)=Df(x)$ and $f \in \mathscr{C}_M$, we have $f_1 \in \mathscr{C}^1_M$. Moreover, it is clear that $\Lambda(f_1)=\Lambda(f)$.

We put $\Omega_1 := \mu(N(\tilde{A}_1, \nu_1/2))$. In the next step, we consider the set $Z_2 := [f_1^{-1}(0) \backslash (M \cup \Omega_1) \cap A$. We claim that $Z_2 \cap \overline{\Omega_1} = \varnothing$. Indeed, assume that $x \in Z_2 \cap \overline{\Omega_1}$; then $x=v+w \in \mu(N(\tilde{A}_1, \nu_1/2))$. Since f_1 satisfies the α_1-normality condition in $\overline{\Omega_1}$, we have $0=f_1(x)=f_1(v)+w$, which implies that $f_1(v)=0$ and $w=0$. Therefore, $x=v$ and $x \in \tilde{A}_1 \cap Z_2 \subset \tilde{A}_1 \cap A$. But this is impossible, because $x \in Z_1 \subset \Omega_1$ and $\Omega_1 \cap Z_2 = \varnothing$. Since all the "new" zeros of f_1 in Z_2 were produced in $N_1=\mu(N(\tilde{D}_1, \nu_1))$, and since they are such that $\gamma_1(x)(f_1(v)+w)+(1-\gamma_1(x))f_1(x)=0$ and that $w \neq 0$, we must have $G_x=G_v \cap G_w = H \cap G_w \subseteq H$. Consequently, the orbit types in Z_2 are larger than α_1. Now we define the following functions:

$$\varepsilon_2(x)=\frac{\varepsilon(x)}{n_{21}}+\frac{\text{dist}(x, \overline{\Omega_1})}{n_{22}}$$

where n_{21} and n_{22} are positive integers to be chosen later. We also put

$$D_2 := \{x \in V \backslash M;\ \text{dist}(x, Z_2)<\varepsilon_2(x)\}$$

$$L_2 := \{x \in V \backslash M;\ \text{dist}(x, Z_2)<2\varepsilon_2(x)\}$$

$$A_2 := \left\{x \in V \backslash M;\ \text{dist}(x, Z_2)<\frac{\varepsilon_2(x)}{2}\right\}$$

We choose $\alpha = \alpha_2$ and consider the sets $\tilde{D}_2 := D_2 \cap V_\alpha \subset L_2$ and $\tilde{A}_2 := A_2 \cap V_\alpha \subset L_2$. If $\tilde{A}_2 = \emptyset$, then we put $f_2 = f_1$. Assume therefore that $\tilde{D}_2 \neq \emptyset$. Then there exists a continuous invariant function $\nu_2 : \tilde{D}_2 \to \mathbb{R}_+$ such that $\nu_2(x) \leq \frac{\varepsilon(x)}{2}$, and the mapping $\mu : N(V_\alpha) \to V$, $\mu(v, w) = v + w$, restricted to the set $N(\tilde{D}_2, v_2)$, is a G-imbedding into L_2. We put $N_2 := \mu(N(\tilde{D}_2, \nu_2))$. Let $\gamma_2 : V \backslash M \to [0, 1]$ be an invariant C^∞-function such that $\gamma_2(x) = 1$ for $x \in \mu(N(\tilde{A}_2, \nu_2/2))$ and $\gamma_2(x) = 0$ for $x \in V \backslash (M \cup N_2)$. We define $f_2 : V \to W$ by

$$f_2(x) := \begin{cases} f_1(x)\,, & \text{if } x \in V \backslash N_2 \\ \gamma_2(x)(f_1(v) + w) + (1 - \gamma_1(x))f_1(x)\,, & \text{if } x = v + w \in N_2 \end{cases}$$

where $x = v + w$ denotes the decomposition with $v \in \tilde{D}_2$ and $w \in N_v(\tilde{D}_2)$. We put $\delta_2 := \max_{x \in \tilde{D}_2} \nu_2(x)$. We may assume that the numbers n_{21} and n_{22} are chosen to be sufficiently large so that the corresponding function $\nu_2(x)$ may be chosen such that

$$\sup_{v + w \in N_2} |f_1(v) - f_1(v + w)| \leq \frac{\eta}{m} - \delta_2$$

Consequently,

$$\begin{aligned} &\sup_{x \in V} |f_2(x) - f(x)| \\ &\quad \leq \sup_{x \in V} |f_2(x) - f_1(x)| + \sup_{x \in V} |f_1(x) - f(x)| \\ &\quad \leq \sup_{x = v + w \in N_2} |\gamma_2(x)(f_1(v) + w) + (1 - \gamma_2(x))f_1(x) - f_1(x)| + \frac{\eta}{m} \\ &\quad \leq \sup_{x = v + w \in N_2} |f_1(v) - f_1(x) + \gamma_2(x) w| + \frac{\eta}{m} \\ &\quad \leq \sup_{x = v + w \in N_2} |f_1(v) - f_1(x)| + \delta_2 + \frac{\eta}{m} \\ &\quad \leq \frac{2\eta}{m} - \delta_2 + \delta_2 = \frac{2\eta}{m} \end{aligned}$$

We can verify, in a similar way to that for f_1, that $f_2 \in \mathscr{C}^1_M$ and that f_2 is a $2\eta/m$-approximation of f such that for every $x \in (f_2^{-1}(0) \backslash M) \cap A$ with $\alpha = (G_v) = \alpha_1$ or α_2, there is $\varepsilon_x > 0$ ($\varepsilon_x = \frac{\nu_1(x)}{2}$ for $\alpha = \alpha_1$ and $\varepsilon_x = \frac{\nu_2(x)}{2}$ for $\alpha = \alpha_2$) such that for all $w \in N_x(V_\alpha)$ with $|w| < \varepsilon_x$, we have $f_2(x + w) = f_2(x) + w$. In other words, the map f_2 satisfies the α-normality condition in A for $\alpha = \alpha_1$ and α_2. We can also verify, in a similar way to that in the first step, that $f_2 \in \mathscr{C}^1_M$. We put $\Omega_2 = \mu(N(\tilde{A}_2, \nu_2/2))$.

In the kth step, we consider the set

$$Z_k := (f_{k-1}^{-1}(0)\backslash(M \cup \Omega_1 \cup \cdots \cup \Omega_{k-1})) \cap A$$

which is disjoint from all the sets $\overline{\Omega_1}, \overline{\Omega_2}, \ldots, \overline{\Omega_{k-1}}$. Moreover, all the orbit types in Z_k are larger than α_{k-1}. Next, we define the function

$$\varepsilon_k(x) := \frac{\varepsilon(x)}{n_{k1}} + \sum_{i=2}^{k} \frac{\operatorname{dist}(x, \Omega_{i-1})}{n_{ki}}$$

where the integers $n_{k1}, \ldots, n_{kk+1}$ are chosen sufficiently large. We put

$$D_k := \{x \in V\backslash M; \operatorname{dist}(x, Z_k) < \varepsilon_k(x)\}$$

$$L_k := \{x \in V\backslash M; \operatorname{dist}(x, Z_2) < 2\varepsilon_k(x)\}$$

$$A_k := \left\{x \in V\backslash M; \operatorname{dist}(x, Z_2) < \frac{\varepsilon_k(x)}{2}\right\}$$

and for $\alpha = \alpha_k$ we consider the sets $\tilde{D}_k := D_k \cap V_\alpha \subset L_k$ and $\tilde{A}_k := A_k \cap V_\alpha \subset L_k$. If $\tilde{A}_k = \emptyset$, then we put $f_k = f_{k-1}$; otherwise, we find a continuous invariant function $\nu_k : \tilde{D}_k \to \mathbb{R}_+$ such that $\nu_k(x) \leq \frac{\varepsilon(x)}{2}$, and the mapping $\mu : N(V_\alpha) \to V$, $\mu(v, w) = v + w$, restricted to the set $N(\tilde{D}_k, \nu_k)$, is a G-imbedding into L_k. We put $N_k := \mu(N(\tilde{D}_k, \nu_k))$. Let $\gamma_k : V\backslash M \to [0, 1]$ be an invariant C^∞-function such that $\gamma_k(x) = 1$ for $x \in \mu(N(\tilde{A}_k, \nu_k/2))$ and $\gamma_k(x) = 0$ for $x \in V\backslash(M \cup N_k)$. We define $f_k : V \to W$ by

$$f_k(x) := \begin{cases} f_{k-1}(x), & \text{if } x \in V\backslash N_k \\ \gamma_k(x)(f_{k-1}(v) + w) + (1 - \gamma_1(x))f_{k-1}(x), & \text{if } x = v + w \in N_k \end{cases}$$

We put $\delta_k = \max_{x \in \tilde{D}_k} \nu_k(x)$ and assume that

$$\sup_{v+w \in N_k} |f_{k-1}(v) - f_{k-1}(v - w)| < \frac{\eta}{m} - \delta_k$$

consequently,

$$\begin{aligned} \sup_{x \in V} |f_k(x) - f(x)| &\leq \sup_{x \in V} |f_k(x) - f_{k-1}(x)| + \frac{(k-1)\eta}{m} \\ &\leq \sup_{x = v + w \in N_k} |f_{k-1}(v) - f_{k-1}(x)| + \delta_k + \frac{(k-1)\eta}{m} \\ &\leq \frac{k\eta}{m} \end{aligned}$$

Next, we show by arguments similar to those above that $f_k \in \mathscr{C}^1_M$. By induction, f_k satisfies the α-normality condition on A for all $\alpha = \alpha_1, \alpha_2, \ldots, a_k$. Moreover, $Df_k(x) = Df(x)$ for all $x \in M$ and, thus, $\Lambda(f_k) = \Lambda(f)$.

For $m = k$, we put $\tilde{f} := f_m$. It is now clear that

$$\sup_{x \in V} |\tilde{f}(x) - f(x)| \le m \frac{\eta}{m} = \eta$$

and that $\tilde{f}$ satisfies all the required properties. $\square$

Assume $f \in \mathscr{C}^1_M$. We consider the bifurcation equation (8.7.1) on M. As was already pointed out, the set of bifurcation points of Eq. (8.7.1) is contained in the set $\Lambda(f)$ of M-singular points of f. Suppose that $x_0 \in M$ is an isolated point of $\Lambda(f)$. We want to use the G-degree to determine if x_0 is a bifurcation point. For this purpose, we consider an open bounded neighborhood D of x_0 in M such that $\overline{D} \cap \Lambda(f) = \{x_0\}$. Since $\overline{D}$ is a compact subset of M, there exists a sufficiently small $\varepsilon > 0$ such that the restriction of the map $\mu : N(M) \to W \oplus \mathbb{R}^n$, $\mu(x, v) = x + v$, to the set $N(\overline{D}, \varepsilon) = \{(x, v) \in N(M);\ x \in \overline{D},\ |v| \le \varepsilon\}$ is an equivariant imbedding. As we may choose $\varepsilon > 0$ arbitrarily small, we can assume that $f(x + v) \ne 0$ for every $x \in \partial D$ and $|v| \le \varepsilon$. Indeed, $\partial D \cap \Lambda(f) = \emptyset$; thus, all points from ∂D are not M-singular points of (8.7.1). We put $\overline{\mathscr{U}} := \mu(N(\overline{D}, \varepsilon))$ and call the set $\overline{\mathscr{U}}$ a *special neighborhood* of the isolated M-singular point $x_0 \in \Lambda(f)$. Let $\varphi : \overline{\mathscr{U}} \to \mathbb{R}$ be an invariant continuous function such that $\varphi(x) < 0$ for all $x \in \overline{D}$ and $\varphi(x) > 0$ for all $x = u + v$ with $u \in \overline{D}$ and $|v| = \varepsilon$. We will again call such a function φ a *complementing function*.

We define the map $f_\varphi : \mathscr{U} \to W \oplus \mathbb{R}$ by

$$f_\varphi(x) = (f(x), \varphi(x)), \qquad x \in \overline{\mathscr{U}}$$

It is clear that $f_\varphi(x) \ne 0$ for all $x \in \partial\mathscr{U}$. Therefore, the G-degree G-$\deg(f_\varphi, \mathscr{U})$ is well defined and we have the following.

Proposition 8.7.3

The G-degree G-$\deg(f_\varphi, \mathscr{U})$ does not depend on the choice of the special neighborhood $\mathscr{U}$ of x_0 nor on the choice of a complementing function φ. Moreover, if G-$\deg(f_\varphi, \mathscr{U}) \ne 0$, then x_0 is a bifurcation point of (8.7.1).

Proof. Suppose that $\varphi_0, \varphi_1 : \overline{\mathscr{U}} \to \mathbb{R}$ are two complementary functions. Then for every $t \in [0, 1]$, the function

$$\varphi_t(x) = t\varphi_1(x) + (1 - t)\varphi_0(x)$$

is also a complementing function. Thus, it follows from the homotopy property that

$$G\text{-}\deg(f_{\varphi_0}, \mathscr{U}) = G\text{-}\deg(f_{\varphi_t}, \mathscr{U}) = G\text{-}\deg(f_{\varphi_1}, \mathscr{U})$$

Suppose now that $\mathscr{U}_0$, $\mathscr{U}_1$ are two special neighborhoods of K such that $\overline{\mathscr{U}}_i := \mu(N(\overline{D}_i, \varepsilon_i))$, where $\overline{D}_0 \subset D_1$. Since the compact set $\overline{D}_1 \backslash D_0$ contains no M-singular point, there exists $\varepsilon > 0$ such that $\varepsilon < \min\{\varepsilon_0, \varepsilon_1\}$ and the set $A := \mu(N(\overline{D}_1 \backslash D_0, \varphi))$ contains no nontrivial solution of (8.7.1). Let $\varphi : W \oplus \mathbb{R}^n \to \mathbb{R}$ be an invariant function such that for $x \in \overline{\mathscr{U}}_1$, $x = u + v$, $u \in \overline{D}_1$, $|v| > \varepsilon/2$, we have $\varphi(u+v) > 0$, and $\varphi(x) < 0$ for $x \in \overline{D}_1$. Then φ is a complementing function for both special neighborhoods $\mathscr{U}_0$ and $\mathscr{U}_1$. Therefore, by using the excision property twice, we obtain the following:

$$\begin{aligned} G\text{-deg}(f_\varphi, \mathscr{U}_1) &= G\text{-deg}(f_\varphi, \mu(N(D_1, \varepsilon))) \\ &= G\text{-deg}(f_\varphi, \mu(N(D_0, \varepsilon))) \\ &= G\text{-deg}(f_\varphi, \mathscr{U}_0) \end{aligned}$$

and thus, $G\text{-deg}(f_\varphi, \mathscr{U})$ does not depend on the choice of a special neighborhood $\mathscr{U}$ of x_0.

Suppose that $G\text{-deg}(f_\varphi, \mathscr{U}) \neq 0$, and let $\mathscr{V}$ be an arbitrary open neighborhood of x_0 in $W \oplus \mathbb{R}$. Then there exists another special neighborhood $\mathscr{U}_1$ of x_0 such that $\mathscr{U}_1 \subset \mathscr{U} \cap \mathscr{V}$. Assume that $\varphi_1 : \overline{\mathscr{U}} \to \mathbb{R}$ is a complementary function. Then $0 \neq G\text{-deg}(f_\varphi, \mathscr{U}) = G\text{-deg}(f_{\varphi_1}, \mathscr{U}_1)$; hence, by the existence property, there exists a solution to the system.

$$\begin{cases} f(x) = 0 \\ \varphi_1(x) = 0 \end{cases}$$

which is evidently a nontrivial solution of (8.7.1). This means that in an arbitrarily small neighborhood of x_0, there is a nontrivial solution of (8.7.1). So, x_0 is a bifurcation point. □

Definition 8.7.4 Let $\alpha \in J(W \oplus \mathbb{R})$. We say that an isolated in $\Lambda(f)$ point x_0 is an *α-essential bifurcation point of (8.7.1)* if for every bounded open neighborhood U of x_0 in $W \oplus \mathbb{R}$, there is $\eta > 0$ such that for all $\tilde{f} \in \mathscr{G}_M(\overline{U})$ satisfying the following:

(i) $\sup_{x \in W \oplus \mathbb{R}} |f(x) - \tilde{f}(x)| < \eta$

(ii) $\Lambda(\tilde{f}) = \Lambda(f)$

the equation

$$\tilde{f}(x) = 0, \qquad x \in W \oplus \mathbb{R} \tag{8.7.2}$$

has a sequence of nontrivial solutions $\{x_n\}$ such that

(a) $\operatorname{dist}(x_n, x_0) \to 0$ as $n \to \infty$

(b) $(G_{x_n}) = \alpha$ for all $n = 1, 2, \ldots$

Lemma 8.7.5

If $x_0 \in \Lambda(f)$ is an α-essential bifurcation point of (8.7.1), then x_0 is a bifurcation point.

Proof. If x_0 is not a bifurcation point, then there exists a neighborhood U of x_0 in $W \oplus \mathbb{R}$ that does not contain a nontrivial solution of (8.7.1). Then the mapping $\tilde{f} = f$ belongs to $\mathcal{G}_M(\overline{U})$ and x_0 is not an α-essential bifurcation point of (8.7.1) for all $\alpha \in J(W \oplus \mathbb{R})$. $\square$

Theorem 8.7.6 (Branching Lemma)

If G-deg$(f_\varphi, \mathcal{U}) = \sum_\alpha n_\alpha \alpha$ is such that $n_\alpha \neq 0$, where $\alpha \in \Phi(G)$, then x_0 is an α-essential bifurcation point of (8.7.1). That is, there exists $\eta > 0$ such that every $\tilde{f} \in \mathcal{G}_M(\overline{\mathcal{U}})$ satisfying

$$\sup_{x \in W \oplus \mathbb{R}} |f(x) - \tilde{f}(x)| < \eta$$

and $\Lambda(\tilde{f}) = \Lambda(f)$, the equation $\tilde{f}(x) = 0$ has a continuum C_α of nontrivial solutions bifurcating from x_0, such that $(G_x) = \alpha$ for all $x \in C_\alpha$ (we say that C_α is an α-branch of nontrivial solutions bifurcating from x_0).

Proof. Let $\mu(N(D, \varepsilon))$ be a special neighborhood $\mathcal{U}$ of the isolated compact M-singular set K and let $\varphi : \mathcal{U} \to \mathbb{R}$ be a complementing function on $\mathcal{U}$. Assume that

$$0 < 2\eta < \inf\{|f(x)|;\ x = u + v,\ u \in \partial D,\ |v| \leq \varepsilon,\ \varphi(x) = 0\}$$

Then for every $\tilde{f} \in \mathcal{G}_M(\mathcal{U})$ such that $\Lambda(\tilde{f}) = \Lambda(f)$ and $\sup_{x \in \mathcal{U}} |\tilde{f}(x) - f(x)| < \eta$, the homotopy

$$h_\varphi(x, t) := (t\tilde{f}(x) + (1 - t)f(x), \varphi(x)), \qquad x \in \overline{\mathcal{U}}, \quad t \in [0, 1]$$

has no zero on $\partial\mathcal{U}$. Indeed, for all $x = u + v$ such that $u \in \partial D$, $|v| \leq \varepsilon$ and $\varphi(x) = 0$, we have

$$\begin{aligned} |t\tilde{f}(x) + (1 - t)f(x)| &\geq |f(x)| - t|\tilde{f}(x) - f(x)| \\ &\geq 2\eta - \eta > 0 \end{aligned}$$

Thus,

$$G\text{-deg}(f_\varphi, \mathcal{U}) = G\text{-deg}(\tilde{f}_\varphi, \mathcal{U})$$

The map $\tilde{f}_\varphi$ is not normal (in the sense of Definition 8.7.1), but it may be "corrected" to a normal map $\tilde{f}_{\tilde{\varphi}}$ such that $\tilde{f}_\varphi^{-1}(0) = \tilde{f}_{\tilde{\varphi}}^{-1}(0)$. For this purpose, we consider the set $Z = \tilde{f}_\varphi^{-1}(0) = \tilde{f}^{-1}(0) \cap \varphi^{-1}(0)$. Since $\tilde{f}$ is a normal bifurcation map, for every $x \in Z$ with $\alpha = (G_x)$, we have

$$\tilde{f}(x+h) = \tilde{f}(x) + h \tag{8.7.3}$$

where $h \in N_\alpha(V_\alpha)$, $|h| < \varepsilon_x$. We may assume, without loss of generality, that the α-normality condition (8.7.3) is satisfied for $x \in U_\alpha$, where U_α is an open invariant neighborhood of $Z \in V_\alpha$ in V_α for all $\alpha \in J(Z)$. We may assume that all the orbit types in $J(Z)$ are totally ordered:

$$\alpha_1 < \alpha_2 < \cdots < \alpha_m$$

Let $\alpha = \alpha_1$. We consider the set $Z_\alpha = Z \cap V_\alpha$. We claim that Z_α is compact. Indeed, if $\{x_n\} \subset Z_\alpha$ is a sequence such that $x_n \to x \not\in Z_\alpha$, then the slice theorem (Theorem 8.1.10) implies $(G_x) < (G_{x_n}) = \alpha$ that is a contradiction of the assumption that α is the minimal orbit type in $J(Z)$.

Consequently, we can find two relatively compact, open, invariant neighborhoods A_1 and B_1 of Z_α in V_α such that

$$Z_\alpha \subset A_1 \subset \overline{A}_1 \subset B_1 \subset \overline{B}_1 \subseteq \overline{U}_\alpha$$

Let $\delta_1 > 0$ be given so $\delta_1 < \varepsilon_x$ for all $x \in B_1$. We put $N_1 = \mu(N(B_1\ \delta_1))$ and denote by $\gamma_1 : V \to [0, 1]$ an invariant C^∞-function such that $\gamma_1(x) = 1$ for $x \in \mu(N(A_1, \delta_1/2))$ and $\gamma_1(x) = 0$ for $x \in V \backslash N_1$. Then we define

$$\tilde{f}_\varphi^1(x) = \begin{cases} \tilde{f}_\varphi(x), & \text{if } x \in V \backslash N_1 \\ \gamma_1(x)[\tilde{f}_\varphi(v) + w] + (1 - \gamma_1(x))\tilde{f}_\varphi(x), & \text{for } x = v + w \in N_1 \end{cases}$$

where $x = v + w$ denotes the decomposition with $v \in B_1$ and $w \in N_v(V_\alpha)$. Note that for $x = v + w \in N_1$, we have

$$\begin{aligned} \tilde{f}_\varphi^1(x) &= (\gamma_1(x)(\tilde{f}(v) + w) + (1 - \gamma_1(x))\tilde{f}(x), \gamma_1(x)\varphi(v) + (1 - \gamma_1(x))\varphi(x)) \\ &= (\tilde{f}(v) + w, \gamma_1(x)\varphi(v) + (1 - \gamma_1(x))\varphi(x)) \\ &= (\tilde{f}(x), \gamma_1(x)\varphi(v) + (1 - \gamma_1(x))\varphi(x)) \end{aligned}$$

Furthermore, for $x \in N_1$, we have $\tilde{f}_\varphi^1(x) = 0$ if and only if $w = 0$, $\tilde{f}(v) = 0$, and $\varphi(v) = 0$. Therefore, $\tilde{f}_\varphi^1(x) = 0$ if and only if $\tilde{f}_\varphi(x) = 0$. We put $\Omega_1 = \mu(N(A_1, \delta_1/2))$. Consequently, the mapping $\tilde{f}_\varphi^1$ is normal in the set Ω_1.

Assume now, for the induction, that we have constructed the mapping $\tilde{f}_\varphi^k$ such that it is normal in the open invariant sets $\Omega_1, \ldots, \Omega_k$ with $Z \cap V_{\alpha_i} \subset \Omega_i$, and that $\tilde{f}_\varphi^k(x) = 0$ if and only if $\tilde{f}_\varphi(x) = 0$. We consider the orbit type $\alpha = \alpha_{k+1}$ and the set $Z_\alpha = Z \cap V_\alpha$. We claim that the set Z_α is compact and

$Z_\alpha \cap \Omega_i = \varnothing$ for all $i = 1, 2, \ldots, k$. Indeed, suppose that $\{x_n\} \subset Z_\alpha$ is a sequence such that $x_n \to x \notin Z_\alpha$. Then $\beta = (G_x) \in \{\alpha_1, \ldots, \alpha_k\}$ and, consequently, $\tilde{f}^k_\varphi$ satisfies the β-normality condition in a neighborhood of x in V_β. But this condition excludes the possibility of the existence of other zeros of $\tilde{f}^k_\varphi$, with the orbit type larger than β in a neighborhood of x. This is a contradiction of the assumption that $x_n \to x$.

Now we define relatively compact, open, and invariant neighborhoods A_{k+1} and B_{k+1} of Z_α in V_α such that

$$Z_\alpha \subset A_{k+1} \subset \overline{A}_{k+1} \subset B_{k+1} \subset \overline{B}_{k+1} \subset U_\alpha$$

Then there is $\delta_{k+1} > 0$ such that $\delta_{k+1} < \varepsilon_x$ for all $x \in B_{k+1}$. We put $N_{k+1} := \mu(N(B_{k+1}, \delta_{k+1}))$. Let $\gamma_{k+1} : V \to [0, 1]$ be an invariant C^∞-function such that $\gamma_{k+1}(x) = 1$ for $x \in \mu\left(N\left(A_{k+1}, \frac{\delta_{k+1}}{2}\right)\right)$ and $\gamma_{k+1}(x) = 0$ for $x \in V \backslash N_{k+1}$. Then we can define

$$\tilde{f}^{k+1}_\varphi(x) = \begin{cases} \tilde{f}^k_\varphi(x), & \text{if } x \in V \backslash N_{k+1} \\ \gamma_{k+1}(x)[\tilde{f}^k_\varphi(v) + w] + (1 - \gamma_{k+1}(x))\tilde{f}^k_\varphi(x), & \text{if } x = v + w \in N_{k+1} \end{cases}$$

and again

$$\tilde{f}^{k+1}_\varphi(x) = (\tilde{f}(x), \gamma_{k+1}(x)\varphi(v) + (1 - \gamma_{k+1}(x)\varphi(x))$$

Thus, for $k = m - 1$, we obtain $\tilde{f}_{\tilde{\varphi}}(x) = \tilde{f}^m_\varphi(x)$, and $\tilde{f}_{\tilde{\varphi}}$ is the required normal mapping such that $\tilde{f}^{-1}_{\tilde{\varphi}}(0) = \tilde{f}^{-1}_\varphi(0)$.

In the above arguments, we have constructed a new complementing function $\tilde{\varphi}$ such that $\tilde{f}_{\tilde{\varphi}}(x) = (\tilde{f}(x), \tilde{\varphi}(x))$ is normal.

To complete the proof, we notice that the normality of $\tilde{f}_{\tilde{\varphi}}$ implies that for every $n_\alpha \neq 0$, there is x such that $\tilde{f}_{\tilde{\varphi}}(x) = 0$ and $(G_x) = \alpha$. Since $\tilde{f}^{-1}_{\tilde{\varphi}}(0) = \tilde{f}^{-1}_\varphi(0)$, the conclusion follows from the fact that φ can be an arbitrary complementing function. □

Now we discuss the global bifurcation problem for (8.7.1). In what follows, we assume that every point of $\Lambda(f)$ is isolated in M. Consequently, for every point $x \in \Lambda(f)$, there is a special neighborhood $\mathcal{U}$ together with a complementing function $\varepsilon : \overline{\mathcal{U}} \to \mathbb{R}$, and we can define the integer $n_\alpha(x)$ by

$$G\text{-deg}(f_\varepsilon, \mathcal{U}) = \sum_\alpha n_\alpha(x)\alpha$$

Let us denote by $\mathfrak{S}(f)$ the closure of the set of all nontrivial solutions of

(8.7.1), that is,

$$\mathfrak{S}(f) := \overline{\{x \in V \backslash M : f(x) = 0\}}$$

Using the same argument as that of Theorem 8.5.4, we obtain the following.

Theorem 8.7.7

Assume that every point of $\Lambda(f)$ is isolated in M. Let $\mathfrak{C}$ be a bounded connected component of the set $\mathfrak{S}(f)$. Then there exists only a finite collection of points $\{x_i\}_{j=1}^{r} \subseteq M$ such that $x_i \in \mathfrak{C}$. Moreover, for all $\alpha \in \Phi(G)$, we have the following relation:

$$\sum_{j=1}^{r} n_\alpha(x_i) = 0 \tag{8.7.4}$$

8.8 STEADY-STATE BIFURCATION SPONTANEOUS SYMMETRY BREAKING

Let G be a compact Lie group and $W = \mathbb{R}^n$ an orthogonal representation of G. Assume that $f : W \times \mathbb{R} \to W$ is a smooth G-equivariant map, that is, $f(gx, \lambda) = gf(x, \lambda)$ for all $g \in G$, $x \in W$, $\lambda \in \mathbb{R}$. Consider the ordinary differential equation

$$\frac{dx}{dt} = f(x, \lambda) \tag{8.8.1}$$

We assume that $f(0, \lambda) = 0$ and $L := D_x F(0, 0)$ is singular. Thus, L has a zero eigenvalue. Due to the presence of symmetries, this zero eigenvalue is generally not simple. In this section, we shall present some particular classes of steady-state bifurcations for the action of the group $SO(3)$.

There is great interest in finding solutions of (8.8.1) with special symmetry such as axisymmetric solutions. The lattice of isotropy subgroups of G, for irreducible representations of G, provides a way to classify solutions by their symmetries. Let $x \in W$ and $H = G_x$ be the isotropy group of x. Then the mapping f preserves the fixed-point subspace W^H, that is, W^H is flow-invariant, that is, any trajectory of (8.8.1) starting in W^H remains in W^H. We would like to know when there exists a *branch of solutions* with isotropy H and bifurcating from the solution $x = 0$. Since this branch of solutions has less symmetry than the solution $x = 0$, this phenomenon of bifurcation is called *spontaneous symmetry breaking*.

Since the isotropy subgroups of points on the same orbit are conjugate, we will talk about an orbit type (H) rather than isotropy groups H when we describe the symmetry property of the orbit.

In this section we will study the case where $V = \mathrm{Ker}(L)$ is an absolutely irreducible representation of G. So, the only linear maps on $\mathrm{Ker}(L)$ that commute with G are scalar multiples of the identity. Therefore, the multiplicity of the zero eigenvalue of L is equal to the dimension m of $\mathrm{Ker}(L)$. Let us denote by $\sigma(L_\lambda)$ the spectrum of the operator $L_\lambda := D_x f(0, \lambda)$, and let σ_- (respectively, σ_+) denote the number of negative eigenvalues of L_λ for $\lambda = -\rho$ (respectively, $\lambda = \rho$) with some sufficiently small $\rho > 0$. We will call the integer $\sigma = \sigma_- - \sigma_+$ the *crossing number* for the problem (8.8.1). We assume that $\sigma = \pm m$, that is, the eigenvalue of L changes sign by passing through zero.

Since we have assumed that $V = \mathrm{Ker}(L)$ is an absolutely irreducible representation of G, $V^G = \{0\}$, and for all $x \in V \backslash \{0\}$, the isotropy group G_x is not equal to G. If (H) is a minimal orbit type in $V \backslash \{0\}$, then the isotropy group H is called *maximal*. All other isotropy groups in $V \backslash \{0\}$ are called *submaximal*.

The following version of the equivariant branching lemma was first proved by Cicogna (1981). Here we present a proof of this result, as an application of the equivariant degree.

Proposition 8.8.1

Suppose that W is an orthogonal representation of the Lie group G, $F : W \times \mathbb{R} \to W$ is a smooth G-equivariant map such that $f(0, \lambda) = 0$, $L_\lambda := D_x f(0, \lambda)$ is nonsingular for $\lambda \neq 0$, and singular for $\lambda = 0$. Moreover, assume that $\mathrm{Ker}(L_0) = V$ *is an absolutely irreducible representation of G such that the crossing number σ associated with (8.8.1) is equal $\pm m$, where $m =$* $\dim V$. *Then for every maximal isotropy group H in V such that* $\dim V^H$ *is odd, there exists a branch of stationary points of (8.8.1) bifurcating from* $(0, 0)$ *with the orbit type exactly* (H).

Proof. Without loss of generality, we can assume that a reduction has been carried out so that $\mathrm{Ker}(L) = \mathbb{R}^n = W$. We will apply Proposition 8.5.1 and Theorem 8.5.2. Therefore, we need to compute

$$G\text{-deg}(f_\varphi, U(r, \rho)) = G\text{-deg}(f_-, \Omega^-(r, \rho)) - G\text{-deg}(f_+, \Omega^+(r, \rho))$$

where $\Omega^\pm(r, \rho) = \{x \in W;\ \|x\| < r\}$, $U(r, \rho)$ is a special neighborhood of $(0, 0)$, $f_\pm : W \to W$ is defined by $f_\pm(x) = f(x, \pm\rho)$, $x \in W$, and φ is a complementary function. By using the linearization at the points $(0, \pm\rho)$, we may assume that $f_\pm(x) = \pm x$ (or $f_\pm = \mp x$). Since $G\text{-deg}(\mathrm{Id}, \Omega^+(r, \rho)) = (G)$, we need only prove that the (H)-component $m_{(H)}$ of $G\text{-deg}(f_-, \Omega^-(r, \rho)) = G\text{-deg}(-\mathrm{Id}, \Omega^-(r, \rho))$ is nonzero. For this purpose, we will apply formula (8.3.2). That is,

$$m_{(H)}(F_-) = [I(F_-^H) - I(F_-^{[H]})]/|W(H)|\,, \qquad F := Id - f$$

Since W^H is an odd-dimensional space and since we can replace f_- by $-\mathrm{Id}$, we have $I(F_-^H) = -1$. By assumption, H is a maximal isotropy group and $W^{[H]} = \{0\}$; thus, $I(F_-^{[H]}) = 1$. Note that $W(H)$ acts freely on the even-dimensional unite sphere in V^H. By Proposition 2.4.8, $|W(H)|$ has to be 1 or 2. Consequently, $m_{(H)} = (-1-1)/|W(H)| = -2$ or -1. This shows that the (H)-component of $G\text{-deg}(f_\varphi, U(r, \rho))$ is equal to -2 or -1, and the conclusion follows from Proposition 8.5.1. □

The method employed in the proof of Proposition 8.8.1 illustrates the importance of computing $G\text{-deg}(-\mathrm{Id}, \Omega)$, where Ω denotes the unit ball in $V = \mathrm{Ker}(L)$, in order to determine the occurrence of bifurcations. In particular, for every nonzero (H)-component $m_{(H)}$ of $G\text{-deg}(-\mathrm{Id}, \Omega) - (G)$, there exists a branch of solutions with orbit type (K) such that $(G) < (K) \leq (H)$. Unfortunately, we are unable to determine if for a submaximal isotropy group H, there is a bifurcating branch with the orbit type (H). Another interesting problem is to construct a parametrized "generic" mapping f, similarly to normal mappings constructed in the definition of the G-degree, which has the property that $m_{(H)} \neq 0$ implies the existence of a branch of solutions with orbit type (H). In other words, the question is: Can we construct some types of parametrized normal mapping f such that if $G\text{-deg}(f_\varepsilon, U(r, \rho)) = \sum_\alpha m_\alpha \cdot (\alpha)$ and $m_\alpha \neq 0$ for some α, then $(0, 0)$ is an α-essential bifurcation point in the sense of Definition 8.7.4?

We mention that our approach can be used to study more complex situations such as the case where $\mathrm{Ker}(L) = V$ is a direct sum of two or more absolutely irreducible representations of G. In this case, the multiplicativity property as well as the multiplication table for $A(G)$ can provide us with the information necessary to compute the degree $G\text{-deg}(f_\varphi, U(r, \rho)) = G\text{-deg}(f_-, \Omega^-(r, \rho)) - G\text{-deg}(f_+, \Omega^+(r, \rho))$. We will illustrate this situation by an example.

Let us now consider the case where $G = SO(3)$. In Example 8.4.8, we computed the G-degrees $SO(3)\text{-deg}(-\mathrm{Id}, \Omega_i)$, where Ω_i denotes the unit ball in the $(2i+1)$-dimensional absolutely irreducible representation V_i, $i = 1, 2, 3, 4, 5$, of the group $SO(3)$. We have obtained the following results:

$$SO(3)\text{-deg}(-\mathrm{Id}, \Omega_1) = (SO(3)) - (SO(2))$$

$$SO(3)\text{-deg}(-\mathrm{Id}, \Omega_2) = (SO(3)) - 2(O(2)) + (\mathbf{V_4})$$

$$SO(3)\text{-deg}(-\mathrm{Id}, \Omega_3) = (SO(3)) - (A_4) - (SO(2)) - (D_3)$$

$$SO(3)\text{-deg}(-\mathrm{Id}, \Omega_4) = (SO(3)) - 2(O(2)) - 2(S_4) + 2(D_4) + 2(D_3) - (\mathbf{V_4})$$

$$SO(3)\text{-deg}(-\mathrm{Id}, \Omega_5) = (SO(3)) - (SO(2)) - (D_5) - (D_4) - (D_3) + (\mathbf{V_4})$$

Example 8.8.2 Consider the steady-state bifurcation problem (8.8.1) in the case where $G = SO(3)$ and $\mathrm{Ker}(L) = V_3 \oplus V_5$. Assume, for example, that $V = V_3 \oplus V_5$, and that $L_{\lambda|V_3} : V_3 \to V_3$ (respectively, $L_{\lambda|V_5} : V_5 \to V_5$) changes its eigenvalue from negative to positive, as λ crosses zero from negative to positive values. Then, in order to determine the "essential" orbit types for bifurcating branches of solutions, that is, the orbit types that have orbit types (H) corresponding to a nonzero component $m_{(H)}$ of the degree $SO(3)\text{-deg}(f_\varphi, U(r,\rho))$, it suffices to notice that

$$\begin{aligned} SO(3)\text{-deg}(f_\varphi, U(r,\rho) &= SO(3)\text{-deg}(-\mathrm{Id}, \Omega) - SO(3)\text{-deg}(\mathrm{Id}, \Omega) \\ &= (SO)(3)\text{-deg}(-\mathrm{Id}, \Omega_3)) \cdot (SO(3)\text{-deg}(-\mathrm{Id}, \Omega_5)) - (SO(3)) \\ &= -(A_4) - (D_5) - (D_4) + (\mathbf{V_4}) \end{aligned}$$

where Ω denotes the unit ball in V. Consequently, we can detect the existence of branches with the orbit types (H), where $H = A_4, D_5, D_4$ (which reflects the fact that the corresponding fixed-point spaces V^H are one-dimensional) and for $H = \mathbf{V_4}$, which is not a maximal orbit type. The interaction between two representations, due to nonlinear coupling, destroys the essential bifurcation of the branches with the orbit types $(SO(2))$ and (D_3); thus, detecting them is difficult. Possibly, there is no normal bifurcations of this type at all. The method developed in this section, although it cannot be applied to detect exact orbit types of bifurcating branches, can be effectively applied to those situations where there are several irreducible components in $\mathrm{Ker}(L)$ to determine what types of bifurcation can surely take place.

EXERCISES

8.8.1 Let m denote a positive integer, and Ω_i the unit ball in the irreducible representation V_i of the group $SO(3)$. Verify that

$$[SO(3)\text{-deg}(-\mathrm{Id}, \Omega_i)]^m = \begin{cases} (SO(3)), & \text{if } m \text{ is even} \\ SO(3)\text{-deg}(-\mathrm{Id}, \Omega_i), & \text{if } m \text{ is odd} \end{cases}$$

for $i = 1, 2, 3, 4, 5$. Do you really need to use the multiplication table? Can you generalize this formula for $[G\text{-deg}(-\mathrm{Id}, \Omega)]^m$, where G is any compact Lie group, and Ω is a unit ball in an absolutely irreducible representation V of G?

8.8.2 Suppose that W is an orthogonal representation of the group $G = SO(3)$, $f: W \times \mathbb{R} \to W$ is a smooth G-equivariant map such that $f(0, \lambda) = 0$, $L_\lambda := D_x f(0, \lambda)$ is nonsingular for $\lambda \neq 0$, and singular for $\lambda = 0$. Assume also that $\mathrm{Ker}(L_0) = V_1^{m_1} \oplus V_2^{m_2} \oplus V_3^{m_3} \oplus V_4^{m_4} \oplus V_5^{m_5}$, where $V_i^m = \underbrace{V_i \oplus \cdots \oplus V_i}_{m \text{ times}}$. Describe, using the G-equivariant degree, all possible steady-state bifurcations from $(0, 0)$. List all the "essential" orbit types for each case.

8.9 BIBLIOGRAPHICAL NOTES

The basic notations and results on the theory of transformation groups are formulated in the first section, and we refer the interested reader to Bredon (1972), Bröcker and tom Dieck (1985), Kawakubo (1991), Murayama (1983), Rubinsztein (1976), Serre (1968), and tom Dieck (1987) for a more detailed discussion. The definition of the Burnside ring introduced here is slightly different from but equivalent to the one given in tom Dieck (1987). The computations of the Burnside ring $A(SO(3))$ are based on the results of tom Dieck (1987) and some computations in Chossat et al. (1990).

For detailed discussions of the equivariant degree and equivariant fixed-point index, we refer the reader to Dold (1983), Komiya (1988), Matsuoka (1989), and Ulrich (1988). The idea of normal approximations presented in Section 8.3 follows the approach developed by Krawcewicz and Xia (1996) and Xia (1994). The original idea of normal approximation should be attributed to Gęba.

Theorem 8.5.2 has been proved in Krawcewicz and Wu (1994). The results on the bifurcation of periodic solutions for time-reversible systems are taken from Krawcewicz and Wu (1994). The branching lemma and the results for steady-state bifurcation of $SO(3)$ are derived from Krawcewicz and Vivi (1995). Also see Cicogna (1981), Golubitsky et al. (1988), Fiedler and Mischaikow (1992), Field and Richardson (1990), Ihring and Golubitsky (1984), Ize (1993), Ize et al. (1989, 1992, 1993), Ruelle (1973) and Smoller (1983) for related work.

References

Abraham, R., Marsden, J. E., and Ratin, T. (1988). *Manifolds, Tensor Analysis, and Applications*, 2nd ed. Springer-Verlag, New York.

Adams, J. F. (1969). *Lectures on Lie Groups*. Benjamin, New York.

Akhmerov, R. R., Kamenskii, M. I., Potapov, A. S., and Rodkina, A. E. (1992). *Measures of Noncompactness and Condensing Operators. Operator Theory: Advances and Applictions*, Vol. 55. Birkhäuser Verlag, Basel.

Alexander, J. C. and Yorke, J. A. (1978). Global bifurcation of periodic orbits. *Amer. J. Math.*, **100**, 263–292.

Alligood, K. T., Mallet-Paret, J., and Yorke, J. A. (1981). Families of periodic orbits: Local continuability does not imply global continuability. *J. Diff. Geom.*, **16**, 483–492.

Alligood, K. T., Mallet-Paret, J., and Yorke, J. A. (1983). An index for the global continuations of relatively isolated sets of periodic orbits. In *Lecture Notes in Mathematics*, No. 1007 (J. Palis, Jr., Ed.). Springer-Verlag, New York, pp. 1–21.

Alligood, K. T. and Yorke, J. A. (1983). Cascades of period-doubling bifurcations: A prerequisite for horseshoes. *Bull. Amer. Math. Soc.*, **9**, 319–322.

Appell, J. (1981). Implicit function, nonlinear integral equations and the measure of noncompactness of superposition operator. *J. Math. Anal. Appl.*, **83**, 251–261.

Appell, J. and Zabrejko, P. (1983). On a theorem of M. A. Krasnosiel'skii. *Nonlinear Anal. TMA*, **7**, 695–706.

Arnol'd, V. I. (1978). *Mathematical Methods of Classical Mechanics. Springer Graduate Texts in Mathematics*, Vol. 60. Springer-Verlag, New York.

Banaś, J. and Goebel, K. (1980). *Measures of Noncompactness in Banach Spaces. Lecture Notes in Pure Applied Mathematics*, Vol. 60. Marcel Dekker, New York.

Bartsch, T. (1993). A simple proof of the degree formula for $(\mathbb{Z}/p)$-equivariant maps. *Math. Zeitschrift*, **12**, 285–292.

Bebernes, J. W. and Schmitt, K. (1973). Periodic boundary value problems for systems of second order differential equations. *J. Diff. Eqns.*, **13**, 33–47.

Bernfeld, S. R., Ladde, G., and Lakshmikantham, V. (1974). Nonlinear boundary value problems and several Lyapunov functions for functional differential equations. *Boll. Un. Mat. Ital.*, **10**, 602–613.

Borisovich, Y. G. and Sapronov, Y. I. (1968). Topological theory of condensing operators. *Dokl. Akad Nauk SSSR*, **183**, 18–20 (in Russian).

Borsuk, K. (1967). *Theory of Retracts*. PWN—Polish Scientific Publishers, Warszawa.

Bott, R. and Tu, L. W. (1982). *Forms in Algebraic Topology*. Springer-Verlag, New York.

Brayton, R. K. (1966). Bifurcation of periodic solutions in a nonlinear difference-differential equation of neutral type. *Quart. Appl. Math.*, **24**, 215–224.

Brayton, R. K. (1967). Nonlinear oscillations in a distributed network. *Quart. Appl. Math.*, **24**, 289–301.

Brayton, R. K. and Krumme, D. W. (1968). Differential-difference equations and nonlinear initial-boundary value problem for linear hyperbolic partial differential equations. *J. Math. Anal. Appl.*, **24**, 372–387.

Brayton, R. K. and Miranker, W. L. (1964). A stability theory for nonlinear mixed initial boundary value problems. *Arch. Rational Mech. Anal.*, **17**, 358–376.

Bredon, G. E. (1972). *Introduction to Compact Transformation Groups*. Academic Press, New York.

Bröcker, T. and Jänich, K. (1982). *Introduction to Differential Topology*. Cambridge University Press, Cambridge.

Bröcker, T. and tom Dieck, T. (1985). *Representations of Compact Lie Groups*. Springer-Verlag, New York.

Brouwer, L. E. J. (1912a). Über Abbildung von Mannigfaltigkeiten. *Math. Ann.*, **71**, 97–115.

Brouwer, L. E. J. (1912b). Invarianz des n-dimensionalen Gebiets. *Math. Ann.*, **71**, 305–313.

Browder, F. E. and Petryshyn, W. V. (1969). Approximation methods and generalized topological degree for nonlinear mappings in Banach spaces. *J. Funct. Anal.*, **3**, 217–245.

Brown, R. F. (1967). *The Lefschetz Fixed Point Theorem*. Foresman and Co., London.

Chafee, N. (1971). A bifurcation problem for a functional differential equation of finitely retarded type. *J. Math. Anal. Appl.*, **35**, 312–348.

Chiappinelli, R. (1989). Remarks on Krasnosiel'skii bifurcation theorem. *Comment. Math. Univ. Carol.*, **30,** 235–241.

Chossat, P., Lauterbach, R., and Melbourne, I. (1990). Steady-state bifurcation with $O(3)$-symmetry. *Arch. Rational Mech. Anal.*, **113**, 313–376.

Chow, S. N. and Hale, J. K. (1982). *Methods of Bifurcation Theory*. Springer-Verlag, New York.

Chow, S. N. and Mallet-Paret, J. (1977). Integral averaging and bifurcation. *J. Diff. Eqns.*, **26**, 112–159.

Chow, S. N. and Mallet-Paret, J. (1978). The Fuller index and global Hopf bifurcation. *J. Diff. Eqns.*, **29**, 66–85.

Chow, S. N., Mallet-Paret, J., and Yorke, J. A. (1978). Global Hopf bifurcation from multiple eigenvalue. *Nonlinear Anal. TMA*, **2**, 753–763.

Chukwu, E. N. (1992). *Stability and Time-optimal Control of Hereditary Systems*. Academic Press, New York.

Cicogna, G. (1981). Symmetry breakdown from bifurcation. *Lettere al Nuovo Cimento*, **31**, 600–602.

Claeyssen, J. R. (1980). The integral-averaging bifurcation method and the general one-delay equation. *J. Math. Anal. Appl.*, **78**, 428–439.

Crandall, M. G. and Rabinowitz, P. H. (1971). Bifurcation from simple eigenvalues. *J. Funct. Anal.*, **8**, 321–340.

Crandall, M. G. and Rabinowitz, P. H. (1974). Bifurcation, perturbation of simple eigenvalues and linearized stability. *Arch. Rational Mech. Anal.*, **52**, 161–180.

Crandall, M. G. and Rabinowitz, P. H. (1977). The Hopf bifurcation theorem in infinite dimensions. *Arch. Rational Mech. Anal.*, **67**, 161–180.

Coddington, E. A. and Levinson, N. (1955). *Theory of Ordinary Differential Equations*. McGraw-Hill, New York.

Cooke, K. L. and Krumme, D. W. (1968). Differential-difference equations and nonlinear initial-boundary value problems for linear hyperbolic partial differential equations. *J. Math. Anal. Appl.*, **24**, 372–387.

Cronin, J. (1964). *Fixed Points and Topological Degree in Nonlinear Analysis. Mathematical Surveys*, Vol. 11. American Mathematical Society, Providence, R.I.

Dancer, E. N. (1980). On the existence of bifurcating solutions in the presence of symmetries. *Proc. Roy. Soc. Edinburgh*, **85A**, 321–336.

Dancer, E. N. (1982). Symmetries, degree, homotopy indices and asymptotically homogeneous problems. *Nonlinear Anal. TMA*, **92**, 667–686.

Dancer, E. N. (1985). A new degree for S^1-invariant gradient mappings and applications. *Ann. Inst. Henri Poincaré, Anal. Non Linéaire*, **2**, 329–370.

Deimling, K. (1984). *Nonlinear Functional Analysis*. Springer-Verlag, New York.

tom Dieck, T., (1987). *Transformation Groups*. de Gruyter, Berlin.

Dinculeanu, N. (1974). *Integration on Locally Compact Groups*. Noordhoff, Leyden, International Publishing.

Dold, A. (1965). Fixed point index and fixed point theorem for Euclidean neighborhood retracts. *Topology*, **4**, 1–8.

Dold, A. (1972). *Lectures on Algebraic Topology*, 2nd ed. Springer-Verlag, New York.

Dold, A. (1982). Fixed point theory and homotopy theory. *Contemp. Math.*, **12**, 105–115.

Dold, A. (1983). Fixed point indices of iterated maps. *Invent. Math.*, **74**, 419–435.

Dold, A. (1984). Combinatorial and geometric fixed point theory. *Riv. Math. Univ. Parma*, **4**, 23–32.

Dubrovin, B. A., Novikov, C. P., and Fomenko, A. T. (1986). *Contemporary Geometry: Methods and Applications*. Nauka, Moscow (in Russian).

Duda, R. (1986). *Introduction to Topology, Part II—Algebraic Topology and Topology of Manifolds*. PWN—Polish Scientific Publishers, Warszawa (in Polish).

Dugundji, J. (1951). An extension of Tietze's theorem. *Pac. J. Math.*, **1**, 353–367.

Dugundji, J. (1966). *Topology*. Allyn and Bacon, New York.

Dugundji, J. and Granas, A. (1982). *Fixed Point Theory, Vol. 1*. PWN—Polish Scientific Publishers, Warszawa.

Dunford, N. and Schwartz, J. (1958). *Linear Operators*. Wiley-Interscience, New York.

Dylawerski, G. (1988). *An S^1-Degree and S^1-Maps Between Representation Spheres, Algebraic Topology and Transformation Groups* (T. tom Dieck, Ed.). *Lecture Notes in Mathematics*, Vol. 1361. Springer-Verlag, New York, pp. 14–28.

Dylawerski, G., Gęba, K. Jodel, J., and Marzantowicz, W. (1991). S^1-equivariant degree and the Fuller index. *Ann. Polon. Math.*, **52**, 243–280.

Elworthy, K. D. and Tromba, A. J. (1970). Degree theory on Banach manifolds. In *Proceedings of Symposia in Pure Mathematics*, Vol. XV. American Mathematical Society, pp. 86–94.

Erbe, L., Gęba, K., and Krawcewicz, W. (1991). Equivariant fixed point index and the period-doubling cascades. *Canadian J. Math.*, **43**, 738–747.

Erbe, L., Gęba, K., Krawcewicz, W., and Wu, J. (1992). S^1-degree and global Hopf bifurcation theory of functional differential equations, *J. Diff. Eqns.*, **98**, 277–298.

Erbe, L. H. and Krawcewicz, W. (1991). Nonlinear boundary value problems for differential inclusions $y'' \in F(t, y, y')$. *Annales Polonici Matematici*, **3**, 195–226.

Erbe, L., Krawcewicz, W., and Wu, J. (1990). Leray–Schauder degree for semilinear Fredholm maps and periodic boundary value problems of neutral equations. *Nonlinear Anal. TMA*, **15**, 747–764.

Erbe, L., Krawcewicz, W., and Wu, J. (1992). Topological transversality in boundary value problems of neutral equations with applications to lossless transmission line theory. *Bull. Soc. Math. Belgique*, **44**, 75–99.

Erbe, L., Krawcewicz, W., and Wu, J. (1993). A composite coincidence degree with applications to boundary value problems of neutral equations. *Trans. Amer. Math. Soc.*, **335**, 459–478.

Erbe, L. and Palamides, P. K. (1987). Boundary value problems for second-order differential systems. *J. Math. Anal. Appl.*, **127**, 80–92.

Ewert, J. (1980). Homotopical properties and the topological degree for γ-contraction vector fields. *Bull. Acad. Polon. Sci.*, **28**, 5–6.

Fiedler, B. (1985). An index for global Hopf bifurcation in parabolic systems. *J. Reine Angew. Math.*, **359**, 1–36.

Fiedler, B. (1988). *Global Bifurcation of Periodic Solutions with Symmetry. Lecture Notes in Mathematics*, No. 1309. Springer-Verlag, New York.

Fiedler, B. and Mischaikow, K. (1992). Dynamics of bifurcations for variational problems with $O(3)$-equivariance: A Conley index approach. *Arch. Rational Mech. Anal.*, **119**, 145–196.

Field, M. and Richardson, R. W. (1990). Symmetry breaking in equivariant bifurcation problems. *Bull. Amer. Math. Soc.*, **22**, 79–84.

Fucks, D. B., Fomenko, A. T., and Gutenmaher, B. L. (1969). *Homotopic Topology*. Moscow Univ. Publications, Moscow (in Russian).

Fuller, B. (1967). An index of fixed point type for periodic orbits. *Amer. J. Math.*, **89**, 133–148.

Gaines, R. E. and Mawhin, J. (1977a). Ordinary differential equations with nonlinear boundary conditions. *J. Diff. Eqns.*, **26**, 200–222.

Gaines, R. E. and Mawhin, J. (1977b). *Coincidence Degree and Nonlinear Differential Equations. Lecture Notes in Mathematics*, No. 568. Springer-Verlag, New York.

Gęba, K. (1968). On the homotopy groups of $GL_c(E)$. *Bull. Acad. Polon. Sci.*, **16**, 699–702.

Gęba, K. Krawcewicz, W., and Wu, J. (1994). An equivariant degree with applications to symmetric bifurcation problems I: Construction of the degree. *Proc. London. Math. Soc.*, **69**, 377–398.

Gęba, K., Massabó, I., and Vignoli, A. (1985). Generalized topological degree and bifurcation. In *Nonlinear Functional Analysis and Its Applications, Proceedings of NATO Advanced Study Institute*. Maratea, Italy, pp. 54–73.

Gęba, K., Massabó, I., and Vignoli, A. (1989). On the Euler characteristic of equivariant gradient vector fields. *Boll. Unione Mat. Italiana*, **4-A**, 243–251.

Gęba, K. and Marzantowicz, W. (1991). Global bifurcation of periodic solutions. *Top. Methods Nonlinear Anal.*, **1**, 67–93.

Golubitsky, M. and Guillemin, V. (1973). *Stable Mappings and Their Singularities*. Springer-Verlag, New York.

Golubitsky, J. M., Stewart, I., and Schaeffer, D. G. (1988). *Singularities and Groups in Bifurcation Theory*, Vol. 2. Springer-Verlag, New York.

Gopalsamy, K. and Zhang, B. G. (1988). On a neutral delay logistic equation. *Dynamics Stability Syst.*, **2**, 183–195.

Granas, A. (1962a). A note on Schauder's theorem on invariance of domain. *Bull. Acad. Polon. Sci.*, **10**, 233–238.

Granas, A. (1962b). The theory of compact vector fields and some of its applications to the topology of functional space (I). *Dissertationes Mathematicae*, **30**, 1–93.

Granas, A. (1972). The Leray–Schauder index and the fixed point theory for arbitrary ANR's. *Bull. Soc. Math. France*, **100**, 209–228.

Granas, A., Guenther, R. B., and Lee, J. W. (1980). Applications of topological transversality to differential equations, I. *Pacific J. Math.*, **89**, 53–67.

Granas, A., Guenther, R. B., and Lee, J. W. (1983). Applications of topological transversality to differential equations, II. *Pacific J. Math.*, **104**, 95–109.

Greenberg, M. J. and Harper, J. R. (1981). *Algebraic Topology, a First Course*. Benjamin/Cummings Pub. Co., New York.

Guillemin, V. and Pollack, A. (1974). *Differential Topology*. Prentice-Hall, Inc., Englewood Cliffs, New Jersey.

Gustafson, G. B. and Schmitt, K. (1974). A note on periodic solutions for delay-differential equations. *Proc. Am. Math. Soc.*, **42**, 161–166.

Gustafson, G. B. and Schmitt, K. (1977). Periodic solutions of hereditary differential equations. *J. Diff. Eqns.*, **13**, 567–587.

Györi, I. and Wu, J. (1991). Neutral equations arising from compartmental systems with pipes. *J. Dynamisc. Diff. Eqns.*, **3**, 289–311.

Haddock, J. R., Krisztin, T., Terjeki, J., and Wu, J. (1994). An invariance principle of Liapunov–Razumikhin type of neutral equations. *J. Diff. Eqns.*, **107**, 395–417.

Hale, J. K. (1971). Critical cases for neutral functional differential equations. *J. Diff. Eqns.*, **10**, 59–82.

Hale, J. K. (1974). Behaviour near constant solution of functional differential equations. *J. Diff. Eqns.*, **15**, 278–294.

Hale, J. K. (1977). *Theory of Functional Differential Equations*. Springer-Verlag, New York.

Hale, J. K. and Mawhin, J. (1974). Coincidence degree and periodic solutions of neutral equations. *J. Diff. Eqns.*, **15**, 295–307.

Hale, J. K. and Meyer, K. R. (1967). A class of functional equations of neutral type. *Memoirs Amer. Math. Soc.*, **76**. American Mathematical Society, Providence.

Hale, J. K. and De Oliviera, J. C. F. (1980). Hopf bifurcation for functional equations. *J. Math. Anal. Appl.*, **74**, 41–59.

Hale, J. K. and Verduyn Lunel, S. M. (1993). *Introduction to Functional Differential Equations*. Springer-Verlag, New York.

Hartman, P. (1964). *Ordinary Differential Equations*. John Wiley & Sons, New York.

Hassard, B., Kazarinoff, N. D., and Wan, Y. (1981). *Theory and Applications of Hopf Bifurcation*. Cambridge University Press, Cambridge.

Hauschild, H. (1974). Bordismentheorie stabl gerahmter G-Mannigfaltigkeiten. *Math Z.*, **139**, 165–171.

Hauschild, H. (1977). Zerspaltung äquivarianter Homotopiemengen. *Math. Ann.*, **230**, 279–292.

Hetzer, G. (1975a). Some remarks on ϕ_t on operators and on the coincidence degree for a Fredholm equation with noncompact nonlinear perturbations. *Ann. Soc. Sci. Bruxelles Ser. I*, **89**, 497–508.

Hetzer, G. (1975b). Boundary value-problems for retarded functional differential equations. *Comment. Math. Univ. Carolina*, **16**, 121–137.

Hetzer, G. (1975c). Some applications of the coincidence degree for set-contractions to functional differential equations of neutral type. *Comment. Math. Univ. Carolina*, **16**, 121–138.

Hetzer, G. (1975d). A note on periodic solutions of second order systems of nonlinear autonomous differential equations. *Annals Soc. Scient. Bruxelles*, **89**, 497–508.

Hirsch, M. W. (1976). *Differential Topology*. Springer-Verlag, New York.

Hochschild, G. (1965). *The Structure of Lie Groups*. Holden-Day, San Francisco, Calif.

Hofbauer, J. and Sigmund, K. (1988). *The Theory of Evolution and Dynamical Systems. London Mathematical Society Students Texts*, Vol. 7. Cambridge University Press, Cambridge.

Hu, S. T. (1965). *Theory of Retracts*. Detroit, Mich.

Husemoller, D. (1966). *Fiber Bundles*, 2nd ed. Springer-Verlag, New York.

Ihring, E. and Golubitsky, M. (1984). Pattern selection with $O(3)$ symmetry. *Physica*, **13D**, 1–33.

Ize, J. (1974). Global bifurcation of periodic orbits. *Communications Technicas C.I.M.A.S.*, **5**, 1–129.

Ize, J. (1976). Bifurcation theory for Fredholm operators. *Mem. Amer. Math. Soc.*, **174**, American Mathematical Society, Providence.

Ize, J. (1985). Obstruction theory and multiparameter Hopf bifurcation. *Trans. Amer. Math. Soc.*, **209**, 757–792.

Ize, J. (1993). Topological bifurcation, *Reportes de Invest. UNAM*, **34**, 1–129.

Ize, J., Massabó, I., and Vignoli, V. (1989). Degree theory for equivariant maps, I. *Trans. Amer. Math. Soc.*, **315**, 433–510.

Ize, J., Massabó, I., and Vignoli, V. (1992). Degree theory for equivariant maps, the general S^1-action. *Memoirs Amer. Math. Soc.*, **481**, American Mathematical Society, Providence.

Ize, J., Massabó, I., and Vignoli, V. (1993). Equivariant degree for abelian actions. Part I: Equivariant homotopy groups, *Reportes de Invest. UNAM*, **30**, 1–48.

Kawakubo, K. (1991). *The Theory of Transformation Groups*. Oxford University Press, Oxford.

Kazarinoff, van den Drissche, P., and Wan, Y. H. (1978). Hopf bifurcation and stability of periodic solutions of differential-difference and integro-differential equations. *J. Inst. Math. Appl.*, **21**, 461–477.

Kirillov, A. A. (1976). *Elements of the Theory of Representations*. Springer-Verlag, New York.

Kolmogorov, A. N. and Fomin, S. V. (1961). *Elements of the Theory of Functions and Functional Analysis*. Grayloc Press, Rochester.

Komiya, K. (1988). Fixed point indices of equivariant maps and Möbius inversion. *Invent. Math.*, **91**, 129–135.

Krasnosiel'skii, M. A. (1965). *Topological Methods in the Theory of Nonlinear Integral Equations*. Pergamon Press, New York.

Krasnosiel'skii, M. A. (1968). The operator of translation along trajectories of ordinary differential equations. American Mathematical Society, Providence, R.I.

Krasnosiel'skii, M. A. and Zabreiko, P. P. (1984). *Geometrical Methods of Nonlinear Analysis*. Springer-Verlag, New York.

Krawcewicz, W. (1988). Contribution à la théorie des équations non linéaires dans les espaces de Banach. *Dissertationes Mathematicae*, **273**, 1–80.

Krawcewicz, W. (1990). Résolution des équations semilinéaires avec la partie linéaire à noyau de dimension infinie via des application A-propres. *Dissertationes Mathematicae*, **295**, 1–67.

Krawcewicz, W. and Vivi, P. (1995). Generic bifurcation and equivariant degree (preprint).

Krawcewicz, W., Vivi, P., and Wu, J. (1996). Computations formulae of an equivariant degree with applications to symmetric bifurcations. *Nonlinear World* (to appear).

Krawcewicz, W. and Wu, J. (1994). Global bifurcation of time-reversible systems (preprint).

Krawcewicz, W., Wu, J., and Xia, H. (1993). Global Hopf bifurcation theory for condensing fields and neutral equations with applications to lossless transmission problems. *Canadian Appl. Math. Quart.*, **1**, 167–220.

Krawcewicz, W. and Xia, H. (1996). Analytic definition of an equivariant degree. *Izvestiya Vuzov Mathematika*, Kazan State University (in press).

Kryszewski, W. and Przeradzki, B. (1986). The topological degree and fixed points of DC-mappings. *Fund. Math.*, **126**, 15–26.

Kuang, Y. (1991). On neutral-delay two-species Lotka–Voltera competitive systems. *J. Austral. Math. Soc. Ser. B*, **32**, 311–326.

Kuiper, N. H. (1965). The homotopy type of the unitary group of Hilbert space. *Topology*, **3**, 19–30.

Ladas, G. E. and Lakshmikanthan, V. (1972). *Differential Equations in Abstract Space*. Academic Press, New York.

Lang, S. (1985). *Differential Manifolds*. Springer-Verlag, New York.

Lasota, A. and Yorke, J. A. (1971). Bounds for periodic solutions of differential equations in Banach spaces, *J. Diff. Eqns.*, **10**, 83–91.

Leray, L. and Schauder, J. (1934). Topologie et équations fonctionnelles. *Ann. Sci. École Norm. Sup.*, **51**, 45–78.

Li, T. Y. (1975). Bounds for the periodic solutions of differential equations. *J. Math. Anal. Appl.*, **49**, 124–129.

Lloyd, N. G. (1978). *Degree Theory*. Cambridge University Press, Cambridge.

Lopes, O. (1975). Forced oscillation in nonlinear neutral differential equations. *SIAM J. Appl. Math.*, **29**, 196–201.

Lyubich, Y. I. (1988). *Introduction to the Theory of Banach Representations of Groups*. Birkhäuser Verlag, Basel.

Makhmudov, A. P. and Aliev, Z. S. (1989). Global bifurcation of solutions of certain nonlinear eigenvalue problems. *Diff. Eqns.*, **25**, 71–76 (English translation).

Mallet-Paret, J. (1977). Generic periodic solutions of functional differential equations. *J. Diff. Eqns.*, **25**, 163–183.

Mallet-Paret, J. and Yorke, J. A. (1982). Snakes: Oriented families of periodic orbits, their sources, sinks and continuation. *J. Diff. Eqns.*, **43**, 419–450.

Marsden, J. E. and McCracken, M. (1976). *The Hopf Bifurcation and Its Applications*. Springer-Verlag, New York.

Matsuoka, T. (1989). The number of periodic points of smooth maps. *Ergod. Th. Dynam. Sys.*, **9**, 153–163.

Matsushima, Y. (1966). *Differentiable Manifolds*. Marcel Dekker, New York.

Mawhin, J. (1971). Periodic solutions of nonlinear functional differential equations. *J. Diff. Eqns.*, **10**, 240–261.

Mawhin, J. (1972). Equivalence theorems for nonlinear operator equations and coincidence degree theorem for some mappings in locally convex topological vector spaces. *J. Diff. Eqns.*, **12**, 610–636.

Mawhin, J. (1979). Topological degree methods in nonlinear boundary value problems. In *CBMS Regional Conference Series in Mathematics*, Vol. 40. American Mathematical Society, Providence, R.I.

Mawhin, J. (1981). *Compacité, Monotinie et Convexité dans l'étude de Problème aux Limites Semi-linéaires. Lecture Notes*, No. 9. Univ. Sherbrooke, Sherbrooke.

Milnor, J. (1959). On spaces having the homotopy type of a CW-complex. *Trans. Amer. Math. Soc.*, **90**, 272–280.

Milnor, J. (1978). *Topology from the Differentiable Viewpoint*. Based on notes by D. W. Weaver, The University Press of Virginia, Charlottesville.

Mischenko, A. S. and Fomenko, A. T. (1980). *Differential Geometry and Topology*. Publications of University of Moscow, Moscow (in Russian).

Montgeomery, D. and Zippin, L. (1955). *Topological Transformation Groups*. Wiley-Interscience, New York.

Muhamadiev, E. M. and Sadovskii, D. B. (1973). Estimation of the spectral radius of a certain operator that is connected with equations of neutral type. *Math. Notes*, **13**, 39–45.

Murayama, M. (1983). On *G*-ANR's and their *G*-homotopy types. *Osaka J. Math.*, **20**, 479–512.

Narasimhan, R. (1968). *Analysis on Real and Complex Manifolds*. North-Holland Publishing Co., New York.

Nirenberg, L. (1974). *Topics in Nonlinear Functional Analysis*. Courant Institute of Mathematical Sciences, New York.

Nussbaum, R. D. (1969a). The fixed point index and fixed point theorems for k-set contractions. Ph.D. Thesis, Univ. of Chicago.

Nussbaum, R. D. (1969b). The fixed point index and asymptotic fixed point theorems for k-set contractions. *Bull. Amer. Math. Soc.*, **75**, 490–495.

Nussbaum, R. D. (1970). The radius of essential spectrum. *Duke Math. J.*, **37**, 473–478.

Nussbaum, R. D. (1971). The fixed point index for local condensing maps. *Ann. Pure Appl.*, **89**, 217–258.

Nussbaum, R. D. (1972). Degree theory for local condensing maps. *J. Math. Anal. Appl.*, **37**, 741–766.

Nussbaum, R. D. (1975). A global bifurcation theorem with application to functional differentail equations. *J. Funct. Anal.*, **19**, 319–338.

Nussbaum, R. D. (1976). Global bifurcation of periodic solutions of some autonomous functional differential equations. *J. Math. Anal. Appl.*, **55**, 699–725.

Nussbaum, R. D. (1978). A Hopf global bifurcation theorem for retarded functional differential equations. *Trans. Amer. Math. Soc.*, **238**, 139–164.

De Oliviera, J. C. F. (1980). Hopf bifurcation for functional differential equations. *Nonlinear Anal. TMA*, **4**, 217–229.

Palais, R. S. (1965). On the homotopy type of certain groups of operators. *Topology*, **3**, 271–279.

Palais, R. S. (1966). Homotopy theory of infinite dimensional manifolds. *Topology*, **5**, 1–16.

Petryshyn, W. V. (1975). On the approximation-solvability of equations involving A-proper and pseudo-A-proper mappings. *Bull. Amer. Math. Soc.*, **8**, 223–312.

Pontriagin, L. S. (1946). *Topological Groups*. Princeton University Press.

Pontriagin, L. S. (1976). *Smooth Manifolds and Their Applications in Homotopy Theory*. Nauka, Moscow (in Russian).

Postnikov, M. M. (1984). *Lectures on Algebraic Topology: Basic Homotopy Theory*. Nauka, Moscow (in Russian).

Quinn, F. (1970). Transversal approximation on Banach manifolds. In *Proceedings of Symposia in Pure Mathematics*. American Mathematical Society, Vol. XV, pp. 213–222.

Rabinowitz, P. (1971). Some global results for nonlinear eigenvalues problems. *J. Funct. Anal.* **7**, 487–513.

Rothe, E. H. (1986). *Introduction to Various Aspects of Degree Theory in Banach Spaces*. Mathematical Surveys and Monographs, Vol. 23. American Mathematical Society, Providence.

Rubinsztein, R. (1976). On the equivariant homotopy of spheres. *Dissertationes Mathematicae*, **134**, 1–48.

Rudin, W. (1976). *Functional Analysis*. McGraw-Hill, New York.

Ruelle, D. (1973). Bifurcation in the presence of a symmetry group. *Arch. Rational Mech. Anal.*, **51**, 136–152.

Rustichini, A. (1989). Hopf bifurcation for functional differential equations of mixed type. *J. Dynamics Diff. Eqns.*, **1**, 145–177.

Sadovskii, B. N. (1968). On measures of noncompactness and concentrative operators. *Probl. Mat. Analiza Sloz. Sistem.*, Voronezh, 89–119.

Sadovskii, B. N. (1970). Some remarks on concentrative operators and measures of noncompactness. *Trudy Mat. Fakult. U.G.U.*, **1**, 112–124.

Salamon, M. D. (1984). *Control and Observation of Neutral Systems*. Pitman Advanced Publishing Program, Boston, Mass.

Sattinger, D. H. (1973). *Topics in Stability and Bifurcation Theory*. Springer-Verlag, New York.

Schwartz, J. T. (1965). *Nonlinear Functional Analysis*. Courant Institute of Mathematical Sciences, New York.

Schmitt, K. (1973). *Equations différentielles et fonctionnelles non linéaires*. Hermann, Paris.

Serre, J.-P. (1968). *Linear Representations of Finite Groups*. Springer-Verlag, New York.

Slemrod, M. (1971). Nonexistence of oscillations in a nonlinear distributed network. *J. Math. Anal. Appl.*, **36**, 22–40.

Smale, S. (1965). An infinite dimensional version of Sard's theorem. *Amer. J. Math.*, **87**, 861–866.

Smoller, J. (1983). *Shock Waves and Reaction-Diffusion Equations*. Springer-Verlag, New York.

Spanier, E. H. (1966). *Algebraic Topology*. Springer-Verlag, New York.

Staffans, O. J. (1987). Hopf bifurcation for functional and functional differential equations with infinite delay. *J. Diff. Eqns.*, **70**, 114–151.

Staffans, O. J. (1983). A neutral FDE with D-operator is retarded. *J. Diff. Eqns.*, **49**, 208–217.

Stech, H. W. (1985a). The Hopf bifurcation for functional differential equations. *J. Math. Anal. Appl.*, **109**, 472–491.

Stech, H. W. (1985b). Nongeneric Hopf bifurcation in functional differential equations. *SIAM J. Math. Anal.*, **16**, 1134–1151.

Stuart, C. A. (1973). Some bifurcation theory for k-set contractions. *Proc. London Math. Soc.*, **27**, 531–550.

Switzer, R. M. (1975). *Algebraic Topology—Homotopy and Homology*. Springer-Verlag, New York.

Thomas, J. W. (1973). A bifurcation theorem for k-set contractions. *Pac. J. Math.*, **44**, 749–756.

Ulrich, H. (1988). *Fixed Point Theory of Parametrized Equivariant Maps. Lecture Notes in Mathematics*, No. 1343. Springer-Verlag, New York.

Vanderbauwhede, A. (1982). Families of periodic solutions for autonomous systems. In *Dynamcial Systems II* (A. Bednarek and L. Cesari, Eds.). Academic Press, New York, pp. 427–446.

Vidossich, G. (1976). On the structure of periodic solutions of differential equations. *J. Diff. Eqns.*, **21**, 263–278.

Waltman, P. and Wong, J. S. W. (1972). Two point boundary value problems for nonlinear functional differential equations. *Trans. Amer. Math. Soc.*, **164**, 39–54.

Wawrzyńczyk, A. (1984). *Group Representations and Special Functions*. D. Riedel, Boston.

Webb, J. R. L. (1971). Remarks on k-set contractions. *Boll. Un. Mat. Ital.*, **4**, 614–629.

Whitney, H. (1944). The selfintersections of a smooth n-manifold in $2n$-space. *Annals Math.*, **45**, 220–246.

Wu, J. (1993a). Global continua of periodic solutions to some differential equations of neutral type. *Tôhoku Math J.*, **45**, 67–88.

Wu, J. (1993b). Delay-induced discrete waves of large amplitude in neural networks with circulant connection matrices. Rept. *FI 93-D802*, Fields Institute.

Wu, J. (1986). The stability of neutral functional differential equations with infinite delay. *Funkcialaj Ekvac.*, **29**, 131–189.

Xia, H. (1994). Equivariant degree and global Hopf bifurcation for NFDEs with symmetry. Ph.D. Thesis, Univ. Alberta, Edmonton, Canada.

Yorke, J. A. (1969). Periods of periodic solutions and Lipschitz constant. *Proc. Amer. Math. Soc.*, **22**, 509–512.

Zeidler, E. (1986). *Nonlinear Functional Analysis, I: Fixed-Point Theorems*. Springer-Verlag, New York.

Zhelobenko, D. P. (1970). *Compact Lie Groups and Their Representations*. Nauka, Moscow (in Russian).

Zhelobenko, D. P. and Shtern, A. I. (1983). *Representations of Lie Groups*. Nauka, Moscow (in Russian).

Index

CANADIAN MATHEMATICAL SOCIETY SERIES OF MONOGRAPHS AND ADVANCED TEXTS

Monographies et Études de la Société Mathématique du Canada

EDITORS: Jonathan M. Borwein and Peter B. Borwein

Frank H. Clarke
**Optimization and Nonsmooth Analysis*
Erwin Klein and Anthony C. Thompson
**Theory of Correspondences: Including Applications to Mathematical Economics*
I. Gohberg, P. Lancaster, and L. Rodman
Invariant Subspaces of Matrices with Applications
Jonathan M. Borwein and Peter B. Borwein
Pi and the AGM—A Study in Analytic Number Theory and Computational Complexity
John H. Berglund, Hugo D. Jünghenn, and Paul Milne
**Analysis of Semigroups: Function Spaces Compactifications Representation*
Subhashis Nag
The Complex Analytic Theory of Teichmüller Spaces
Manfred Kracht and Erwin Kreyszig
**Methods of Complex Analysis in Partial Differential Equations with Applications*
Ernest J. Kani and Robert A. Smith
The Collected Papers of Hans Arnold Heilbronn
Victor P. Snaith
**Topological Methods in Galois Representation Theory*
Kalathoor Varadarajan
The Finiteness Obstruction of CTC Wall
F. A. Sherk, P. McMullen, A. Thompson, and A. Weiss
Kaleidoscopes: Selected Writings of H. S. M. Coxeter
Robert V. Moody and Arturo Pianzola
Lie Algebras with Triangular Decompositions
Peter A. Fillmore
A User's Guide to Operator Algebras
Alf van der Poorten
Notes on Fermat's Last Theorem

*Indicates an out-of-print title.

Peter Lancaster and Kęstutis Šalkauskas
Transform Methods in Applied Mathematics
Ole A. Nielsen
An Introduction to Integration and Measure Theory
Wieslaw Krawcewicz and Jianhong Wu
Theory of Degrees with Application to Bifurcations and Differential Equations